Acta Numerica 1998

Acta Numerica

Volume 7 1998

CAMBRIDGE
UNIVERSITY PRESS

CAMBRIDGE UNIVERSITY PRESS
Cambridge, New York, Melbourne, Madrid, Cape Town, Singapore,
São Paulo, Delhi, Dubai, Tokyo, Mexico City

Cambridge University Press
The Edinburgh Building, Cambridge CB2 8RU, UK

Published in the United States of America by Cambridge University Press, New York

www.cambridge.org
Information on this title: www.cambridge.org/9780521157650

First published 1998
First paperback edition 2010

A catalogue record for this publication is available from the British Library

ISBN 978-0-521-64316-0 Hardback
ISBN 978-0-521-15765-0 Paperback

Contents

Acta Numerica (1998), pp. 1–49

Monte Carlo and quasi-Monte Carlo methods

Russel E. Caflisch*
Mathematics Department, UCLA,
Los Angeles, CA 90095-1555, USA
E-mail: caflisch@math.ucla.edu

Monte Carlo is one of the most versatile and widely used numerical methods. Its convergence rate, $O(N^{-1/2})$, is independent of dimension, which shows Monte Carlo to be very robust but also slow. This article presents an introduction to Monte Carlo methods for integration problems, including convergence theory, sampling methods and variance reduction techniques. Accelerated convergence for Monte Carlo quadrature is attained using quasi-random (also called low-discrepancy) sequences, which are a deterministic alternative to random or pseudo-random sequences. The points in a quasi-random sequence are correlated to provide greater uniformity. The resulting quadrature method, called quasi-Monte Carlo, has a convergence rate of approximately $O((\log N)^k N^{-1})$. For quasi-Monte Carlo, both theoretical error estimates and practical limitations are presented. Although the emphasis in this article is on integration, Monte Carlo simulation of rarefied gas dynamics is also discussed. In the limit of small mean free path (that is, the fluid dynamic limit), Monte Carlo loses its effectiveness because the collisional distance is much less than the fluid dynamic length scale. Computational examples are presented throughout the text to illustrate the theory. A number of open problems are described.

* Research supported in part by the Army Research Office under grant number DAAH04-95-1-0155.

CONTENTS

1. Introduction

Monte Carlo provides a direct method for performing simulation and integration. Because it is simple and direct, Monte Carlo is easy to use. It is also robust, since its accuracy depends on only the crudest measure of the complexity of the problem. For example, Monte Carlo integration converges at a rate $O(N^{-1/2})$ that is independent of the dimension of the integral. For this reason, Monte Carlo is the only viable method for a wide range of high-dimensional problems, ranging from atomic physics to finance.

The price for its robustness is that Monte Carlo can be extremely slow. The order $O(N^{-1/2})$ convergence rate is decelerating, since an additional factor of 4 increase in computational effort only provides an additional factor of 2 improvement in accuracy. The result of this combination of ease of use, wide range of applicability and slow convergence is that an enormous amount of computer time is spent on Monte Carlo computations.

This represents a great opportunity for researchers in computational science. Even modest improvements in the Monte Carlo method can have substantial impact on the efficiency and range of applicability for Monte Carlo methods. Indeed, much of the effort in the development of Monte Carlo has been in construction of variance reduction methods which speed up the computation. A description of some of the most common variance reduction methods is given in Section 4.

Variance reduction methods accelerate the convergence rate by reducing the constant in front of the $O(N^{-1/2})$ for Monte Carlo methods using random or pseudo-random sequences. An alternative approach to acceleration is to change the choice of sequence. Quasi-Monte Carlo methods use quasi-random (also known as low-discrepancy) sequences instead of random or pseudo-random. Unlike pseudo-random sequences, quasi-random sequences do not attempt to imitate the behaviour of random sequences. Instead, the elements of a quasi-random sequence are correlated to make them more uniform than random sequences. For this reason, Monte Carlo integration using quasi-random points converges more rapidly, at a rate $O(N^{-1}(\log N)^k)$, for some constant k. Quasi-random sequences are described in Sections 5 and 6.

In spite of their importance in applications, Monte Carlo methods receive relatively little attention from numerical analysts and applied mathematicians. Instead, it is number theorists and statisticians who design the pseudo-random, quasi-random and other types of sequence that are used in Monte Carlo, while the innovations in Monte Carlo techniques are developed mainly by practitioners, including physicists, systems engineers and statisticians.

The reasons for the near neglect of Monte Carlo in numerical analysis and applied mathematics are related to its robustness. First, Monte Carlo methods require less sophisticated mathematics than other numerical methods. Finite difference and finite element methods, for example, require careful mathematical analysis because of possible stability problems, but stability is not an issue for Monte Carlo. Instead, Monte Carlo nearly always gives an answer that is qualitatively correct, but acceleration (error reduction) is always needed. Second, Monte Carlo methods are often phrased in non-mathematical terms. In rarefied gas dynamics, for example, Monte Carlo allows for direct simulation of the dynamics of the gas of particles, as described in Section 7. Finally, it is often difficult to obtain definitive results on Monte Carlo, because of the random noise. Thus computational improvements often come more from experience than from a particular insightful calculation.

This article is intended to provide an introduction to Monte Carlo methods for numerical analysts and applied mathematicians. In spite of the reasons cited above, there are ample opportunities for this community to make significant contributions to Monte Carlo. First of all, any improvements can have a big impact, because of the prevalence of Monte Carlo computations. Second, the methodology of numerical analysis and applied mathematics, including well controlled computational experiments on canonical problems, is needed for Monte Carlo. Finally, there are some outstanding problems on which a numerical analysis or applied mathematics viewpoint is clearly needed; for example:

- design of Monte Carlo simulation for transport problems in the diffusion limit (Section 7)
- formulation of a quasi-Monte Carlo method for the Metropolis algorithm (Section 6)
- explanation of why quasi-Monte Carlo behaves like standard Monte Carlo when the dimension is large and the number of simulation is of moderate size (Section 6).

Some older, but still very good, general references on Monte Carlo are Kalos and Whitlock (1986) and Hammersley and Handscomb (1965).

The focus of this article is on Monte Carlo for integration problems. Integration problems are simply stated, but they can be extremely challenging. In addition, integration problems contain most of the difficulties that are found in more general Monte Carlo computations, such as simulation and optimization.

The next section formulates the Monte Carlo method for integration and describes its convergence. Section 3 describes random number generators and sampling methods. Variance reduction methods are discussed in Section 4 and quasi-Monte Carlo methods in Section 5. Effective use of quasi-Monte Carlo requires some modification of standard Monte Carlo techniques, as described in Section 6. Monte Carlo methods for rarefied gas dynamics are described in Section 7, with emphasis on the loss of effectiveness for Monte Carlo in the fluid dynamic limit.

2. Monte Carlo integration

The integral of a Lebesgue integrable function $f(x)$ can be expressed as the average or *expectation* of the function f evaluated at a random location. Consider an integral on the one-dimensional unit interval

$$I[f] = \int_0^1 f(x)\,\mathrm{d}x = \bar{f}. \tag{2.1}$$

Let x be a random variable that is uniformly distributed on the unit interval. Then

$$I[f] = E[f(x)]. \tag{2.2}$$

For an integral on the unit cube $I^d = [0,1]^d$ in d dimensions,

$$I[f] = E[f(\boldsymbol{x})] = \int_{I^d} f(\boldsymbol{x})\,\mathrm{d}\boldsymbol{x}, \tag{2.3}$$

in which $\boldsymbol{x}$ is a uniformly distributed vector in the unit cube.

The Monte Carlo quadrature formula is based on the probabilistic interpretation of an integral. Consider a sequence $\{x_n\}$ sampled from the uniform distribution. Then an empirical approximation to the expectation is

$$I_N[f] = \frac{1}{N}\sum_{n=1}^{N} f(x_n). \tag{2.4}$$

According to the Strong Law of Large Numbers (Feller 1971), this approximation is convergent with probability one; that is,

$$\lim_{N\to\infty} I_N[f] \to I[f]. \tag{2.5}$$

In addition, it is unbiased, which means that the average of $I_N[f]$ is exactly $I[f]$ for any N; that is,

$$E[I_N[f]] = I[f], \tag{2.6}$$

in which the average is over the choice of the points $\{x_n\}$.

In general, define the Monte Carlo integration error

$$\epsilon_N[f] = I[f] - I_N[f] \tag{2.7}$$

so that the bias is $E[\epsilon_N[f]]$ and the root mean square error (RMSE) is

$$E[\epsilon_N[f]^2]^{1/2}. \tag{2.8}$$

2.1. Accuracy of Monte Carlo

The Central Limit Theorem (CLT) (Feller 1971) describes the size and statistical properties of Monte Carlo integration errors.

Theorem 2.1 For N large,

$$\epsilon_N[f] \approx \sigma N^{-1/2} \nu \tag{2.9}$$

in which ν is a standard normal ($N(0,1)$) random variable and the constant $\sigma = \sigma[f]$ is the square root of the variance of f; that is,

$$\sigma[f] = \left(\int_{I^d} (f(x) - I[f])^2 \, \mathrm{d}x \right)^{1/2}. \tag{2.10}$$

A more precise statement is that

$$\begin{aligned} \lim_{N\to\infty} \operatorname{Prob}\left(a < \frac{\sqrt{N}}{\sigma} \epsilon_N < b \right) &= \operatorname{Prob}(a < \nu < b) \\ &= \int_a^b (2\pi)^{-1/2} \mathrm{e}^{-x^2/2} \, \mathrm{d}x. \end{aligned} \tag{2.11}$$

This says that the error in Monte Carlo integration is of size $O(N^{-1/2})$ with a constant that is just the variance of the integrand f. Moreover, the statistical distribution of the error is approximately a normal random variable. In contrast to the usual results of numerical analysis, this is a probabilistic result. It does not provide an absolute upper bound on the error; rather it says that the error is of a certain size with some probability. On the other hand, this result is an equality, so that the bounds it provides are tight. The use of this result will be discussed at the end of this section.

Now we present a partial proof of the Central Limit Theorem, which proves that the error size is $O(N^{-1/2})$. Derivation of the Gaussian distribution for the error is more difficult (Feller 1971). First define $\xi_i = \sigma^{-1}(f(x_i) - \bar{f})$ for x_i uniformly distributed. Then

$$\begin{aligned} E[\xi_i] &= 0, \\ E[\xi_i^2] &= \int \sigma^{-2}(f(x_i) - \bar{f})^2 \, \mathrm{d}x = 1, \\ E[\xi_i \xi_j] &= 0 \quad \text{if } i \neq j. \end{aligned} \tag{2.12}$$

The last equality is due to the independence of the $x_i's$.

Now consider the sum

$$S_N = N^{-1} \sum_1^N \xi_i = \sigma^{-1} \varepsilon_N. \tag{2.13}$$

Its variance is

$$\begin{aligned}
E[S_N^2]^{1/2} &= E[N^{-2} \left(\sum_{i=1}^N \xi_i \right)^2]^{1/2} \\
&= N^{-1} \left\{ E\left[\sum_{i=1}^N \xi_i^2 \right] + E\left[\sum_{i=1}^N \sum_{j \neq i} \xi_i \xi_j \right] \right\}^{1/2} \\
&= N^{-1} \left\{ \sum_{i=1}^N 1 + 0 \right\}^{1/2} \\
&= N^{-1/2}.
\end{aligned} \tag{2.14}$$

Therefore

$$E[\varepsilon_N^2] = \sigma N^{-1/2}, \tag{2.15}$$

which shows that the RMSE is of size $O(\sigma N^{-1/2})$.

The converse of the Central Limit Theorem is useful for determining the size of N required for a particular computation. Since the error bound from the CLT is probabilistic, the precision of the Monte Carlo integration method can only be ensured within some confidence level. To ensure an error of size at most ϵ with confidence level c requires the number of sample points N to be

$$N = \epsilon^{-2} \sigma^2 s(c), \tag{2.16}$$

in which s is the confidence function for a normal variable; that is,

$$\begin{aligned}
c &= \int_{-s(c)}^{s(c)} \mathrm{e}^{-x^2/2} \, \mathrm{d}x / \sqrt{2\pi} \\
&= \operatorname{erf}(s(c)/\sqrt{2}).
\end{aligned} \tag{2.17}$$

For example, 95 per cent confidence in the error size requires that $s = 2$, approximately.

In an application, the exact value of the variance is unknown (it is as difficult to compute as the integral itself), so the formula (2.16) cannot be directly used. There is an easy way around this, which is to determine the empirical error and variance (Hogg and Craig 1995). Perform M computations using independent points x_i for $1 \le i \le MN$. For each j obtain values

$I_N^{(j)}$ for $1 \leq j \leq M$. The empirical RMSE is then $\tilde{\varepsilon}_N$, given by

$$\tilde{\varepsilon}_N = \left(M^{-1} \sum_{j=1}^{M} \left(I_N^{(j)} - \bar{I}_N \right)^2 \right)^{1/2}, \tag{2.18}$$

in which

$$\bar{I}_N = M^{-1} \sum_{j=1}^{M} I_N^{(j)}. \tag{2.19}$$

The empirical variance is $\tilde{\sigma}$ given by

$$\tilde{\sigma} = N^{1/2} \tilde{\varepsilon}_N. \tag{2.20}$$

This value can be used for σ in (2.16) to determine the value of N for a given precision level ε and a given confidence level c.

2.2. Comparison to grid-based methods

Most people who see Monte Carlo for the first time are surprised that it is a viable method. How can a random array be better than a grid? There are several ways to answer this question. First, compare the convergence rate of Monte Carlo with that of a grid-based integration method such as Simpson's rule. The convergence rate for grid-based quadrature is $O(N^{-k/d})$ for an order k method in dimension d, since the grid with N points in the unit cube has spacing $N^{-1/d}$. On the other hand, the Monte Carlo convergence rate is $O(N^{-1/2})$ independent of dimension. So Monte Carlo beats a grid in high-dimension d, if

$$k/d < 1/2. \tag{2.21}$$

On the other hand, for an analytic function on a periodic domain, the value of k is infinite, so that this simple explanation fails. A more realistic explanation for the robustness of Monte Carlo is that it is practically impossible to lay down a grid in high dimension. The simplest cubic grid in d dimensions requires at least 2^d points. For $d = 20$, which is not particularly large, this requires more than a million points. Moreover, it is practically impossible to refine a grid in a high dimension, since a refinement requires increasing the number of points by factor 2^d. In contrast to these difficulties for a grid in high dimension, the accuracy of Monte Carlo quadrature is nearly independent of dimension and each additional point added to the Monte Carlo quadrature formula provides an incremental improvement in its accuracy. To be sure, the value of N at which the $O(N^{-1/2})$ error estimate becomes valid (that is, the length of the transient) is difficult to predict, but experience shows that, for problems of moderate complexity in moderate dimension (for instance $d = 20$), the $O(N^{-1/2})$ error size is typically attained for moderate values of N.

Two additional interpretations of Monte Carlo quadrature are worthwhile. Consider the Fourier representation of a periodic function with period one

$$f(x) = \sum_{k=-\infty}^{\infty} \hat{f}(k)e^{2\pi ikx}. \tag{2.22}$$

The integral $I[f]$ is just $\hat{f}(0)$; that is, the contributions to the integral are 0 from all wave-numbers $k \neq 0$. For a grid of spacing $1/n$, the grid-based quadrature formula is

$$I_n^{(g)}[f] = n^{-1}\sum_{i=1}^{n} f(i/n). \tag{2.23}$$

The contributions to this sum are 0 (as they should be) from wave-numbers $k \neq mn$ for some integer m, and the contributions are $\hat{f}(k)$ for wave-numbers $k = mn$. That is, the accuracy of grid-based quadrature is 100 per cent for $k \neq mn$, but 0 per cent for $k = mn$ (with $m \neq 0$). Monte Carlo quadrature using a random array is partially accurate for all k, which is superior to a grid-based method if the Fourier coefficients decay slowly.

Finally, insight into the relative performance of grid-based and Monte Carlo methods is gained by considering the points themselves in a high dimension. For a regular grid, the change from one point to the next is only a variation of one component at a time, that is, $(0, 0, \ldots, 0, 0) \mapsto (0, 0, \ldots, 0, 1/n)$. In many problems, this is an inefficient use of the unvaried components. In a random array, all components are varied in each point, so that the state space is sampled more fully. This accords well with the global nature of Monte Carlo: each point of a Monte Carlo integration formula is an estimate of the integral over the entire domain.

3. Generation and sampling methods

3.1. Random number generators

The numbers generated by computers are not random, but pseudo-random, which means that they are made to have many of the properties of random number sequences (Niederreiter 1992). While this is a well-developed subject, occasional problems still occur, mostly with very long sequences ($N \geq 10^9$). The methods used to generate pseudo-random numbers are mostly linear congruential methods. There is a series of reliable pseudo-random number generators in the popular book *Numerical Recipes* (Press, Teukolsky, Vettering and Flannery 1992). It is important to use the routines *ran0, ran1, ran2, ran3, ran4* from the *second* edition (for instance, *ran1* is recommended for $N < 10^8$); the routines *RAN0, RAN1, RAN2, RAN3* from the first edition of this book had some deficiencies (see Press and Teukolsky (1992)).

For very large problems requiring extremely large values of N (as high as 10^{13}!), reliable sequences can be obtained from the project Scalable Parallel Random Number Generators Library for Parallel Monte Carlo Computations (`http://www.ncsa.uiuc.edu/Apps/SPRNG/`).

3.2. Sampling methods

Standard (pseudo-) random number generators produce uniformly distributed variables. Non-uniform variables can be sampled through transformation of uniform variables. For a non-uniform random variable with density $p(x)$, the expectation of a function $f(x)$ is

$$E[f] = I[f] = \int f(x)p(x)\,\mathrm{d}x. \tag{3.1}$$

For a sequence of random numbers $\{x_n\}$ distributed according to the density p, the empirical approximation to the expectation is

$$I_N[f] = \frac{1}{N}\sum_{n=1}^{N} f(x_n) \tag{3.2}$$

and the resulting quadrature error is

$$\epsilon_N[f] = I[f] - I_N[f]. \tag{3.3}$$

As in the one-dimensional case, the Central Limit Theorem says that

$$\epsilon_N[f] \approx N^{-1/2}\sigma\nu \tag{3.4}$$

in which ν is $N(0,1)$ and

$$\sigma^2 = \int (f-\bar{f})^2 p(x)\,\mathrm{d}x. \tag{3.5}$$

Next we discuss methods for generating non-uniform random variables starting from uniform random variables.

3.3. Transformation method

This is a general method for producing a random variable x with density $p(x)$, through transformation of a uniform random variable. Let y be a uniform variable and look for a function $X(y)$, so that $x = X(y)$ has the desired density $p(x)$.

Define the cumulative distribution function

$$P(x) = \int_{-\infty}^{x} p(x')\,\mathrm{d}x'. \tag{3.6}$$

Determination of the mapping $X(y)$ is through the following computation of the expectation. For any function f,

$$E_p[f(x)] = E_{\mathrm{unif}}[f(X(y))], \tag{3.7}$$

so that, using a change of variables,

$$\begin{aligned}\int f(x)p(x)\,\mathrm{d}x &= \int f(X(y))\,\mathrm{d}y \\ &= \int f(x)(\mathrm{d}y/\mathrm{d}x)\,\mathrm{d}x. \end{aligned} \tag{3.8}$$

This implies that $p(x) = \mathrm{d}y/\mathrm{d}x = 1/X'(y)$ so that $\int^{X(y)} p(x)\,\mathrm{d}x = y$; *i.e.*,

$$\begin{aligned}P(X(y)) &= y \\ X(y) &= P^{-1}(y). \end{aligned} \tag{3.9}$$

This formulation is convenient and explicit but not necessarily easy to implement, as it may be difficult to compute the inverse of the function P.

3.4. Gaussian random variables

As an example of the transformation method, a Gaussian (normal) random variable has density p and cumulative distribution function P given by

$$\begin{aligned}p(x) &= (2\pi)^{-1/2}\mathrm{e}^{-x^2/2}, \\ P(x) &= \frac{1}{\sqrt{2\pi}}\int_{-\infty}^{x} \mathrm{e}^{-t^2/2}\,\mathrm{d}t \\ &= \tfrac{1}{2} + \tfrac{1}{2}\,\mathrm{erf}\,(x/\sqrt{2}), \end{aligned} \tag{3.10}$$

in which *erf* is the error function defined by

$$\mathrm{erf}\,(z) = \frac{2}{\sqrt{\pi}}\int_0^z \mathrm{e}^{-t^2}\,\mathrm{d}t. \tag{3.11}$$

One way to sample from a Gaussian distribution is to apply the inverse of the Gaussian cumulative distribution function P to a uniform random variable y. Sample a normal variable x, starting from a uniform variable y, by

$$y = P(x) = \tfrac{1}{2} + \tfrac{1}{2}\,\mathrm{erf}\,(x/\sqrt{2}), \tag{3.12}$$

that is,

$$x = \sqrt{2}\,\mathrm{erf}^{-1}(2y-1). \tag{3.13}$$

Approximate formulas for $P^{(-1)}(y)$ or erf^{-1} are found in Kennedy and Gentle (1980).

For the Gaussian distribution, as well as for a number of other distributions, special transformations are a useful alternative to the general transformation method. The simplest of these is the Box–Muller method. It provides a direct way of generating normal random variables without inverting the error function. Starting with two uniform variables y_1, y_2, two normal variables x_1, x_2 are obtained by

$$\begin{aligned} x_1 &= \sqrt{-2\log(y_1)}\cos(2\pi y_2), \\ x_2 &= \sqrt{-2\log(y_1)}\sin(2\pi y_2). \end{aligned} \tag{3.14}$$

Box–Muller is based on the following observation. First change from rectangular to polar coordinates, that is,

$$(x_1, x_2) = (r\cos\theta, r\sin\theta), \tag{3.15}$$

so that

$$\mathrm{d}x_1\,\mathrm{d}x_2 = r\,\mathrm{d}r\,\mathrm{d}\theta. \tag{3.16}$$

The corresponding transformation of the probability density function is

$$(2\pi)^{-1}\mathrm{e}^{-(x_1^2+x_2^2)/2}\,\mathrm{d}x_1\,\mathrm{d}x_2 = (2\pi)^{-1}\mathrm{e}^{-r^2/2}r\,\mathrm{d}r\,\mathrm{d}\theta. \tag{3.17}$$

This shows that the angular variable $y_1 = \theta/(2\pi)$ is uniformly distributed over the unit interval. Next the variable r is easily sampled, since it has density $re^{-r^2/2}$ and corresponding cumulative distribution function

$$P(r) = \int_0^r \mathrm{e}^{-r'^2/2}r'\,\mathrm{d}r' = 1 - \mathrm{e}^{-r^2/2}. \tag{3.18}$$

Therefore r can be sampled by

$$r = P^{-1}(y_2) = \sqrt{-2\log(1-y_2)}. \tag{3.19}$$

After replacing y_2 by $1 - y_2$, the resulting transform

$$(y_1, y_2) \to (r, \theta) \to (x_1, x_2) \tag{3.20}$$

is given in (3.14).

The only disadvantages of the Box–Muller method are that it requires evaluation of transcendental functions and that it generates two random variables when only one may be needed. See Marsaglia (1991) for examples of efficient use of Box–Muller.

3.5. Acceptance–rejection method

Another general way of producing random variables from a given density $p(x)$ is based on a probabilistic approach. This method shows the power and flexibility of stochastic methods. For a given density function $p(x)$, suppose that we know a function $q(x)$ satisfying

$$q(x) \geq p(x), \tag{3.21}$$

and that we have a way to sample from the probability density

$$\hat{q}(x) = q(x)/I[q], \tag{3.22}$$

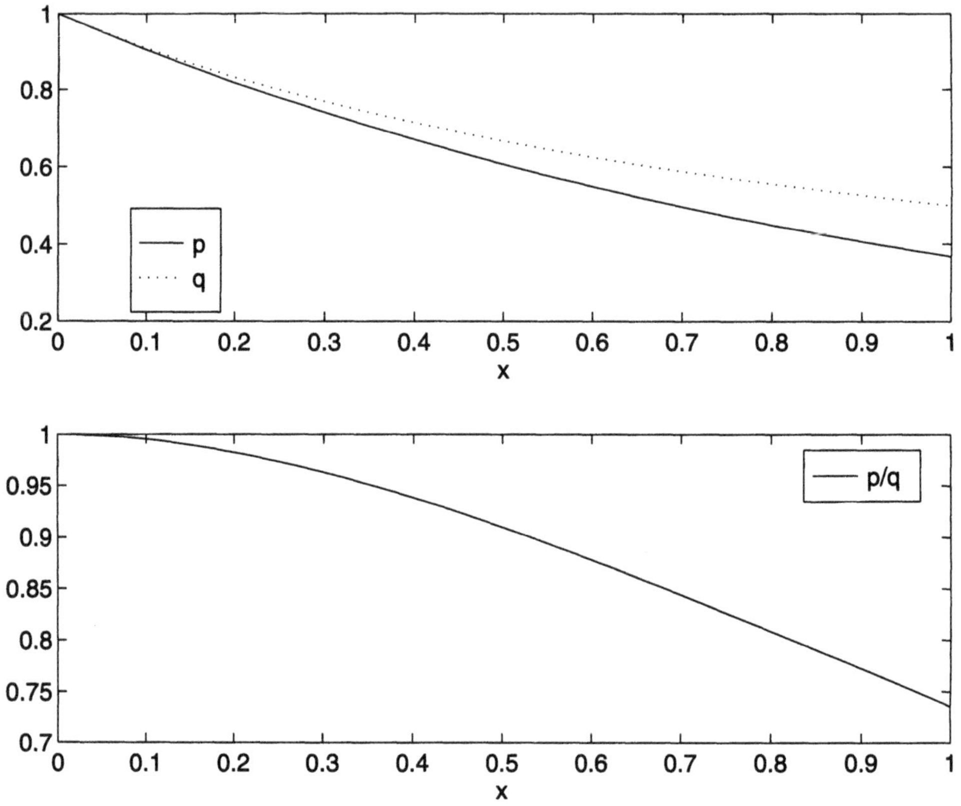

Fig. 1. Typical choice of p, q and p/q. Accept for $y < p/q$ and reject for $y > p/q$

in which $I[q] = \int q(x')\,\mathrm{d}x'$. In practice, q is often chosen to be a constant.

The acceptance–rejection procedure goes as follows. Pick two random variables, x', y, in which x' is a trial variable chosen according to the probability density $\hat{q}(x')$, and y is a decision variable chosen according to the uniform density on $0 < y < 1$. The decision is to

- *accept* if $0 < y < p(x')/q(x')$
- *reject* if $p(x')/q(x') < y < 1$.

This procedure is repeated until a value x' is accepted. Once a value x' is accepted, take

$$x = x'. \tag{3.23}$$

This procedure is depicted graphically in Figure 1.

Here is a derivation of the acceptance–rejection method. Since $0 \le p \le q$,

$$\begin{aligned} p(x) &= \frac{p(x)}{q(x)}\hat{q}(x)I[q] \\ &= \int_0^1 \chi\left(\frac{p(x)}{q(x)} > y\right)\,\mathrm{d}y\;\hat{q}(x)I[q]. \end{aligned} \tag{3.24}$$

So,

$$\begin{aligned}\int f(x)p(x)\,\mathrm{d}x &= \int\int_0^1 f(x)\,\chi\left(\frac{p(x)}{q(x)} > y\right)\hat{q}(x)\,\mathrm{d}y\,\mathrm{d}x\; I[q] \\ &\approx N'^{-1} \sum_{p(x'_n)/q(x'_n)>y_n} f(x'_n)\; I[q] \\ &\approx N^{-1} \sum_{\text{accepted pts}} f(x_n), \end{aligned} \tag{3.25}$$

in which

$$\begin{aligned} N' &= \text{total number of trial points,} \\ N &= \text{total number of accepted points} \\ &\approx N'/I[q]. \end{aligned} \tag{3.26}$$

The acceptance–rejection method has some obvious advantages that often make it the method of choice for generating random variables with a given distribution. It does not require inversion of the cumulative distribution function P. In addition, it works even if the density function p has not been normalized to have integral 1. One disadvantage of the method is that it may be inefficient, requiring many trials before acceptance. In practice this is often not a serious deficiency, since the largest share of the computation is on the integrand rather than on the random point selection.

An extension of the acceptance–rejection method called the Metropolis algorithm (Kalos and Whitlock 1986) is used to find the equilibrium distribution for a stochastic process.

4. Variance reduction

In Monte Carlo integration, the error ϵ and the number N of samples are related by

$$\epsilon = O\left(\sigma N^{-1/2}\right), \tag{4.1}$$

$$N = O(\sigma/\epsilon)^2. \tag{4.2}$$

The computational time required for the method is proportional to the sample number N and thus is of size $O(\sigma/\epsilon)^2$, which shows that computational time grows rapidly as the desired accuracy is tightened. There are two options for acceleration (error reduction) of Monte Carlo. The first is variance reduction, in which the integrand is transformed in a way that reduces the constant variance σ. The second is to modify the statistics, that is, to replace the random variables by an alternative sequence which improves the exponent 1/2. In this section, various strategies for variance reduction are described. In Section 5, we discuss quasi-random variables, an example of the second approach. One note of caution is that the acceleration methods described here may require extra computational time, which must be balanced against savings gained by reduced N. In most examples, however, the savings due to variance reduction are quite significant.

4.1. *Antithetic variables*

Antithetic variables is one of the simplest and most widely used variance reduction methods. The method is as follows: for each (multi-dimensional) sample value $\boldsymbol{x}$, also use the value $-\boldsymbol{x}$. The resulting Monte Carlo quadrature rule is

$$I_N[f] = \frac{1}{2N} \sum_{n=1}^{N} \{f(\boldsymbol{x}_n) + f(-\boldsymbol{x}_n)\}. \tag{4.3}$$

For example, the vector $\boldsymbol{x}$ could be the discrete states in a random walk, so that the dimension would be the number of time-steps in the walk. When antithetic variables are used for a problem involving a random walk, then for each path $\boldsymbol{x} = (x_1, \ldots, x_k)$, we also use the path $-\boldsymbol{x} = (-x_1, \ldots, -x_k)$.

The use of antithetic variables is motivated by an expansion for small values of the variance. Consider, for example, the expectation $E[f(x)]$ in which x is an $N(0, \sigma^2)$ random variable with σ small. Set

$$x = \sigma \hat{x}. \tag{4.4}$$

The Taylor expansion of $f = f(\sigma\hat{x})$ (for small σ) is

$$f = f(0) + f'(0)\sigma\hat{x} + O(\sigma^2). \tag{4.5}$$

Since the distribution of $\hat{x}$ is symmetric about 0, the average $E[\hat{x}]$ of the linear term is zero. In a standard Monte Carlo integral of f, these terms do not cancel exactly, so that the Monte Carlo error is proportional to σ. With antithetic variables, on the other hand, the linear terms cancel exactly, so that the remaining error is proportional to σ^2.

4.2. *Control variates*

The idea of control variates is to use an integrand g, which is similar to f and for which the integral $I[g] = \int g(x)\,\mathrm{d}x$ is known. The integral $I[f]$ is then written as

$$\int f(x)\,\mathrm{d}x = \int (f(x) - g(x))\,\mathrm{d}x + \int g(x)\,\mathrm{d}x. \tag{4.6}$$

Monte Carlo quadrature is used on the integral of $f - g$ to obtain the formula

$$I_n[f] = \frac{1}{N} \sum_{n=1}^{N} (f(x_n) - g(x_n)) + I[g]. \tag{4.7}$$

The resulting integration error $\epsilon_N[f] = I[f] - I_N[f]$ is of size

$$\epsilon_N[f] \approx \sigma_{f-g} N^{-1/2}, \tag{4.8}$$

in which the relevant variance is

$$\sigma_{f-g}^2 = \int \left(\tilde{f}(x) - \tilde{g}(x)\right)^2 \mathrm{d}x \tag{4.9}$$

using the notation

$$\tilde{f}(x) = f(x) - I[f], \qquad \tilde{g}(x) = g(x) - I[g]. \tag{4.10}$$

This shows that the control variate method is effective if

$$\sigma_{f-g} \ll \sigma_f. \tag{4.11}$$

Optimal use of a control variate is made by introducing a multiplier λ. For a given function g, write the integral of f as

$$\int f(x)\,\mathrm{d}x = \int (f(x) - \lambda g(x))\,\mathrm{d}x + \lambda \int g(x)\,\mathrm{d}x. \tag{4.12}$$

The error in Monte Carlo quadrature of the first integral is proportional to the variance

$$\sigma_{f-\lambda g}^2 = \int \left(\tilde{f}(x) - \lambda\tilde{g}(x)\right)^2 \mathrm{d}x. \tag{4.13}$$

The optimal value of λ is found by minimizing $\sigma_{f-\lambda g}^2$ to obtain

$$\begin{aligned} \lambda &= E[\tilde{f}\tilde{g}]/E[\tilde{g}^2] \\ &= \left(\int \tilde{f}\tilde{g}\,\mathrm{d}x\right) / \left(\int \tilde{g}^2\,\mathrm{d}x\right). \end{aligned} \tag{4.14}$$

4.3. Matching moments method

Monte Carlo integration error is partly due to statistical sampling error, that is, differences between the desired density function $p(x)$ and the empirical density function of the sampled points $\{x_n\}_{n=1}^N$. Part of this difference can be seen directly through comparison of the moments of the two distributions. Define the first and second moments m_1 and m_2 of p as

$$m_1 = \int x p(x)\,\mathrm{d}x, \qquad m_2 = \int x^2 p(x)\,\mathrm{d}x. \tag{4.15}$$

The first and second moments μ_1 and μ_2 of the sample $\{x_n\}_{n=1}^N$ are

$$\mu_1 = N^{-1}\sum_{n=1}^{N} x_n, \qquad \mu_2 = N^{-1}\sum_{n=1}^{N} x_n^2. \tag{4.16}$$

The moment error is due to the inequality of these moments, that is,

$$\mu_1 \neq m_1, \qquad \mu_2 \neq m_2. \tag{4.17}$$

A partial correction to the statistical sampling error is to make the moments exactly match. This can be done by a simple transformation of the sample points. To match the first moment of the sample with that of p, replace x_n by

$$y_n = (x_n - \mu_1) + m_1. \tag{4.18}$$

This satisfies

$$N^{-1} \sum y_n = m_1 \tag{4.19}$$

so that the first moment is exactly right. To match the first two moments, replace x_n by

$$y_n = (x_n - \mu_1)/c + m_1, \qquad c = \sqrt{\frac{m_2 - m_1^2}{\mu_2 - \mu_1^2}}. \tag{4.20}$$

Then

$$N^{-1} \sum y_n = m_1, \qquad N^{-1} \sum y_n^2 = m_2. \tag{4.21}$$

These transformed sequences with the correct moments must be used with some caution. Since the transformation involves μ_1 and μ_2, the new sample points y_n are no longer independent, and Monte Carlo error estimates are not so straightforward. For example, the Central Limit Theorem is not directly applicable, and the method may be biased. Actually, this is an example of the second approach to acceleration of Monte Carlo, through modification of the properties of the random sequence. It is presented here along with variance reduction methods, however, because the resulting improvement in Monte Carlo accuracy is comparable to that gained through variance reduction.

Pullin (1979) has formulated a method in which the sample points are independent and have a prescribed empirical mean and variance, but come from a Gaussian distribution with a randomly chosen mean and variance.

4.4. Stratification

Stratification combines the benefits of a grid with those of random variables. In the simplest case, stratification is based on a regular grid with uniform density in one dimension. Split the integration region $\Omega = [0, 1]$ into M pieces Ω_k given by

$$\Omega_k = \left[\frac{(k-1)}{M}, \frac{k}{M}\right]. \tag{4.22}$$

Then, each subset has the same measure, $|\Omega_k| = 1/M$. Define the averages over each Ω_k by

$$\bar{f}(x) = \bar{f}_k = |\Omega_k|^{-1} \int_{\Omega_k} f(x)\, dx \qquad \text{for } x \in \Omega_k. \tag{4.23}$$

For each k, sample $N_k = N/M$ points $\{x_i^{(k)}\}$ uniformly distributed in Ω_k. Then the stratified quadrature formula is just the sum of the quadratures over each subset, that is,

$$I_N = N^{-1} \sum_{k=1}^{M} \sum_{i=1}^{N/M} f\left(x_i^{(k)}\right). \tag{4.24}$$

The Monte Carlo quadrature error for this stratified sum is

$$\begin{aligned} \epsilon &\approx N^{-1/2}\sigma_s, \\ \sigma_s^2 &= \int \left(f(x) - \bar{f}(x)\right)^2 \,\mathrm{d}x \\ &= \sum_{k=1}^{M} \int_{\Omega_k} \left(f(x) - \bar{f}_k\right)^2 \,\mathrm{d}x. \end{aligned} \tag{4.25}$$

For this stratified quadrature rule, there is a simple result. Stratified Monte Carlo quadrature always beats the unstratified quadrature, since

$$\sigma_s \le \sigma. \tag{4.26}$$

The proof of this inequality is straightforward. For each k, $c = \bar{f}_k$ is the minimizer of

$$\int_{\Omega_k} \left(f(x) - c\right)^2 \,\mathrm{d}x. \tag{4.27}$$

In particular,

$$\int_{\Omega_k} \left(f(x) - \bar{f}_k\right)^2 \,\mathrm{d}x \le \int_{\Omega_k} \left(f(x) - \bar{f}\right)^2 \,\mathrm{d}x. \tag{4.28}$$

Add this over all k to get

$$\begin{aligned} \sigma_s^2 &= \sum_{k=1}^{M} \int_{\Omega_k} \left(f(x) - \bar{f}_k\right)^2 \,\mathrm{d}x \\ &\le \sum_{k=1}^{M} \int_{\Omega_k} \left(f(x) - \bar{f}\right)^2 \,\mathrm{d}x \\ &= \sigma^2. \end{aligned} \tag{4.29}$$

Stratification can be phrased more generally as follows. Split the integration region Ω into M pieces Ω_k with

$$\Omega = \cup_{k=1}^{M} \Omega_k. \tag{4.30}$$

Take N_k random variables in each piece Ω_k with

$$\sum_{k=1}^{M} N_k = N. \tag{4.31}$$

In each subset Ω_k choose points $x_n^{(k)}$ distributed with density $p^{(k)}(x)$ in which

$$\begin{aligned} p^{(k)}(x) &= p(x)/\bar{p}_k, \\ \bar{p}_k &= \int_{\Omega_k} p(x)\,\mathrm{d}x. \end{aligned} \tag{4.32}$$

The stratified quadrature formula is the following sum over k:

$$I_N[f] = \sum_{k=1}^{M} \frac{\bar{p}_k}{N_k} \sum_{n=1}^{N_k} f\left(x_n^{(k)}\right). \tag{4.33}$$

The resulting integration error is

$$\begin{aligned} \epsilon_N[f] &= I[f] - I_N[f] \\ &= \sum_{k=1}^{M} \epsilon_{N_k}^{(k)}[f]. \end{aligned} \tag{4.34}$$

The components of this error are

$$\begin{aligned} \epsilon_{N_k}^{(k)}[f] &\approx N_k^{-1/2} \bar{p}_k \left(\int_{\Omega_k} (f(x) - \bar{f}_k)^2 p^{(k)}(x)\,\mathrm{d}x \right)^{1/2} \\ &= (\bar{p}_k/N_k)^{1/2} \sigma^{(k)}, \end{aligned} \tag{4.35}$$

in which the variances are

$$\begin{aligned} \sigma^{(k)} &= (\bar{p}_k)^{1/2} \left(\int_{\Omega_k} (f(x) - \bar{f}_k)^2 p^{(k)}(x)\,\mathrm{d}x \right)^{1/2} \\ &= \left(\int_{\Omega_k} (f(x) - \bar{f}_k)^2 p(x)\,\mathrm{d}x \right)^{1/2}, \end{aligned} \tag{4.36}$$

and the averages are

$$\bar{f}_k = \int_{\Omega_k} f(x) p(x)\,\mathrm{d}x / \bar{p}_k. \tag{4.37}$$

Stratification always lowers the integration error if the distribution of points is balanced. The balance condition is that, for all k,

$$\bar{p}_k/N_k = 1/N, \tag{4.38}$$

that is, the number of points in set Ω_k is proportional to its weighted size $\bar{p}_k$. The resulting error for stratified quadrature is

$$\epsilon_N \approx N^{-1/2}\sigma_s, \qquad \sigma_s^2 = \sum_{k=1}^{M} {\sigma^{(k)}}^2. \tag{4.39}$$

Since the variance over a subset is always less than the variance over the whole set, that is,

$$\sigma_s \le \sigma, \tag{4.40}$$

then stratification always lowers the integration error. Actually, a better choice than the balance condition may be to put more points where f has largest variation. An adaptive method for stratification was formulated by Lepage (1978) and is described by Press et al. (1992).

4.5. Importance sampling

Importance sampling is probably the most widely used Monte Carlo variance reduction method. Consider the simple integral $\int f(x)\,\mathrm{d}x$ and rewrite it by introducing a density p as

$$\begin{aligned} I[f] &= \int f(x)\,\mathrm{d}x \\ &= \int \frac{f(x)}{p(x)} p(x)\,\mathrm{d}x. \end{aligned} \tag{4.41}$$

Now think of this as an integral with density function p. Sample points x_n from the density $p(x)$ and form the Monte Carlo estimate

$$I_N[f] = \frac{1}{N}\sum_{n=1}^{N} \frac{f(x_n)}{p(x_n)}. \tag{4.42}$$

The resulting error $\epsilon_N[f] = I[f] - I_N[f]$ has size

$$\epsilon_N[f] \approx \sigma_p N^{-1/2}, \tag{4.43}$$

in which the variance σ_p is

$$\sigma_p = \int \left(\frac{f(x)}{p(x)} - I\right)^2 p(x)\,\mathrm{d}x. \tag{4.44}$$

So importance sampling will reduce the Monte Carlo quadrature error, if

$$\sigma_p \ll \sigma. \tag{4.45}$$

Importance sampling is an effective method when f/p is nearly constant, so that σ_p is small. One difficulty in this method is that sampling from the density p is required, but this can be performed using acceptance–rejection, if necessary. One use of importance sampling is to emphasize rare but important events, *i.e.*, small regions of space in which the integrand f is large.

4.6. Russian roulette

Some Monte Carlo computations involve infinite sums, for instance, due to iteration of an integral equation. The Russian roulette method converts an infinite sum into a sum that is of finite length for each sample. Consider the sum

$$S = \sum_{n=0}^{\infty} a_n \tag{4.46}$$

and suppose that the terms a_n are exponentially decreasing, that is,

$$|a_n| \le c\lambda^n, \tag{4.47}$$

in which $0 < \lambda < 1$. Choose $\lambda < \kappa < 1$, and let M be chosen according to a discrete exponential distribution, so that

$$\mathrm{Prob}(M \ge n) = \kappa^n. \tag{4.48}$$

Define the random sum

$$\tilde{S} = \sum_{n=0}^{M} \kappa^{-n} a_n. \tag{4.49}$$

Since $|\kappa^{-n} a_n| < (\lambda/\kappa)^n$ and $|\lambda/\kappa| < 1$, then this sum is uniformly bounded for all M. Then

$$\begin{aligned} E[\tilde{S}] &= \sum_{n=0}^{\infty} \mathrm{Prob}(M \ge n)\kappa^{-n} a_n \\ &= S. \end{aligned} \tag{4.50}$$

This formula leads to the following Monte Carlo method for computation of the infinite sum (4.46):

$$S_N = N^{-1} \sum_{i=0}^{N} \sum_{n=0}^{M_i} \kappa^{-n} a_n, \tag{4.51}$$

in which the values M_i are chosen according to the probability distribution (4.48). In this method, the terms a_n could also be computed by a Monte Carlo integral.

4.7. Example of variance reduction

As an example of the use of variance reduction consider the three-dimensional Gaussian integral

$$I[u] = (2\pi)^{-3/2} \int_{-\infty}^{\infty}\int_{-\infty}^{\infty}\int_{-\infty}^{\infty} u e^{-(x_1^2+x_2^2+x_3^2)/2}\, dx_1\, dx_2\, dx_3, \tag{4.52}$$

in which the integrand u is defined by

$$\begin{aligned} u &= (1+r_0)^{-1}(1+r_1)^{-1}(1+r_2)^{-1}(1+r_3)^{-1}, \\ r_1 &= r_0 e^{\sigma x_1 - \sigma^2/2}, \\ r_2 &= r_1 e^{\sigma x_2 - \sigma^2/2}, \\ r_3 &= r_2 e^{\sigma x_3 - \sigma^2/2}. \end{aligned} \tag{4.53}$$

This integral can be interpreted as the present value of a payment of \$1 four years from now, for an interest rate of r_i in year i. The interest rates are allowed to evolve according to a lognormal model with variance σ, that is,

$$r_i = r_{i-1} e^{\sigma x_i - \sigma^2/2}, \tag{4.54}$$

in which the $x_i's$ are independent $N(0,1)$ Gaussian random variables. The factors $(1+r_i)^{-1}$ are the discount factors. For the computations below, we take $r_0 = 0.10$ and $\sigma = 0.1$.

We evaluate $I[u]$ by sampling x_i from an $N(0,1)$ distribution using the transformation method, that is, by direct inversion of the cumulative distribution function for a normal random variable. Then we apply antithetic variables and control variates to this problem.

Approximate $(1+r_i)^{-1}$ by $1-r_i$ for $i \geq 1$. Then form the control variate as

$$g = (1+r_0)^{-1}(1-r_1)(1-r_2)(1-r_3). \tag{4.55}$$

Since g consists of a sum of linear exponentials, its integral can be performed exactly, for instance

$$\begin{aligned} \int_{-\infty}^{\infty} e^{\lambda x} e^{-x^2/2}\, dx &= e^{\lambda^2/2} \int_{-\infty}^{\infty} e^{-(x-\lambda)^2/2}\, dx \\ &= e^{\lambda^2/2} \int_{-\infty}^{\infty} e^{-y^2/2}\, dy \\ &= \sqrt{2\pi} e^{\lambda^2/2}. \end{aligned} \tag{4.56}$$

Numerical results are presented in Figure 2. This compares the quadrature error from standard Monte Carlo, antithetic variables and control variates, using both standard pseudo-random points and the quasi-random (Sobol') points that will be discussed in the next section. In order to make a meaningful comparison, we have performed an average (root mean square) of the error over 20 runs for each value of N. The computations are all independent, that is, the same random number is never used twice. The figure also shows the results from a single run without averaging or variance reduction. The results show the following points.

- Both control variates and antithetic variables provide significant error reduction.
- Control variates and antithetic variables together perform better than either one alone. This shows that the error reduction from the two methods are different.
- Quasi-Monte Carlo, which will be discussed in Section 5, has smaller error and a faster rate of convergence.
- Control variates and antithetic variables, both separately and together, provide further error reduction when used with quasi-Monte Carlo.

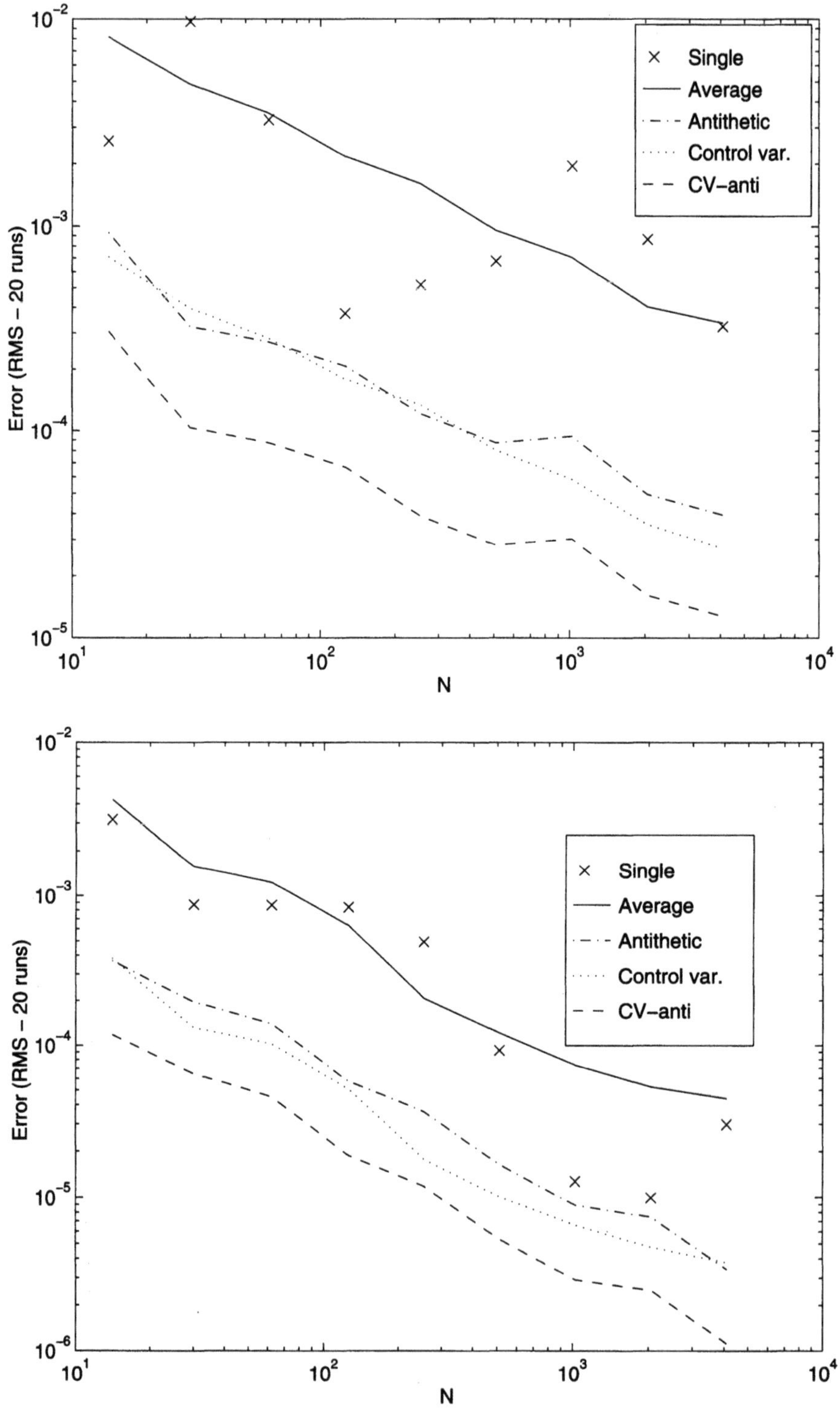

Fig. 2. Discounted cashflow, Monte Carlo (top) and quasi-Monte Carlo (bottom)

5. Quasi-random numbers

Quasi-random sequences are a deterministic alternative to random or pseudo-random sequences (Kuipers and Niederreiter 1974, Hua and Wang 1981, Niederreiter 1992, Zaremba 1968). In contrast to pseudo-random sequences, which try to mimic the properties of random sequences, quasi-random sequences are designed to provide better uniformity than a random sequence, and hence faster convergence for quadrature formulas. Uniformity of a sequence is measured in terms of its *discrepancy*, which is defined in Section 5.2 below, and for this reason quasi-random sequences are also called low-discrepancy sequences. Some have objected to the name quasi-random, since these sequences are intentionally not random. We continue to use this name, however, because of the unpredictability of quasi-random sequences.

5.1. Monte Carlo versus quasi-Monte Carlo

The difference between standard Monte Carlo and quasi-Monte Carlo is best illustrated by Figure 3, which compares a pseudo-random sequence with a quasi-random sequence (a Sobol' sequence) in two dimensions. Standard Monte Carlo methods use pseudo-random sequences and provide a convergence rate of $O(N^{-1/2})$ for Monte Carlo quadrature using N samples. In addition to integration problems, standard Monte Carlo is applicable to optimization and simulation problems.

The limiting factor in accuracy of standard Monte Carlo is the clumping that occurs in the points of a random or pseudo-random sequence. Clumping of points, as well as spaces that have no points, is clearly seen in the pseudo-random sequence in Figure 3. The reason for this clumping is that the points are independent. Since different points know nothing about each other, there is some small chance that they will lie very close together. A simple argument shows that about $\sqrt{N}$ out of N points lie in clumps.

Quasi-Monte Carlo methods use quasi-random sequences, which are deterministic, with correlations between the points to eliminate clumping. The resulting convergence rate is $O((\log N)^k N^{-1})$. Because of the correlations, quasi-random sequences are less versatile than random or pseudo-random sequences. They are designed for integration, rather than simulation or optimization. On the other hand, the desired result of a simulation can often be written as an expectation, which is an integral, so that quasi-Monte Carlo is then applicable. As discussed below, this often introduces high dimensionality, which can limit the effectiveness of quasi-Monte Carlo sequences.

5.2. Discrepancy

Quasi-random sequences were invented by number theorists who were interested in the uniformity properties of numerical sequences (Kuipers and Niederreiter 1974, Hua and Wang 1981). An important first step is formu-

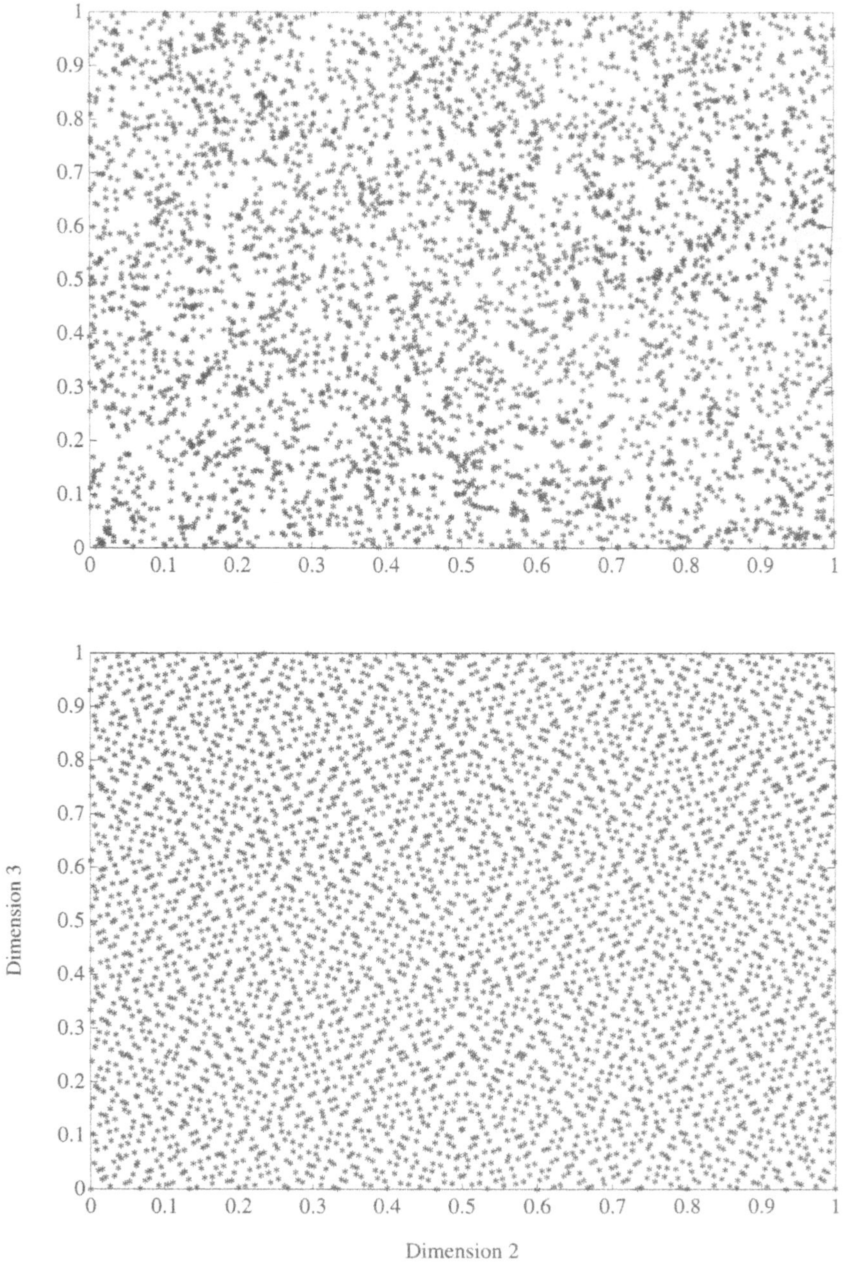

Fig. 3. Two-dimensional projection of a pseudo-random sequence (top) and a Sobol' sequence (bottom)

lating a quantitative measure of uniformity. Uniformity of a sequence of points is measured in terms of its *discrepancy.*

For a sequence of N points $\{\boldsymbol{x}_n\}$ in the unit cube I^d, define

$$R_N(J) = \frac{1}{N}\#\{\boldsymbol{x}_n \in J\} - m(J) \tag{5.1}$$

for any subset J of I^d. $R_N(J)$ is just the Monte Carlo quadrature error in measuring the volume of J. The discrepancy is then defined by some norm applied to $R_N(J)$. First, restrict J to be a rectangular set, and define the set of all such subsets to be E. Since a rectangular set can be defined by two antipodal vertices, a rectangle J can be identified as $J = J(\boldsymbol{x}, \boldsymbol{y})$ in which the points $\boldsymbol{x}$ and $\boldsymbol{y}$ are antipodal vertices. Now define two discrepancies: D_N, which is an L^∞ norm on J, and T_N, which is an L^2 norm, defined by

$$\begin{aligned} D_N &= \sup_{J\in E} |R_N(J)| \\ T_N &= \left[\int_{(\boldsymbol{x},\boldsymbol{y})\in I^{2d}, x_i<y_i} R_N(J(\boldsymbol{x},\boldsymbol{y}))^2 \, \mathrm{d}\boldsymbol{x} \, \mathrm{d}\boldsymbol{y}\right]^{\frac{1}{2}}. \end{aligned} \tag{5.2}$$

It is useful to also define the discrepancy over rectangular sets with one vertex at 0, as

$$\begin{aligned} D_N^* &= \sup_{J\in E^*} |R_N(J)| \\ T_N^* &= \left[\int_{I^d} R_N(J(0,\boldsymbol{x}))^2 \, \mathrm{d}\boldsymbol{x}\right]^{\frac{1}{2}} \end{aligned} \tag{5.3}$$

in which E^* is the set $\{J(0, \boldsymbol{x})\}$. Other definitions include terms from lower-dimensional projections of the sequence (Hickernell 1998).

5.3. Koksma–Hlawka inequality

Quasi-random sequences are useful for integration because they lead to much smaller error than standard Monte Carlo quadrature. The basis for analysing quasi-Monte Carlo quadrature error is the Koksma–Hlawka inequality.

Consider the integral

$$I[f] = \int_{I^d} f(x) \, \mathrm{d}x \tag{5.4}$$

and the Monte Carlo approximation

$$I_N[f] = \frac{1}{N}\sum_{n=1}^{N} f(x_n), \tag{5.5}$$

and define the quadrature error

$$\varepsilon[f] = |I[f] - I_N[f]|. \tag{5.6}$$

Define the variation (in the Hardy–Krause sense) of f, a function of a single variable, as

$$V[f] = \int_0^1 \left| \frac{\mathrm{d}f}{\mathrm{d}t} \right| \mathrm{d}t. \tag{5.7}$$

In d dimensions, the variation is defined as

$$V[f] = \int_{I^d} \left| \frac{\partial^d f}{\partial t_1 \cdots \partial t_d} \right| \mathrm{d}t_1 \cdots \mathrm{d}t_d + \sum_{i=1}^{d} V[f_1^{(i)}] \tag{5.8}$$

in which $f_1^{(i)}$ is the restriction of the function f to the boundary $x_i = 1$. Since these restrictions are functions of $d-1$ variables, this definition is recursive.

Theorem 5.1 (Koksma–Hlawka theorem) For any sequence $\{x_n\}$ and any function f with bounded variation, the integration error ε is bounded as

$$\varepsilon[f] \leq V[f]\, D_N^*. \tag{5.9}$$

It is instructive to compare the Koksma–Hlawka result to the root mean square error (RMSE) of standard Monte Carlo integration using a random sequence, that is,

$$E[\varepsilon[f]^2]^{1/2} = \sigma[f] N^{-1/2}, \tag{5.10}$$

in which

$$\begin{aligned} \sigma[f] &= \left(\int_{I^d} (f(x) - I[f])^2 \, \mathrm{d}x \right)^{1/2}, \\ \varepsilon[f] &= \int_{I^d} f(x) \, \mathrm{d}x - \frac{1}{N} \sum_{n=1}^{N} f(x_n). \end{aligned} \tag{5.11}$$

In comparing the Koksma–Hlawka inequality (5.9) and the RMSE for standard Monte Carlo integration (5.10), note the following points.

- Both error bounds are a product of one factor that depends on the sequence (*i.e.*, D_N for Koksma–Hlawka and $N^{-1/2}$ for RMSE) and a factor that depends on the function f (*i.e.*, $V[f]$ for Koksma–Hlawka and $\sigma[f]$ for RMSE).
- The Koksma–Hlawka inequality is a worst-case bound, whereas the RMSE (5.10) is a probabilistic bound.
- The RMSE (5.10) is an equality, and so it is tight, whereas Koksma–Hlawka is an upper bound.
- Our experience is that $V[f]$ in Koksma–Hlawka is usually a gross overestimate, while the discrepancy is indicative of actual performance.

A further, more subtle point is that for standard Monte Carlo each point is an estimate of the entire integral. A typical quasi-random sequence, on the other hand, has a hierarchical structure, so that the initial points sample the integrand on a coarse scale and later points sample it on a finer scale. This is the source of the $\log N$ terms in the discrepancy bounds cited below.

5.4. Proof of the Koksma–Hlawka inequality

First consider functions f that vanish on the boundary of the unit cube. Note that

$$R(\boldsymbol{x}) = \left\{ \frac{1}{N} \sum_{n=1}^{N} \delta(\boldsymbol{x} - \boldsymbol{x}_n) - 1 \right\} \mathrm{d}\boldsymbol{x}, \tag{5.12}$$

in which $R(\boldsymbol{x}) = R_N(J(0, \boldsymbol{x}))$ as defined in Section 5.2. Since there are no boundary terms, then

$$\begin{aligned} \varepsilon[f] &= \left| \int_{I^d} f(\boldsymbol{x}) \,\mathrm{d}\boldsymbol{x} - \frac{1}{N} \sum_{n=1}^{N} f(\boldsymbol{x}_n) \right| \\ &= \left| \int_{I^d} \left\{ 1 - \frac{1}{N} \sum_{n=1}^{N} \delta(\boldsymbol{x} - \boldsymbol{x}_n) \right\} f(\boldsymbol{x}) \,\mathrm{d}\boldsymbol{x} \right| \\ &= \left| \int_{I^d} R(\boldsymbol{x}) \,\mathrm{d}f(\boldsymbol{x}) \right| \\ &\le \left(\sup_{\boldsymbol{x}} R(\boldsymbol{x}) \right) \int_{I^d} |\mathrm{d}f(\boldsymbol{x})| \\ &= D_N^* V[f]. \end{aligned} \tag{5.13}$$

For f that is nonzero on the boundary of the unit cube, the terms from integration by parts are bounded by the boundary terms in $V[f]$.

5.5. Average integration error

Most computational experience confirms that discrepancy is a good indicator of quasi-Monte Carlo performance, while variation is not a typical indicator. In 1990, Woźniakowski (Woźniakowski 1991) proved the following surprising result.

Theorem 5.2 We have

$$E[\varepsilon_N[f]^2]^{1/2} = T_N^*, \tag{5.14}$$

in which the expectation E is taken with respect to functions f distributed according to the Brownian sheet measure.

The Brownian sheet measure is a measure on function space. It is a natural generalization of simple Brownian motion $b(x)$ to multi-dimensional 'time' $\boldsymbol{x}$. With probability one, a function $f(\boldsymbol{x})$, chosen from the Brownian sheet measure, is approximately 'half-differentiable'. In particular, its variation is infinite. In this context, we can definitely say that variation is not a good indicator of integration error. Since Woźniakowski's identity (5.14) is an equality, it follows that the L^2 discrepancy T_N^* does agree with the typical size of quasi-Monte Carlo integration error. Our computational experience is that T_N^* and D_N are usually of the same size.

Denote $\boldsymbol{x}' = (x_i')_{i=1}^d$ in which $x_i' = 1 - x_i$; also denote the finite difference operator $D_i f(\boldsymbol{x}') = f(\boldsymbol{x}' + \Delta_i \hat{e}_i') - f(\boldsymbol{x}')$, in which $\hat{e}_i'$ is the ith coordinate vector. The Brownian sheet is based at the point $\boldsymbol{x}' = 0$, that is, $\boldsymbol{x} = (1, \ldots, 1)$, and has $f(\boldsymbol{x}' = 0) = f(1, \ldots, 1) = 0$. For any point $\boldsymbol{x}'$ in $I^d$and any set of positive lengths Δ_i (with $x_i' + \Delta_i < 1$), the multi-dimensional difference $D_1 \ldots D_d f(\boldsymbol{x})$ is a normally distributed random variable with mean zero and variance

$$E[(D_1 \ldots D_d f(\boldsymbol{x}))^2] = \Delta_1 \ldots \Delta_d. \tag{5.15}$$

This implies that

$$E[\,\mathrm{d}f(\boldsymbol{x})\,\mathrm{d}f(\boldsymbol{y})] = \delta(\boldsymbol{x} - \boldsymbol{y})\,\mathrm{d}\boldsymbol{x}\,\mathrm{d}\boldsymbol{y}, \tag{5.16}$$

in which $\mathrm{d}f$ is understood in the sense of the Itô calculus (Karatzas and Shreve 1991). Moreover $f(\boldsymbol{x}') = 0$ if $x_i' = 0$ for any i. For any $\boldsymbol{x}$ in I^d, $f(\boldsymbol{x})$ is normally distributed with mean zero and variance

$$E[f(\boldsymbol{x})^2] = \prod_{i=1}^{d} x_i'. \tag{5.17}$$

Woźniakowski's proof of (5.14) was a straightforward calculation of both sides of the equality. The following proof, derived by Morokoff and Caflisch (1994), is based on the properties of the Brownian sheet measure and follows the same lines as the proof of the Koksma–Hlawka inequality.

Proof of Woźniakowski's identity. First rewrite the integration error $E[f]$ using integration by parts, following the steps of the proof of the Koksma–Hlawka inequality. Note that

$$R(\boldsymbol{x}) = \left\{ \frac{1}{N} \sum_{n=1}^{N} \delta(\boldsymbol{x} - \boldsymbol{x}_n) - 1 \right\} \mathrm{d}\boldsymbol{x}, \tag{5.18}$$

in which $R(\boldsymbol{x}) = R_N(J(0, \boldsymbol{x}))$ as defined in Section 5.2. Also, $R(\boldsymbol{x}) = 0$ if $x_i = 0$, and $f(\boldsymbol{x}) = 0$ if $x_i = 1$, for any i. This implies that the boundary terms all disappear in the following integration by parts:

$$\begin{aligned}
\varepsilon[f] &= \left| \int_{I^d} f(\boldsymbol{x})\,\mathrm{d}\boldsymbol{x} - \frac{1}{N}\sum_{n=1}^{N} f(\boldsymbol{x}_n) \right| \\
&= \left| \int_{I^d} \left\{ 1 - \frac{1}{N}\sum_{n=1}^{N} \delta(\boldsymbol{x} - \boldsymbol{x}_n) \right\} f(\boldsymbol{x})\,\mathrm{d}\boldsymbol{x} \right| \\
&= \left| \int_{I^d} R(\boldsymbol{x})\,\mathrm{d}f(\boldsymbol{x}) \right|.
\end{aligned}$$

The quantity $\mathrm{d}f$ in this identity is defined here through the Itô calculus, even though $V[f] = \infty$ with probability one.

It follows from (5.16) that the average square error is

$$\begin{aligned}
E[\varepsilon[f]^2] &= E\left[\left(\int_{I^d} R(\boldsymbol{x})\,\mathrm{d}f(\boldsymbol{x}) \right)^2 \right] \\
&= \int_{I^d \times I^d} R(\boldsymbol{x})R(\boldsymbol{y})E[\,\mathrm{d}f(\boldsymbol{x})\,\mathrm{d}f(\boldsymbol{y})] \\
&= \int_{I^d} R(\boldsymbol{x})^2\,\mathrm{d}\boldsymbol{x} \\
&= (T_N^*)^2.
\end{aligned} \tag{5.19}$$

5.6. Quasi-random number generators

An infinite sequence $\{\boldsymbol{x}_n\}$ is *uniformly distributed* if

$$\lim_{N\to\infty} D_N = 0. \tag{5.20}$$

The sequence is *quasi-random* if

$$D_N \le c(\log N)^k N^{-1}, \tag{5.21}$$

in which c and k are constants that are independent of N, but may depend on the dimension d. In particular, it is possible to construct sequences for which $k = d$. It is also common to say that a sequence is quasi-random only if the exponent k is equal to the dimension d of the sequence.

The simplest example of a quasi-random sequence is the Van der Corput sequence in one dimension ($d = 1$) (Niederreiter 1992). It is described most simply as follows. Write out n in base 2. Then obtain the nth point x_n by reversion of the bits of n around the decimal point, that is,

$$\begin{aligned}
n &= a_m a_{m-1} \dots a_1 a_0 \quad \text{(base 2)} \\
x_n &= 0.a_0 a_1 \dots a_{m-1} a_m \quad \text{(base 2)}.
\end{aligned} \tag{5.22}$$

Halton sequences (Halton 1960) are a generalization of this procedure. In the kth dimension, expand n in base p_k, the kth prime, and form the kth component by reversion of this expansion around the decimal point. The discrepancy of a Halton sequence is bounded by

$$D_N(\text{Halton}) \leq c_d(\log N)^d N^{-1} \tag{5.23}$$

in which c_d is a constant depending on d. On the other hand, the average discrepancy of a random sequence is

$$E[T_N^2(\text{random})]^{1/2} = c_d N^{-1/2}. \tag{5.24}$$

Additional sequences go by the names of Haselgrove (Haselgrove 1961), Faure (Faure 1982), Sobol' (Sobol' 1967, 1976), Niederreiter (Niederreiter 1992, Xing and Niederreiter 1995), and Owen (Owen 1995, 1997, 1998, Hickernell 1996). Niederreiter has formulated a general theory for quasi-random sequences in terms of (t, s) nets (Niederreiter 1992). All of these sequences satisfy bounds like (5.23), but the more recent constructions have much better constants c_d.

An algorithm for construction of a Sobol' sequence is found in Press et al. (1992) and for Niederreiter sequences in Bratley, Fox and Niederreiter (1994). Various quasi-random number generators are found in the software collection ACM-Toms at `http://www.netlib.org/toms/index.html`.

For quasi-random sequences with discrepancy of size $O((\log N)^d N^{-1})$, the Koksma–Hlawka inequality implies that the integration error is of this same size, that is,

$$\varepsilon[f] \leq cV[f](\log N)^d N^{-1}. \tag{5.25}$$

Note that the Koksma–Hlawka inequality applies for any sequence. For quasi-random sequences, the discrepancy is small, so that the integration error is also small.

5.7. Limitations: smoothness and dimension

The Koksma–Hlawka inequality and discrepancy bounds for a quasi-random sequence together imply that quasi-Monte Carlo quadrature converges much faster than standard Monte Carlo quadrature. On the other hand, there are several distinct limitations to the effectiveness of quasi-Monte Carlo methods (Morokoff and Caflisch 1994, 1995).

First, there is no theoretical basis for empirical estimates of accuracy of quasi-Monte Carlo methods. Recall from Section 2 that the Central Limit Theorem can be used to test the accuracy of standard Monte Carlo quadrature as the computation proceeds, and then to predict the required number of samples N for a given accuracy level. Since there is no Central Limit Theorem for quasi-random sequences, there is no corresponding empirical error estimate for quasi-Monte Carlo. On the other hand, confidence in the accuracy of quasi-Monte Carlo integration comes from refining a set of computations.

Second, quasi-Monte Carlo methods are designed for integration and are not directly applicable to simulation. This is because of the correlations between the points of a quasi-random sequence. However, in many simulations the desired result is an expectation of some quantity, which can then be written as a high-dimensional integral on which quasi-Monte Carlo can be used. In fact, we can think of the different dimensions of a quasi-random point as independent, so that quasi-random sequences can represent a simulation by allocating one dimension per time-step or decision. This approach often requires a high-dimensional quasi-random sequence.

Third, quasi-Monte Carlo integration is found to lose its effectiveness when the dimension of the integral becomes large. This can be anticipated from the bound $(\log N)^d N^{-1}$ on discrepancy. For large dimension d, this bound is dominated by the $(\log N)^d$ term unless

$$N > 2^d. \tag{5.26}$$

Fourth, quasi-Monte Carlo integration is found to lose its effectiveness if the integrand f is not smooth. The factor $V[f]$ in the Koksma–Hlawka inequality (5.9) is an indicator of this dependence, although we actually find that a much smaller amount of smoothness, somewhere between continuity and differentiability, is usually enough. The limitation on smoothness is significant, because many Monte Carlo methods involve decisions of some sort, which usually correspond to multiplication by 0 or 1.

In the next subsection, some computational evidence for limitations on dimension and smoothness will be presented. Techniques for overcoming these limitations will be discussed in Section 6.

An important lesson from our computational experience is that quasi-Monte Carlo integration is almost always as accurate as standard Monte Carlo integration. So the 'loss of effectiveness' cited above means that quasi-Monte Carlo performs no better than standard Monte Carlo. The reasons for this are not clear, but it is consistent with the discrepancy computations presented in the next subsection.

5.8. Dependence on dimension

To demonstrate how quasi-random sequences depend on dimension, we will present results from two computational tests. The first is a computation of L^2 discrepancy for a quasi-random sequence. The second is an examination of the one- and two-dimensional projections of quasi-random sequences.

The L^2 discrepancy T_N can be computed directly (Morokoff and Caflisch 1994) by an explicit formula. Figure 4 shows computation of T_N for a Sobol' sequence for dimension 4 and 16. In addition to T_N, each graph shows a plot of $N^{-1/2}$ with a constant coming from the expected discrepancy of a random sequence. Each graph also shows a plot of N^{-1} (with a constant of

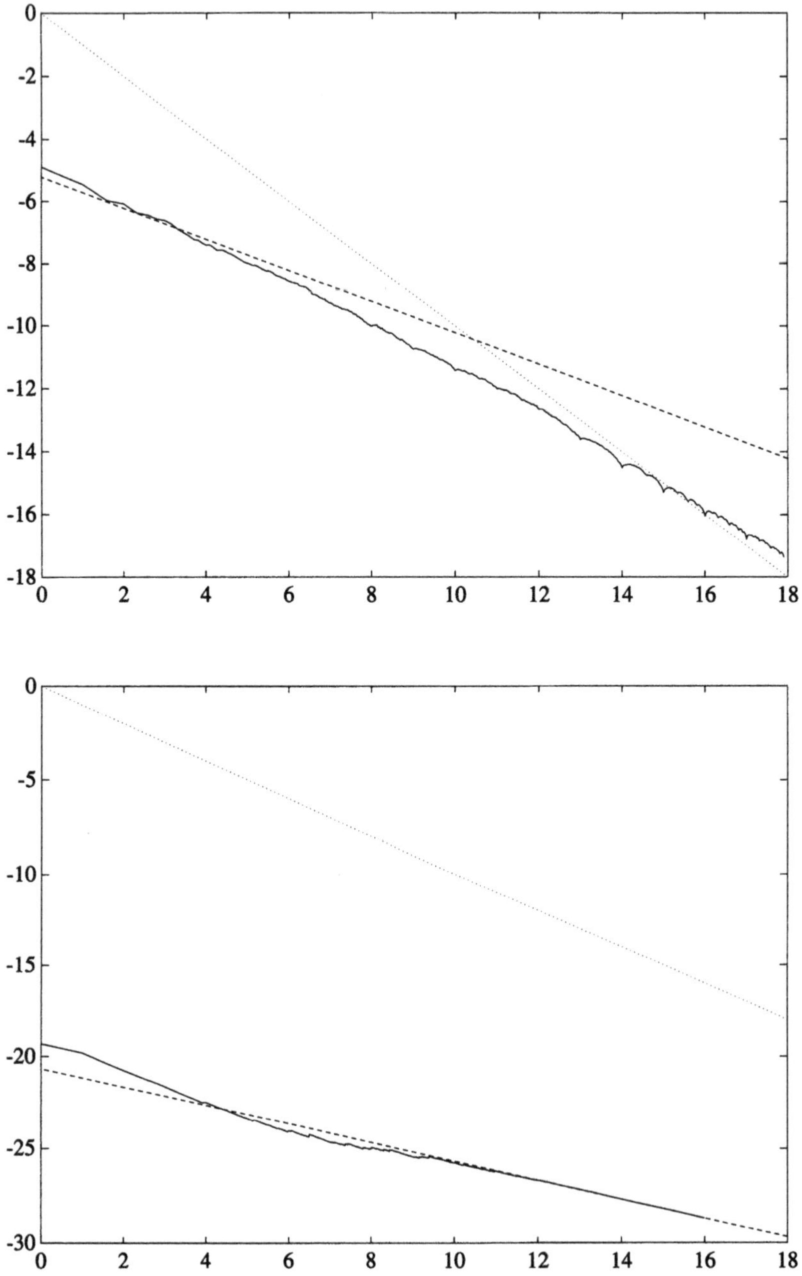

Fig. 4. Log-log (base 10) plot of discrepancy T_N for a Sobol' sequence in 4 dimensions (top) and for 16 dimensions (bottom)

no significance). Since the plots are on a log-log scale, these reference curves are lines of slope $-1/2$ and -1.

The figure for small dimension (4) and for large N, shows that $T_N \approx O(N^{-1})$. We have not tried to match the $\log N$ terms, since we do not seem to be able to detect them reliably. On the other hand, the graph for large dimension (16), shows that for moderate sized N, $T_N \approx O(N^{-1/2})$, with a constant that agrees with the discrepancy of a random sequence. Eventually, as N gets very large, T_N must be nearly of size $O(N^{-1})$. The agreement between discrepancy of quasi-random and random sequences for large d and moderate N has not been explained.

Next we consider one- and two-dimensional projections of a quasi-random sequence. Many, but not all, quasi-random sequences have the following special property. All one-dimensional projections of the sequence (*i.e.*, all single components) are equally well distributed. The Sobol' sequence, for example, has this property. This implies that any function f that consists of a sum of one-dimensional functions will be well integrated by quasi-Monte Carlo, in other words, functions of the form

$$f(\boldsymbol{x}) = \sum_{n=1}^{d} f_n(x_n). \tag{5.27}$$

In particular, quasi-Monte Carlo performs very well for linear functions

$$f(\boldsymbol{x}) \approx \sum_{i=1}^{d} a_i x_i. \tag{5.28}$$

Now consider two-dimensional projections. Figure 5 shows a 'good' and a 'bad' two-dimensional projection of a Sobol' sequence. The top plot of the 'good' projection is very uniform, while the bottom plot of the 'bad' projection shows gaps and a repetitive structure to the points. For this projection, the holes would be filled in as the further points of the sequence are laid down. At certain special values of N, the sequence would be quite uniform. We have observed patterns of this sort in a variety of quasi-random sequences (Morokoff and Caflisch 1994), but they are not expected to occur in the randomized quasi-random points of Owen (1995, 1997).

6. Quasi-Monte Carlo techniques

In this section, two techniques are described for overcoming the smoothness and high-dimension limitations of quasi-Monte Carlo. The first method is a smoothing method and will be demonstrated on the acceptance–rejection method, which involves a characteristic function in its standard form, coming from the decision to accept or reject. The second technique is a rearrangement of dimension, so as to put most of the weight of an integral into the leading dimensions of the quasi-random sequence.

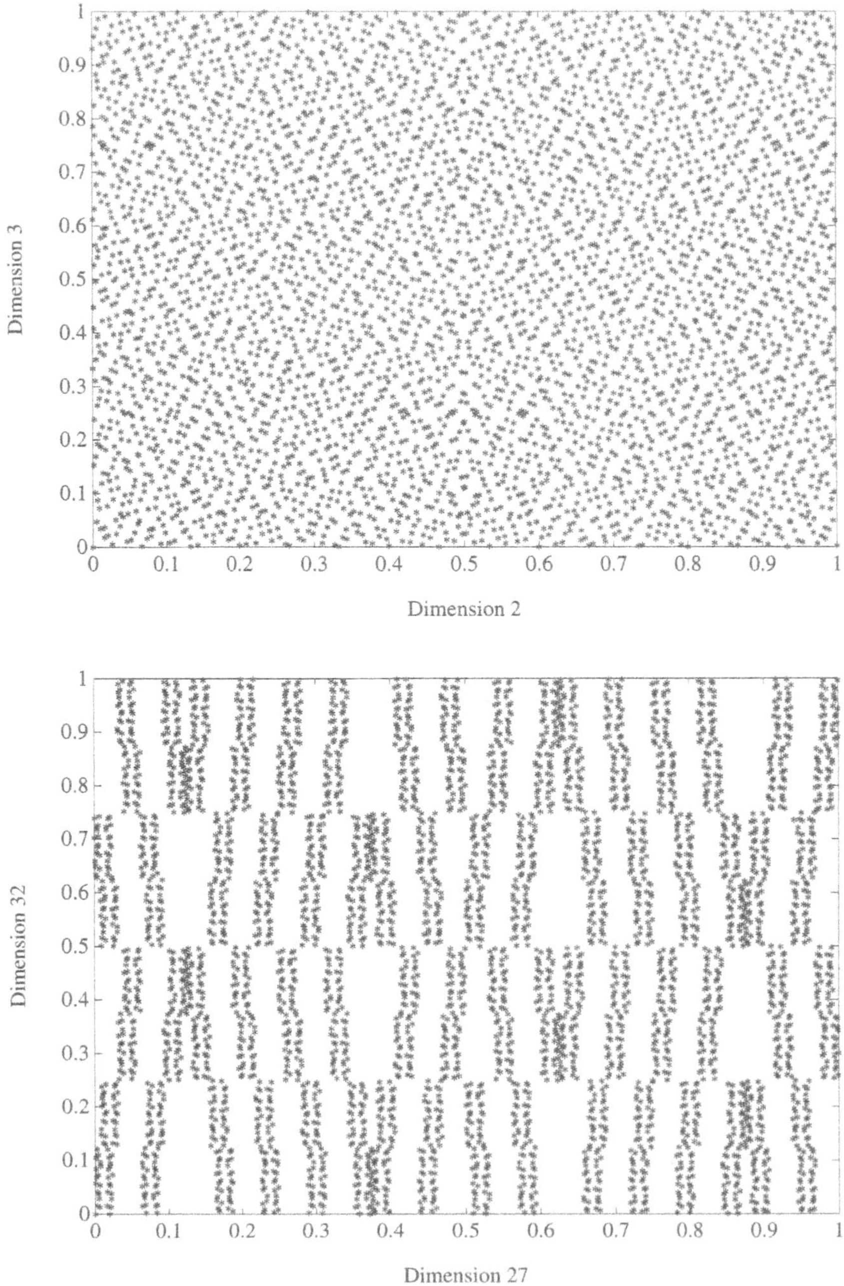

Fig. 5. Good and bad two-dimensional projections of a Sobol' sequence

6.1. *Smoothing*

The effectiveness of quasi-Monte Carlo is often lost on problems involving discontinuities, such as decisions in the acceptance–rejection method. One way to overcome this difficulty is to smooth out the integrand without changing the value of the integral. Here we describe a smoothed version of the acceptance–rejection method and compare it with standard acceptance–rejection, when used with a quasi-random sequence.

Standard acceptance–rejection can be expressed in an integration problem in the following way. Consider the integral of a function f multiplying a function p on the unit interval, with $0 \leq p(x) \leq 1$. Rewrite the integral as

$$\int_0^1 f(x)p(x)\,\mathrm{d}x = \int_0^1 \int_0^1 f(x)\chi(y < p(x))\,\mathrm{d}y\,\mathrm{d}x. \tag{6.1}$$

The acceptance–rejection method is a slight modification of the usual Monte Carlo formula for this integral, that is,

- sample (x, y) uniformly
- weight $w = \chi(y < p(x)) = 1$ (accept), if $y < p(x)$
- weight $w = \chi(y < p(x)) = 0$ (reject), if $y > p(x)$
- repeat until number of accepted points is N.

The Monte Carlo quadrature formula is then

$$I_N = N^{-1} \sum_{i=1}^{M} w_i f(x_i), \tag{6.2}$$

in which

$$N \approx \sum_{i=1}^{M} w_i. \tag{6.3}$$

The smoothed acceptance–rejection method (Spanier and Maize 1994, Moskowitz and Caflisch 1996) is formulated by replacing the discontinuous function $\chi(y < p(x))$ by a smooth function $q(x, y)$ satisfying

$$\int_0^1 q(x, y)\,\mathrm{d}y = p(x). \tag{6.4}$$

For example, one choice of q is a piecewise linear function:

$$q(x, y) = \begin{cases} 1 & \text{if } y < p(x) - \epsilon, \\ 0 & \text{if } y > p(x) + \epsilon, \\ \text{linear} & p(x) - \epsilon < y < p(x) + \epsilon. \end{cases} \tag{6.5}$$

The integral can then be rewritten as

$$\int_0^1 f(x)p(x)\,\mathrm{d}x = \int_0^1 \int_0^1 f(x)q(x, y)\,\mathrm{d}y\,\mathrm{d}x, \tag{6.6}$$

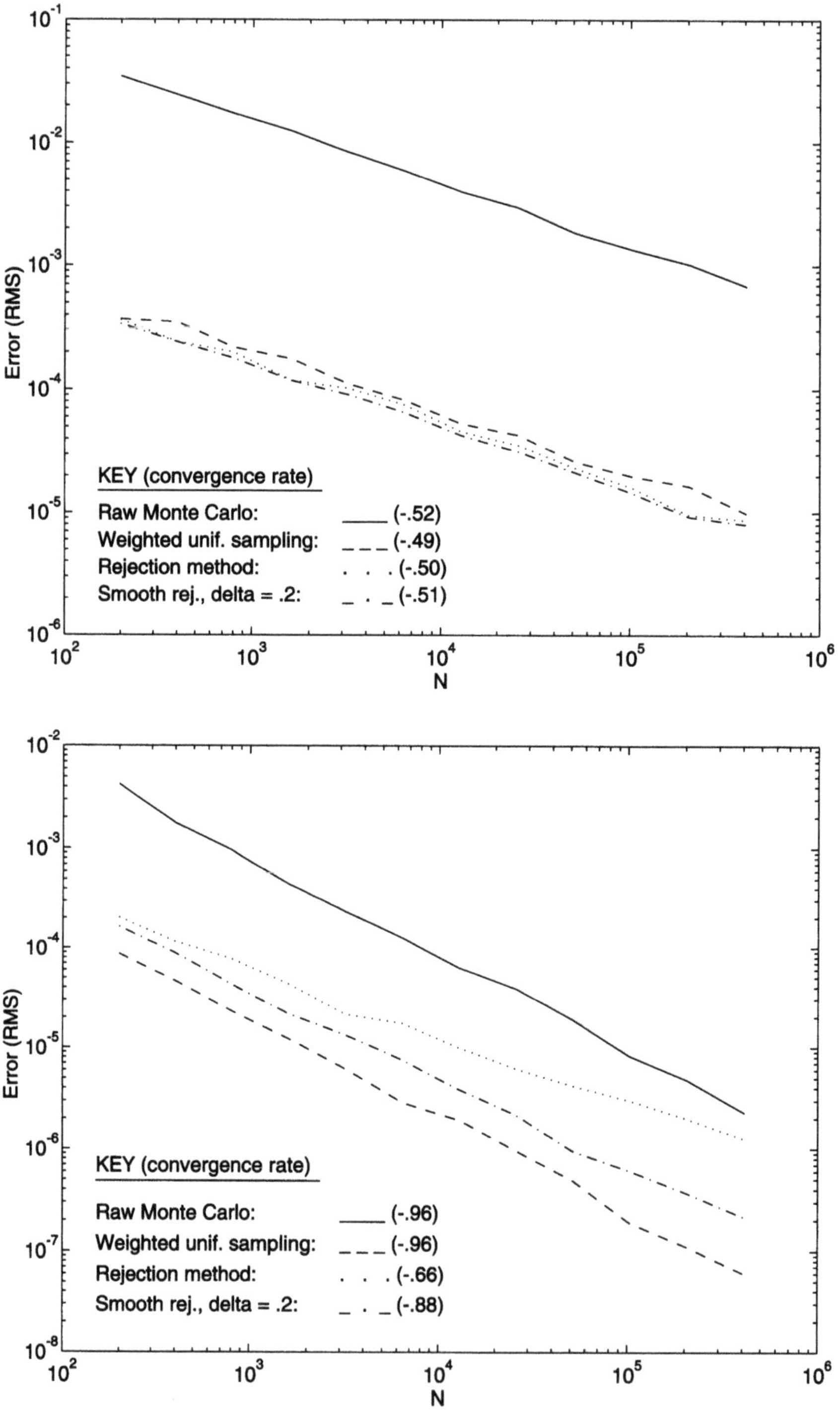

Fig. 6. Smoothed acceptance–rejection for Monte Carlo (top) and quasi-Monte Carlo (bottom)

and the sampling procedure is the following:

- sample (x, y) uniformly
- set the weight $w = q(x, y)$
- repeat until

$$\sum_{i=1}^{M} q(x_i, y_i) \approx N. \tag{6.7}$$

The quadrature formula, using smoothed acceptance–rejection, is then

$$J_N = N^{-1} \sum_{i=1}^{M} q(x_i, y_i) f(x_i), \tag{6.8}$$

in which

$$\sum_{i=1}^{M} q(x_i, y_i) \approx N. \tag{6.9}$$

Figure 6 shows the results of standard and smoothed acceptance–rejection for Monte Carlo using both a pseudo-random sequence and a quasi-random sequence. The integral evaluated here (Moskowitz and Caflisch 1996) has the form

$$I = \int_{I^7} f(\boldsymbol{x}) p(\boldsymbol{x}) \, \mathrm{d}\boldsymbol{x}, \tag{6.10}$$

in which $I^7 = [0, 1]^7$ and

$$\begin{aligned} p(\boldsymbol{x}) &= \exp\left\{-\left(\sin^2\left(\frac{\pi}{2}x_1\right) + \sin^2\left(\frac{\pi}{2}x_2\right) + \sin^2\left(\frac{\pi}{2}x_3\right)\right)\right\}, \\ f(\boldsymbol{x}) &= \mathrm{e} \arcsin\left(\sin(1) + \left[\frac{x_1 + \cdots + x_7}{200}\right]\right). \end{aligned} \tag{6.11}$$

The integral I is evaluated in several ways. First, by 'raw Monte Carlo', by which we mean evaluation of the product $f(x)p(x)$ at points from a uniformly distributed sequence. Second, points are sampled from the (unnormalized) density function p using acceptance–rejection, then f is evaluated at these points. Third, standard acceptance–rejection is replaced by smoothed acceptance–rejection. Finally, all three of these methods are performed both with pseudo-random and quasi-random points. These numerical results show the following.

- Importance sampling, using acceptance–rejection, reduces variance and leads to smaller errors for both Monte Carlo and quasi-Monte Carlo.
- For Monte Carlo, smoothing has little effect on errors, since it does not change the variance.
- Without importance sampling, the quasi-Monte Carlo method converges at a rapid rate, but with a constant that is larger than for the method with importance sampling.

- For quasi-Monte Carlo, the error for unsmoothed acceptance–rejection decreases at a slower rate, because of discontinuities involved in the acceptance–rejection decision.
- For quasi-Monte Carlo, the rapid convergence rate is regained using smoothing.

Although smoothed acceptance–rejection regains much of the effectiveness of quasi-Monte Carlo, it entails a loss of efficiency. Since fractional weights are involved, more accepted samples are required to get total weight N. Moreover, we do not have an effective quasi-Monte Carlo method for the Metropolis algorithm (Kalos and Whitlock 1986), which is a stochastic process involving acceptance–rejection.

6.2. Dimension reduction: Brownian bridge method

For problems in a dimension of moderate size, quasi-Monte Carlo provides a significant speed-up over standard Monte Carlo. Many of the most important applications of Monte Carlo, however, involve integrals of a very high dimension. One class of examples comprises path integrals involving Brownian motion $b(t)$. A typical example is a Feynman–Kac integral (Karatzas and Shreve 1991) of the form

$$I = E\left[\int_0^T f(b(t)) \exp\left\{\int_0^t \lambda(b(s), s)\,\mathrm{d}s\right\}\,\mathrm{d}t\right]. \tag{6.12}$$

In order to evaluate this expectation by Monte Carlo, we need to discretize the time in the Brownian motion to obtain a random walk with M steps. Then the integral becomes an integral over the M steps of a random walk, each of which is distributed by a normal distribution. An accurate representation of the integral requires a large number M of time-steps, which results in an integral of large-dimension M. Because of the high dimension, we find that quasi-Monte Carlo loses much of its effectiveness on such a problem. The main point of this section is to show how a rearrangement of the dimensions can regain the effectiveness of quasi-Monte Carlo for this problem.

The standard discretization of a random walk is to represent the position $b(t + \Delta t)$ in terms of the previous position $b(t)$ by the formula

$$b(t + \Delta t) = b(t) + \sqrt{\Delta t}\ \nu, \tag{6.13}$$

in which ν is an $N(0, 1)$ random variable. Using a sequence of independent samples of ν, we can generate the random walk sequentially by

$$y_0 = 0,\ \ y_1 = b(\Delta t),\ \ y_2 = b(2\Delta t), \ldots. \tag{6.14}$$

This representation leads to the M-dimensional integral discussed above.

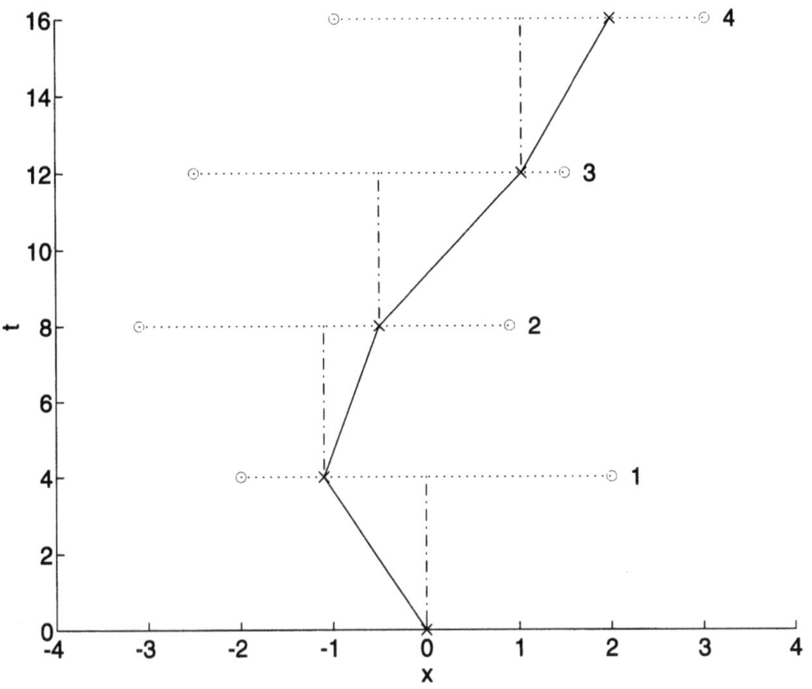

Fig. 7. Standard discretization of random walk

The effectiveness of quasi-Monte Carlo will be regained using an alternative representation of the random walk, the Brownian bridge discretization, which was first introduced as a quasi-Monte Carlo technique by Moskowitz and Caflisch (1996). This representation relies on the following Brownian bridge formula (Karatzas and Shreve 1991) for $b(t + \Delta t_1)$, given $b(t)$ and $b(T = t + \Delta t_1 + \Delta t_2)$:

$$b(t + \Delta t_1) = ab(t) + (1 - a)b(T) + c\nu, \tag{6.15}$$

in which

$$a = \Delta t_2/(\Delta t_1 + \Delta t_2), \qquad c = \sqrt{a\Delta t_1}. \tag{6.16}$$

Using this representation, the random walk can be generated by successive subdivision. Suppose for simplicity that M is a power of 2. Then generate the random walk in the following order:

$$y_0 = 0,\ y_M,\ y_{M/2},\ y_{M/4},\ y_{3M/4},\ \ldots. \tag{6.17}$$

The standard discretization and Brownian bridge discretization are represented schematically in Figures 7 and 8.

The significance of this representation is that it first chooses the large time-steps over which the changes in $b(t)$ are large. Then it fills in the small time-steps in between, in which the changes in $b(t)$ are quite small. The

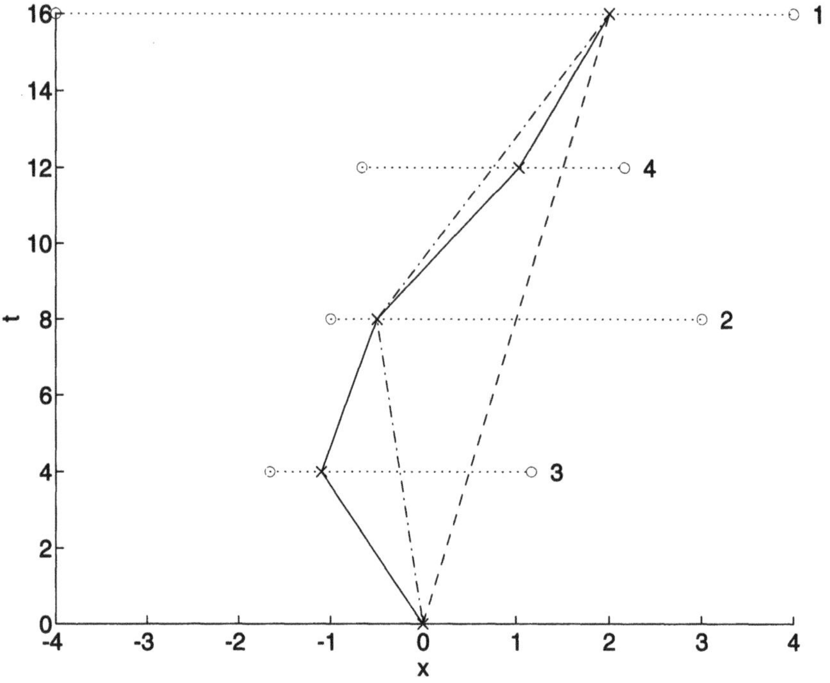

Fig. 8. Brownian bridge discretization of random walk

advantage of this representation is that it concentrates the variance into the early, large time-steps. This is much like a principal component analysis (Acworth, Broadie and Glasserman 1997).

Although the actual dimension of the problem is not changed, in some sense the effective dimension of the problem is lowered, so that quasi-Monte Carlo retains its effectiveness. To make this statement quantitative, suppose that, at some given value of N, the discrepancy is of size N^{-1} for dimension d (omitting logarithmic terms for simplicity), but is of size $N^{-1/2}$ for the remaining dimensions. We expect that the integration error is roughly of the size of the variance times the discrepancy. Using the Brownian bridge discretization, the variance over the first d dimensions is σ_0, which is about the same size as the original value of σ, whereas the variance over the remaining $M-d$ dimensions is σ_1, which is much smaller, that is,

$$\sigma_1 \ll \sigma_0 \approx \sigma. \tag{6.18}$$

Denote ε_s and ε_{bb} to be the errors for quasi-Monte Carlo using the standard discretization and the Brownian bridge discretization, respectively. Then, approximately,

$$\varepsilon_s = \sigma N^{-1/2}, \qquad \varepsilon_{bb} = \sigma_0 N^{-1} + \sigma_1 N^{-1/2}. \tag{6.19}$$

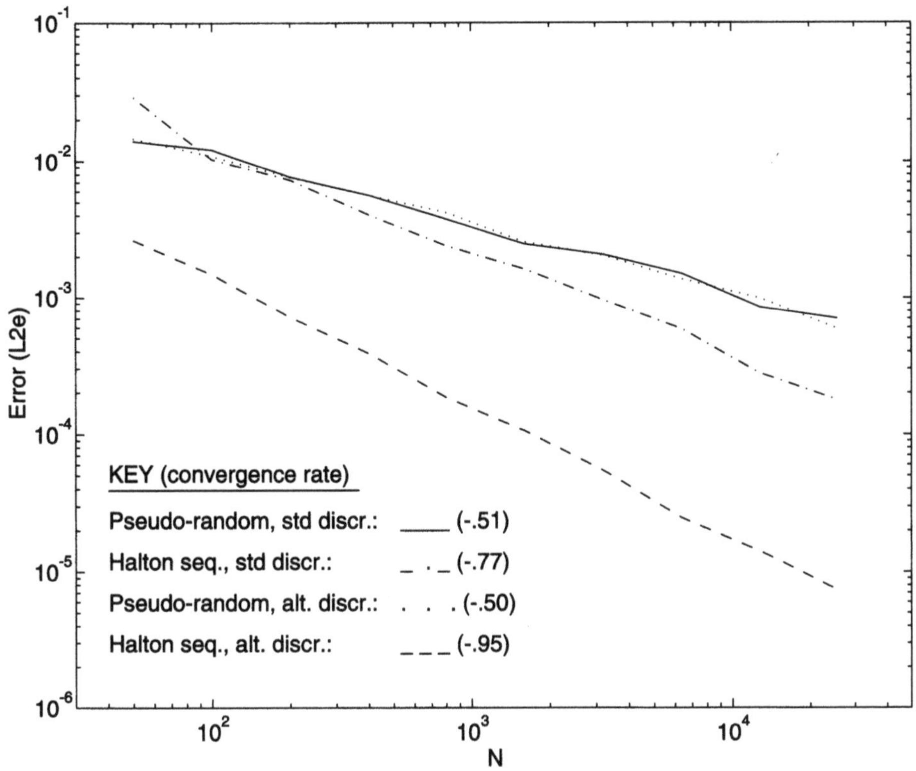

Fig. 9. Log-log plot for Feynman–Kac integral, 75 runs, $T = 0.08$, $m = 32$

The ordering (6.18) then implies that

$$\varepsilon_{bb} \ll \varepsilon_s. \tag{6.20}$$

This shows that the Brownian bridge discretization can provide a significant improvement in quasi-Monte Carlo integration for path integral problems.

As an example, consider an integral of the form (6.12), in which

$$\begin{aligned} \lambda(x,t) &= \left(\frac{1}{t+1} + \frac{1}{x^2+1} - \frac{4x^2}{(x^2+1)^2}\right), \\ f(x) &= \frac{1}{x^2+1}. \end{aligned} \tag{6.21}$$

According to the Feynman–Kac formula (Karatzas and Shreve 1991), the integral $I = F(x,t)$, in which x is the starting point of the Brownian motion, solves the following linear parabolic differential equation:

$$\frac{\partial F}{\partial t}(x,t) = \frac{1}{2}\frac{\partial^2 F}{\partial x^2}(x,t) + \lambda(x,t)\frac{\partial F}{\partial x}(x,t), \quad \text{with } F(x,0) = f(x). \tag{6.22}$$

The exact solution is $F(x,t) = (t+1)(x^2+1)^{-1}$.

Computational results for this problem of Moskowitz and Caflisch (1996) are presented in Figure 9, at time $T = 0.08$ with $M = 32$. The results show the following.

- For Monte Carlo using pseudo-random numbers, the error is the same for the standard and Brownian bridge discretizations. This is because the variance of the two is the same.
- For quasi-Monte Carlo with the standard discretization, the error is only a little less than for standard Monte Carlo; *i.e.*, the effectiveness of quasi-Monte Carlo has been lost for this high-dimension ($M = 32$).
- With the Brownian bridge discretization, the error for quasi-Monte Carlo is substantially reduced, in terms of both the rate of convergence and the constant; *i.e.*, the effectiveness of quasi-Monte Carlo has been regained.

It is necessary to use the transformation method (Section 3.3) rather than the Box–Muller method (Section 3.4), since the latter method has large gradients (Morokoff and Caflisch 1993). Similar results have been obtained on problems from computational finance by Caflisch, Morokoff and Owen (1997*b*).

7. Monte Carlo methods for rarefied gas dynamics

Computations for rarefied gas dynamics, as occur in the outer parts of the atmosphere, are difficult since the gas is represented by a distribution function over space, time and molecular velocity. For general problems, the only effective method has been a Monte Carlo method, as described below. One reason for including this application of Monte Carlo in this survey is that formulation of an effective Monte Carlo method in the fluid dynamic limit is an open problem.

7.1. The Boltzmann equation

The kinetic theory of gases describes the behaviour of a gas in which the density is not too large, so that the only interactions between gas particles are binary collisions. The resulting nonlinear Boltzmann equation,

$$\frac{\partial}{\partial t}F + \boldsymbol{\xi}\cdot\frac{\partial}{\partial \mathbf{x}}F = \varepsilon^{-1}Q(F,F), \tag{7.1}$$

for the molecular distribution function $F(\mathbf{x},\boldsymbol{\xi},t)$, is a basic equation of nonequilibrium statistical mechanics (Cercignani 1988, Chapman and Cowling 1970). The collision operator is

$$Q(F,F)(\boldsymbol{\xi}) = \int (F(\boldsymbol{\xi}_1')F(\boldsymbol{\xi}') - F(\boldsymbol{\xi}_1)F(\boldsymbol{\xi}))B(\omega,|\boldsymbol{\xi}_1-\boldsymbol{\xi}|)\,\mathrm{d}\omega\,\mathrm{d}\boldsymbol{\xi}_1, \tag{7.2}$$

in which $\boldsymbol{\xi}, \boldsymbol{\xi}_1$ represent velocities before a collision, $\boldsymbol{\xi}', \boldsymbol{\xi}_1'$ represent velocities after a collision, and the collision parameters are represented by $\omega \in S^2$, where S^2 is the unit sphere in $\mathbb{R}^3$. The parameter ε is a measure of the 'mean free time', which is defined as the characteristic time between collisions of a particle.

For a molecular distribution F, the macroscopic (or fluid dynamic) variables are the density ρ, velocity $\boldsymbol{u}$ and temperature T defined by

$$\begin{aligned} \rho &= \int F \,\mathrm{d}\boldsymbol{\xi}, \\ \boldsymbol{u} &= \rho^{-1} \int \boldsymbol{\xi} F \,\mathrm{d}\boldsymbol{\xi}, \\ T &= (3\rho)^{-1} \int |\boldsymbol{\xi} - \boldsymbol{u}|^2 F \,\mathrm{d}\boldsymbol{\xi}. \end{aligned} \tag{7.3}$$

The importance of these moments of F is that they correspond to conserved quantities, namely, mass, momentum and energy, for the collisional process, which is expressed as follows:

$$\begin{aligned} \int Q(F,F) \,\mathrm{d}\boldsymbol{\xi} &= 0, \\ \int \boldsymbol{\xi} Q(F,F) \,\mathrm{d}\boldsymbol{\xi} &= 0, \\ \int |\boldsymbol{\xi}|^2 Q(F,F) \,\mathrm{d}\boldsymbol{\xi} &= 0. \end{aligned} \tag{7.4}$$

These also provide the degrees of freedom for the equilibrium case, the Maxwellian distributions

$$M(\boldsymbol{\xi},\ \rho,\ \boldsymbol{u},\ T) = \rho(2\pi T)^{-3/2} \exp\left(-|\boldsymbol{\xi} - \boldsymbol{u}|^2/2T\right). \tag{7.5}$$

In the limit $\varepsilon \to 0$, the solution F of (7.1) is given by the Hilbert (or Chapman–Enskog) expansion, in which the leading term is a Maxwellian distribution of the form (7.5), in which $\rho,\ \boldsymbol{u},\ T$ satisfy the compressible Euler (or Navier–Stokes) equations of fluid dynamics (Chapman and Cowling 1970). The Euler equations are

$$\begin{aligned} \rho_t + \boldsymbol{\nabla} \cdot (\boldsymbol{u}\rho) &= 0, \\ (\rho \boldsymbol{u})_t + \boldsymbol{\nabla} \cdot (\boldsymbol{u}\boldsymbol{u}\rho) + \boldsymbol{\nabla}(\rho T) &= 0, \\ \left(\frac{1}{2}\rho|\boldsymbol{u}|^2 + \frac{3}{2}\rho T\right)_t + \boldsymbol{\nabla} \cdot \left(\boldsymbol{u}\left(\frac{1}{2}\rho|\boldsymbol{u}|^2 + \frac{5}{2}\rho T\right)\right) &= 0. \end{aligned} \tag{7.6}$$

This expansion is not valid in layers around shocks, boundaries and non-equilibrium initial data, where special boundary, shock or initial layer expansions can be constructed (Caflisch 1983). By varying this limit, taking the velocity $\boldsymbol{u}$ to be size ε as well, the compressible equations are replaced by the incompressible fluid equations (Bardos, Golse and Levermore 1991, 1993, De Masi, Esposito and Lebowitz 1989).

7.2. Particle methods

In particle methods for transport theory, the distribution $F(\mathbf{x}, \boldsymbol{\xi}, t)$ is represented as the sum

$$F_N(\mathbf{x}, \boldsymbol{\xi}, t) = \sum_{n=1}^{N} \delta(\boldsymbol{\xi} - \boldsymbol{\xi}_n(t))\delta(\mathbf{x} - \mathbf{x}_n(t)), \tag{7.7}$$

and the positions $\mathbf{x}_n(t)$ and velocities $\boldsymbol{\xi}_n(t)$ are evolved in time to simulate the effects of convection and collisions. The most common particle methods use random collisions between a reasonable number of particles (*e.g.*, 10^3–10^6) to simulate the dynamics of many particles (*e.g.*, 10^{23}).

In the direct simulation Monte Carlo (DSMC) method pioneered by Bird (1976, 1978), the numerical method is designed to simulate the physical processes as closely as possible. This makes it easy to understand and to insert new physics; it is also numerically robust. First, space and time are discretized into spatial cells of size Δx^3 and time-steps of duration Δt. In each time-step the evolution is divided into two steps: transport and collisions. For the transport step each particle is moved from position $\mathbf{x}_n(t)$ to $\mathbf{x}_n(t + \Delta t) = \mathbf{x}_n(t) + \Delta t\, \boldsymbol{\xi}_n(t)$.

In the collision step, random collisions are performed between particles within each spatial bin. In each collision, particles $\boldsymbol{\xi}_n$ and $\boldsymbol{\xi}_m$ are chosen randomly from the full set of particles in a bin, with probability p_{mn} given by

$$p_{mn} = \frac{S(|\boldsymbol{\xi}_m - \boldsymbol{\xi}_n|)}{\sum_{1 \le i \le j \le N_0} S(|\boldsymbol{\xi}_i - \boldsymbol{\xi}_j|)}, \tag{7.8}$$

in which N_0 is the number of particles in the spatial bin, and

$$S(|\boldsymbol{\xi}_i - \boldsymbol{\xi}_j|) = \int_{S^2} B(\omega, |\boldsymbol{\xi}_m - \boldsymbol{\xi}_n|)\, \mathrm{d}\omega \tag{7.9}$$

is the total collision rate between particles of velocity $\boldsymbol{\xi}_i$ and $\boldsymbol{\xi}_j$. As written, this choice requires $O(N^2)$ operations to evaluate the sum in the denominator in (7.8). By a standard acceptance–rejection scheme, however, each choice can be made in $O(1)$ steps, so that the total method requires only $O(N)$ steps.

Next the collision parameters ω are randomly chosen from a uniform distribution on S^2. The outcome of the collision is two new velocities $\boldsymbol{\xi}'_n$ and $\boldsymbol{\xi}'_m$ which replace the old velocities $\boldsymbol{\xi}_n$ and $\boldsymbol{\xi}_m$.

The number of collisions performed in each time-step has been determined by several methods. In the original 'time-counter' (TC) method, a collision time Δt_c is determined. It is equal to one over the frequency for that collision type, that is,

$$\Delta t_c = 2(nN_0S(|\boldsymbol{\xi}_m - \boldsymbol{\xi}_n|))^{-1}, \tag{7.10}$$

in which n is the number density of particles. This is added to the time-counter $t_c = \sum \Delta t_c$. In the time interval of length Δt beginning at t, collisions are continued until t_c exceeds the final time, that is, until $t_c \geq t + \Delta t$. For N particles this method has operation count $O(N)$.

The unlikely possibility of choosing a collision with very small frequency can result in a large collisional time-step that may cause relatively large errors. To remove such errors, Bird (1976) has developed a 'no time-counter' (NTC) method. This method uses a maximum collision probability $S_{\max}$. The number of collisions to be performed in time-step $\mathrm{d}t$ is chosen as if the collision frequency were exactly $S_{\max}$ for all collision pairs. For each selection of a collision pair, the collision between $\boldsymbol{\xi}_m$ and $\boldsymbol{\xi}_n$ is then performed with probability $S(|\boldsymbol{\xi}_m - \boldsymbol{\xi}_n|)/S_{\max}$, as in an acceptance–rejection scheme. The DSMC method with the time-counter or no-time-counter algorithm has been enormously successful. A rigorous convergence result for DSMC was proved by Wagner (1992).

Several related methods have been developed by other researchers, including Koura (1986), Nanbu (1986), and Goldstein, Sturtevant and Broadwell (1988). Additional modifications of Nanbu's method, including use of quasi-random sequences, have been developed by Babovsky, Gropengiesser, Neunzert, Struckmeier and Wiesen (1990).

7.3. Methods with the correct diffusion limit

One of the most difficult problems of transport theory is that there may be wide variation of mean free paths within a single problem. In regions where the mean free path is large, there are few collisions, so that large numerical time-steps may be taken, whereas in regions where the mean free path is small, the time- and space-steps must be small. Thus the highly collisional regions determine the numerical time-step, which may make computations impractically slow. On the other hand, much of the extra effort in the collisional regions seems wasted, since in those regions the gas will be nearly in fluid dynamic equilibrium, so that a fluid dynamic description (7.6) should be valid.

A partial remedy to this problem is to use a numerical method that converts to a numerical method for the correct fluid equations in regions where the mean free time is small. This allows large, fluid dynamic time-steps in the collisional region and large collisional time-steps in the large mean free time region.

Consider a discretization of the Boltzmann equation (7.1) with discrete space and time scales Δx and Δt. If Δx, Δt are much smaller than ε, then the Boltzmann solution is well resolved. On the other hand, in regions of small mean free path, we want to let the discretization scale be much larger than the collision scale ε. So we consider the limit in which the numerical parameters Δx, Δt are held fixed, while $\varepsilon \to 0$. We say that the discretized equation has the correct diffusion limit, if in this limit the solution of the discrete equations for (7.1) goes to the solution for a discretization of the fluid equation (7.6).

In the context of linear neutron transport, Larsen and co-workers (Larsen, Morel and Miller 1987, Borgers, Larsen and Adams 1992) have investigated the diffusion limit for a variety of difference schemes and have found that many of them have the correct diffusion limit only in special regimes. They have constructed some alternative methods that always have the correct diffusion limit. Jin and Levermore (1993) have applied a similar procedure to an interface problem to get a constraint on the quadrature set for the discretization of a scattering integral. Another class of methods that uses information from the diffusion limit to improve a numerical transport methods is the family of 'diffusion synthetic acceleration methods' (Larsen 1984). For the Broadwell model of the Boltzmann equation, finite difference numerical methods with the correct diffusion limit have been developed (Caflisch, Jin and Russo 1997*a*, Jin, Pareschi and Toscani 1998). These use implicit differencing for the collision step, because it is a stiff problem with rate ε^{-1}.

For the nonlinear Boltzmann equation, on the other hand, Monte Carlo methods are the most practical because of the large number of degrees of freedom. Application of the methods would require some kind of implicit step to handle the collisions, but no such method has been formulated so far. We believe that a method that uses fluid dynamic information to improve the collisional step for rarefied gas dynamic computations would have great impact. Some important partial steps in this direction have been taken (Gabetta, Pareschi and Toscani 1997), but they have not yet been extended to particle methods such as DSMC.

REFERENCES

P. Acworth, M. Broadie and P. Glasserman (1997), A comparison of some Monte Carlo and quasi Monte Carlo techniques for option pricing, in *Monte Carlo and Quasi-Monte Carlo Methods 1996* (G. Larcher, H. Niederreiter, P. Hellekalek and P. Zinterhof, eds), Springer.

H. Babovsky, F. Gropengiesser, H. Neunzert, J. Struckmeier and J. Wiesen (1990), 'Application of well-distributed sequences to the numerical simulation of the Boltzmann equation', *J. Comput. Appl. Math.* **31**, 15–22.

C. Bardos, F. Golse and C. D. Levermore (1991), 'Fluid dynamic limits of kinetic equations: I. Formal derivations', *J. Statist. Phys.* **63**, 323–344.

C. Bardos, F. Golse and C. D. Levermore (1993), 'Fluid dynamic limits of kinetic equations: II. Convergence proofs for the Boltzmann equation', *Comm. Pure Appl. Math.* **46**, 667–753.

G. A. Bird (1976), *Molecular Gas Dynamics*, Oxford University Press.

G. A. Bird (1978), 'Monte Carlo simulation of gas flows', *Ann. Rev. Fluid Mech.* **10**, 11–31.

C. Borgers, E. W. Larsen and M. L. Adams (1992), 'The asymptotic diffusion limit of a linear discontinuous discretization of a two-dimensional linear transport equation', *J. Comput. Phys.* **98**, 285–300.

P. Bratley, B. L. Fox and H. Niederreiter (1994), 'Algorithm 738 – Programs to generate Niederreiter's discrepancy sequences', *ACM Trans. Math. Software* **20**, 494–495.

R. E. Caflisch (1983), Fluid dynamics and the Boltzmann equation, in *Nonequilibrium Phenomena I: The Boltzmann equation* (E. W. Montroll and J. L. Lebowitz, eds), Vol. 10 of *Studies in Statistical Mechanics*, North Holland, pp. 193–223.

R. E. Caflisch, S. Jin and G. Russo (1997*a*), 'Uniformly accurate schemes for hyperbolic systems with relaxation', *SIAM J. Numer. Anal.* **34**, 246–281.

R. E. Caflisch, W. Morokoff and A. B. Owen (1997*b*), 'Valuation of mortgage-backed securities using Brownian bridges to reduce effective dimension', *J. Comput. Finance* **1**, 27–46.

C. Cercignani (1988), *The Boltzmann Equation and its Applications*, Springer.

S. Chapman and T. G. Cowling (1970), *The Mathematical Theory of Non-Uniform Gases*, Cambridge University Press.

A. De Masi, R. Esposito and J. L. Lebowitz (1989), 'Incompressible Navier–Stokes and Euler limits of the Boltzmann equation', *Comm. Pure Appl. Math.* **42**, 1189–1214.

H. Faure (1982), 'Discrépance de suites associées à un système de numération (en dimension s)', *Acta Arithmetica* **41**, 337–351.

W. Feller (1971), *An Introduction to Probability Theory and its Applications: Vol. I*, Wiley.

E. Gabetta, L. Pareschi and G. Toscani (1997), 'Relaxation schemes for nonlinear kinetic equations', *SIAM J. Numer. Anal.* **34**, 2168–2194.

D. Goldstein, B. Sturtevant and J. E. Broadwell (1988), Investigations of the motion of discrete-velocity gases, in *Rarefied Gas Dynamics: Theoretical and Computational Techniques* (E. P. Muntz, D. P. Weaver and D. H. Campbell, eds), Vol. 118 of *Progress in Aeronautics and Astronautics*, Proceedings of the 16th International Symposium on Rarefied Gas Dynamics, pp. 100–117.

J. H. Halton (1960), 'On the efficiency of certain quasi-random sequences of points in evaluating multi-dimensional integrals', *Numer. Math.* **2**, 84–90.

J. M. Hammersley and D. C. Handscomb (1965), *Monte Carlo Methods*, Wiley, New York.

C. Haselgrove (1961), 'A method for numerical integration', *Math. Comput.* **15**, 323–337.

F. J. Hickernell (1996), 'The mean square discrepancy of randomized nets', *ACM Trans. Model. Comput. Simul.* **6**, 274–296.

F. J. Hickernell (1998), 'A generalized discrepancy and quadrature error bound', *Math. Comput.* **67**, 299–322.

R. V. Hogg and A. T. Craig (1995), *Introduction to Mathematical Statistics*, Prentice Hall.

L. K. Hua and Y. Wang (1981), *Applications of Number Theory to Numerical Analysis*, Springer, Berlin/New York.

S. Jin and C. D. Levermore (1993), 'Fully-discrete numerical transfer in diffusive regimes', *Transp. Theory Statist. Phys.* **22**, 739–791.

S. Jin, L. Pareschi and G. Toscani (1998), 'Diffusive relaxation schemes for discrete-velocity kinetic equations'. Preprint.

M. H. Kalos and P. A. Whitlock (1986), *Monte Carlo Methods, Vol. I: Basics*, Wiley, New York.

I. Karatzas and S. E. Shreve (1991), *Brownian Motion and Stochastic Calculus*, Springer, New York.

W. J. Kennedy and J. E. Gentle (1980), *Statistical Computing*, Dekker, New York.

K. Koura (1986), 'Null-collision technique in the direct-simulation Monte Carlo method', *Phys. Fluids* **29**, 3509–3511.

L. Kuipers and H. Niederreiter (1974), *Uniform Distribution of Sequences*, Wiley, New York.

E. W. Larsen (1984), 'Diffusion-synthetic acceleration methods for discrete-ordinates problems', *Transp. Theory Statist. Phys.* **13**, 107–126.

E. W. Larsen, J. E. Morel and W. F. Miller (1987), 'Asymptotic solutions of numerical transport problems in optically thick, diffusive regimes', *J. Comput. Phys.* **69**, 283–324.

G. Lepage (1978), 'A new algorithm for adaptive multidimensional integration', *J. Comput. Phys.* **27**, 192–203.

G. Marsaglia (1991), 'Normal (Gaussian) random variables for supercomputers', *J. Supercomputing* **5**, 49–55.

W. Morokoff and R. E. Caflisch (1993), 'A quasi-Monte Carlo approach to particle simulation of the heat equation', *SIAM J. Numer. Anal.* **30**, 1558–1573.

W. Morokoff and R. E. Caflisch (1994), 'Quasi-random sequences and their discrepancies', *SIAM J. Sci. Statist. Comput.* **15**, 1251–1279.

W. Morokoff and R. E. Caflisch (1995), 'Quasi-Monte Carlo integration', *J. Comput. Phys.* **122**, 218–230.

B. Moskowitz and R. E. Caflisch (1996), 'Smoothness and dimension reduction in quasi-Monte Carlo methods', *J. Math. Comput. Modeling* **23**, 37–54.

K. Nanbu (1986), Theoretical basis of the direct simulation Monte Carlo method, in *Proceedings of the 15th International Symposium on Rarefied Gas Dynamics* (V. Boffi and C. Cercignani, eds), Vol. I, pp. 369–383.

H. Niederreiter (1992), *Random Number Generation and Quasi-Monte Carlo Methods*, SIAM, Philadelphia, PA.

A. B. Owen (1995), Randomly permuted (t, m, s)-nets and (t, s)-sequences, in *Monte Carlo and Quasi-Monte Carlo Methods in Scientific Computing* (H. Niederreiter and P. J.-S. Shiue, eds), Springer, New York, pp. 299–317.

A. B. Owen (1997), 'Monte Carlo variance of scrambled net quadrature', *SIAM J. Numer. Anal.* **34**, 1884–1910.

A. B. Owen (1998), 'Scrambled net variance for integrals of smooth functions', *Ann. Statist.* In press.

W. H. Press and S. A. Teukolsky (1992), 'Portable random number generators', *Comput. Phys.* **6**, 522–524.

W. H. Press, S. A. Teukolsky, W. T. Vettering and B. P. Flannery (1992), *Numerical Recipes in C: The Art of Scientific Computing*, 2nd edn, Cambridge University Press.

D. I. Pullin (1979), 'Generation of normal variates with given sample and mean variance', *J. Statist. Comput. Simul.* **9**, 303–309.

I. M. Sobol' (1967), 'The distribution of points in a cube and the accurate evaluation of integrals', *Zh. Vychisl. Mat. Mat. Fiz.* **7**, 784–802. In Russian.

I. M. Sobol' (1976), 'Uniformly distributed sequences with additional uniformity property', *USSR Comput. Math. Math. Phys.* **16**, 1332–1337.

J. Spanier and E. H. Maize (1994), 'Quasi-random methods for estimating integrals using relatively small samples', *SIAM Rev.* **36**, 18–44.

W. Wagner (1992), 'A convergence proof for Bird's Direct Simulation Monte Carlo method for the Boltzmann equation', *J. Statist. Phys.* **66**, 1011–1044.

H. Woźniakowski (1991), 'Average case complexity of multivariate integration', *Bull. Amer. Math. Soc.* **24**, 185–194.

C. P. Xing and H. Niederreiter (1995), 'A construction of low-discrepancy sequences using global function fields', *Acta Arithmetica* **73**, 87–102.

S. K. Zaremba (1968), 'The mathematical basis of Monte Carlo and quasi-Monte Carlo methods', *SIAM Rev.* **10**, 303–314.

Acta Numerica (1998), pp. 51–150

Nonlinear approximation

Ronald A. DeVore*
Department of Mathematics,
University of South Carolina,
Columbia, SC 29208, USA
E-mail: devore@math.sc.edu

This is a survey of nonlinear approximation, especially that part of the subject which is important in numerical computation. Nonlinear approximation means that the approximants do not come from linear spaces but rather from nonlinear manifolds. The central question to be studied is what, if any, are the advantages of nonlinear approximation over the simpler, more established, linear methods. This question is answered by studying the rate of approximation which is the decrease in error versus the number of parameters in the approximant. The number of parameters usually correlates well with computational effort. It is shown that in many settings the rate of nonlinear approximation can be characterized by certain smoothness conditions which are significantly weaker than required in the linear theory. Emphasis in the survey will be placed on approximation by piecewise polynomials and wavelets as well as their numerical implementation. Results on highly nonlinear methods such as optimal basis selection and greedy algorithms (adaptive pursuit) are also given. Applications to image processing, statistical estimation, regularity for PDEs, and adaptive algorithms are discussed.

* This research was supported by Office of Naval Research Contract N0014-91-J1343 and Army Research Office Contract N00014-97-1-0806.

CONTENTS

1. Nonlinear approximation: an overview

The fundamental problem of approximation theory is to resolve a possibly complicated function, called the *target function*, by simpler, easier to compute functions called the *approximants*. Increasing the resolution of the target function can generally only be achieved by increasing the complexity of the approximants. The understanding of this trade-off between resolution and complexity is the main goal of constructive approximation. Thus the goals of approximation theory and numerical computation are similar, even though approximation theory is less concerned with computational issues. The differing point in the two subjects lies in the information assumed to be known about the target function. In approximation theory, one usually assumes that the values of certain simple linear functionals applied to the target function are known. This information is then used to construct an approximant. In numerical computation, information usually comes in a different, less explicit form. For example, the target function may be the solution to an integral equation or boundary value problem and the numerical analyst needs to translate this into more direct information about the target function. Nevertheless, the two subjects of approximation and computation are inexorably intertwined and it is impossible to understand fully the possibilities in numerical computation without a good understanding of the elements of constructive approximation.

It is noteworthy that the developments of approximation theory and numerical computation followed roughly the same line. The early methods utilized approximation from finite-dimensional linear spaces. In the beginning, these were typically spaces of polynomials, both algebraic and trigonometric. The fundamental problems concerning order of approximation were solved in this setting (primarily by the Russian school of Bernstein, Chebyshev, and their mathematical descendants). Then, starting in the late 1950s came the development of piecewise polynomials and splines and their incorporation into numerical computation. We have in mind the finite element methods (FEM) and their counterparts in other areas such as numerical quadrature, and statistical estimation.

It was noted shortly thereafter that there was some advantage to be gained by not limiting the approximations to come from linear spaces, and therein emerged the beginnings of nonlinear approximation. Most notable in this regard was the pioneering work of Birman and Solomyak (1967) on adaptive approximation. In this theory, the approximants are not restricted to come from spaces of piecewise polynomials with a fixed partition; rather, the partition was allowed to depend on the target function. However, the number of pieces in the approximant is controlled. This provides a good match with numerical computation since it often represents closely the cost of computation (number of operations). In principle, the idea was simple: we should use a finer mesh where the target function is not very smooth (singular) and a coarser mesh where it is smooth. The paramount question remained, however, as to just how we should measure this smoothness in order to obtain definitive results.

As is often the case, there came a scramble to understand the advantages of this new form of computation (approximation) and, indeed, rather exotic spaces of functions were created (Brudnyi 1974, Bergh and Peetre 1974), to define these advantages. But to most, the theory that emerged seemed too much a tautology and the spaces were not easily understood in terms of classical smoothness (derivatives and differences). But then came the remarkable discovery of Petrushev (1988) (preceded by results of Brudnyi (1974) and Oswald (1980)) that the efficiency of nonlinear spline approximation could be characterized (at least in one variable) by classical smoothness (Besov spaces). Thus the advantage of nonlinear approximation became crystal clear (as we shall explain later in this article).

Another remarkable development came in the 1980s with the development of multilevel techniques. Thus, there were the roughly parallel developments of multigrid theory for integral and differential equations, wavelet analysis in the vein of harmonic analysis and approximation theory, and multiscale filterbanks in the context of image processing. From the viewpoint of approximation theory and harmonic analysis, the wavelet theory was important on several counts. It gave simple and elegant unconditional bases (wavelet bases) for function spaces (Lebesgue, Hardy, Sobolev, Besov, Triebel–Lizorkin) that simplified some aspects of Littlewood–Paley theory (see Meyer (1990)). It provided a very suitable vehicle for the analysis of the core linear operators of harmonic analysis and partial differential equations (Calderón–Zygmund theory). Moreover, it allowed the solution of various functional analytic and statistical extremal problems to be made directly from wavelet coefficients.

Wavelet theory provides simple and powerful decompositions of the target function into a series of building blocks. It is natural, then, to approximate the target function by selecting terms of this series. If we take partial sums of this series we are approximating again from linear spaces. It was easy to establish that this form of linear approximation offered little, if any, advantage over the already well established spline methods. However, it is also possible to let the selection of terms to be chosen from the wavelet series depend on the target function f and keep control only over the number of terms to be used. This is a form of nonlinear approximation which is called *n-term approximation*. This type of approximation was introduced by Schmidt (1907). The idea of n-term approximation was first utilized for multivariate splines by Oskolkov (1979).

Most function norms can be described in terms of wavelet coefficients. Using these descriptions not only simplifies the characterization of functions with a specified approximation order but also makes transparent strategies for achieving good or best n-term approximations. Indeed, it is enough to retain the n terms in the wavelet expansion of the target function that are largest relative to the norm measuring the error of approximation. Viewed in another way, it is enough to threshold the properly normalized wavelet coefficients. This leads to approximation strategies based on what is called *wavelet shrinkage* by Donoho and Johnstone (1994). Wavelet shrinkage is used by these two authors and others to solve several extremal problems in statistical estimation, such as the recovery of the target function in the presence of noise.

Because of the simplicity in describing n-term wavelet approximation, it is natural to try to incorporate a good choice of basis into the approximation problem. This leads to a double stage nonlinear approximation problem where the target function is used both to choose a good (or best) basis from a given library of bases and then to choose the best n-term approximation relative to the good basis. This is a form of *highly nonlinear approximation*. Other examples are greedy algorithms and adaptive pursuit for finding an n-term approximation from a redundant set of functions. Our understanding of these highly nonlinear methods is quite fragmentary. Describing the functions that have a specified rate of approximation with respect to highly nonlinear methods remains a challenging problem.

Our goal in this paper is to be tutorial rather than complete in our description of nonlinear approximation. We spare the reader some of the finer aspects of the subject in search of clarity. In this vein, we begin in Section 2 by considering approximation in a Hilbert space. In this simple setting the problems of linear and nonlinear approximation are easily settled and the distinction between the two subjects is readily seen.

In Section 3, we consider approximation of univariate functions by piecewise constants. This form of approximation is the prototype of both spline approximation and wavelets. Understanding linear and nonlinear approximation by piecewise constants will make the transition to the fuller aspects of splines (Section 6) and wavelets (Section 7) more digestible.

In Section 8, we treat highly nonlinear methods. Results in this subject are in their infancy. Nevertheless, the methods are already in serious numerical use, especially in image processing.

As noted earlier, the thread that runs through this paper is the following question: what properties of a function determine its rate of approximation by a given nonlinear method? The final solution of this problem, when it is known for a specific method of approximation, is most often in terms of Besov spaces. However, we try to postpone the full impact of Besov spaces until the reader has, we hope, developed significant feeling for smoothness conditions and their role in approximation. Nevertheless, it is impossible to understand this subject fully without finally coming to grips with Besov spaces. Fortunately, they are not too difficult when viewed via moduli of smoothness (Section 4) or wavelet coefficients (Section 7).

Nonlinear approximation is used significantly in many applications. Perhaps the greatest success for this subject has been in image processing. Nonlinear approximation explains the thresholding and quantization strategies used in compression and noise removal. It also explains how quantization and thresholding may be altered to accommodate other measures of error. It is also noteworthy that it explains precisely which images can be compressed well by certain thresholding and quantization strategies. We discuss some applications of nonlinear methods to image processing in Section 10.

Another important application of nonlinear approximation lies in the solution of operator equations. Most notable, of course, are the adaptive finite element methods for elliptic equations (see Babuška and Suri (1994)) as well as the emerging nonlinear wavelet methods in the same subject (see Dahmen (1997)). For hyperbolic problems, we have the analogous developments of moving grid methods. Applications of nonlinear approximation in PDEs are touched upon in Section 10.

In approximation theory, one measures the complexity of the approximation process by the number of parameters needed to specify the approximant. This agrees in principle with the concepts of complexity in information theory. However, it does not necessarily agree with computational complexity, which measures the number of computations necessary to render the approximant. This is particularly the case when the target function is not explicitly available and must be computed through a numerical process such as in the numerical solution of integral or differential equations. We shall not touch on this finer notion of computational complexity in this survey. Good references for computational complexity in the framework of linear and nonlinear approximation is given in the book of Traub, Wasilkowski and Woźniakowski (1988), the paper of E. Novak (1996), and the references therein.

Finally, we close this introduction with a couple of helpful remarks about notation. Constants appearing in inequalities will be denoted by C and may vary at each occurrence, even in the same formula. Sometimes we will indicate the parameters on which the constant depends. For example, $C(p)$ (respectively, $C(p,\alpha)$) means the constant depends only on p (respectively, p and α). However, usually the reader will have to consult the text to understand the parameters on which C depends. More ubiquitous is the notation

$$A \asymp B, \tag{1.1}$$

which means there are constants $C_1, C_2 > 0$ such that $C_1 A \le B \le C_2 A$. Here A and B are two expressions depending on other variables (parameters). When there is any chance of confusion, we will indicate in the text the parameters on which C_1 and C_2 depend.

2. Approximation in a Hilbert space

The problems of approximation theory are simplest when they take place in a Hilbert space $\mathcal{H}$. Yet the results in this case are not only illuminating but very useful in applications. It is worthwhile, therefore, to begin with a brief discussion of linear and nonlinear approximation in this setting.

Let $\mathcal{H}$ be a separable Hilbert space with inner product $\langle \cdot, \cdot \rangle$ and norm $\|\cdot\|_H$ and let η_k, $k = 1, 2, \ldots$, be an orthonormal basis for $\mathcal{H}$. We shall consider two types of approximation corresponding to the linear and nonlinear settings.

For linear approximation, we use the linear space $\mathcal{H}_n := \operatorname{span}\{\eta_k : 1 \le k \le n\}$ to approximate an element $f \in \mathcal{H}$. We measure the approximation error by

$$E_n(f)_{\mathcal{H}} := \inf_{g \in \mathcal{H}_n} \|f - g\|_{\mathcal{H}}. \tag{2.1}$$

As a counterpart in nonlinear approximation, we have *n-term approximation*, which replaces $\mathcal{H}_n$ by the space Σ_n consisting of all elements $g \in \mathcal{H}$ that can be expressed as

$$g = \sum_{k \in \Lambda} c_k \eta_k, \tag{2.2}$$

where $\Lambda \subset \mathbb{N}$ is a set of indices with $\#\Lambda \le n$.[1] Notice that, in contrast to $\mathcal{H}_n$, the space Σ_n is not linear. A sum of two elements in Σ_n will in general

[1] We use $\mathbb{N}$ to denote the set of natural numbers and $\#S$ to denote the cardinality of a finite set S.

need $2n$ terms in its representation by the η_k. Analogous to E_n, we have the *error of n-term approximation*

$$\sigma_n(f)_{\mathcal{H}} := \inf_{g \in \Sigma_n} \|f - g\|_{\mathcal{H}}. \tag{2.3}$$

We pose the following question. Given a real number $\alpha > 0$, for which elements $f \in \mathcal{H}$ do we have

$$E_n(f)_{\mathcal{H}} \leq M n^{-\alpha}, \quad n = 1, 2, \ldots, \tag{2.4}$$

for some constant $M > 0$? Let us denote this class of f by $\mathcal{A}^\alpha((\mathcal{H}_n))$, where our notation reflects the dependence on the sequence $(\mathcal{H}_n)$, and define $|f|_{\mathcal{A}^\alpha((\mathcal{H}_n))}$ as the infimum of all M for which (2.4) holds. $\mathcal{A}^\alpha$ is called an *approximation space*: it gathers under one roof all $f \in \mathcal{H}$ which have a common approximation order. We denote the corresponding class for (Σ_n) by $\mathcal{A}^\alpha((\Sigma_n))$.

We shall see that it is easy to describe the above approximation classes in terms of the coefficients in the orthogonal expansion

$$f = \sum_{k=1}^{\infty} \langle f, \eta_k \rangle \eta_k. \tag{2.5}$$

Let us use in this section the abbreviated notation

$$f_k := \langle f, \eta_k \rangle, \quad k = 1, 2, \ldots. \tag{2.6}$$

Consider first the case of linear approximation. The best approximation to f from $\mathcal{H}_n$ is given by the projection

$$P_n f := \sum_{k=1}^{n} f_k \eta_k \tag{2.7}$$

onto $\mathcal{H}_n$ and the approximation error satisfies

$$E_n(f)_{\mathcal{H}}^2 = \sum_{k=n+1}^{\infty} |f_k|^2. \tag{2.8}$$

We can characterize $\mathcal{A}^\alpha$ in terms of the dyadic sums

$$F_m := \left(\sum_{k=2^{m-1}+1}^{2^m} |f_k|^2 \right)^{1/2}, \quad m = 1, 2, \ldots. \tag{2.9}$$

Indeed, it is almost a triviality to see that $f \in \mathcal{A}^\alpha((\mathcal{H}_n))$ if and only if

$$F_m \leq M 2^{-m\alpha}, \quad m = 1, 2, \ldots, \tag{2.10}$$

and the smallest M for (2.10) is equivalent to $|f|_{\mathcal{A}^\alpha((\mathcal{H}_n))}$. To some, (2.10) may not seem so pleasing since it is so close to a tautology. However, it usually serves to characterize the approximation spaces $\mathcal{A}_\alpha((\mathcal{H}_n))$ in concrete settings.

It is more enlightening to consider a variant of $\mathcal{A}^\alpha$. Let $\mathcal{A}_2^\alpha((\mathcal{H}_n))$ denote the set of all f such that

$$|f|_{\mathcal{A}_2^\alpha((\mathcal{H}_n))} := \left(\sum_{n=1}^{\infty}[n^\alpha E_n(f)_{\mathcal{H}}]^2 \frac{1}{n}\right)^{1/2} \tag{2.11}$$

is finite. From the monotonicity of $E_k(f)_{\mathcal{H}}$, it follows that

$$|f|_{\mathcal{A}_2^\alpha((\mathcal{H}_n))} \asymp \left(\sum_{k=0}^{\infty} 2^{2k\alpha} E_{2^k}(f)_{\mathcal{H}}^2\right)^{1/2} . \tag{2.12}$$

The condition for membership in $\mathcal{A}_2^\alpha$ is slightly stronger than membership in $\mathcal{A}^\alpha$. The latter requires that the sequence $(n^\alpha E_n)$ is bounded while the former requires that it is square summable with weight $1/n$.

The space $\mathcal{A}_2^\alpha((\mathcal{H}_n))$ is characterized by

$$\sum_{k=1}^{\infty} k^{2\alpha}|f_k|^2 \le M^2 \tag{2.13}$$

and the smallest M satisfying (2.13) is equivalent to $|f|_{\mathcal{A}_2^\alpha((\mathcal{H}_n))}$. We shall give the simple proof of this fact since the ideas in the proof are used often. First of all, note that (2.13) is equivalent to

$$\sum_{m=1}^{\infty} 2^{2m\alpha} F_m^2 \le (M')^2 \tag{2.14}$$

with M of (2.13) and M' of (2.14) comparable. Now, we have

$$2^{2m\alpha} F_m^2 \le 2^{2m\alpha} E_{2^{m-1}}(f)_{\mathcal{H}}^2,$$

which, when using (2.12), gives one of the implications of the asserted equivalence. On the other hand,

$$2^{2m\alpha} E_{2^m}(f)_{\mathcal{H}}^2 = 2^{2m\alpha} \sum_{k=m+1}^{\infty} F_k^2$$

and therefore

$$\sum_{m=0}^{\infty} 2^{2m\alpha} E_{2^m}(f)_{\mathcal{H}}^2 \le \sum_{m=0}^{\infty} 2^{2m\alpha} \sum_{k=m+1}^{\infty} F_k^2 \le C \sum_{k=1}^{\infty} 2^{2k\alpha} F_k^2,$$

which gives the other implication of the asserted equivalence.

Let us digest these results with the following example. We take for $\mathcal{H}$ the space $L_2(\mathbb{T})$ of 2π-periodic functions on the unit circle $\mathbb{T}$ which has the Fourier basis $\{(2\pi)^{-\frac{1}{2}}e^{ikx} : k \in \mathbb{Z}\}$. (Note here the indexing of the basis functions on $\mathbb{Z}$ rather than $\mathbb{N}$.) The space $\mathcal{H}_n := \text{span}\{e^{ikx} : |k| \le n\}$ is the space $\mathcal{T}_n$ of trigonometric polynomials of degree $\le n$. The coefficients with respect to this basis are the Fourier coefficients $\hat{f}(k)$ and therefore (2.13) states that $\mathcal{A}_2^\alpha((\mathcal{T}_n))$ is characterized by the condition

$$\sum_{k\in\mathbb{Z}\setminus\{0\}} |k|^{2\alpha}|\hat{f}(k)|^2 \le M. \tag{2.15}$$

If α is an integer, (2.15) describes the Sobolev space $W^\alpha(L_2(\mathbb{T}))$ of all 2π-periodic function with their αth derivative in $L_2(\mathbb{T})$ and the sum in (2.15) is the square of the semi-norm $|f|_{W_2^\alpha(L_2(\mathbb{T}))}$. For noninteger α, (2.15) characterizes, by definition, the fractional order Sobolev space $W^\alpha(L_2(\mathbb{T}))$. One should note that one half of the characterization (2.15) of $\mathcal{A}_2^\alpha((\mathcal{T}_n))$ gives the inequality

$$\left(\sum_{n=1}^{\infty}[n^\alpha E_n(f)_{\mathcal{H}}]^2\frac{1}{n}\right)^{1/2} \le C|f|_{W^\alpha(L_2(\mathbb{T}))} \tag{2.16}$$

which is slightly stronger than the inequality

$$E_n(f)_{\mathcal{H}} \le Cn^{-\alpha}|f|_{W^\alpha(L_2(\mathbb{T}))}, \tag{2.17}$$

which is more frequently found in the literature.

Using (2.10), it is easy to prove that the space $\mathcal{A}^\alpha((\mathcal{T}_n))$ is identical with the Besov space $B_\infty^\alpha(L_2(\mathbb{T}))$ and, for noninteger α, this is the Lipschitz space $\text{Lip}(\alpha, L_2(\mathbb{T}))$. (We introduce and discuss amply the Besov and Lipschitz spaces in Sections 3.2 and 4.5.)

Let us return now to the case of a general Hilbert space $\mathcal{H}$ and nonlinear approximation from Σ_n. We can characterize the space $\mathcal{A}^\alpha((\Sigma_n))$ by using the rearrangement of the coefficients f_k. We denote by $\gamma_k(f)$ the kth largest of the numbers $|f_j|$. We first want to observe that $f \in \mathcal{A}^\alpha((\Sigma_n))$ if and only if

$$\gamma_n(f) \le Mn^{-\alpha-1/2} \tag{2.18}$$

and the infimum of all M which satisfy (2.18) is equivalent to $|f|_{\mathcal{A}^\alpha((\Sigma_n))}$. Indeed, we have

$$\sigma_n(f)_{\mathcal{H}}^2 = \sum_{k>n} \gamma_k(f)^2. \tag{2.19}$$

Therefore, if f satisfies (2.18), then clearly

$$\sigma_n(f)_{\mathcal{H}} \le CMn^{-\alpha},$$

so that $f \in \mathcal{A}^\alpha((\Sigma_n))$ and we have one of the implications in the asserted characterization. On the other hand, if $f \in A^\alpha((\Sigma_n))$, then

$$\gamma_{2n}(f)^2 \le n^{-1} \sum_{m=n+1}^{2n} \gamma_m(f)^2 \le n^{-1}\sigma_n(f)_{\mathcal{H}}^2 \le |f|^2_{A^\alpha((\Sigma_n))} n^{-2\alpha-1}.$$

Since a similar inequality holds for $\gamma_{2n+1}(f)$, we have the other implication of the asserted equivalence.

It is also easy to characterize other approximation classes such as the $\mathcal{A}_2^\alpha((\Sigma_n))$, which is the analogue of $\mathcal{A}_2^\alpha((\mathcal{H}_n))$. We shall formulate such results in Section 5.

Let us return to our example of trigonometric approximation. Approximation by Σ_n is n-term approximation by trigonometric sums. It is easy to see the distinction between linear and nonlinear approximation in this case. Linear approximation corresponds to a certain decay in the Fourier coefficients $\hat{f}(k)$ as the frequency k increases, whereas nonlinear approximation corresponds to a decay in the rearranged coefficients. Thus, nonlinear approximation does not recognize the frequency location of the coefficients. If we reassign the Fourier coefficients of a function $f \in \mathcal{A}^\alpha$ to new frequency locations, the resulting function is still in $\mathcal{A}^\alpha$. Thus, in the nonlinear case there is no correspondence between rate of approximation to classical smoothness as there was in the linear case. It is possible to have large coefficients at high frequency just as long as there are not too many of them. For example, the functions e^{ikx} are obviously in all of the spaces $\mathcal{A}^\alpha$ even though their derivatives are large when k is large.

3. Approximation by piecewise constants

For our next taste of nonlinear approximation, we shall consider in this section several types of approximation by piecewise constants corresponding to linear and nonlinear approximation. Our goal is to see in this very simple setting the advantages of nonlinear methods. We begin with a target function f defined on $\Omega := [0, 1)$ and approximate it in various ways by piecewise constants with n pieces. We shall be interested in the efficiency of such approximation, that is, how the error of approximation decreases as n tends to infinity. We shall see that, in many cases, we can characterize the functions f which have certain approximation orders (for instance $O(n^{-\alpha})$, $0 < \alpha \le 1$). Such characterizations will illuminate the distinctions between linear and nonlinear approximation.

3.1. Linear approximation by piecewise constants

We begin by considering approximation by piecewise constants on partitions of Ω which are fixed in advance. This will be our reference point for comparisons with nonlinear approximations that follow. This form of linear approximation is also important in numerical computation since it is the simplest setting for FEM and other numerical methods based on approximation by piecewise polynomials. We shall see that there is a complete understanding in this case of the properties of the target function needed to guarantee certain approximation rates. As we shall amplify below, this theory explains what we should be able to achieve with proper numerical methods and also tells us what form good numerical estimates should take.

Let N be a positive integer and let $T := \{0 =: t_0 < t_1 < \cdots < t_N := 1\}$ be an ordered set of points in Ω. These points determine a partition $\Pi := \Pi(T) := \{I_k\}_{k=1}^N$ of Ω into N disjoint intervals $I_k := [t_{k-1}, t_k)$, $1 \le k \le N$. Let $\mathcal{S}^1(T)$ denote the space of piecewise constant functions relative to this partition. The characteristic functions $\{\chi_I : I \in \Pi\}$ form a basis for $\mathcal{S}^1(T)$: each function $S \in \mathcal{S}^1(T)$ can be represented uniquely by

$$S = \sum_{I \in \Pi} c_I \chi_I. \tag{3.1}$$

Thus $\mathcal{S}^1(T)$ is a linear space of dimension N.

For $0 < p \le \infty$, we introduce the error in approximating a function $f \in L_p[0,1)$ by the elements of $\mathcal{S}^1(T)$:

$$s(f,T)_p := \inf_{S \in \mathcal{S}^1(T)} \|f - S\|_{L_p[0,1)}. \tag{3.2}$$

We would like to understand what properties of f and T determine $s(f,T)_p$. For the moment, we shall restrict our discussion to the case $p = \infty$ which corresponds to uniformly continuous functions f on $[0,1)$ to be approximated in the uniform norm (L_∞-norm) on $[0,1)$. The quality of approximation that $S'(T)$ provides is related to the mesh length

$$\delta_T := \max_{0 \le k < N} |t_{k+1} - t_k|. \tag{3.3}$$

We shall first give estimates for $s(f,T)_\infty$ and then later ask in what sense these estimates are best possible. We recall the definition of the Lipschitz spaces $\operatorname{Lip}\alpha$. For each $0 \le \alpha \le 1$ and $M > 0$, we let $\operatorname{Lip}_M \alpha$ denote the set of all functions f on Ω such that

$$|f(x) - f(y)| \le M|x-y|^\alpha.$$

Then $\operatorname{Lip}\alpha := \cup_{M>0} \operatorname{Lip}_M \alpha$. The infimum of all M for which $f \in \operatorname{Lip}_M \alpha$ is by definition $|f|_{\operatorname{Lip}\alpha}$. In particular, $f \in \operatorname{Lip} 1$ if and only if f is absolutely continuous and $f' \in L_\infty$; moreover, $|f|_{\operatorname{Lip} 1} = \|f'\|_{L_\infty}$.

If the target function $f \in \operatorname{Lip}_M \alpha$, then

$$s(f,T)_\infty \le M(\delta_T/2)^\alpha. \tag{3.4}$$

Indeed, we define the piecewise constant function $S \in \mathcal{S}^1(T)$ by

$$S(x) := f(\xi_I), \quad x \in I,\ I \in \Pi_n,$$

with ξ_I the midpoint of I. Then, $|x - \xi_I| \le \delta_T/2$, $x \in I$, and hence

$$\|f - S\|_{L_\infty[0,1)} \le M(\delta_T/2)^\alpha, \tag{3.5}$$

which gives (3.4).

We turn now to the question of whether the estimate (3.4) is the best we can do. We shall see that this is indeed the case in several senses. First, suppose that for a function f we know that

$$s(f,T)_\infty \le M\delta_T^\alpha, \tag{3.6}$$

for *every* partition T. Then, we can prove that f is in $\mathrm{Lip}\,\alpha$ and moreover $|f|_{\mathrm{Lip}\,\alpha} \le M$. Results of this type are called inverse theorems in approximation theory whereas results like (3.4) are called direct theorems.

To prove the inverse theorem, we need to estimate the smoothness of f from the approximation errors $s(f,T)_\infty$. In the case at hand, the proof is very simple. Let S_T be a best approximation to f from $\mathcal{S}^1(T)$ in the $L_\infty(\Omega)$-norm. (A simple compactness argument shows the existence of best approximants.) If x, y are two points from Ω that are in the same interval $I \in \Pi(T)$, then from (3.6)

$$\begin{aligned} |f(x) - f(y)| &\le |f(x) - S_T(x)| + |f(y) - S_T(y)| \\ &+ |S_T(x) - S_T(y)| \le 2s(f,T)_\infty \le 2M\delta_T^\alpha \end{aligned} \tag{3.7}$$

because $S_T(x) = S_T(y)$ (S_T is constant on I). Since we can choose T so that δ_T is arbitrarily close to $|x-y|$, we obtain

$$|f(x) - f(y)| \le 2M(\delta_T)^\alpha \le 2M|x-y|^\alpha \tag{3.8}$$

which shows that $f \in \mathrm{Lip}\,\alpha$ and $|f|_{\mathrm{Lip}\,\alpha} \le 2M$.

Here is one further observation on the above analysis. If f is a function for which $s(f,T)_\infty = o(\delta_T)$ holds for all T, then the above argument gives that $f(x+h) - f(x) = o(h)$, $h \to 0$, for each $x \in \Omega$. Thus f is constant (its derivative is 0 everywhere). This is called a saturation theorem in approximation theory. Only trivial functions can be approximated with order better than $O(\delta_T)$.

The above discussion is not completely satisfactory for numerical analysis. In numerical algorithms, we usually have only a sequence of partitions. However, with some massaging, the above arguments can be applied in this case as well. Consider, for example, the case where

$$\Delta_n := \{k/n : 0 \le k \le n\} \tag{3.9}$$

consists of n equally spaced points from Ω (with spacing $1/n$). Then, for each $0 < \alpha \le 1$, a function f satisfies

$$s_n(f)_\infty := s(f, \Delta_n)_\infty = O(n^{-\alpha}) \tag{3.10}$$

if and only if $f \in \mathrm{Lip}\,\alpha$ (see DeVore and Lorentz (1993)). The saturation result holds as well. If $s_n(f)_\infty = o(n^{-1})$ then f is constant. Of course the direct estimates in this setting follow from (3.4). The inverse estimates are a little more subtle and use the fact that the sets Δ_n mix; that is, each point $x \in (0,1)$ falls in the 'middle' of many intervals from the partitions associated to Δ_n. If we consider partitions that do not mix then, while direct estimates are equally valid, the inverse estimates generally fail. A case in point are the dyadic partitions whose sets of breakpoints Δ_{2^n} are nested. A piecewise constant function from $S^1(\Delta_{2^n})$ will be approximated exactly by elements from $S^1(\Delta_{2^m})$, $m \geq n$, and yet these functions are not even continuous.

An analysis similar to that given above holds for approximation in L_p, for $1 \leq p < \infty$, and even for $0 < p < 1$. To explain these results, we define the space $\mathrm{Lip}(\alpha, L_p(\Omega))$, $0 < \alpha \leq 1$, $0 < p \leq \infty$, which is the set of all functions $f \in L_p(\Omega)$ for which

$$\|f(\cdot + h) - f\|_{L_p[0,1-h)} \leq Mh^\alpha, \quad 0 < h < 1. \tag{3.11}$$

Again, the smallest $M \geq 0$ for which (3.11) holds is $|f|_{\mathrm{Lip}(\alpha, L_p(\Omega))}$.

By analogy with (3.4), there are $S_T \in \mathcal{S}^1(T)$ such that

$$s(f,T)_p \leq \|f - S_T\|_{L_p(\Omega)} \leq C_p |f|_{\mathrm{Lip}(\alpha, L_p(\Omega))} \delta_T^\alpha \tag{3.12}$$

with the constant C_p depending at most on p. Indeed, for $p \geq 1$, we can define S_T by

$$S_T(x) := a_I(f), \quad x \in I, \ I \in \Pi(T), \tag{3.13}$$

with[2]

$$a_I(f) := \frac{1}{|I|} \int_I f \, dx$$

the average of f over I. With this definition of S_T one easily derives (3.12); see Section 2 of Chapter 12 in DeVore and Lorentz (1993). When $0 < p < 1$, we replace $a_I(f)$ by the median of f on the interval I (see Brown and Lucier (1994)).

Inverse estimates follow the same lines as the case $p = \infty$ discussed above. We limit further discussion to the case Δ_n of equally spaced breakpoints given by (3.9). Then, if f satisfies

$$s_n(f)_p := s(f, \Delta_n)_p \leq Mn^{-\alpha}, \quad n = 1, 2, \ldots, \tag{3.14}$$

for some $0 < \alpha \leq 1$, $M > 0$, then $f \in \mathrm{Lip}(\alpha, L_p(\Omega))$ and

$$|f|_{\mathrm{Lip}(\alpha, L_p(\Omega))} \leq C_p M.$$

[2] We shall use the notation $|E|$ to denote the Lebesgue measure of a set E throughout this paper.

The saturation theorem is also valid: if $s_n(f)_p = o(n^{-1})$, $n \to \infty$, then f is constant.

In summary, we know precisely when a function satisfies $s_n(f)_p = O(n^{-\alpha})$, $n = 1, 2, \ldots$; it should be in the space $\mathrm{Lip}(\alpha, L_p(\Omega))$. This provides a guide to the construction and analysis of numerical methods based on approximation by piecewise constants. For example, suppose that we are using $\mathcal{S}^1(\Delta_n)$ to generate a numerical approximation $A_n u$ to a function u which is known to be in $\mathrm{Lip}(1, L_p(\Omega))$. The values of u would not be known to us but would be generated by our numerical method. The estimates (3.4) or (3.12) tell us what we could expect of the numerical method in the best of all worlds. If we are able to prove that our numerical method satisfies

$$\|u - A_n u\|_{L_p(\Omega)} \le C_p |f|_{\mathrm{Lip}(1, L_p(\Omega))} n^{-1}, \quad n = 1, 2, \ldots, \tag{3.15}$$

we can rest assured that we have done the best possible (save for the numerical constant C_p). If we cannot prove such an estimate then we should try to understand why. Moreover, (3.15) is the correct form of error estimates based on approximation by piecewise constants on uniform partitions.

There are numerous generalizations of the results given in this section. First of all, piecewise constants can be replaced by piecewise polynomials of degree r with r arbitrary but fixed (see Section 6.2). One can require that the piecewise polynomials have smoothness C^{r-2} at the breakpoints with an identical theory. Of course, inverse theorems still require some mixing condition. Moreover, all of these results hold in the multivariate case as is discussed in Section 6.2. We can also do a more subtle analysis of approximation orders where $O(n^{-\alpha})$ is replaced by a more general statement on the rate of decay of the error. This is important for a fuller understanding of approximation theory and its relationship to function spaces. We shall discuss these issues in Section 4 after the reader has more familiarity with more fundamental approximation concepts.

3.2. Nonlinear approximation by piecewise constants

In linear approximation by piecewise constants, the partitions are chosen in advance and are independent of the target function f. The question arises whether there is anything to be gained by allowing the partition to depend on f. This brings us to try to understand approximation by piecewise constants where the number of pieces is fixed but the actual partition can vary with the target function. This is the simplest case of what is called variable knot spline approximation. It is also one of the simplest and most instructive examples of nonlinear approximation.

If T is a finite set of points $0 =: t_0 < t_1 < \cdots < t_n := 1$ from Ω, we denote by $\mathcal{S}^1(T)$ the functions S which are piecewise constant with breakpoints from T. Let $\Sigma_n := \cup_{\#T=n+1} \mathcal{S}^1(T)$, where $\#T$ denotes the cardinality of T.

Each function in Σ_n is piecewise constant with at most n pieces. Note that Σ_n is not a linear space; for example, adding two functions in Σ_n results in a piecewise constant function but with as many as $2n$ pieces. Given $f \in L_p(\Omega)$, $0 < p \le \infty$, we introduce

$$\sigma_n(f)_p := \inf_{S \in \Sigma_n} \|f - S\|_{L_p(\Omega)}, \tag{3.16}$$

which is the L_p-error of nonlinear piecewise constant approximation to f.

As noted earlier, we would like to understand what properties of f determine the rate of decrease of $\sigma_n(f)_p$. We shall begin our discussion with the case $p = \infty$, which corresponds to approximating the continuous function f in the uniform norm. We shall show the following result of Kahane (1961). For a function $f \in C(\Omega)$ we have

$$\sigma_n(f)_\infty \le \tfrac{M}{2n}, \quad n = 1, 2, \ldots, \tag{3.17}$$

if and only if $f \in \mathrm{BV}$, *i.e.*, f, is of bounded variation on Ω and $|f|_{\mathrm{BV}} := \mathrm{Var}_\Omega(f)$ is identical with the smallest constant M for which (3.17) holds.

We sketch the proof of Kahane's result since it is quite simple and instructive. Suppose first that $f \in \mathrm{BV}$ with $M := \mathrm{Var}_\Omega(f)$. Since f is, by assumption, continuous, we can find $T := \{0 =: t_0, \ldots, t_n := 1\}$ such that $\mathrm{Var}_{[t_{k-1}, t_k)} f \le M/n$, $k = 1, \ldots, n$. If a_k is the median value of f on $[t_{k-1}, t_k]$, and $S_n(x) := a_k$, $x \in [t_{k-1}, t_k)$, $k = 1, \ldots, n$, then $S_n \in \Sigma_n$ and satisfies

$$\|f - S_n\|_{L_\infty(\Omega)} \le M/2n, \tag{3.18}$$

which shows (3.17).

Conversely, suppose that (3.17) holds for some $M > 0$. Let $S_n \in \Sigma_n$ satisfy $\|f - S_n\|_{L_\infty(\Omega)} \le (M + \epsilon)/(2n)$ with $\epsilon > 0$. If $x_0 := 0 < x_1 < \cdots < x_m := 1$ is an arbitrary partion for Ω and ν_k is the number of values that S_n attains on $[x_{k-1}, x_k)$, then one easily sees that

$$|f(x_k) - f(x_{k-1})| \le 2\nu_k \|f - S_n\|_{L_\infty(\Omega)} \le \tfrac{\nu_k(M+\epsilon)}{n}, \quad k = 1, 2, \ldots, m. \tag{3.19}$$

Since $\sum_{k=1}^m \nu_k \le m + n$, we have

$$\sum_{k=1}^m |f(x_k) - f(x_{k-1})| \le \sum_{k=1}^m \tfrac{\nu_k(M+\epsilon)}{n} \le (M + \epsilon)(1 + \tfrac{m}{n}). \tag{3.20}$$

Letting $n \to \infty$ and then $\epsilon \to 0$ we find

$$\sum_{k=1}^m |f(x_k) - f(x_{k-1})| \le M, \tag{3.21}$$

which shows that $\mathrm{Var}_\Omega(f) \le M$.

There are elements of the above proof that are characteristic of nonlinear approximation. Firstly, the partition providing (3.17) depends on f.

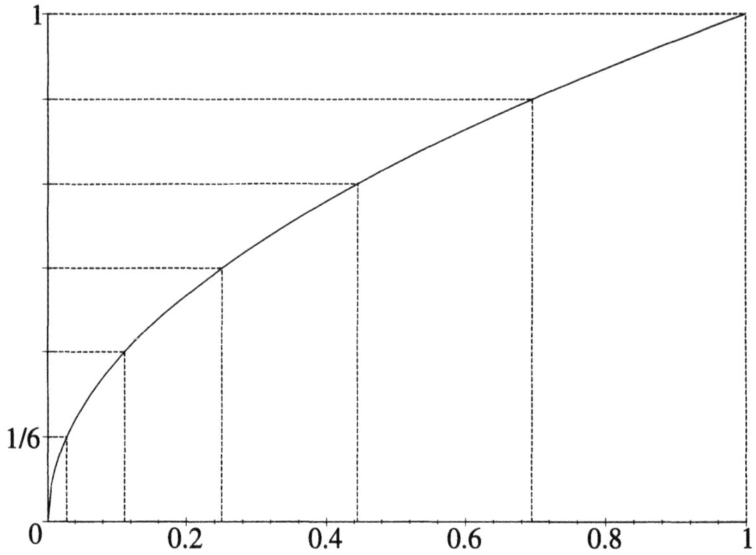

Fig. 1. Best selection of breakpoints for $f(x) = x^{1/2}$ when $n = 6$

Secondly, this partition is obtained by balancing the variation of f over the intervals I in this partition. In other types of nonlinear approximation, $\mathrm{Var}_I(f)$ will be replaced by some other expression $B(f, I)$ defined on intervals I (or other sets in more general settings).

Let us pause now for a moment to compare Kahane's result with what we know about linear approximation by piecewise constants in the uniform norm. In both cases, we can characterize functions which can be approximated with efficiency $O(n^{-1})$. In the case of linear approximation from $\mathcal{S}^1(T_n)$ (as described in the previous section), this is the class of functions $\mathrm{Lip}\,(1, L_\infty(\Omega))$ or, equivalently, functions f for which $f' \in L_\infty(\Omega)$. On the other hand, for nonlinear approximation, it is the class BV of functions of bounded variation. It is well known that $\mathrm{BV} = \mathrm{Lip}(1, L_1(\Omega))$ with equivalent norms. Thus in both cases the function is required to have one order of smoothness but measured in quite different norms. For linear approximation the smoothness is measured in L_∞, the same norm as the underlying approximation. For nonlinear approximation the smoothness is measured in L_1. Thus, in nonlinear approximation, the smoothness is measured in a weaker norm. What is the significance of L_1? The answer lies in the Sobolev embedding theorem. Among the spaces $\mathrm{Lip}(1, L_p(\Omega))$, $0 < p \le \infty$, $p = 1$ is the smallest value for which this space is embedded in $L_\infty(\Omega)$. In other words, the functions in $\mathrm{Lip}(1, L_1(\Omega))$ barely get into $L_\infty(\Omega)$ (the space in which we measure error) and yet we can approximate them quite well.

An example might be instructive. Consider the function $f(x) = x^\alpha$ with $0 < \alpha < 1$. This function is in $\mathrm{Lip}(\alpha, L_\infty(\Omega))$ and in no higher-order Lipschitz space. It can be approximated by elements of $\mathcal{S}^1(T_n)$ with order

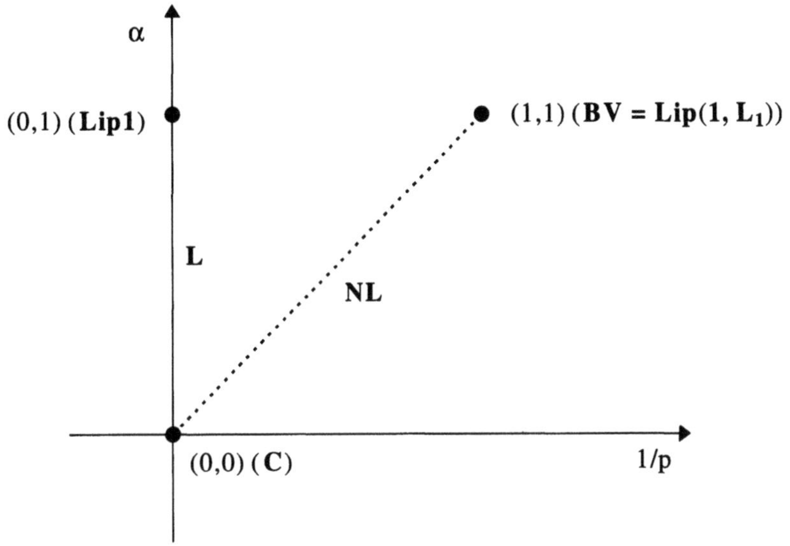

Fig. 2. Linear and nonlinear approximation in C

exactly $O(n^{-\alpha})$. On the other hand, this function is clearly of bounded variation (being monotone) and hence can be approximated by the elements of Σ_n to order $O(n^{-1})$. It is easy to see how to construct such an approximant. Consider the graph of f as depicted in Figure 1. We divide the range of f (which is the interval $[0,1)$) on the y-axis into n pieces corresponding to the y values $y_k := k/n$, $k = 0, 1, \dots, n$. The preimage of these points is the set $\{x_k := (k/n)^{1/\alpha} : 0 \le k \le n\}$, which forms our set T of breakpoints for the best piecewise polynomial approximant from Σ_n.

It will be useful to have a way of visualizing spaces of functions as they occur in our discussion of approximation. This will give us a simple way to keep track of various results and also add to our understanding. We shall do this by using points in the upper right quadrant of the plane. The x-axis will correspond to the L_p spaces except that L_p is identified with $x = 1/p$ not with $x = p$. The y axis will correspond to the order of smoothness. For example $y = 1$ will mean a space of smoothness order one (or one time differentiable, if you like). Thus $(1/p, \alpha)$ corresponds to a space of smoothness α measured in the L_p-norm. For example, we could identify this point with the space $\mathrm{Lip}(\alpha, L_p)$ although when we get to finer aspects of approximation theory we may want to vary this interpretation slightly.

Figure 2 gives a summary of our knowledge so far. The vertical line segment (marked L) connecting $(0,0)$ (L_∞) to $(0,1)$ $(\mathrm{Lip}(1, L_\infty))$ correspond to the spaces we engaged when we characterized approximation order for linear approximation (approximation from $\mathcal{S}^1(T_n)$). For example, $(0,1)$ $(\mathrm{Lip}(1, L_\infty))$ was the space of functions with approximation order $O(n^{-1})$.

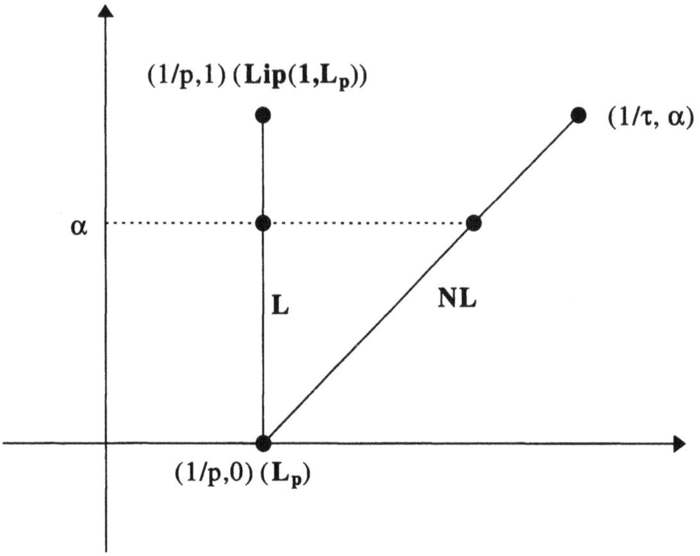

Fig. 3. Linear and nonlinear approximation in L_p

On the other hand, for nonlinear approximation from Σ_n, we saw that the point $(1,1)$ $(\text{Lip}(1, L_1))$ describes the space of functions which are approximated with order $O(n^{-1})$. We shall see later (Section 4) that the point (α, α) on the line connecting $(0,0)$ to $(1,1)$ (marked NL) describes the space of functions approximated with order $O(n^{-\alpha})$ (a few new wrinkles come in here which is why we are postponing a precise discussion).

More generally, approximation in L_p, $0 < p \leq \infty$, is depicted in Figure 3. The spaces corresponding to linear approximation lie on the vertical line segment (marked L) connecting $(1/p, 0)$ (L_p) to $(1/p, 1)$ $(\text{Lip}(1, L_p))$, whereas the line segment (marked NL) emanating from $(1/p, 0)$ with slope one will describe the nonlinear approximation spaces. The points on this line are of the form $(1/\tau, \alpha)$ with $1/\tau = \alpha + 1/p$. Again, this line segment in nonlinear approximation corresponds to the limiting spaces in the Sobolev embedding theorem. Spaces to the left of this line segment are embedded into L_p; those to the right are not.

There are various generalizations of nonlinear piecewise constant approximation which we shall address in due course. For univariate approximation, we can replace piecewise constant functions by piecewise polynomials of fixed degree r with n free knots with a similar theory (Section 6.3). However, multivariate approximation by piecewise polynomials leads to new difficulties, as we shall see in Section 6.5.

Approximation by piecewise constants (or more generally piecewise polynomials) with free knots is used in numerical PDEs. It is particularly useful when the solution is known to develop singularities. An example would be a nonlinear transport equation in which shocks appear (see Section 10). The significance of the above results to the numerical analyst is that it clarifies what is the optimal performance that can be obtained by such methods. Once the norm has been chosen in which the error is to be measured, then we understand the minimal smoothness that will allow a given approximation rate. We also understand what form error estimates should take. For example, consider numerically approximating a function u by a piecewise constant function $A_n u$ with n free knots. We have seen that, in the case of uniform approximation, the correct form of the error estimate is

$$\|u - A_n u\|_{L_\infty(\Omega)} \le C\frac{|u|_{\mathrm{BV}}}{n}. \tag{3.22}$$

This is in contrast to the case of fixed knots where $|u|_{\mathrm{BV}}$ is replaced by $\|u'\|_{L_\infty(\Omega)}$. A similar situation exists when error is measured in other L_p-norms, as will be developed in Section 6.

The above theory of nonlinear piecewise constant approximation also tells us the correct form for local error estimators. Approximating in L_∞, we should estimate local error by local variation. Approximating in L_p, the variation will be replaced by other set functions obtained from certain Besov or Sobolev norms (see Section 6.1).

3.3. Adaptive approximation by piecewise constants

One disadvantage of piecewise constant approximation with free knots is that it is not always easy to find partitions that realize the optimal approximation order. This is particularly true in the case of numerical approximation when the target function is not known to us but is only approximated as we proceed numerically. One way to ameliorate this situation is to generate partitions adaptively. New breakpoints are added as new information is gained about the target function. We shall discuss this type of approximation in this section with the goal of understanding what is lost in terms of accuracy of approximation when adaptive partitions are used in place of free partitions. Adaptive approximation is also important because it generalizes readily to the multivariate case when intervals are replaced by cubes.

The starting point for adaptive approximation is a function $\mathcal{E}(I)$ which is defined for each interval $I \subset \Omega$ and estimates the approximation error on I. Namely, let $E(f, I)_p$ be the local error in approximating f by constants in the $L_p(I)$-norm:

$$E(f, I)_p := \inf_{c \in \mathbb{R}} \|f - c\|_{L_p(I)}. \tag{3.23}$$

Then, we assume that $\mathcal{E}$ satisfies

$$E(f, I)_p \le \mathcal{E}(I). \tag{3.24}$$

In numerical settings, $\mathcal{E}(I)$ is an upper bound for $E(f,I)_p$ obtained from the information at hand. It is at this point that approximation theory and numerical analysis sometimes part company. Approximation theory assumes enough about the target function to have an effective error estimator $\mathcal{E}$, a property not always verifiable for numerical estimators.

To retain the spirit of our previous sections, let us assume for our illustration that $p = \infty$ so that we are approximating continuous functions in the $L_\infty(\Omega)$ norm. In this case, a simple upper bound for $E(f,I)_\infty$ is provided by

$$E(f,I)_\infty \le \mathrm{Var}_I(f) \le \int_I |f'(x)|\,\mathrm{d}x, \tag{3.25}$$

which holds whenever these quantities are defined for the continuous function f (*i.e.*, f should be in BV for the first estimate, $f' \in L_1$ for the second). Thus, we could take for $\mathcal{E}$ any of the three quantities appearing in (3.25). A common feature of each of these error estimators is that

$$\mathcal{E}(I_1) + \mathcal{E}(I_2) \le \mathcal{E}(I_1 \cup I_2), \quad I_1 \cap I_2 = \emptyset. \tag{3.26}$$

We shall restrict our attention to adaptive algorithms that create partitions of Ω consisting of dyadic intervals. Our development parallels completely the standard treatment of adaptive numerical quadrature. We shall denote by $D := D(\Omega)$ the set of all dyadic intervals in Ω; for specificity we take these intervals to be closed on the left end-point and open on the right. Each interval $I \in D$ has two *children*. These are the intervals $J \in D$ such that $J \subset I$ and $|J| = |I|/2$. If J is a child of I then I is called the *parent* of J. Intervals $J \in D$ such that $J \subset I$ are *descendants* of I those with $I \subset J$ are *ancestors* of I.

A typical adaptive algorithm proceeds as follows. We begin with our target function f, an error estimator $\mathcal{E}$, and a target tolerance ϵ which relates to the final approximation error we want to attain. At each step of the algorithm we have a set $\mathcal{G}$ of good intervals (on which the local error meets the tolerance) and a set $\mathcal{B}$ of bad intervals (on which we do not meet the tolerance). Good intervals become members of our final partition. Bad intervals are further processed: they are halved and their children are checked for being good or bad.

Initially, we check $\mathcal{E}(\Omega)$. If $\mathcal{E}(\Omega) \le \epsilon$ then we define $\mathcal{G} = \{\Omega\}$, $\mathcal{B} := \emptyset$ and we terminate the algorithm. On the other hand, if $\mathcal{E}(\Omega) > \epsilon$, we define $\mathcal{G} = \emptyset$, $\mathcal{B} := \{\Omega\}$ and proceed with the following general step of the algorithm.

General step. Given any interval I in the current set $\mathcal{B}$ of bad intervals, we process it as follows. For each of the two children J of I, we check $\mathcal{E}(J)$. If $\mathcal{E}(J) \le \epsilon$, then J is added to the set of good intervals. If $E(J) > \epsilon$, then J is added to the set of bad intervals. Once a bad interval is processed it is removed from $\mathcal{B}$.

The algorithm terminates when $\mathcal{B} = \emptyset$, and the final set of good intervals is denoted by $\mathcal{G}_\epsilon := \mathcal{G}_\epsilon(f)$. The intervals in $\mathcal{G}_\epsilon$ form a partition of Ω, that is, they are pairwise disjoint and their union is all of Ω. We define

$$S_\epsilon := \sum_{I \in \mathcal{G}_\epsilon} c_I \chi_I, \tag{3.27}$$

where c_I is a constant that satisfies

$$\|f - c_I\|_{L_\infty(I)} \le \mathcal{E}(I) \le \epsilon, \quad I \in \mathcal{G}_\epsilon.$$

Thus, S_ϵ is a piecewise constant function approximating f to tolerance ϵ:

$$\|f - S_\epsilon\|_{L_\infty(\Omega)} \le \epsilon. \tag{3.28}$$

The approximation efficiency of the adaptive algorithm depends on the number $N_\epsilon(f) := \#\mathcal{G}_\epsilon(f)$ of good intervals. We are interested in estimating N_ϵ so that we can compare adaptive efficiency with free knot spline approximation. For this we recall the space $L \log L$, which consists of all integrable functions for which

$$\|f\|_{L \log L} := \int_\Omega |f(x)|(1 + \log |f(x)|)\, dx$$

is finite. This space contains all spaces L_p, $p > 1$, but is strictly contained in $L_1(\Omega)$. We have shown in DeVore (1987) that any of the three estimators of (3.25) satisfy

$$N_\epsilon(f) \le C \frac{\|f'\|_{L \log L}}{\epsilon}. \tag{3.29}$$

We shall give the proof of (3.29), which is not difficult. It will allow us to introduce some concepts that are useful in nonlinear approximation and numerical estimation, such as the use of maximal functions. The Hardy–Littlewood maximal function Mf is defined for a function in $L_1(\Omega)$ by

$$Mf(x) := \sup_{I \ni x} \frac{1}{|I|} \int_I |f(y)|\, dy, \tag{3.30}$$

where the sup is taken over all intervals $I \subset \Omega$ which contain x. Thus $Mf(x)$ is the smallest number that bounds all of the averages of $|f|$ over intervals which contain x. The maximal function Mf is at the heart of differentiability of functions (see Chapter 1 of Stein (1970)). We shall need the fact (see pages 243–246 of Bennett and Sharpley (1988)) that

$$\|f\|_{L \log L} \asymp \int_\Omega Mf(y)\, dy. \tag{3.31}$$

We shall use Mf to count N_ϵ. We assume that $\mathcal{G}_\epsilon \neq \{\Omega\}$. Suppose that $I \in \mathcal{G}_\epsilon$. Then the parent J of I satisfies

$$\epsilon < \mathcal{E}(J) \leq \int_J |f'(y)|\,\mathrm{d}y \leq |J| Mf'(x), \tag{3.32}$$

for all $x \in J$. In particular, we have

$$\epsilon \leq |J| \inf_{x\in I} Mf'(x) \leq \frac{|J|}{|I|}\int_I Mf'(y)\,\mathrm{d}y = 2\int_I Mf'(y)\,\mathrm{d}y. \tag{3.33}$$

Since the intervals in $\mathcal{G}_\epsilon$ are disjoint, we have

$$N_\epsilon \epsilon \leq 2\sum_{I\in\mathcal{G}_\epsilon}\int_I Mf'(y)\,\mathrm{d}y = 2\int_\Omega Mf'(y)\,\mathrm{d}y \leq C\|f'\|_{L\log L},$$

where the last inequality uses (3.31). This proves (3.29).

In order to compare adaptive approximation with free knot splines, we introduce the adaptive approximation error

$$a_n(f)_\infty := \inf\{\epsilon : N_\epsilon(f) \leq n\}. \tag{3.34}$$

Thus, with the choice $\epsilon = (C\|f'\|_{L\log L})/n$, and C the constant in (3.29), our adaptive algorithm generates a partition $\mathcal{G}$ with at most n dyadic intervals and, from (3.28), we have

$$a_n(f)_\infty \leq \|f - S_\epsilon\|_{L_\infty(\Omega)} \leq C\frac{\|f'\|_{L\log L}}{n}. \tag{3.35}$$

Let's compare $a_n(f)_\infty$ with the error $\sigma_n(f)_\infty$ for free knot approximation. In free knot splines we obtained the approximation rate $\sigma_n(f)_\infty = O(n^{-1})$ if and only if $f \in \mathrm{BV}$. This condition is slightly weaker than requiring that f' is in $L_1(\Omega)$ (the derivative of f should be a Borel measure). On the other hand, assuming that f satisfies the slightly stronger condition $f' \in L\log L$, we find $a_n(f)_\infty \leq C/n$. Thus, the cost in using adaptive algorithms is slight from the viewpoint of the smoothness condition required on f to produce the order $O(n^{-1})$.

It is much more difficult to prove error estimates for numerically based adaptive algorithms. What is needed is a comparison (from above and below) of the error estimator $\mathcal{E}(I)$ with the local approximation error $E(f,I)_p$ or one of the good estimators like $\int_I |f'|$. Nevertheless, the above results are useful in that they give the form such error estimators $\mathcal{E}(I)$ should take and also give the form the error analysis should take.

There is a comparable theory for adaptive approximation in other L_p-norms and even in several variables (Birman and Solomyak 1967).

3.4. n-term approximation: a first look

There is another view toward the results we have obtained thus far, which is important because it generalizes readily to a variety of settings. In each of the three types of approximation (linear, free knot, and adaptive), we have constructed an approximant of the form

$$S = \sum_{I \in \Lambda} c_I \chi_I, \tag{3.36}$$

where Λ is a set of intervals and the c_I are constants. Thus, a general approximation problem that would encompass all three of the above is to approximate using sums (3.36) where $\#\Lambda \le n$. This is called *n-term approximation*. We formulate this problem more formally as follows.

Let Σ_n^* be the set of all piecewise constant functions that can be written as in (3.36) with $\#\Lambda \le n$. Then, Σ_n^* is a nonlinear space. As in our previous considerations, we define the L_p-approximation error

$$\sigma_n^*(f)_p := \inf_{S \in \Sigma_n^*} \|f - S\|_{L_p(\Omega)}. \tag{3.37}$$

Note that we do not require that the intervals of Λ form a disjoint partition; we allow possible overlap in the intervals.

It is easy to see that the approximation properties of n-term approximation is equivalent to that of free knot approximation. Indeed, $\Sigma_n \subset \Sigma_n^* \subset \Sigma_{2n}$, $n = 1, 2, \ldots$, and therefore

$$\sigma_{2n}(f)_p \le \sigma_n^*(f)_p \le \sigma_n(f)_p. \tag{3.38}$$

Thus, for example, a function f satisfies $\sigma_n^*(f)_p = O(n^{-\alpha})$ if and only if $\sigma_n^*(f)_p = O(n^{-\alpha})$.

The situation with adaptive algorithms is more interesting and enlightening. In analogy to the above, one defines Σ_n^a as the set of functions S which can be expressed as in (3.36), but now with $\Lambda \subset D$ and σ_n^a defined accordingly. The analogue of (3.38) would compare σ_n^a and a_m. Of course, $\sigma_n^a \le a_n$, $n \ge 1$. But no comparison $a_{cn} \le \sigma_n^a$, $n = 1, 2, \ldots$, is valid for any fixed constant $c \ge 1$. The reason is that adaptive algorithms do not create arbitrary functions in Σ_n^a. For example, the adaptive algorithm cannot have a partition with just one small dyadic interval; it automatically carries with it a certain entourage of intervals. We can explain this in more detail by using binary trees.

Consider any of the adaptive algorithms of the previous section. Given an $\epsilon > 0$, let $\mathcal{B}_\epsilon$ be the collection of all $I \in D$ such that $\mathcal{E}(I) > \epsilon$ (the collection of bad intervals). Then, whenever $I \in \mathcal{B}_\epsilon$, its parent is too. Thus $\mathcal{B}_\epsilon$ is a binary tree with root Ω. The set of dyadic intervals $\mathcal{G}_\epsilon$ is precisely the set of good intervals I (*i.e.*, $\mathcal{E}(I) \le \epsilon$) whose parent is bad. The inefficiency of the adaptive algorithm occurs when $\mathcal{B}_\epsilon$ contains a long chain of intervals

$I_1 \supset I_2 \supset \cdots \supset I_m$ with I_k the parent of I_{k+1} with the property that the other child of I_k is always good, $k = 1, \ldots, m-1$. This occurs, for example, when the target function f has a singularity at some point $x_0 \in I_m$ but is smooth otherwise. The partition $\mathcal{G}_\epsilon$ will contain one dyadic interval at each level (the sibling J_k of I_k). Using free knot partitions, we would zoom in faster on this singularity and thereby avoid this entourage of intervals J_k.

There are ways of modifying the adaptive algorithm to make it comparable to approximation from Σ_n^a, which we now briefly describe. If we are confronted with a long chain $I_0 \supset I_1 \supset \cdots \supset I_m$ of bad intervals from $\mathcal{B}_\epsilon$, the adaptive algorithm would place each of the sibling intervals J_k of I_k, $k = 0, \ldots, m$, into the good partition. We can decrease the number of intervals needed in the following way. We find the shortest subchain $I_0 = I_{j_0} \supset I_{j_1} \supset \cdots \supset I_{j_\ell} = I_m$ for which $\mathcal{E}(I_{j-1} \setminus I_j) < \epsilon$, $j = 1, \ldots, \ell$. Then, it is sufficient to use the intervals I_{j_i}, $i = 0, \ldots, \ell$, in place of the intervals J_k, $k = 0, \ldots, m$, in the construction of an approximant from Σ_n^a (see DeVore and Popov (1987) or Cohen, DeVore, Petrushev and Xu (1998) for a further elaboration on these ideas).

3.5. Wavelets: a first look; the Haar system

The two topics of approximating functions and representing them are closely related. For example, approximation by trigonometric sums is closely related to the theory of Fourier series. Is there an analogue in approximation by piecewise constants? The answer is yes. There are in fact several representations of a given function f using a basis of piecewise constant functions. The most important of these is the Haar basis, which we shall now describe.

Rather than simply introducing the Haar basis and giving its properties, we prefer to present this topic from the viewpoint of multiresolution analysis (MRA) since this is the launching point for the construction of wavelet bases, which we shall discuss in more detail in Section 7. Wavelets and multilevel methods are increasingly coming into favour in numerical analysis.

Let us return to the linear spaces $\mathcal{S}^1(\Delta_n)$ of piecewise constant functions on the partition of Ω with spacing $1/n$. We shall only need the case $n = 2^k$ and we denote this space by $\mathcal{S}_k := \mathcal{S}^1(\Delta_{2^k})$. The characteristic functions χ_I, $I \in D_k(\Omega)$, are a basis for $\mathcal{S}_k$. If we approximate well a smooth function f by a piecewise constant function $S = \sum_{I \in D_k} c_I \chi_I$ from $\mathcal{S}_k$, then the coefficients c_I will not change much: c_I will be close to c_J if I is close to J. We would like to take advantage of this fact to find a more compact representation for S. That is, we should be able to find a more favourable basis for $\mathcal{S}_k$ for which the coefficients of S are either zero or small.

The spaces $\mathcal{S}_k$ form a ladder: $\mathcal{S}_k \subset \mathcal{S}_{k+1}$, $k = 0, 1, \ldots$. We let $W_k := \mathcal{S}_{k+1} \ominus \mathcal{S}_k$ be the orthogonal complement of $\mathcal{S}_k$ in $\mathcal{S}_{k+1}$. This means that

W_k consists precisely of the functions in $w \in \mathcal{S}_{k+1}$ orthogonal to $\mathcal{S}_k$:

$$\int_\Omega w(x)S(x)\,\mathrm{d}x = 0, \quad \text{for all } S \in \mathcal{S}_k.$$

We then have

$$S_{k+1} = S_k \oplus W_k, \quad k = 0, 1, \ldots. \tag{3.39}$$

Thus W_k represents the *detail* that must be added to $\mathcal{S}_k$ in order to obtain $\mathcal{S}_{k+1}$.

The spaces W_k have a very simple structure. Consider, for example, $W := W_0$. Since $S_1 = S_0 + W_0$, and S_1 has dimension 2 and S_0 dimension 1, the space W_1 will be spanned by a single function from $\mathcal{S}_1$. Orthogonality gives us that this function is a nontrivial multiple of

$$H(x) := \chi_{[0,1/2)} - \chi_{[1/2,1)} = \begin{cases} 1, & 0 \le x < 1/2, \\ -1, & 1/2 \le x < 1. \end{cases} \tag{3.40}$$

H is called the Haar function. More generally, it is easy to see that W_k is spanned by the following (normalized) shifted dilates of H:

$$H_{j,k}(x) := 2^{k/2} H(2^k x - j), \quad j = 0, \ldots, 2^k - 1. \tag{3.41}$$

The function $H_{j,k}$ is a scaled version of H fitted to the interval $2^{-k}[j, j+1)$ which has $L_2(\Omega)$-norm one: $\|H_{j,k}\|_{L_2(\Omega)} = 1$.

From (3.39), we find

$$\mathcal{S}_m = \mathcal{S}_0 \oplus W_0 \oplus \cdots \oplus W_{m-1}. \tag{3.42}$$

It follows that χ_Ω together with the functions $H_{j,k}$, $j = 0, \ldots, 2^k - 1$, $k = 0, \ldots, m-1$, form an orthonormal basis for $\mathcal{S}_m$ which is, in many respects, better than the old basis χ_I, $I \in D_m$. But, before taking up that point, we want to see that we can take $m \to \infty$ in (3.42) and thereby obtain a basis for $L_2(\Omega)$.

It will be useful to have an alternative notation for the Haar functions $H_{j,k}$. Each j, k corresponds to the dyadic interval $I := 2^{-k}[j, j+1)$. We shall write

$$H_I := H_{j,k} = |I|^{-1/2} H(2^k \cdot -j). \tag{3.43}$$

From (3.42) we see that each $S \in \mathcal{S}_m$ has the representation

$$S = \langle S, \chi_\Omega \rangle \chi_\Omega + \sum_{I \in \cup_{0 \le k < m} D_k} \langle S, H_I \rangle H_I, \tag{3.44}$$

where

$$\langle f, g \rangle := \int_\Omega f(x) g(x)\,\mathrm{d}x \tag{3.45}$$

is the inner product in $L_2(\Omega)$.

Let P_m denote the orthogonal projector onto $\mathcal{S}_m$. Thus, $P_m f$ is the best $L_2(\Omega)$-approximation to f from $\mathcal{S}_m$. It is the unique element in $\mathcal{S}_m$ such that $f - P_m f$ is orthogonal to $\mathcal{S}_m$. Using the orthonormal basis of (3.44), we see that

$$P_m f = \langle f, \chi_\Omega \rangle \chi_\Omega + \sum_{I \in \cup_{0 \le k < m} D_k} \langle f, H_I \rangle H_I. \tag{3.46}$$

Since $\operatorname{dist}(f, \mathcal{S}_m)_{L_2(\Omega)} \to 0$, $m \to \infty$, we can take the limit in (3.46) to obtain

$$f = \langle f, \chi_\Omega \rangle \chi_\Omega + \sum_{I \in D} \langle f, H_I \rangle H_I \tag{3.47}$$

In other words, χ_Ω together with the functions H_I, $I \in D$, form an orthonormal basis, called the *Haar basis*, for $L_2(\Omega)$.

Some of the advantages of the Haar basis for $\mathcal{S}_m$ over the standard basis $(\chi_I, I \in D_m)$ are obvious. If we wish to increase our resolution of the target function by approximating from $\mathcal{S}_{m+1}$ rather than $\mathcal{S}_m$, we do not need to recompute our approximant. Rather, we merely add a layer of the decomposition (3.47) to the approximant corresponding to the wavelet space W_{m+1}. Of course, the orthogonality of W_m to S_m means that this new information is independent of our previous information about f. It is also clear that the coefficients of the basis function H_I, $I \in D_m$, tend to zero as $m \to \infty$. Indeed, we have

$$\|f\|^2_{L_2(\Omega)} = |\langle f, \chi_\Omega \rangle|^2 + \sum_{I \in D} |\langle f, H_I \rangle|^2. \tag{3.48}$$

Therefore, this series converges absolutely.

3.6. n-term approximation: a second look

We shall next consider n-term approximation using the Haar basis. This is a special case of n-term wavelet approximation considered in more detail in Section 7.4. Let Σ_n^H denote the collection of all functions S of the form

$$S = c\chi_\Omega + \sum_{I \in \Lambda} c_I H_I, \tag{3.49}$$

where $\Lambda \subset D$ is a set of dyadic intervals with $\#\Lambda \le n$. As before, we let

$$\sigma_n^H(f)_p := \inf_{S \in \Sigma_n^H} \|f - S\|_{L_p(\Omega)} \tag{3.50}$$

be the error of n-term approximation.

We shall consider first the case of approximation in $L_2(\Omega)$ where the matter is completely transparent. In fact, in this case, in view of the norm equivalence (3.48), we see that a best approximation from Σ_n^H is given by

$$S = \langle f, \chi_\Omega \rangle \chi_\Omega + \sum_{I \in \Lambda} \langle f, H_I \rangle H_I, \tag{3.51}$$

where $\Lambda \subset D$ is a set corresponding to the n biggest Haar coefficients. Since there may be coefficients of equal absolute values, best approximation from Σ_n^H is not necessarily unique.

Since we are dealing with an orthonormal system, we can apply the results of Section 2 to characterize the class of functions f which satisfy

$$\sigma_n^H(f)_2 \le M n^{-\alpha}, \quad n = 1, 2, \ldots. \tag{3.52}$$

Namely, let $\gamma_n := \gamma_n(f)$ be the absolute of the nth largest Haar coefficient. It follows from the characterization (2.18) that, for any $\alpha > 0$, a function f satisfies (3.52) if and only if

$$\gamma_n(f) \le \frac{M'}{n^{\alpha+1/2}}. \tag{3.53}$$

Moreover, the smallest constant M in (3.52) is equivalent (independently of f) to the smallest constant M' in (3.53).

It is interesting to note that the above characterization holds for any $\alpha > 0$; it is not necessary to assume that $\alpha \le 1$. It is not apparent how the characterization (3.53) relates directly to the smoothness of f. We shall see later, when we develop n-term wavelet approximation in more detail, that, for $0 < \alpha < 1$, (3.53) is tantamount to requiring that f have α orders of smoothness in L_τ, where τ is defined by $1/\tau = \alpha + 1/2$. We recall our convention for interpreting smoothness spaces as points in the upper right quadrant of $\mathbb{R}^2$, as described in Section 3.2. The point $(1/\tau, \alpha)$ lies on the line with slope one which passes through $(1/2, 0)$ ($L_2(\Omega)$). Thus, the characterization of n-term Haar approximation (in $L_2(\Omega)$) is the same as the previous characterizations of free knot approximation.

The study of n-term Haar approximation in $L_2(\Omega)$ benefited greatly from the characterization of $L_2(\Omega)$ in terms of wavelet coefficients. The situation for approximation in $L_p(\Omega)$, $1 < p < \infty$, can also be treated, although the computation of $L_p(\Omega)$ norms is more subtle (see (7.27)). It turns out that a norm close to the L_p norm is given by

$$\|f\|_{B_p}^p := |\langle f, \chi_\Omega \rangle|^p + \sum_{I \in D} \|\langle f, H_I \rangle H_I\|_{L_p(\Omega)}^p, \tag{3.54}$$

which is known as the B_p norm. For approximation in the B_p norm, the theory is almost identical to $L_2(\Omega)$. Now, a best approximation from Σ_n^H is given by

$$S = \langle f, \chi_\Omega \rangle \chi_\Omega + \sum_{I \in \Lambda} \langle f, H_I \rangle H_I, \tag{3.55}$$

where $\Lambda \subset D$ is a set corresponding to the n biggest terms $\|\langle f, H_I\rangle H_I\|_{L_p(\Omega)}$. This selection procedure, to build the set Λ, depends on p because

$$\|H_I\|_{L_p(\Omega)} = |I|^{1/p-1/2}.$$

In other words, the coefficients are scaled depending on their dyadic level before we select the largest coefficients.

This same selection procedure works for approximation in L_p (DeVore, Jawerth and Popov 1992); however, now the proof is more involved and will be discussed in Section 7.4 when we treat the more general case of wavelets.

3.7. Optimal basis selection: wavelet packets

We have shown in Section 2 that, in the setting of a Hilbert space, it is a simple matter to determine a best n-term approximation to a target function f using elements of an orthonormal basis. A basis is good for f if the absolute value of the coefficients of f, when they are reordered according to decreasing size, tend rapidly to zero. We can increase our approximation efficiency by finding such a good basis for f. Thus, we may want to include in our approximation process a search over a given collection (usually called a *library*) of orthonormal bases in order to choose one which is good for our target function f. This leads to another degree of nonlinearity in our approximation process since now we have the choice of basis in addition to the choice of best n terms with respect to that basis. From a numerical perspective, however, we must be careful that this process can be implemented computationally. In other words, we cannot allow too many bases in our selection: our library of bases must be computationally implementable. In the case of piecewise constant approximation, such a library of bases was given by Coifman and Wickerhauser (1992) and is a special case of what are known as *wavelet packet libraries.*

We introduce some notation which will simplify our description of wavelet packet libraries. If g is a function from $L_2(\mathbb{R})$, we let

$$g_I(x) := |I|^{-1/2} g(2^n x - k), \quad I = 2^{-n}[k, k+1). \tag{3.56}$$

If g is supported on $\Omega = [0,1)$, then g_I will be supported on the dyadic interval I. We also introduce the following scaling operators which appear in the construction of multiresolution analysis for the Haar function. For a function $g \in L_2(\mathbb{R})$, we define

$$A_0 g := g(2\cdot) + g(2\cdot -1)); \qquad A_1 g := g(2\cdot) - g(2\cdot -1)). \tag{3.57}$$

If g is supported on Ω, the functions $A_0 g$, $A_1 g$ are also supported on Ω and have the same L_2 norm as g. Also, the functions $A_0 g$ and $A_1 g$ are orthogonal, that is,

$$\int_\Omega A_0 g A_1 g = 0.$$

Let $\gamma_0 := \chi_\Omega$ and $\gamma_1 := H$ be the characteristic and Haar functions. They satisfy

$$\gamma_0 = A_0\gamma_0; \qquad \gamma_1 = A_1\gamma_0. \tag{3.58}$$

In the course of our development of wavelet packets we will apply the operators A_0 and A_1 to generate additional functions. It is most convenient to index these functions on binary strings b. Such a b is a string of 0s and 1s. For such a string b, let $b0$ be the new string obtained from b by appending 0 to the end of b and let $b1$ be the corresponding string obtained by appending 1 to the end of b. Then, we inductively define

$$\gamma_{b0} := A_0\gamma_b; \qquad \gamma_{b1} := A_1\gamma_b. \tag{3.59}$$

In particular, (3.58) gives that $\gamma_{00} := A_0\gamma_0 = \chi_\Omega$ and $\gamma_{01} := A_1\gamma_0 = H$. Note that there is redundancy in that two binary strings b and b' represent the same integer in base 2 if and only if $\gamma_b = \gamma_{b'}$.

We can now describe the wavelet packet bases for S_m with $m \geq 1$, a fixed integer. We associate to each binary string b its length $\#b$, and the space

$$\Gamma_b := \text{span}\{(\gamma_b)_I : I \in D_{m-\#b}\}. \tag{3.60}$$

The functions $(\gamma_b)_I$ form an orthonormal basis for Γ_b. While the two functions γ_b and $\gamma_{b'}$ may be identical for $b \neq b'$, the subspaces Γ_b and $\Gamma_{b'}$ are not the same because b and b' will have different lengths. For any binary string b, we have

$$\Gamma_b = \Gamma_{b0} \oplus \Gamma_{b1}, \tag{3.61}$$

and the union of the two bases (given by (3.60)) for Γ_{b0} and Γ_{b1} give an alternative orthonormal basis for Γ_b.

The starting point of multiresolution analysis and our construction of the Haar wavelet was the decomposition $\mathcal{S}_m = \mathcal{S}_{m-1} \oplus W_{m-1}$ given in (3.42). In our new notation, this decomposition is

$$\Gamma_0 = \Gamma_{00} \oplus \Gamma_{01}. \tag{3.62}$$

In multiresolution analysis, the process is continued by decomposing $\mathcal{S}_{m-1} = \mathcal{S}_{m-2} \oplus W_{m-2}$ or, equivalently, $\Gamma_{00} = \Gamma_{000} \oplus \Gamma_{001}$. We take $W_{m-2} = \Gamma_{001}$ in our decomposition and continue. Our new viewpoint is that we can apply the recipe (3.57) to further decompose $\Gamma_{01} = W_{m-1}$ into two orthogonal subspaces as described in (3.61). Continuing in this way, we get other orthogonal decompositions of $\mathcal{S}_m$ and other orthonormal bases which span this space.

We can depict these orthogonal decompositions by a binary tree as given in Figure 4. Each node of the tree can be indexed by a binary string b. The number of digits k in b corresponds to its depth in the tree. Associated to b are the function γ_b and the space Γ_b, which has an orthonormal basis

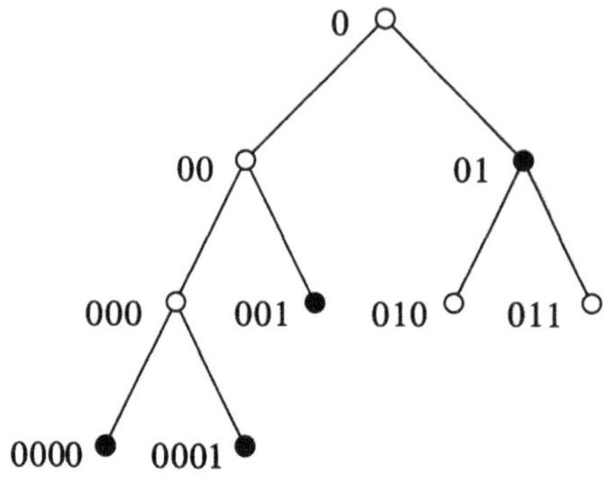

Fig. 4. The binary tree for wavelet packets

consisting of the functions $(\gamma_b)_I$, $I \in D_{m-k}$. If we move down and to the left from b we append digit 0 to b, while if we move down one level on the right branch we append digit 1. The tree stops when we reach level m.

The above construction generates many orthonormal bases of $\mathcal{S}_m$. We can associate each binary b to the dyadic interval I_b whose left end-point is $b/2\#b$ and whose length is $2^{-\#b}$, where $\#b$ is the number of digits in b (the level of b in the tree). If we take a collection B of such b, such that the I_b, $b \in B$, are a disjoint cover of Ω, then $\mathcal{S}_m = \Gamma_0 = \bigoplus_{b \in B} \Gamma_b$. The union of all the bases for these spaces form an orthonormal basis for $\mathcal{S}_m$. For example, in Figure 4, the solid nodes correspond to such a cover. The same story applies to any node b of the tree. The portion of the tree starting at b has the same structure as the entire tree and we obtain many bases for Γ_b by using interval decompositions of I_b as described above.

Several of the bases for $\mathcal{S}_m$ are noteworthy. By choosing just Ω in the interval decomposition of Ω, we obtain just the space Γ_0 and its basis $(\gamma_0)_I = \chi_I$, $I \in D_m$. The choice $B = \{00\cdots0\} \cup \{01,\ 001, \ldots\}$ corresponds to the dyadic intervals $2^{-m}[0,1)$, $2^{-m+1}[1/2,1], \ldots$, $[1/2,1)$ and gives the Haar basis. We can also take all the nodes at the lowest level (level m) of the tree. These nodes each correspond to spaces of dimension one. The basis obtained in this way is the Walsh basis from Fourier analysis.

It is important to note that we can efficiently compute the coefficients of a function $S \in \mathcal{S}_m$ with respect to all of the spaces Γ_b by using (3.57). For example, let γ_b be the generator of Γ_b. Then, $\Gamma_b = \Gamma_{b0} \oplus \Gamma_{b1}$. If $S \in \mathcal{S}_m$ and $c_{b,I} := \langle S, (\gamma_b)_I \rangle$, $I \in D_{m-k}$ are the coefficients of S with respect to these functions, then, for $I \in D_{m-k-1}$,

$$c_{b0,I} = \frac{1}{\sqrt{2}}(c_{b,I_0} + c_{b,I_1}), \qquad c_{b1,I} = \frac{1}{\sqrt{2}}(c_{b,I_0} - c_{b,I_1}), \tag{3.63}$$

where I_0 and I_1 are the left and right halves of I. Similarly, we can obtain the coefficients c_{b,I_0}, c_{b,I_1} from the coefficients $c_{b0,I}, c_{b1,I}$. Thus, for example, starting with the coefficients for the basis at the top (or bottom) of the tree, we can compute all other coefficients with $O(m2^m)$ operations.

For all numerical applications, the above construction is sufficient. One chooses m sufficiently large and considers all bases of $\mathcal{S}_m$ given as above. For theoretical reasons, however, one may want bases for $L_2(\Omega)$. This can be accomplished by letting $m \to \infty$ in the above depiction, thereby obtaining an infinite tree.

A typical adaptive basis selection algorithm, for approximating the target function f, chooses a coefficient norm that measures the spread of coefficients, and finds a basis that minimizes this norm. As we have seen in Section 2, n-term approximation efficiency using orthonormal bases is related to ℓ_τ norms of the coefficients. Thus, a typical algorithm would begin by fixing a sufficiently large value of m for the desired numerical accuracy, choosing $\tau > 0$, and finding a basis for the ℓ_τ norm, as we shall now describe.

If f is our target function, we let $S = P_m f$ be the orthogonal projection of f onto $\mathcal{S}_m$. The coefficients $\langle f, (\gamma_b)_I \rangle = \langle S, (\gamma_b)_I \rangle$ can be computed efficiently as described above. Let B be any orthonormal subcollection of the functions $(\gamma_b)_I$ and define

$$N_\tau(B) := N_\tau(f, B) := \sum_B |\langle f, (\gamma_b)_I \rangle|^\tau. \tag{3.64}$$

We want to find a basis B for Γ_0 which minimizes (3.64). To do this, we begin at the bottom of the tree and work our way up, at each step exchanging the current basis for a new one if the new basis gives a smaller N_τ.

For each node b at the bottom of the tree (*i.e.*, at level m), the space Γ_b has dimension one and has the basis $\{\gamma_b\}$. A node occurring at level $m-1$ corresponds to the space Γ_b. It has two bases from our collection. The first is $\{(\gamma_b)_I\}_{I \in D_1}$; the second is $\{\gamma_{b0}, \gamma_{b1}\}$. We compare these two bases and choose the one, which will be denoted by B_b, that minimizes $N_\tau(B)$. We do this for every node b at level $m-1$. We then proceed up the tree. If bases have been chosen for every node at level k, and if b is a node at level $k-1$, we compare $N_\tau(\{((\gamma_b)_I)_{I \in D_{k-1}}\})$ with $N_\tau(B_{b0} \cup B_{b1})$. The basis that minimizes N_τ is denoted by B_b and is our best basis for node b. At the conclusion, we shall have the best basis B_0 for node 0, that is, the basis which gives the smallest value of $N_\tau(B)$ among all wavelet packet bases for $\mathcal{S}_m$. This algorithm requires $O(m2^m)$ computations.

4. The elements of approximation theory

To move into the deeper aspects of nonlinear approximation, it will be necessary to call on some of the main tools of approximation theory. We have seen in the study of piecewise constant approximation that a prototypical theorem characterizes approximation efficiency in terms of the smoothness of the target function. For other methods of nonlinear approximation, it is not always easy to decide the appropriate measure of smoothness which characterizes approximation efficiency. There are, however, certain aids which make our search for this connection easier. The most important of these is the theory of interpolation of function spaces and the role of Jackson and Bernstein inequalities. This section will introduce the basics of interpolation theory and relate it to the study of approximation rates and smoothness. In the process, we shall engage three types of spaces: approximation spaces, interpolation spaces, and smoothness spaces. These three topics are intimately connected and it is these connections which give us insight on how to solve our approximation problems.

4.1. Approximation spaces

In our analysis of piecewise constant approximation, we have repeatedly asked the question: which functions are approximated at a given rate like $O(n^{-\alpha})$? It is time to put questions like this into a more formal framework. We shall consider the following general setting in this section. There will be a normed space $(X, \|\cdot\|_X)$, in which approximation takes place. Our approximants will come from spaces $X_n \subset X$, $n = 0, 1, \ldots$, and we introduce the approximation error

$$E_n(f)_X := \operatorname{dist}(f, X_n)_X := \inf_{g \in X_n} \|f - g\|_X. \tag{4.1}$$

In the case of linear approximation, n will usually be the dimension of X_n, or a quantity closely related to $\dim X_n$. In nonlinear approximation, n relates to the number of free parameters. For example, n might be the number of knots (breakpoints) in piecewise constant approximation with free knots. The X_n can be quite general spaces; in particular, they do not have to be linear. But we shall make the following assumptions (some only for convenience):

(i) $X_0 := \{0\}$
(ii) $X_n \subset X_{n+1}$
(iii) $aX_n = X_n$, $a \in \mathbb{R}$, $a \neq 0$
(iv) $X_n + X_n \subset X_{cn}$ for some integer constant $c \geq 1$ independent of n
(v) each $f \in X$ has a best approximation from X_n
(vi) $\lim_{n \to \infty} E_n(f)_X = 0$ for all $f \in X$.

Assumptions (iii), (iv), and (vi) are the most essential. The others can be eliminated or modified with a similar theory.

It follows from (ii) and (vi) that $E_n(f)_X$ monotonically decreases to 0 as n tends to ∞.

We wish to gather under one roof all functions which have a common approximation rate. In analogy with the results of the previous section, we introduce the space $\mathcal{A}^\alpha := \mathcal{A}^\alpha(X)$, which consists of all functions $f \in X$ for which

$$E_n(f)_X = O(n^{-\alpha}), \quad n \to \infty. \tag{4.2}$$

Our goal, as always, is to characterize $\mathcal{A}^\alpha$ in terms of something we know, such as a smoothness condition. It turns out that we shall sometimes need to consider finer statements about the decrease of the error $E_n(f)_X$. This will take the form of slight variants to (4.2), which we now describe.

Let $\mathbb{N}$ denote the set of natural numbers. For each $\alpha > 0$ and $0 < q < \infty$, we define the approximation space $\mathcal{A}^\alpha_q := \mathcal{A}^\alpha_q(X, (X_n))$ as the set of all $f \in X$ such that

$$|f|_{\mathcal{A}^\alpha_q} := \begin{cases} \left(\sum_{n=1}^\infty [n^\alpha E_n(f)_X]^q \frac{1}{n}\right)^{1/q}, & 0 < q < \infty, \\ \sup_{n \geq 1} n^\alpha E_n(f)_X, & q = \infty, \end{cases} \tag{4.3}$$

is finite, and further define $\|f\|_{\mathcal{A}^\alpha_q} := |f|_{\mathcal{A}^\alpha_q} + \|f\|_X$. Thus, the case $q = \infty$ is the space $\mathcal{A}^\alpha$ described by (4.2). For $q < \infty$, the requirement for membership in $\mathcal{A}^\alpha_q$ gets stronger as q decreases:

$$\mathcal{A}^\alpha_q \subset \mathcal{A}^\alpha_p, \quad 0 < q < p \leq \infty.$$

However, all of these spaces correspond to a decrease in error like $O(n^{-\alpha})$.

Because of the monotonicity of the sequence $(E_n(f)_X)$, we have the equivalence

$$|f|_{\mathcal{A}^\alpha_q} \asymp \begin{cases} \left(\sum_{k=0}^\infty [2^{k\alpha} E_{2^k}(f)_X]^q\right)^{1/q}, & 0 < q < \infty, \\ \sup_{k \geq 0} 2^{k\alpha} E_{2^k}(f)_X, & q = \infty. \end{cases} \tag{4.4}$$

It is usually more convenient to work with (4.4) than (4.3).

The next sections will develop some general principles which can be used to characterize the approximation spaces $\mathcal{A}^\alpha_q$.

4.2. Interpolation spaces

Interpolation spaces arise in the study of the following problem of analysis. Given two spaces X and Y, for which spaces Z is it true that each linear operator T mapping X and Y boundedly into themselves automatically maps Z boundedly into itself? Such spaces Z are called *interpolation spaces* for the pair X, Y and the problem is to construct and, more ambitiously, to characterize the spaces Z. The classical result in this direction is the Riesz–Thorin theorem, which states that the spaces L_p, $1 < p < \infty$, are interpolation spaces for the pair L_1, L_∞ and the Calderón–Mitjagin theorem, which characterizes all the interpolation spaces for this pair as the rearrangement of invariant function spaces (see Bennett and Sharpley (1988)). There are two primary methods for constructing interpolation spaces Z: the complex method as developed by Calderón (1964a) and the real method of Lions and Peetre (see Peetre (1963)). We shall only need the latter in what follows.

Interpolation spaces arise in approximation theory in the following way. Consider our problem of characterizing the approximation spaces $\mathcal{A}^\alpha(X)$ for a given space X and approximating subspaces X_n. If we obtain information about $\mathcal{A}^\alpha(X)$ for a given value of α, is it possible to parlay that information into statements about other approximation spaces $\mathcal{A}^\beta(X)$, with $\beta \neq \alpha$? The answer is yes: we can interpolate this information. Using these ideas, we can usually characterize approximation spaces as interpolation spaces between X and a suitably chosen second space Y. Thus, our goal of characterizing approximation spaces gets reduced to that of characterizing certain interpolation spaces. Fortunately, much effort has been put into the problem of characterizing interpolation spaces, and characterizations (usually as smoothness spaces) are known for most classical pairs of spaces X, Y. Thus, our approximation problem is solved.

An example might motivate the reader. In our study of approximation by piecewise constants, we saw that $\mathrm{Lip}(1, L_p(\Omega))$ characterizes the functions which are approximated with order $O(n^{-1})$ in $L_p(\Omega)$ by linear approximation from $\mathcal{S}^1(\Delta_n)$. Interpolation gives that the spaces $\mathrm{Lip}(\alpha, L_p(\Omega))$ characterize the functions which are approximated with order $O(n^{-\alpha})$, $0 < \alpha < 1$. A similar situation exists in nonlinear approximation.

Our description of how to solve the approximation problem is a little unfair to approximation theory. It makes it sound as if we reduce the approximation problem to the interpolation problem and then call upon the interpolation theory for the final resolution. In fact, one can go both ways, that is, one can also think of characterizing interpolation spaces by approximation spaces. Indeed, this is often how interpolation spaces are characterized. Thus, both theories shed considerable light on the other, and this is the view we shall adopt in what follows.

As mentioned, we shall restrict our development to the real method of interpolation using the Peetre K-functional, which we now describe. Let X, Y be a pair of normed linear spaces. We shall assume that Y is continuously embedded in X ($Y \subset X$ and $\|\cdot\|_X \leq C\|\cdot\|_Y$). (There are a few applications in approximation theory where this is not the case and one can make simple modifications in what follows to handle those cases as well.) For any $t > 0$, we define the K-functional

$$K(f,t) := K(f,t;X,Y) := \inf_{g \in Y} \|f - g\|_X + t|g|_Y, \tag{4.5}$$

where $\|\cdot\|_X$ is the norm on X and $|\cdot|_Y$ is a semi-norm on Y. We shall also meet cases where $|\cdot|_Y$ is only a quasi-semi-norm, which means that the triangle inequality is replaced by $|g_1 + g_2|_Y \leq C(|g_1|_Y + |g_2|_Y)$ with an absolute constant C. To spare the reader, we shall ignore this distinction in what follows.

The function $K(f,\cdot)$ is defined on $\mathbb{R}_+$ and is monotone and concave (being the pointwise infimum of linear functions). Notice that, for each $t>0$, $K(f,t)$ describes a type of approximation. We approximate f by functions g from Y with the penalty term $t|g|_Y$. The role of the penalty term is paramount. As we vary $t>0$, we gain additional information about f.

K-functionals have many uses. As noted earlier, they were originally introduced as a means of generating interpolation spaces. To see that application, let T be a linear operator which maps X and Y into themselves with a norm not exceeding M in both cases. Then, for any $g\in Y$, we have $Tf=T(f-g)+Tg$ and therefore

$$K(Tf,t)\le \|T(f-g)\|_X+t|Tg|_Y\le M(\|f-g\|_X+t|g|_Y). \tag{4.6}$$

Taking an infimum over all g, we have

$$K(Tf,t)\le MK(f,t),\quad t>0. \tag{4.7}$$

Suppose further that $\|\cdot\|$ is a function norm defined for real-valued functions on $\mathbb{R}_+$. We can apply this norm to (4.7) and obtain

$$\|K(Tf,\cdot)\|\le M\|K(f,\cdot)\|. \tag{4.8}$$

Each function norm $\|\cdot\|$ can be used in (4.8) to define a space of functions (those functions for which the right side of (4.8) is finite) and this space will be an interpolation space. We shall restrict our attention to the most common of these, which are the θ,q norms. They are analogous to the norms we used in defining approximation spaces. If $0<\theta<1$ and $0<q\le\infty$, then the interpolation space $(X,Y)_{\theta,q}$ is defined as the set of all functions $f\in X$ such that

$$|f|_{(X,Y)_{\theta,q}}:=\begin{cases}\left(\int_0^\infty[t^{-\theta}K(f,t)]^q\frac{dt}{t}\right)^{1/q}, & 0<q<\infty,\\ \sup_{t>0}t^{-\theta}K(f,t), & q=\infty,\end{cases} \tag{4.9}$$

is finite.

The spaces $(X,Y)_{\theta,q}$ are interpolation spaces. The usefulness of these spaces depends on understanding their nature for a given pair (X,Y). This is usually accomplished by characterizing the K-functional for the pair. We shall give several examples of this in Sections 4.4–4.5.

Here is a useful remark which we shall have need for later. We can apply the θ,q method for generating interpolation spaces to any pair (X,Y). In particular, we can apply the method to a pair of θ,q spaces. The question is whether we get anything new and interesting. The answer is no: we simply get θ,q spaces of the original pair (X,Y). This is called the reiteration theorem of interpolation. Here is its precise formulation. Let $X':=(X,Y)_{\theta_1,q_1}$

and $Y' := (X, Y)_{\theta_2, q_2}$. Then, for all $0 < \theta < 1$ and $0 < q \le \infty$, we have

$$(X', Y')_{\theta,q} = (X, Y)_{\alpha,q}, \quad \alpha := (1-\theta)\theta_1 + \theta\theta_2. \tag{4.10}$$

We make two observations which can simplify the norm in (4.9). Firstly, using the fact that Y is continuously embedded in X, we obtain an equivalent norm by taking the integral in (4.9) over $[0, 1]$. Secondly, since $K(f, .)$ is monotone, the integral over $[0, 1]$ can be discretized. This gives that the norm of (4.9) is equivalent to

$$|f|_{(X,Y)_{\theta,q}} \asymp \begin{cases} \left(\sum_{k=0}^{\infty} [2^{k\theta} K(f, 2^{-k})]^q\right)^{1/q}, & 0 < q < \infty, \\ \sup_{k \ge 0} 2^{k\theta} K(f, 2^{-k}), & q = \infty \end{cases} \tag{4.11}$$

(see Chapter 6 of DeVore and Lorentz (1993) for details).

In this form, the definitions of interpolation spaces and approximation spaces are almost identical: we have replaced E_{2^k} by $K(f, 2^{-k})$. It should therefore come as no surprise that one space can often be characterized by the other. What is needed for this is a comparison between the error $E_n(f)$ and the K-functional K. Of course, this can only be achieved if we make the right choice of the space Y in the definition of K. But how can we decide what Y should be? This is the role of the Jackson and Bernstein inequalities given in the next subsection.

4.3. Jackson and Bernstein inequalities

In this subsection, we shall make a considerable simplification in the search for a characterization of approximation spaces and bring out fully the connection between approximation and interpolation spaces. We assume that X is the space in which approximation takes place and assume that we can find a positive number $r > 0$ and a second space Y continuously embedded in X for which the following two inequalities hold.

Jackson inequality: $E_n(f)_X \le Cn^{-r}|f|_Y, \quad f \in Y, \quad n = 1, 2, \ldots.$

Bernstein inequality: $|S|_Y \le Cn^r \|S\|_X, \quad S \in X_n, \quad n = 1, 2, \ldots.$

Whenever these two inequalities hold, we can draw a comparison between $E_n(f)_X$ and $K(f, n^{-r}, X, Y)$. For example, assume that the Jackson inequality is valid and let $g \in Y$ be such that

$$\|f - g\|_X + n^{-r}|g|_Y = K(f, n^{-r}).$$

(In fact we do not know of the existence of such a g, and so an ϵ should be added into this argument, but to spare the reader we shall not insist upon such precision in this survey.) If S is a best approximation to g from X_n, then

$$\begin{aligned} E_n(f)_X &\le \|f - S\|_X \le \|f - g\|_X + \|g - S\|_X \\ &\le \|f - g\|_X + Cn^{-r}|g|_Y \le CK(f, n^{-r}), \end{aligned} \tag{4.12}$$

where the last inequality makes use of the Jackson inequality.

By using the Bernstein inequality, we can reverse (4.12) in a certain weak sense (see Theorem 5.1 of Chapter 7 in DeVore and Lorentz (1993)). From this one derives the following relation between approximation spaces and interpolation spaces.

Theorem 1 If the Jackson and Bernstein inequalities are valid, then for each $0 < \gamma < r$ and $0 < q \leq \infty$ the following relation holds between approximation spaces and interpolation spaces

$$A_q^\gamma(X) = (X, Y)_{\gamma/r,q} \tag{4.13}$$

with equivalent norms.

Thus, Theorem 1 will solve our problem of characterizing the approximation spaces if we know two ingredients:

(i) an appropriate space Y for which the Jackson and Bernstein inequalities hold

(ii) a characterization of the interpolation spaces $(X, Y)_{\theta,q}$.

The first step is the difficult one from the viewpoint of approximation (especially in the case of nonlinear approximation). Fortunately, step (ii) is often provided by classical results in the theory of interpolation. We shall mention some of these in the next sections and also relate these to our examples of approximation by piecewise constants. But for now we want to make a very general and useful remark concerning the relation between approximation and interpolation spaces by stating the following elementary result of DeVore and Popov (1988*b*).

Theorem 2 For any space X and spaces X_n, as well as for any $r > 0$ and $0 \leq s \leq \infty$, the spaces X_n, $n = 1, 2, \ldots$, satisfy the Jackson and Bernstein inequalities for $Y = \mathcal{A}_s^r(X)$. Therefore, for any $0 < \alpha < r$ and $0 < q \leq \infty$, we have

$$\mathcal{A}_q^\alpha(X) = (X, \mathcal{A}_s^r(X))_{\alpha/r,q}. \tag{4.14}$$

In other words, the approximation family $\mathcal{A}_q^\alpha(X)$ is an interpolation family.

We also want to expand on our earlier remark that approximation can often be used to characterize interpolation spaces. We shall point out that, in certain cases, we can realize the K-functional by an approximation process.

We continue with the above setting. We say a sequence (T_n), $n = 1, 2, \ldots$, of (possibly nonlinear) operators, with T_n mapping X into X_n, provides *near best approximation* if there is an absolute constant $C > 0$ such that

$$\|f - T_n f\|_X \leq C E_n(f)_X, \quad n = 1, 2, \ldots.$$

We say this family is stable on Y if

$$|T_n f|_Y \le C|f|_Y, \quad n = 1, 2, \ldots,$$

with an absolute constant $C > 0$.

Theorem 3 Let X, Y, (X_n) be as above and suppose that (X_n) satisfies the Jackson and Bernstein inequalities. Suppose further that the sequence of operators (T_n) provides near best approximation and is stable on Y. Then, T_n realizes the K-functional, that is,

$$\|f - T_n f\|_X + n^{-r}|T_n f|_Y \le CK(f, n^{-r}, X, Y)$$

with an absolute constant C.

For a proof and further results of this type, we refer the reader to Cohen, DeVore and Hochmuth (1997).

4.4. Interpolation for L_1, L_∞

The utility of the K-functional rests on our ability to characterize it and thereby characterize the interpolation spaces $(X, Y)_{\theta,q}$. Much effort was put forward in the 1970s and 1980s to establish such characterizations for classical pairs of spaces. The results were quite remarkable in that the characterizations that ensued were always in terms of classical entities that have a long-standing place in analysis. We shall give several examples of this. In the present section, we limit ourselves to the interpolation of Lebesgue spaces, which are classical to the theory. In later sections, we shall discuss interpolation of smoothness spaces, which are more relevant to our approximation needs.

Let us begin with the pair $L_1(A, \mathrm{d}\mu), L_\infty(A, \mathrm{d}\mu)$ with $(A, \mathrm{d}\mu)$ a given sigma-finite measure space. Hardy and Littlewood recognized the importance of the *decreasing rearrangement* f^* of a μ-measurable function f. The function f^* is a nonnegative, nonincreasing function defined on $\mathbb{R}_+$ which is equimeasurable with f:

$$\mu(f, t) := \mu\{x : |f(x)| > t\} = |\{s : f^*(s) > t\}|, \quad t > 0, \tag{4.15}$$

where we recall our notation for $|E|$ to denote the Lebesgue measure of a set E. The rearrangement f^* can be defined directly via

$$f^*(t) := \inf\{y : \mu(f, t) \le y\}. \tag{4.16}$$

Thus, f^* is essentially the inverse function to $\mu(f, t)$. We have the following beautiful formula for the K-functional for this pair (see Chapter 6 of DeVore and Lorentz (1993)):

$$K(f, t, L_1, L_\infty) = \int_0^t f^*(s)\,\mathrm{d}s, \tag{4.17}$$

which holds whenever $f \in L_1 + L_\infty$. From the fact that

$$\int_A |f|^p \,\mathrm{d}\mu = \int_0^\infty (f^*(s))^p \,\mathrm{d}s$$

it is easy to deduce from (4.17) the Riesz–Thorin theorem for this pair.

With the K-functional in hand, we can easily describe the (θ, q) interpolation spaces in terms of Lorentz spaces. For each $0 < p < \infty$, $0 < q \le \infty$, the Lorentz space $L_{p,q}(A, \mathrm{d}\mu)$ is defined as the set of all μ-measurable f such that

$$\|f\|_{L_{p,q}} := \begin{cases} \left(\int_0^\infty [t^{1/p} f^*(t)]^q \frac{\mathrm{d}t}{t}\right)^{1/q}, & 0 < q < \infty, \\ \sup t^{1/p} f^*(t), & q = \infty, \end{cases} \tag{4.18}$$

is finite. Of course, the form of the integral in (4.18) is quite familiar to us. If we replace f^* by $\frac{1}{t}\int_0^t f^*(s)\,\mathrm{d}s = K(f,t)/t$ and use the Hardy inequalities (see Chapter 6 of DeVore and Lorentz (1993) for details) we obtain that

$$(L_1(A, \mathrm{d}\mu), L_\infty(A, \mathrm{d}\mu))_{1-1/p,q} = L_{p,q}(A, \mathrm{d}\mu), \quad 1 < p < \infty,\ 0 < q \le \infty. \tag{4.19}$$

Several remarks are in order. The space $L_{p,\infty}$ is better known as weak L_p and can be equivalently defined by the condition

$$\mu\{x : |f(x)| > y\} \le M^p y^{-p}. \tag{4.20}$$

The smallest M for which (4.20) is valid is equivalent to the norm in $L_{p,\infty}$.

The above results include the case when $\mathrm{d}\mu$ is purely atomic. This will be useful for us in what follows, in the following context. Let $\mathbb{N}$ be the set of natural numbers and let $\ell_p = \ell_p(\mathbb{N})$ be the collection of all sequences $x = (x(n))_{n\in\mathbb{N}}$ for which

$$\|x\|_{\ell_p} := \begin{cases} \left(\sum_{n=1}^\infty |x(n)|^p\right)^{1/p}, & 0 < p < \infty, \\ \sup_{n\in\mathbb{N}} |x(n)|, & p = \infty, \end{cases} \tag{4.21}$$

is finite. Then, $\ell_p(\mathbb{N}) = L_p(\mathbb{N}, \mathrm{d}\mu)$ with μ the counting measure. Hence, the above results apply. The Lorentz spaces in this case are denoted by $\ell_{p,q}$. The space $\ell_{p,\infty}$ (weak ℓ_p) consists of all sequences that satisfy

$$x^*(n) \le M n^{-1/p} \tag{4.22}$$

with $(x^*(n))$ the decreasing rearrangement of $(|x(n)|)$. This can equivalently be stated as

$$\#\{n : |x(n)| > y\} \le M^p y^{-p}. \tag{4.23}$$

The interpolation theory for L_p spaces applies to more than the pair (L_1, L_∞). We formulate this only for the spaces $\ell_{p,q}$ which we shall use later. For any $0 < p_1 < p_2 < \infty$, $0 < q_1, q_2 \le \infty$, we have

$$(\ell_{p_1,q_1}, \ell_{p_2,q_2})_{\theta,q} = \ell_{p,q}, \quad 1/p := \frac{1-\theta}{p_1} + \frac{\theta}{p_2}, \quad 0 < q \le \infty, \tag{4.24}$$

with equivalent norms. For $1 \le p_1, p_2 \le \infty$, this follows from (4.19) by using the reiteration theorem (4.10). The general case needs slight modification (see Bergh and Löfström (1976)).

Interpolation for the pair (L_1, L_∞) is rather unusual in that we have an exact identity for the K-functional. Usually we only get an equivalent characterization of K. One other case where an exact identity is known is interpolation between C and Lip 1, in which case

$$K(f,t;C,\mathrm{Lip}\,1) = \frac{1}{2}\bar{\omega}(f,2t), \quad t>0,$$

where ω is the modulus of continuity (to be defined in the next section) and $\bar{\omega}$ is its concave majorant (see Chapter 6 of DeVore and Lorentz (1993)).

4.5. Smoothness spaces

We have introduced various smoothness spaces in the course of discussing approximation by piecewise constants. In this section, we want to be a bit more systematic and describe the full range of smoothness spaces that we shall need in this survey. There are two important ways to describe smoothness spaces. One is through notions such as differentiability and moduli of smoothness. Most smoothness spaces were originally introduced into analysis in this fashion. A second way is to expand functions into a series of building blocks (for instance Fourier or wavelet) and describe smoothness as decay conditions on the coefficients in such expansions. That these descriptions are equivalent is at the heart of the subject. We shall give both descriptions. The first is given here in this section; the second in Section 7 when we discuss wavelet decompositions.

We begin with the most important and best known smoothness spaces, the Sobolev spaces. Suppose that $1 \le p \le \infty$ and $r > 0$ is an integer. If $\Omega \subset \mathbb{R}^d$ is a domain (for us this will mean an open, connected set), we define $W^r(L_p(\Omega))$ as the collection of all measurable functions f defined on Ω which have all their distributional derivatives $D^\nu f$, $|\nu| \le r$, in $L_p(\Omega)$. Here $|\nu| := |\nu_1| + \cdots + |\nu_d|$ when $\nu = (\nu_1, \ldots, \nu_d)$. The semi-norm for $W^r(L_p(\Omega))$ is defined by

$$|f|_{W^r(L_p(\Omega))} := \sum_{|\nu|=r} \|D^\nu f\|_{L_p(\Omega))}, \tag{4.25}$$

and their norm by $\|f\|_{W^r(L_p(\Omega))} := |f|_{W^r(L_p(\Omega))} + \|f\|_{L_p(\Omega)}$. Thus, Sobolev spaces measure smoothness of order r in L_p when r is a positive integer and $1 \le p \le \infty$. Their deficiency is that they do not immediately apply when r is nonintegral or when $p < 1$. We have seen several times already the need for smoothness spaces for these extended parameters when engaging nonlinear approximation.

We have seen in the Lipschitz spaces that one way to describe smoothness of fractional order is through differences. We have previously used only first differences; now we shall need differences of arbitrary order which we presently define. For $h \in \mathbb{R}^d$, let T_h denote the translation operator which is defined for a function f by $T_h f := f(\cdot + h)$ and let I denote the identity operator. Then, for any positive integer r, $\Delta_h^r := (T_h - I)^r$ is the rth *difference operator with step h*. Clearly $\Delta_h^r = \Delta_h(\Delta_h^{r-1})$. Also,

$$\Delta_h^r(f, x) := \sum_{k=0}^{r} (-1)^{r-k} \binom{r}{k} f(x + kh). \tag{4.26}$$

Here and later we use the convention that $\Delta_h^r(f, x)$ is defined to be zero when any of the points $x, \ldots, x + rh$ are not in Ω.

We can use Δ_h^r to measure smoothness. If $f \in L_p(\Omega)$, $0 < p \leq \infty$,

$$\omega_r(f, t)_p := \sup_{|h| \leq t} \|\Delta_h^r(f, \cdot)\|_{L_p(\Omega)} \tag{4.27}$$

is the rth *order modulus of smoothness of f in $L_p(\Omega)$*. In the case $p = \infty$, $L_\infty(\Omega)$ is replaced by $C(\Omega)$, the space of uniformly continuous functions on Ω. We always have that $\omega_r(f, t)_p \to 0$ monotonically as $t \to 0$. The faster this convergence to 0 the smoother is f.

We create smoothness spaces by bringing together all functions whose moduli of smoothness have a common behaviour. We shall particularly need this idea with the Besov spaces which are defined as follows. There will be three parameters in our description of Besov spaces. The two primary parameters are α, which gives the order of smoothness (for instance the number of derivatives), and p, which gives the L_p space in which smoothness is to be measured. A third parameter q, which is secondary to the two primary parameters, will allow us to make subtle distinctions in smoothness spaces with the same primary parameters.

Let $\alpha > 0$, $0 < p \leq \infty$, and $0 < q \leq \infty$. We take $r := [\alpha] + 1$ (the smallest integer larger than α). We say f is in the Besov space $B_q^\alpha(L_p(\Omega))$ if

$$|f|_{B_q^\alpha(L_p(\Omega))} := \begin{cases} \left(\int_0^\infty [t^{-\alpha}\omega_r(f, t)_p]^q \frac{dt}{t}\right)^{1/q}, & 0 < q < \infty, \\ \sup_{t>0} t^{-\alpha}\omega_r(f, t)_p, & q = \infty, \end{cases} \tag{4.28}$$

is finite. This expression defines the semi-norm on $B_q^\alpha(L_p(\Omega))$; the Besov norm is given by $\|f\|_{B_q^\alpha(L_p(\Omega))} := |f|_{B_q^\alpha(L_p(\Omega))} + \|f\|_{L_p(\Omega)}$. Here, we have complete analogy with the definitions (4.3) and (4.9) of approximation and interpolation spaces.

The Besov spaces give a full range of smoothness in that α can be any positive number, and p can range over $(0, \infty]$. As noted earlier, q is a secondary index which gives finer gradations of smoothness with the same primary indices.

We shall next make some further remarks which will help clarify Besov spaces, especially their relationship to other smoothness spaces such as the Sobolev and Lipschitz spaces. We assume from here on that the domain Ω is a Lipschitz domain (see Adams (1975)) – slightly weaker conditions on Ω suffice for most of the following statements.

We have taken r as the smallest integer larger than α. Actually, any choice of $r > \alpha$ will define the same space with an equivalent norm (see Chapter 2 of DeVore and Lorentz (1993)). If we take $\alpha < 1$ and $q = \infty$, the Besov space $B^\alpha_\infty(L_p(\Omega))$ is the same as $\mathrm{Lip}(\alpha, L_p(\Omega))$ with an identical semi-norm and norm. However, when $\alpha = 1$, we get a different space because the Besov space uses ω_2 in its definition but $\mathrm{Lip}(1, L_p(\Omega))$ uses ω_1. In this case, the Besov space is larger since $\omega_2(f,t)_p \leq 2^{\max(1/p,1)}\omega_1(f,t)_p$. Sometimes $B^1_\infty(C(\Omega))$ is called the Zygmund space.

For the same reason that Lip 1 is not a Besov space, the Sobolev space $W^r_p(L_p(\Omega))$, $1 \leq p \leq \infty$, $p \neq 2$, r an integer, is not the same as the Besov space $B^r_\infty(L_p(\Omega))$. The Besov space is slightly larger. We could describe the Sobolev space $W^r_p(L_p(\Omega))$, $1 < p \leq \infty$, by replacing ω_{r+1} by ω_r in the definition of $B^r_\infty(L_p(\Omega))$. When α is nonintegral, the fractional order Sobolev space $W^\alpha(L_p(\Omega))$ is defined to be $B^\alpha_p(L_p(\Omega))$. Two special cases are noteworthy. When $p = 2$, the Besov space $B^r_2(L_2(\Omega))$ is the same as the Sobolev space $W^r(L_2(\Omega))$; this is an anomaly that only holds for $p = 2$. The Lipschitz space $\mathrm{Lip}(1, L_1(\Omega))$ is the same as BV when Ω is an interval in $\mathbb{R}^1$. In higher dimensions, we use $\mathrm{Lip}(1, L_1(\Omega))$ as the definition of $\mathrm{BV}(\Omega)$; it coincides with some but not all of the many other definitions of BV.

Increasing the secondary index q in Besov spaces gives a larger space, *i.e.*,

$$B^\alpha_{q_1}(L_p(\Omega)) \subset B^\alpha_{q_2}(L_p(\Omega)), \quad q_1 < q_2.$$

However, the distinctions between these spaces are small.

The Sobolev embedding theorem gives much additional information about the relationship between Besov spaces with different values of the parameters. It is easiest to describe these results pictorially. As earlier, we identify a Besov space with primary indices p and α with the point $(1/p, \alpha)$ in the upper right quadrant of $\mathbb{R}^2$. The line with slope d passing through $(1/p, 0)$ is the demarcation line for embeddings of Besov spaces into $L_p(\Omega)$ (see Figure 5). Any Besov space with primary indices corresponding to a point above that line is embedded into $L_p(\Omega)$ (regardless of the secondary index q). Besov spaces corresponding to points on the demarcation line may or may not be embedded in $L_p(\Omega)$. For example the Besov spaces $B^\alpha_\tau(L_\tau(\Omega))$ with $1/\tau = \alpha/d + 1/p$ correspond to points on the demarcation line and they are embedded in $L_p(\Omega)$. Points below the demarcation line are never embedded in $L_p(\Omega)$.

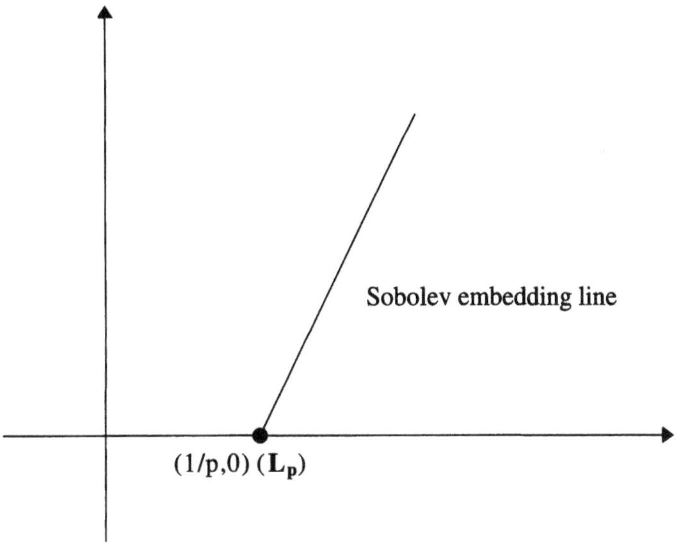

Fig. 5. Sobolev embedding

4.6. Interpolation of smoothness spaces

There is a relatively complete description of interpolation between Sobolev or Besov spaces. We shall point out the results most important for our later use.

Let us first consider interpolation between $L_p(\Omega)$ and a Sobolev space $W^r(L_p(\Omega))$. Interpolation for this pair appears often in linear approximation. One way to describe the interpolation spaces for this pair is to know its K-functional. The remarkable fact (proved in the case of domains by Johnen and Scherer (1977)) is that

$$K(f, t^r, L_p(\Omega), W^r(L_p(\Omega)) \asymp \omega_r(f, t)_p, \quad t > 0. \tag{4.29}$$

This brings home the point we made earlier that K-functionals can usually be described by some classical entity (in this case the modulus of smoothness). From (4.29), it is a triviality to deduce that

$$(L_p(\Omega), W^r(L_p(\Omega)))_{\theta,q} = B_q^{\theta r}(L_p(\Omega)), \quad 0 < \theta < 1, \ 0 < q \le \infty, \tag{4.30}$$

with equivalent norms. From the reiteration Theorem (4.10) for interpolation we deduce that, for $\alpha_1 < \alpha_2$ and any $0 < q_1, q_2 \le \infty$, we have for any $0 < \theta < 1$, $0 < q \le \infty$,

$$(B_{q_1}^{\alpha_1}(L_p(\Omega)), B_{q_2}^{\alpha_2}(L_p(\Omega)))_{\theta,q} = B_q^{\alpha}(L_p(\Omega)), \quad \alpha := (1-\theta)\alpha_1 + \theta\alpha_2. \tag{4.31}$$

We can also replace $B_{q_1}^{\alpha_1}(L_p(\Omega))$ by $L_p(\Omega)$ and obtain

$$(L_p(\Omega), B_r^{\alpha}(L_p(\Omega)))_{\theta,q} = B_q^{\theta\alpha}(L_p(\Omega)), \quad 0 < \theta < 1, \ 0 < q \le \infty, \tag{4.32}$$

for any $0 < r \le \infty$.

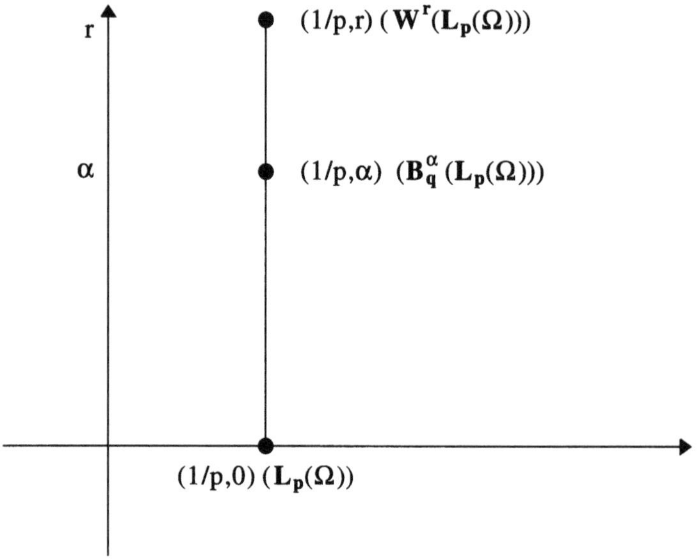

Fig. 6. Graphical interpretation of interpolation between $L_p(\Omega)$ and $W^r(L_p(\Omega))$

We can interpret these results pictorially as in Figure 6. The space $L_p(\Omega)$ corresponds to the point $(1/p, 0)$, and $W^r(L_p(\Omega))$ corresponds to the point $(1/p, r)$. Thus, (4.30) states that the interpolation spaces for this pair correspond to the Besov spaces on the (vertical) line segment connecting the points $(1/p, 0)$ and $(1/p, r)$. A similar picture interprets (4.31) and (4.32).

This pictorial interpretation is very instructive. When we want to interpolate between a pair of spaces (X, Y), we identify them with their corresponding points in the upper quadrant of $\mathbb{R}^2$. The points on the line segment connecting them are the interpolation spaces and, in fact, given the parameter θ, the interpolation space corresponds to the point on this line segment which divides the segment by the ratio $\theta : 1 - \theta$. However, care should be taken in this interpretation regarding the second parameter q, since it does not enter into the picture. In some cases, we can take any value of q, as is the case for the examples considered so far. However, in some cases that we shall see shortly, this interpretation only holds for certain q appropriately chosen.

Let us consider another example, which corresponds to interpolation in a case where the line segment is horizontal. DeVore and Scherer (1979) have shown that, if $1 \le p_1 < p_2 \le \infty$, then the (θ, p) interpolation between Sobolev spaces $W^r(L_{p_1}(\Omega))$ and $W^r(L_{p_2}(\Omega))$ gives Sobolev spaces $W^r(L_p(\Omega))$ when $\frac{1}{p} = \frac{1-\theta}{p_1} + \frac{\theta}{p_2}$, while changing θ, p into the more general θ, q gives the modified Sobolev spaces $W^r(L_{p,q}(\Omega))$ which use the Lorentz spaces $L_{p,q}(\Omega)$ in their definition (which we do not give).

There are characterizations for the θ, q interpolation spaces for many other pairs of Besov spaces (see Bergh and Löfström (1976) or Cohen, DeVore and Hochmuth (1997), for example). However, we shall restrict our further discussion to the following special case, which occurs in nonlinear approximation. We fix a value of $p \in (0, \infty)$ and consider the Besov spaces $B_\tau^\alpha(L_\tau(\Omega))$ where τ and α are related by

$$\frac{1}{\tau} = \frac{\alpha}{d} + \frac{1}{p}. \tag{4.33}$$

These spaces all correspond to points on the line segment with slope d passing through $(1/p, 0)$ (which corresponds to $L_p(\Omega)$). We have the following interpolation result for the pair $(L_p(\Omega), B_\tau^\alpha(L_\tau(\Omega)))$:

$$(L_p(\Omega), B_\tau^\alpha(L_\tau(\Omega)))_{\theta,q} = B_q^{\theta\alpha}(L_q(\Omega)), \quad \text{provided } \frac{1}{q} = \frac{\theta\alpha}{d} + \frac{1}{p}. \tag{4.34}$$

In other words, interpolating between two Besov spaces corresponding to points on this line, we get another Besov space corresponding to a point on this line provided we choose the secondary indices in a suitable way.

We shall obtain more information about Besov spaces and their interpolation properties in Section 7 when we discuss their characterization by wavelet decompositions.

5. Nonlinear approximation in a Hilbert space: a second look

Let us return to the example of approximation in a Hilbert space which began our discussion in Section 2. We continue with the discussion and notation of that section.

We have seen that for nonlinear (n-term) approximation in $\mathcal{H}$ we could characterize $\mathcal{A}_\infty^r((\mathcal{H}_n))$ for any $r > 0$ by the condition

$$\gamma_n(f) \leq Mn^{-r-1/2}, \tag{5.1}$$

with $\gamma_n(f)$ the rearranged coefficients. We now see that (5.1) states that the sequence $f_k := \langle f, \eta_k \rangle$ is in weak $\ell_{\tau(r)}$ $(\ell_{\tau(r),\infty})$, with $\tau(r)$ defined by

$$\frac{1}{\tau(r)} = r + \frac{1}{2},$$

and the smallest M for which (5.1) holds is equivalent to the weak ℓ_τ norm.

We can now characterize all of the approximation spaces $\mathcal{A}_q^\alpha(\mathcal{H})$ in terms of the coefficients f_k. Recall that Theorem 2 shows that, for any $r > 0$, the nonlinear spaces $\Sigma_n(\mathcal{H})$, satisfy Jackson and Bernstein inequalities for the space $Y := \mathcal{A}_\infty^r(\mathcal{H})$ and

$$\mathcal{A}_q^\alpha(\mathcal{H}) = (\mathcal{H}, \mathcal{A}_\infty^r(\mathcal{H}))_{\alpha/r,q}. \tag{5.2}$$

The mapping $f \to (f_k)$ is invertible and gives an isometry between $\mathcal{H}$ and $\ell_2(\mathbb{N})$ and also between $\mathcal{A}^r_\infty$ and $\ell_{\tau,\infty}(\mathbb{N})$. We can interpolate, and deduce that this mapping is an isometry between $\mathcal{A}^\alpha_q(\mathcal{H})$ and $\ell_{\tau(\alpha),q}(\mathbb{N})$ with τ defined by $1/\tau = \alpha + 1/2$. Hence, we have the following complete characterization of the approximation spaces for nonlinear n-term approximation.

Theorem 4 For nonlinear n-term approximation in a Hilbert space $\mathcal{H}$, a function f is in $\mathcal{A}^\alpha_q(\mathcal{H}))$ if and only if its coefficients are in $\ell_{\tau(\alpha),q}$, $\tau(\alpha) := (\alpha + 1/2)^{-1}$, and $|f|_{\mathcal{A}^\alpha_q(\mathcal{H})} \asymp \|(f_k)\|_{\ell_{\tau(\alpha),q}}$.

6. Piecewise polynomial approximation

Now that we have the tools of approximation firmly in hand, we shall survey the main developments of nonlinear approximation, especially as they apply to numerical computation. We shall begin in this section with piecewise polynomial approximation. The reader should keep in mind the case of piecewise constant approximation that we used to motivate nonlinear approximation.

6.1. Local approximation by polynomials

As the name suggests, piecewise polynomial approximation pieces together local polynomial approximants. Therefore, we need to have a good understanding of local error estimates for polynomial approximation. This is an old and well-established chapter in approximation and numerical computation, which we shall briefly describe in this section.

For each positive integer r, we let $\mathcal{P}_r$ denote the space of polynomials in d variables of total degree $< r$ (polynomials of order r). Let $0 < p \le \infty$ and let I be a cube in $\mathbb{R}^d$. If $f \in L_p(I)$, the local approximation error is defined by

$$E_r(f, I)_p := \inf_{P \in \mathcal{P}_r} \|f - P\|_{L_p(I)}. \tag{6.1}$$

The starting point for estimating the efficiency of piecewise polynomial approximation in L_p is to have good estimates for $E_r(f, I)_p$. Perhaps the simplest of these is the estimate

$$E_r(f, I)_p \le C|I|^{r/d} |f|_{W^r(L_p(I))}, \tag{6.2}$$

which holds for $1 \le p \le \infty$, $|\cdot|_{W^r(L_p(I))}$ the Sobolev semi-norm of Section 4.5, and the constant C depending only on r. This is sometimes known as the Deny–Lions lemma in numerical analysis. There are several proofs of this result available in the literature (see, for instance, Adams (1975)), usually by constructing a bounded projector from L_p onto $\mathcal{P}_r$. It can also be proved indirectly (see DeVore and Sharpley (1984)).

The estimate (6.2) remains valid when I is replaced by a more general domain. Suppose, for example, that $\mathcal{O}$ is a domain that satisfies the uniform cone condition (see Adams (1975)) and is contained in a cube I with $|I|^{1/d} \leq C\mathrm{diam}(\mathcal{O})$. If $f \in W^r(L_p(\mathcal{O}))$, then it can be extended to a function on I with comparable norm (Adams (1975) or DeVore and Sharpley (1984)). Applying (6.2) for I we deduce its validity on $\mathcal{O}$ with a constant C now depending on r and $\mathcal{O}$. We shall use this in what follows for polyhedral domains. The constant C then depends on r and the smallest angle in $\mathcal{O}$. Similar remarks apply to the other estimates for $E_r(f,I)_p$ that follow.

Using the ideas of interpolation introduced in Section 4 (see (4.29)), one easily derives from (6.2) that

$$E_r(f,I)_p \leq C_r\omega_r(f,|I|,I)_p, \tag{6.3}$$

with ω_r the rth order modulus of smoothness of f introduced in Section 4.5. This is called Whitney's theorem in approximation and this estimate is equally valid in the case $p < 1$. The advantage of (6.3) over (6.2) is that it applies to any $f \in L_p(I)$ and it also implies (6.2) because of elementary properties of ω_r.

Whitney's estimate is not completely satisfactory when it is necessary to add local estimates over varying cubes I. A more suitable form is obtained by replacing $\omega_r(f,|I|,I)_p$ by

$$w_r(f,I)_p := \left(\frac{1}{|I|}\int_{|s|\leq|I|^{1/d}}\int_I |\Delta_s^r(f,x)|^p \,\mathrm{d}x\,\mathrm{d}s\right)^{1/p}. \tag{6.4}$$

Then, we have (see, for instance, DeVore and Popov (1988*a*))

$$E_r(f,I)_p \leq C_r w_r(f,I)_p, \tag{6.5}$$

which holds for all $r \geq 1$ and all $0 < p \leq \infty$ (with the obvious change in norms for $p = \infty$).

It is also possible to bound $E_r(f,I)_p$ in terms of smoothness measured in spaces L_q, $q \neq p$. Such estimates are essentially embedding theorems and are important in nonlinear approximation. For example, in analogy with (6.2), we have for each $1 \leq q,p \leq \infty$ and $r > d(1/q - 1/p)_+$,

$$E_r(f,I)_p \leq C_r|I|^{r/d+1/p-1/q}|f|_{W^r(L_q(I))}. \tag{6.6}$$

We shall sketch a simple idea for proving such estimates, which is at the heart of proving embedding theorems. We consider $q \leq p$ (the other case is trivial). It is enough to prove (6.6) in the case $I = [0,1]^d$ since it follows for other cubes by a linear change of variables. For each dyadic cube $J \subset I$, let P_J be a polynomial in $\mathcal{P}_r$ that satisfies

$$\|f - P_J\|_{L_q(J)} \leq E_r(f,J)_q,$$

and define $S_k := \sum_{J\in D_k(I)} P_J\chi_J$, where $D_k(I)$ is the collection of all dyadic subcubes of I of side length 2^{-k}. Then, $S_0 = P_I$ and $S_k \to f$ in $L_p(I)$. Therefore,

$$E_r(f,I)_p \le \|f - P_I\|_{L_p(I)} \le \sum_{k=0}^{\infty} \|S_{k+1} - S_k\|_{L_p(I)}. \tag{6.7}$$

Now, for each polynomial $P \in \mathcal{P}_r$ and each cube J, we have $\|P\|_{L_p(J)} \le C|J|^{1/p-1/q}\|P\|_{L_q(J)}$ with the constant depending only on r (see Lemma 3.1 of DeVore and Sharpley (1984) for the simple proof). From this, it follows that

$$\begin{aligned} \|S_{k+1} - S_k\|^p_{L_p(I)} &= \sum_{J\in D_{k+1}(I)} \|S_{k+1} - S_k\|^p_{L_p(J)} \\ &\le C2^{-kd(1-p/q)} \sum_{J\in D_{k+1}(I)} \|S_{k+1} - S_k\|^p_{L_q(J)}. \end{aligned}$$

Now on J, we have $S_{k+1} - S_k = P_{J'} - P_J$ where J' is the parent of J. We write $P_{J'} - P_J = P_{J'} - f + f - P_J$ and use (6.2) (with p replaced by q) on each difference to obtain

$$\begin{aligned} \|S_{k+1} - S_k\|^p_{L_p(I)} &\le C2^{-kd(rp/d+1-p/q)} \sum_{J\in D_{k+1}(I)} |f|^p_{W^r(L_q(J'))} \\ &\le C2^{-kd(rp/d+1-p/q)} \left(\sum_{J\in D_{k+1}(I)} |f|^q_{W^r(L_q(J'))} \right)^{p/q} \\ &= C2^{-kd(rp/d+1-p/q)} |f|^p_{W^r(L_q(I))}. \end{aligned}$$

Here we used the facts that $\|\cdot\|_{\ell_p} \le \|\cdot\|_{\ell_q}$ if $q \le p$ and that a point $x \in I$ appears at most 2^d times in a cube J', as J runs over the cubes in $D_k(I)$. If we use this estimate in (6.7), we arrive at (6.6).

We can also allow $q < 1$ and nonintegral r in (6.6) if we use the Besov spaces. Namely, if $r > 0$ satisfies $r \ge d(1/q - 1/p)_+$, then

$$E_r(f,I)_p \le C_r|I|^{r/d+1/p-1/q}|f|_{B^r_q((L_q(I))}. \tag{6.8}$$

Notice that we can allow $r/d + 1/p - 1/q = 0$ in (6.8), which corresponds to the embedding of $B^r_q(L_q(I))$ into $L_p(I)$. The case $r/d > (1/q - 1/p)_+$ can be proved as above using the set subadditivity of $|\cdot|^q_{B^r_q(L_q(J))}$. For proofs of these results for Besov spaces see DeVore and Popov (1988*a*).

Finally, as we have remarked earlier, by using extensions, these results can be established for more general domains such as domains with a uniform cone condition. In particular, for any polyhedron $\mathcal{C}$, we have

$$E_r(f,\mathcal{C})_p \le C_r \mathrm{diam}(\mathcal{C})^{r/d+1/p-1/q}|f|_{W^r(L_q(\mathcal{C}))}, \tag{6.9}$$

with the constant depending only on r, d, the number of vertices of $\mathcal{C}$, and the smallest angle in $\mathcal{C}$. Similarly, we have the extension of (6.8) to polyhedra.

6.2. Piecewise polynomial approximation: the linear case

For the purpose of orienting the results on nonlinear approximation which follow, we shall in this section consider approximation by piecewise polynomials on fixed partitions. These results will be the analogue of approximation by piecewise constants on uniform partitions given in Section 3.1. For convenience, we shall consider approximation on the unit cube $\Omega := [0,1]^d$. The following results can be established for more general domains by using extension theorems similar to what we have mentioned earlier in this section.

By a *partition* of Ω, we mean a finite collection $\Delta := \{\mathcal{C}\}$ of polyhedrons $\mathcal{C}$ which are pairwise disjoint and union to Ω. Given such a collection, we define the *partition diameter*

$$\mathrm{diam}(\Delta) := \max_{\mathcal{C}} \mathrm{diam}(\mathcal{C}). \tag{6.10}$$

We assume that the number of vertices of each cell $\mathcal{C}$ is bounded independently of $\mathcal{C} \in \Delta$.

Let $\mathcal{S}^r(\Delta)$ denote the space of piecewise polynomials of order r relative to Δ. That is, a function S is in $\mathcal{S}^r(\Delta)$ if and only if it is a polynomial of order r on each cell $\mathcal{C} \in \Delta$. For $0 < p \le \infty$, we let

$$s_\Delta(f)_p := \inf_{S \in \mathcal{S}^r(\Delta)} \|f - S\|_{L_p(\Omega)}. \tag{6.11}$$

We shall fix $1 \le p \le \infty$ and estimate $s_\Delta(f)_p$. A similar analysis holds for $p < 1$ with Sobolev norms replaced by Besov norms.

We assume that each cell $\mathcal{C}$ is contained in a cube $J \subset I$ with $|J|^{1/d} \le C\mathrm{diam}(\mathcal{C})$ with C depending only on c_Δ. Hence, by extending f to this cube (if it is not already defined there) we see that, for each $\mathcal{C} \in \Delta$, there is a polynomial $P_\mathcal{C} \in \mathcal{P}_r$ which satisfies (6.9):

$$\|f - P_\mathcal{C}\|_{L_p(\mathcal{C})} \le C\mathrm{diam}(\Delta)^r |f|_{W^r(L_p(\mathcal{C}))}. \tag{6.12}$$

If we raise the estimates in (6.12) to the power p (in the case $p < \infty$) and add them, we arrive at

$$s_\Delta(f)_p \le C\mathrm{diam}\,\Delta^r |f|_{W^r(L_p(\Omega))}. \tag{6.13}$$

Of course, (6.12) is well known in both approximation and numerical circles. It is the proper form for numerical estimates based on piecewise polynomials of order r. It is the *Jackson inequality* for this type of approximation. By interpolation (as described in Section 4.2), we obtain the following estimate

$$s_\Delta(f)_p \le C\omega_r(f, \operatorname{diam}\Delta)_p, \tag{6.14}$$

where $\omega_r(f, \dot{)}_p = \omega_r(f, \cdot, \Omega)_p$ is the rth order modulus of smoothness of f in $L_p(\Omega)$ as introduced in Section 4.5. The advantage of (6.14) is that it does not require that f is in $W^r(L_p(\Omega))$ and in fact applies to any $f \in L_p(\Omega)$. For example, if $f \in \operatorname{Lip}(\alpha, L_p(\Omega))$, then (6.14) implies

$$s_\Delta(f)_p \le C|f|_{\operatorname{Lip}(\alpha, L_p(\Omega))}|\operatorname{diam}\Delta|^\alpha. \tag{6.15}$$

We would now like to understand to what extent estimates like (6.15) are best possible. It is not difficult to prove that, if $f \in L_p(\Omega)$ is a function for which

$$s_\Delta(f)_p \le M|\operatorname{diam}\Delta|^\alpha \tag{6.16}$$

holds *for every partition* Δ, then $f \in \operatorname{Lip}(\alpha, L_p(\Omega))$ and the smallest M for which (6.16) holds is equivalent to $|f|_{\operatorname{Lip}(\alpha, L_p(\Omega))}$. Indeed, for each $h \in \mathbb{R}^d$ and each $x \in \Omega$ such that the line segment $[x, x + rh) \subset \Omega$, there is a partition Δ with $\operatorname{diam}(\Delta) \le |h|$ and $\operatorname{dist}(x, \partial\mathcal{C}) \ge \text{const}\,|h|$ for every $\mathcal{C} \in \Delta$. This allows an estimate for $|\Delta_h^r(f, x)|$ by using ideas similar to the inverse estimates for piecewise constant approximation given in Section 3.1.

We note that the direct and inverse theorems relating approximation order to smoothness take the same form as those in Section 3.1. Using our interpretation of smoothness spaces given in Figure 6, we see that the approximation spaces for this form of linear approximation correspond to points on the vertical line segment joining $(1/p, 0)$ (L_p) to $(1/p, r)$ $(\operatorname{Lip}(r, L_p)$. Thus the only distinction from the piecewise constant case considered in Section 3.1 is that we can allow α to range over the larger interval $[0, r]$ because we are using piecewise polynomials of order r. Also, note that to achieve approximation order $O(n^{-\alpha})$ we would need spaces $\mathcal{S}^r(\Delta_n)$ of linear space dimension $\approx n^d$, that is, we have the curse of dimensionality.

More generally, if we only know (6.16) for a specific sequence of partitions (Δ_n), we can still prove that $f \in \operatorname{Lip}(\alpha, L_p(\Omega))$ provided the partitions mix sufficiently so that each x falls in the middle of sufficiently many $\mathcal{C}$. We do not formulate this precisely but refer readers to Section 2 of Chapter 12 of DeVore and Lorentz (1993) for a precise formulation in the univariate case.

Mixing conditions are not valid in most numerical settings. Indeed, the typical numerical case is where approximation takes place from a sequence $\mathcal{S}^r(\Delta_n)$, where Δ_n is a refinement of Δ_{n-1}. This means that the spaces are nested: $\mathcal{S}^r(\Delta_{n-1}) \subset \mathcal{S}^r(\Delta_n)$, $n = 1, 2, \ldots$. In this case, one can prove the inverse theorems only for a smaller range of α. It is easy to see that restrictions are needed on α. For example, functions f in $\mathcal{S}^r(\Delta_n)$ will be approximated exactly for $m \geq n$. But functions in $\mathcal{S}^r(\Delta_n)$ do not have much smoothness because they are discontinuous across the faces of the partition. This can be remedied by considering approximation by elements of $\mathcal{S}^r(\Delta_n)$ which have additional smoothness across the faces of the partition. We do not formulate inverse theorems in this case but refer the reader to Section 3 of Chapter 12 in DeVore and Lorentz (1993) where similar univariate results are proved.

We should mention, however, that considering splines with smoothness brings out new questions concerning direct estimates of approximation like (6.12). It is not easy to understand the dimension of spaces of smooth multivariate piecewise polynomials, let alone their approximation power (see Jia (1983)).

As the reader can now see, there are still interesting open questions concerning the approximation power of splines on general partitions, which relate the smoothness of the splines to the approximation power. These are difficult problems and have to a large extent been abandoned with the advent of box splines and, later, wavelets. These two developments shifted the viewpoint of spline approximation away from partitions and more toward the spanning functions. We shall get into this topic more in Section 7 when we discuss wavelet approximation.

6.3. Free knot piecewise polynomial approximation

To begin our development of nonlinear approximation by piecewise polynomials we shall consider the case of approximating a univariate function f defined on $\Omega = [0,1]$ by piecewise polynomials of fixed order r. The theory here is the analogue of piecewise constants discussed in Section 3.2.

Let the natural number r be fixed and for each $n = 1, 2, \ldots$, let $\Sigma_n := \Sigma_{n,r}$ be the space of piecewise polynomials of order r with n pieces on Ω. Thus, for each element $S \in \Sigma_n$ there is a partition Λ of Ω consisting of n disjoint intervals $I \subset \Omega$ and polynomials $P_I \in \mathcal{P}_r$ such that

$$S = \sum_{I \in \Lambda} P_I \chi_I. \tag{6.17}$$

For each $0 < p \leq \infty$, we define the error of approximation

$$\sigma_n(f)_p := \sigma_{n,r}(f)_p := \inf_{S \in \Sigma_{n,r}} \|f - S\|_{L_p(\Omega)}. \tag{6.18}$$

The case $p = \infty$ is sufficiently different that we shall restrict our discussion to the case $p < \infty$ and refer the reader to DeVore and Lorentz (1993) or Petrushev and Popov (1987) for the case $p = \infty$.

We can characterize the functions f that can be approximated with an order like $O(n^{-\alpha})$. We recall the approximation spaces

$$\mathcal{A}_q^\alpha(L_p(\Omega)) = \mathcal{A}_q^\alpha(L_p(\Omega), (\Sigma_n)).$$

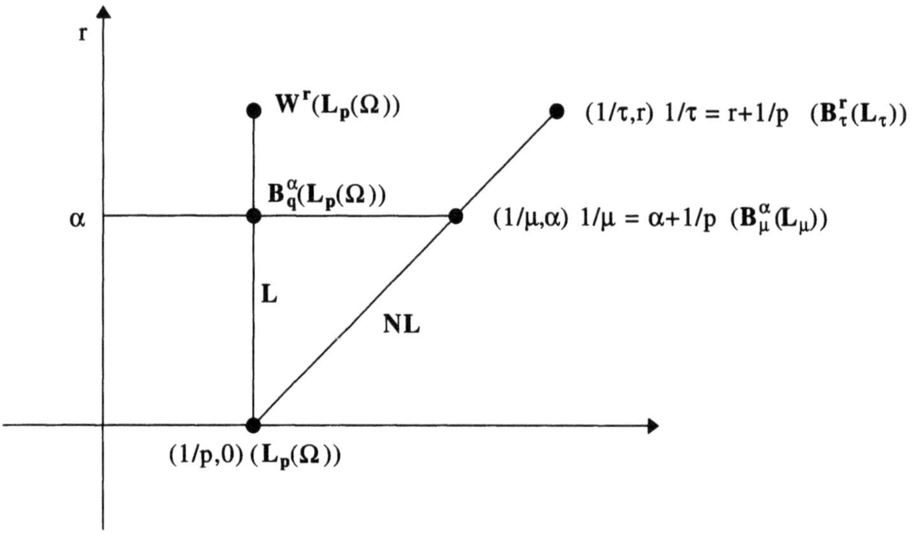

Fig. 7. Nonlinear approximation in L_p

According to the theory in Section 4, we can characterize these approximation spaces if we establish Jackson and Bernstein inequalities for this type of approximation. We fix the space $L_p(\Omega)$ in which approximation is going to take place. The space Y will be the Besov space $B^r_\tau(L_\tau(\Omega))$, $1/\tau = r + 1/p$ which was defined in Section 4.5. To understand this space, we return to our picture of smoothness spaces in Figure 7. The space $L_p(\Omega)$ of course corresponds to the point $(1/p, 0)$. The space $B^r_\tau(L_\tau(\Omega))$ corresponds to the point $(1/\tau, r)$, which lies on the line with slope one that passes through $(1/p, 0)$. As we have noted several times before, this line corresponds to the limiting case of the Sobolev embedding theorem. Thus, we are in complete analogy with the case of piecewise constant approximation described in Section 3.2.

The following inequalities were established by Petrushev (1988)

$$\sigma_n(f)_p \leq Cn^{-r}|f|_{B^r_\tau(L_\tau(\Omega))} \tag{6.19}$$

$$|S|_{B^r_\tau(L_\tau(\Omega))} \leq Cn^r\|f\|_{L_p(\Omega)} \tag{6.20}$$

with the constants C depending only on r. The first of these is the Jackson inequality and the second the companion Bernstein inequality.

Let us say a few words about how one proves these inequalities, since the techniques for doing so appear often in nonlinear approximation. To prove the Jackson inequality, for each $f \in B^\alpha_\tau(L_\tau(\Omega))$, we must find a favourable partition of Ω into n disjoint intervals. This is done by balancing $\Phi(I) := |f|^\tau_{B^\alpha_\tau(L_\tau(I))}$. The key here is that, with a proper renormalization of the Besov norm, Φ is set subadditive. Thus, we can find intervals I_j, $j = 1, \ldots, n$, so that $\Phi(I_j) = \Phi(\Omega)/n$. This gives our desired partition. We then use (6.8) to bound the local approximation error on each I_j, add these and arrive at (6.19) (see Chapter 12 of DeVore and Lorentz (1993) for more details). Therefore, as was the case in our introduction of nonlinear approximation by piecewise constants, we find our optimal partitions by a balancing suitable set function, in the present case Φ.

The proof of the Bernstein inequality is also very instructive. If $S \in \Sigma_n$, then $S = \sum_{I\in\Lambda} \gamma_I$ where Λ is a partition of Ω into n intervals and $\gamma_I = P_I\chi_I$ with $P_I \in \mathcal{P}_r$. For each such γ_I, it is not difficult to calculate its Besov norm and find

$$|\gamma_I|_{B^\alpha_\tau(L_\tau(\Omega))} \le C\|\gamma_I\|_{L_p(I)}, \tag{6.21}$$

with C an absolute constant. Then, using the subadditivity of $|\cdot|^\tau_{B^\alpha_\tau(L_\tau(\Omega))}$, we find that

$$\begin{aligned} |S|^\tau_{B^\alpha_\tau(L_\tau(\Omega))} &\le \sum_{I\in\Lambda} |\gamma_I|^\tau_{B^\alpha_\tau(L_\tau(\Omega))} \\ &\le \sum_{I\in\Lambda} \|\gamma_I\|^\tau_{L_p(\Omega)} \\ &\le Cn^{1-\tau/p}\left(\sum_{I\in\Lambda} \|\gamma_I\|^p_{L_p(\Omega)}\right)^{\tau/p} = Cn^{\alpha\tau}\|S\|^\tau_{L_p(\Omega)}. \end{aligned}$$

With the Jackson and Bernstein inequalities in hand, we can now refer to our general theory in Section 4 and obtain the following characterization of the approximation spaces: for each $0 < \alpha < r$, $0 < q \le \infty$, $0 < p < \infty$,

$$\mathcal{A}^\alpha_q(L_p(\Omega)) = (L_p(\Omega), B^r_\tau(L_\tau(\Omega)))_{\alpha/r,q}. \tag{6.22}$$

Therefore, we have a solution to our problem of characterizing the approximation spaces $\mathcal{A}^\alpha_q(L_p(\Omega))$ to the extent that we understand the interpolation spaces appearing in (6.22). Fortunately, we know a lot about these interpolation spaces. For example, for each $0 < \alpha < r$, there is one value of q for which this interpolation space is a Besov space. Namely, if $1/q = \alpha + 1/p$, then

$$\mathcal{A}^\alpha_q(L_p(\Omega)) = (L_p(\Omega), B^r_\tau(L_\tau(\Omega))_{\alpha/r,q} = B^\alpha_q(L_q(\Omega)). \tag{6.23}$$

For other values of q these interpolation spaces can be described in various ways. We defer a discussion of this until we treat nonlinear wavelet approximation where these interpolation spaces will reappear.

Returning to our picture of smoothness spaces, we see that the approximation spaces $\mathcal{A}^\alpha_q(L_p(\Omega))$ correspond to the point $(1/\tau, \alpha)$ with $1/\tau = \alpha + 1/p$. Thus, these spaces lie on the line with slope one passing through $(1/p, 0)$. In other words, we have the same interpretation as in nonlinear approximation by piecewise constants except that now α can range over the larger interval $(0, r)$ corresponding to the order r of the piecewise polynomials.

We have emphasized that the Besov spaces $B^\alpha_\tau(L_\tau(\Omega))$, $1/\tau = \alpha + 1/p$, which occur in characterizing free knot spline approximation, lie on the demarcation line in the Sobolev embedding theorem. This is an indication that these spaces are quite large when compared to the Besov spaces $B^\alpha_q(L_p(\Omega))$ which appear in characterizing linear approximation. Some examples might further drive this point home. Any function f which is a piecewise polynomial (with a finite number of pieces) is in all of these spaces, that is, we can take α arbitrarily large. Indeed, f can be approximated exactly once n and r are large enough and hence this result follows from (6.22). A simple argument shows that this remains true for any piecewise analytic function f. Hence, any such function can be approximated to accuracy $O(n^{-\alpha})$ for any $\alpha > 0$ with nonlinear piecewise polynomial approximation. Another instructive example is the function $f(x) = x^\gamma$, $\gamma > -1/p$ (so that $f \in L_p(\Omega)$). This function satisfies (see de Boor (1973))

$$\sigma_{n,r}(f)_p = O(n^{-r}).$$

This can be proved by balancing the approximation errors.

6.4. *Free knots and free degree*

There are many variants of piecewise polynomial approximation. One of the most important is to allow not only the partition to vary with f but also the orders (degrees) of the polynomial pieces. Approximation of this type occurs in the *h-p* method in FEM which has been introduced and studied by Babuška and his collaborators (see Babuška and Suri (1994)). While the theory for this type of approximation is far from complete, it will be useful to mention a few facts that separate it from the free knot case discussed above.

Let Σ^*_n denote the set of all piecewise polynomials

$$S = \sum_{I \in \Delta} P_I \chi_I, \qquad (6.24)$$

where Δ is a partition and for each $I \in \Delta$ there is a polynomial P_I of order r_I with $\sum_{I\in\Delta} r_I \le n$. As usual, we let

$$\sigma^*_n(f)_p := \inf_{S\in\Sigma^*_n} \|f - S_n\|_{L_p(\Omega)}. \qquad (6.25)$$

Clearly, for each $r = 1, 2, \ldots$, we have $\sigma^*_{nr}(f)_p \le \sigma_{n,r}(f)_p$ because $\Sigma_{n,r} \subset \Sigma^*_{nr}$. To see that σ^*_n can be considerably better than $\sigma_{n,r}$, we consider the following example, which was studied in DeVore and Scherer (1980). Let $f(x) = x^\beta$ with $\beta > 0$. We have seen that $\sigma_{n,r}(f)_p \asymp n^{-r}$. On the other hand, it is shown in the above reference that

$$\sigma^*_n(f) \le Ce^{-c\sqrt{n}}, \quad c := \sqrt{2} - 1 \qquad (6.26)$$

and that this estimate cannot be improved in the sense of a better exponential rate.

6.5. Free partition splines: the multivariate case

Up to this point our discussion of nonlinear approximation has been almost entirely limited to approximating univariate functions. The question arises, for example, whether the results of the previous section on free knot spline approximation can be extended to the multivariate case.

For the moment, we restrict our discussion to the bivariate case and approximation on $\Omega := [0,1]^2$. In this case, we consider the space $\Sigma^{\#}_{n,r}$ consisting of all functions

$$S = \sum_{T \in \Delta} P_T \chi_T \tag{6.27}$$

with $\Delta = \{T\}$ a partition of Ω consisting of n triangles and the P_T polynomials of total order r on T for each $T \in \Delta$. Let

$$\sigma^{\#}_{n,r}(f)_p := \inf_{S \in \Sigma^{\#}_{n,r}} \|f - S\|_{L_p(\Omega)}. \tag{6.28}$$

Here # is used to make a distinction from the univariate case.

There is no known characterization of $\mathcal{A}^\alpha_q(L_p(\Omega), (\Sigma^{\#}_{n,r}))$ for any values of α, p, q. This remains one of the most interesting and challenging problems in nonlinear approximation. We shall mention some of the difficulties encountered in trying to characterize these approximation classes, since this has influenced developments in multivariate nonlinear approximation.

A first remark is that the space $\Sigma^{\#}_N$ does not satisfy assumption (iv) of Section 4.1: that is, for no constant c do we have $\Sigma^{\#}_n + \Sigma^{\#}_n \subset \Sigma^{\#}_{cn}$. For instance, consider a partition Δ_1 of Ω consisting of n vertical strips of equal size, each divided into two triangles, and the corresponding partition Δ_2 made from horizontal strips. Let S_1 be a piecewise polynomial relative to Δ_1 and S_2 another piecewise polynomial relative to Δ_2. Then the sum $\Delta_1 + \Delta_2$ will be a piecewise polynomial which in general requires $4n^2$ triangles in its partition.

Even more relevant to our problem is a result (communicated to us by Jonathan Goodman) that constructs functions $f(x)$ and $g(y)$ which individually can be approximated with order $O(1/n)$ by the elements of $\Sigma^{\#}_n$ but whose sum can only be approximated to order $O(1/\sqrt{n})$. Thus, the approximation spaces $\mathcal{A}^\alpha_q(L_p(\Omega))$ are not linear. This precludes their characterization by classical smoothness spaces, which are always linear.

Here is another relevant comment. The starting point for proving direct estimates for nonlinear piecewise polynomial approximation are good local error estimates for polynomial approximation, such as those given in Section 6.1. The appropriate local error estimators for polynomial approximation on general triangles are not known. They should take into consideration the shape and orientation of the triangles. For example, less smoothness of the target function should be required in directions where the triangle is thin, more in directions where the triangle is fat. While one may guess appropriate error estimators, none have been utilized successfully in nonlinear schemes.

Given the situation described above concerning nonlinear piecewise polynomial approximation, it comes as no surprise that other avenues were explored to handle nonlinearity in the multivariate case. The most successful of these has been n-term approximation, which took the following viewpoint. In the univariate case the elements in the space Σ_n can also be desribed as a sum of n (or perhaps Cn) fundamental building blocks. In the case of piecewise constants these are simply the characteristic functions χ_I of intervals I. In the general case of univariate, nonlinear piecewise polynomial approximation the building blocks are B-splines. Therefore, one generalization of Σ_n to the multivariate case would take the form of n-term approximation using multivariate building blocks. The first examples were for box splines (DeVore and Popov 1987) but this was later abandoned for the more computationally favourable wavelets. We shall discuss wavelets in Section 7.

6.6. *Rational approximation*

Another natural candidate for nonlinear approximation is the set of rational functions. Let $\mathcal{R}_n(\mathbb{R}^d)$ denote the space of rational functions in d variables. Thus, an element R in $\mathcal{R}_n$ is the quotient, $R = P/Q$, of two polynomials P,Q (in d variables) of total degree $\leq n$. We define the approximation error

$$r_n(f)_p := \inf_{R\in\mathcal{R}_n} \|f - R\|_{L_p(\Omega)}. \tag{6.29}$$

The status of rational approximation is more or less the same as for piecewise polynomials. In one variable, we have

$$\mathcal{A}_q^\alpha(L_p(\Omega), (\mathcal{R}_n)) = \mathcal{A}_q^\alpha(L_p(\Omega), (\Sigma_{n,r})), \quad 0 < \alpha < r. \tag{6.30}$$

Thus, on the one hand the approximation problem is solved but on the other hand the news is somewhat depressing since there is nothing to gain or lose (in the context of the approximation classes) in choosing rational functions over piecewise polynomials.

The characterizations (6.30) were proved by Pekarski (1986) and Petrushev (1988) by comparing σ_n to r_n. A typical comparison is given by the inequalities

$$r_n(f)_p \leq Cn^{-\alpha} \sum_{k=1}^{n} k^{\alpha-1}\sigma_{k,r}(f)_p, \quad n \geq r, \tag{6.31}$$

which hold for all $1 \le p < \infty$, $\alpha > 0$ and approximation on an interval. Similar inequalities reverse the roles of $\sigma_{n,r}(f)_p$ and $r_n(f)_p$. Thus the approximation classes for univariate rational approximation coincide with Besov spaces when $1/q = \alpha + 1/p$ (see (6.23)). In a strong sense, rational approximation can be viewed as piecing together local polynomial approximants similar to piecewise polynomials.

We should also mention the work of Peller (1980), who characterized the approximation classes for rational approximation in the BMO metric (which can be considered as a slight variant of L_∞). In the process, Peller characterized interpolation spaces between BMO and the Besov space $B_1^1(L_1)$ and found the trace classes for Hankel operators, thus unifying three important areas of analysis.

There are some direct estimates for multivariate rational approximation (see, for example, DeVore and Yu (1990)) but they fall far short of being optimal. The characterization of approximation spaces for multivariate rationals has met the same resistance as piecewise polynomials for more or less the same reasons.

There have been several other important developments in rational approximation. One of these was Newman's theorem (see Newman (1964)) which showed that the function $f(x) = |x|$ satisfies $r_n(f)_\infty = O(e^{-c\sqrt{n}})$ (a very stunning result at the time). Subsequently, similar results were proved for other special functions (such as $e^{-|x|^\beta}$) and even asymptotics for the error $r_n(f)$ were found. A mainstay technique in these developments was Padé approximation. This is to rational functions what Taylor expansions are to polynomials. A first reference for Padé approximation is the book of Baker (1975).

7. Wavelets

Wavelets were ripe for discovery in the 1980s. Multigrid methods in numerical computation, box splines in approximation theory, and the Littlewood–Paley theory in harmonic analysis all pointed to multilevel decompositions. However, the great impetus came from two discoveries: the multiresolution analysis of Mallat and Meyer (see Mallat (1989)) and most of all the discovery by Daubechies (1988) of compactly supported orthogonal wavelets with arbitrary smoothness.

Wavelets are tailor-made for nonlinear approximation and certain numerical applications. Computation is fast and simple, and strategies for generating good nonlinear approximations are transparent. Since wavelets provide unconditional bases for a myriad of function spaces and smoothness spaces, the characterization of approximation classes is greatly simplified. Moreover, wavelets generalize readily to several dimensions.

There are many excellent accounts of multiresolution and wavelet theory. We shall introduce only enough of the theory to set our notation and provide us with the vehicle we need for our development of nonlinear approximation. The Haar function is a wavelet (albeit not a very smooth one) and (3.47) is typical of wavelet decompositions. We shall begin our discussion of multiresolution by considering approximation from shift invariant spaces which provides the *linear theory* for wavelet approximation.

In the development of wavelets and multiresolution analysis, one needs to make modest assumptions on the refinement function φ so the theory develops smoothly. We shall not stress these assumptions, and in fact in many cases not even mention them, in order to keep our exposition short and to the point. The reader needs to consult one of the following references to find precise formulations of the results we state here: Daubechies (1992), Meyer (1990), DeVore and Lucier (1992).

7.1. Shift invariant spaces

In multiresolution analysis, there are two fundamental operations we perform on functions: shift and dilation. If f is defined on $\mathbb{R}^d$ and $j \in \mathbb{Z}^d$, then $f(\cdot - j)$ is the (integer) shift of f by j. Meanwhile, if $a > 0$ is a real number then $f(a\cdot)$ is the dilate of f by a. In this section, we consider spaces invariant under shifts. We then dilate them to create new and finer spaces. The main goal is to understand the approximation properties of these dilated spaces.

We shall not discuss *shift invariant spaces* in their full generality in order to move more directly to multiresolution analysis. The results stated below have many extensions and generalizations (see de Boor, DeVore and Ron (1993) and the references therein).

Let φ be a compactly supported function in $L_2(\mathbb{R}^d)$. We define $\tilde{\mathcal{S}}(\varphi)$ as the set of all finite linear combinations of the shifts of φ. The space $\mathcal{S} := \mathcal{S}(\varphi)$ is defined to be the closure of $\tilde{\mathcal{S}}(\varphi)$ in $L_2(\mathbb{R}^d)$. We say that $\mathcal{S}$ is the *principal shift invariant* (PSI) *space* generated by φ.

For each $k \geq 0$, the space $\mathcal{S}_k := \mathcal{S}_k(\varphi)$ is defined to be the dilate of $\mathcal{S}$ by 2^k. A function T is in $\mathcal{S}_k$ if and only if $T = S(2^k\cdot)$ with $S \in \mathcal{S}(\varphi)$. The space $\mathcal{S}_k$ is invariant under the shifts $j2^{-k}$, $j \in \mathbb{Z}^d$. We shall be interested in the approximation properties (in the $L_2(\mathbb{R}^d)$-norm) of $\mathcal{S}_k$ as $k \to \infty$. We let

$$E_k(f) := E_k(f)_2 := \inf_{S \in \mathcal{S}_k} \|f - S\|_{L_2(\mathbb{R}^d)}, \quad k = 0, 1, \ldots. \tag{7.1}$$

The approximation properties of $\mathcal{S}_k$ are related to polynomial reproduction in $\mathcal{S}$. It was Schoenberg (1946) who first recognized that polynomial reproduction could be described by the Fourier transform $\hat{\varphi}$ of φ; subsequently, Strang and Fix (1973) used Fourier transforms to describe approximation properties. We say φ satisfies the Strang–Fix condition of order $r \in \mathbb{N}$ if

$$\hat{\varphi}(0) \neq 0, \text{ and } D^j\hat{\varphi}(2k\pi) = 0, \quad k \in \mathbb{Z}^d \setminus \{0\}, |j| < r. \tag{7.2}$$

When φ satisfies the Strang–Fix condition of order r then $\mathcal{S}(\phi)$ locally contains all polynomials of order r (degree $< r$). (Actually, this and results stated below require a little more about φ in terms of smoothness, which we choose not to formulate exactly.) Moreover, it is easy to prove the Jackson inequality: for all f in the Sobolev space $W^r(L_2(\mathbb{R}^d))$, we have

$$E_k(f) \leq C2^{-kr}|f|_{W^r(L_2(\mathbb{R}^d))}, \quad k = 0, 1, \ldots. \tag{7.3}$$

The companion Bernstein inequality to (7.3) is

$$|S|_{W^r(L_2(\mathbb{R}^d))} \leq C2^{kr}\|S\|_{L_2(\mathbb{R}^d)}, \quad S \in \mathcal{S}_k. \tag{7.4}$$

It is valid if φ is in $W^r(L_2(\mathbb{R}^d))$. Under these conditions on φ, we can use the general results of Section 4.3 to obtain the following characterization of approximation spaces:

$$\mathcal{A}^\alpha_q(L_2(\mathbb{R}^d)) = B^\alpha_q(L_2(\mathbb{R}^d)), \quad 0 < \alpha < r, \ 0 < q \leq \infty. \tag{7.5}$$

Notice that this is exactly the same characterization as for the other types of linear approximation we have discussed earlier. There is a similar theory for approximation in $L_p(\mathbb{R}^d)$, $1 \leq p \leq \infty$, and even $0 < p < 1$.

7.2. Multiresolution and wavelet decompositions

Multiresolution adds one essential new ingredient to the setting of the previous section. We require that the spaces $\mathcal{S}_k$ are nested, that is, $\mathcal{S}_k \subset \mathcal{S}_{k+1}$, which is of course equivalent to $\mathcal{S}_0 \subset \mathcal{S}_1$. This is in turn equivalent to requiring that φ is in $\mathcal{S}_1$.

We shall limit our discussion to the multiresolution analysis that leads to the biorthogonal wavelets of Cohen, Daubechies and Feauveau (1992). These are the wavelets used most often in applications. Accordingly, we start with the univariate case and assume that φ is a function for which the spaces $\mathcal{S}_k = \mathcal{S}_k(\varphi)$ of the previous section provide approximation:

$$\operatorname{dist}(f, \mathcal{S}_k)_{L_2(\mathbb{R})} \to 0. \tag{7.6}$$

We know that this will hold, for example, if φ satisfies the Strang–Fix condition for some order $r > 0$. We assume further that the shifts $\varphi(\cdot - j)$, $j \in \mathbb{Z}$, are a Riesz basis for $\mathcal{S}$ and that the dual basis is given by the shifts of a compactly supported function $\tilde{\varphi}$ whose dilated spaces $\mathcal{S}_k(\tilde{\varphi})$ also form a multiresolution analysis. Duality means that

$$\int_{\mathbb{R}} \varphi(x - j)\tilde{\varphi}(x - k)\,\mathrm{d}x = \delta_{jk}. \tag{7.7}$$

with δ_{jk} the Kronecker delta.

The fact that $\varphi \in \mathcal{S}_1$ implies that φ is refinable:

$$\varphi(x) = \sum_{k\in\mathbb{Z}} c_k \varphi(2x - k). \tag{7.8}$$

The compact support of φ implies that there is only a finite number of nonzero coefficents c_k in (7.8). They are called the refinement mask for φ (in image processing they are called the (low pass) filter coefficients). The dual function $\tilde{\varphi}$ satisfies a corresponding refinement equation with mask coefficients $\tilde{c}_k$.

Let $\langle \cdot, \cdot \rangle$ denote the inner product in $L_2(\mathbb{R})$ and let P be the projector

$$Pf := \sum_{j\in\mathbb{Z}} \langle f, \tilde{\varphi}(\cdot - j)\rangle \varphi(\cdot - j) \tag{7.9}$$

which maps $L_2(\mathbb{R})$ onto $\mathcal{S}$. By dilation, we obtain the corresponding projectors P_k which map $L_2(\mathbb{R})$ onto $\mathcal{S}_k$, $k \in \mathbb{Z}$. We are particularly interested in the projector $Q := P_1 - P_0$ which maps $L_2(\mathbb{R})$ onto a subspace W of $\mathcal{S}_1$. The space W is called a wavelet space; it represents the *detail* which, when added to $\mathcal{S}_0$, gives $\mathcal{S}_1$ via the formula $S = PS + QS$, $S \in \mathcal{S}_1$. One of the main results of wavelet/multiresolution theory is that W is a PSI space generated by the function

$$\psi(x) = \sum_{k\in\mathbb{Z}} d_k \tilde{\varphi}(2x - k), \quad d_k := (-1)^k \tilde{c}_{1-k}. \tag{7.10}$$

Also, the shifts $\psi(\cdot - j)$, $j \in \mathbb{Z}$, form a Riesz basis for W whose dual functionals are represented by $\tilde{\psi}(\cdot - j)$ where $\tilde{\psi}$ is obtained from φ in the same way ψ was obtained from $\tilde{\varphi}$. In other words,

$$Qf = \sum_{j\in\mathbb{Z}} 2^{-k} \langle f, \tilde{\psi}(\cdot - j)\rangle \psi(\cdot - j). \tag{7.11}$$

Of course, by dilation, we obtain the spaces W_k, the projectors Q_k and the representation

$$Q_k f = \sum_{j\in\mathbb{Z}} 2^k \langle f, \tilde{\psi}(2^k \cdot - j)\rangle \psi(2^k \cdot - j). \tag{7.12}$$

From (7.6), we know that $P_k f \to f$, $k \to \infty$. It can also be shown that $P_k f \to 0$, $k \to -\infty$, and therefore we have

$$f = \sum_{k=-\infty}^{\infty} (P_{k+1} f - P_k f) = \sum_{k\in\mathbb{Z}} \sum_{j\in\mathbb{Z}} 2^k \langle f, \tilde{\psi}(2^k \cdot - j)\rangle \psi(2^k \cdot - j). \tag{7.13}$$

The factor 2^k multiplying the inner product arises from scaling. This is the biorthogonal wavelet decomposition of an arbitrary $f \in L_2(\mathbb{R})$. We would like to simplify the wavelet notation and better expose the nature of the representation (7.13). For this we shall use the following convention. To $j \in \mathbb{Z}^d$, $k \in \mathbb{Z}$, we associate the dyadic cube $I = 2^{-k}(j+\Omega)$ with $\Omega := [0,1]^d$, the unit cube in $\mathbb{R}^d$. To each function η defined on $\mathbb{R}^d$, we let

$$\eta_I(x) := |I|^{-1/2}\eta(2^k \cdot -j). \tag{7.14}$$

The cube I roughly represents the support of η_I; in the case that $\eta = \chi_\Omega$ or $\eta = H$ with the H the Haar function, then I is precisely the support of η_I.

Let D be the set of all dyadic intervals in $\mathbb{R}$ and D_k those dyadic intervals of length 2^{-k}. We can now rewrite (7.13) as

$$f = \sum_{I \in D} c_I(f)\psi_I, \quad c_I(f) := \langle f, \tilde{\psi}_I \rangle. \tag{7.15}$$

The Riesz basis property of the ψ_I gives that

$$\|f\|_{L_2(\mathbb{R})} \asymp \left(\sum_{I \in D} |c_I(f)|^2 \right)^{1/2}. \tag{7.16}$$

The special case of orthogonal wavelets is noteworthy. In this case, one begins with a scaling function φ whose shifts are an orthonormal system for $\mathcal{S}(\varphi)$. Thus $\tilde{\varphi} = \varphi$ and the space W is orthogonal to $\mathcal{S}_0$: each function $S \in W$ satisfies

$$\int_{\mathbb{R}} S S_0 \, dx = 0, \quad S_0 \in \mathcal{S}_0. \tag{7.17}$$

The decomposition $\mathcal{S}_1 = \mathcal{S}_0 \oplus W$ is orthogonal and the functions ψ_I, $I \in D$ are an orthonormal basis for $L_2(\mathbb{R})$.

We turn now to the construction of wavelet bases in several dimensions. There are several possibilities. The most often used construction is the following. Let φ be a univariate scaling function and ψ its corresponding wavelet. We define $\psi^0 := \varphi$, $\psi^1 := \psi$. Let E' denote the collection of vertices of the unit cube $[0,1]^d$ and E the set of nonzero vertices. For each vertex $e = (e_1, \ldots, e_d) \in E'$, we define the multivariate functions

$$\psi^e(x_1, \ldots, x_d) := \psi^{e_1}(x_1) \cdots \psi^{e_d}(x_d) \tag{7.18}$$

and define $\Psi := \{\psi^e : e \in E\}$. If $D = D(\mathbb{R}^d)$ is the set of dyadic cubes in $\mathbb{R}^d$, then the collection of functions

$$\{\psi_I^e, \quad I \in D, \ e \in E\} \tag{7.19}$$

forms a Riesz basis for $L_2(\mathbb{R}^d)$; an orthonormal basis if ψ is an orthogonal wavelet. The dual basis functions $\tilde{\psi}_I^e$ have an identical construction starting with $\tilde{\varphi}$ and $\tilde{\psi}$. Thus, each $f \in L_2(\mathbb{R}^d)$ has the wavelet expansion

$$f = \sum_{I \in D} \sum_{e \in E} c_I^e(f)\psi_I^e, \quad c_I^e(f) := \langle f, \tilde{\psi}_I^e \rangle. \tag{7.20}$$

Another construction of multivariate wavelet bases is to simply take the tensor products of the univariate basis ψ_I. This gives the basis

$$\psi_R(x_1,\dots,x_d) := \psi_{I_1}(x_1)\cdots\psi_{I_d}(x_d), \quad R := I_1 \times \cdots I_d, \tag{7.21}$$

where the R are multidimensional parallelepipeds. Notice that the support of the function ψ_R corresponds to R and is nonisotropic. It can be long in one direction and short in another. This is in contrast to the previous bases whose supports are isotropic. We shall be almost exclusively interested in the first basis.

7.3. Characterization of function spaces by wavelet coefficients

Wavelet coefficients provide simple characterizations of most function spaces. The norm in the function space is equivalent to a sequence norm applied to the wavelet coefficients. We shall need such characterizations for the case of L_p spaces and Besov spaces.

It is sometimes convenient in the characterizations that follow to choose different normalizations for the wavelets, and hence coefficients, appearing in the decomposition (7.20). In (7.20) we have normalized the wavelets and dual functions in $L_2(\mathbb{R}^d)$. We can also normalize the wavelets in $L_p(\mathbb{R}^d)$, $0 < p \le \infty$, by taking

$$\psi_{I,p}^e := |I|^{-1/p+1/2}\psi_I^e, \quad I \in D,\ e \in E, \tag{7.22}$$

with a similar definition for the dual functions. Then, we can rewrite (7.20) as

$$f = \sum_{I\in D}\sum_{e\in E} c_{I,p}^e(f)\psi_{I,p}^e, \quad c_{I,p}^e(f) := \langle f, \tilde{\psi}_{I,p'}^e\rangle, \tag{7.23}$$

with $1/p + 1/p' = 1$. We also define

$$c_{I,p}(f) := \left(\sum_{e\in E} |c_{I,p}^e(f)|^p\right)^{1/p}. \tag{7.24}$$

One should note that it is easy to go from one normalization to another. For example, for any $0, p, q \le \infty$, we have

$$\psi_{I,p} = |I|^{1/q-1/p}\psi_{I,q}, \quad c_{I,p}(f) = |I|^{1/p-1/q}c_{I,q}(f). \tag{7.25}$$

The characterization of L_p spaces by wavelet coefficients comes from the Littlewood–Paley theorem of harmonic analysis. One cannot simply characterize the L_p spaces by ℓ_p norms of the wavelet coefficients. Rather, one must go through the square function

$$S(f,x) := \left(\sum_{I\in D} c_{I,2}(f)^2|I|^{-1}\chi_I(x)\right)^{1/2} = \left(\sum_{I\in D} c_{I,p}(f)^2|I|^{-2/p}\chi_I(x)\right)^{1/2} \tag{7.26}$$

which incorporates the interaction between dyadic levels. Here, as earlier, χ_I is the characteristic function of the interval I. For $1 < p < \infty$, one has

$$\|f\|_{L_p(\mathbb{R}^d)} \asymp \|S(f,\cdot)\|_{L_p(\mathbb{R})} \tag{7.27}$$

with the constants of equivalency depending only on p. Notice that, in the case $p = 2$, (7.27) reduces to (7.16). One can find proofs of (7.27) (which use techniques of harmonic analysis such as maximal functions) in Meyer (1990) or DeVore, Konyagin and Temlyakov (1998).

The equivalence (7.27) can be extended to the range $p \le 1$ if the space L_p is replaced by the Hardy space H_p and more assumptions are made of the wavelet ψ. In this sense, most of the theory of approximation given below can be extended to this range of p.

We have introduced the Besov spaces $B_q^\alpha(L_p(\mathbb{R}^d))$ for $0 < q, p \le \infty$, $\alpha > 0$, in Section 4.5. The following is the wavelet characterization of these spaces:

$$|f|_{B_q^\alpha(L_p(\mathbb{R}^d))} \asymp \begin{cases} \left(\sum_{k=-\infty}^{\infty} 2^{k\alpha q}\left(\sum_{I\in D_k} c_{I,p}(f)^p\right)^{q/p}\right)^{1/q}, & 0 < q < \infty, \\ \sup_{k\in\mathbb{Z}} 2^{k\alpha}\left(\sum_{I\in D_k} c_{I,p}(f)^p\right)^{1/p}, & q = \infty. \end{cases} \tag{7.28}$$

Several remarks are in order to explain (7.28).

Remark 7.1 Other normalizations for the coefficients $c_I(f)$ are frequently used. The form of (7.28) then changes by the introduction of a factor $|I|^\beta$ into each term, with β a fixed constant.

Remark 7.2 We can define spaces of functions for all $\alpha > 0$ by using the right side of (7.28). However, these spaces will coincide with Besov spaces only for a certain range of α and p that depend on the wavelet ψ. In the case $1 \le p \le \infty$, we need that

(a) $\psi \in B_q^\beta(L_p(\mathbb{R}^d))$, for some $\beta > \alpha$,
(b) ψ has r vanishing moments with $r > \alpha$.

When $p < 1$, we also need that $r > d/p - d$ (see the following remark).

Remark 7.3 When $p < 1$, (7.28) characterizes the space $B_q^\alpha(H_p(\mathbb{R}^d))$ (with the correct range of parameters) where this latter Besov space can be defined by replacing the L_p modulus of smoothness by the H_p modulus of smoothness (see Kyriazis (1996)). However, if $\alpha > d/p - d$, this space is the same as $B_q^\alpha(L_p(\mathbb{R}^d))$.

Remark 7.4 For a fixed value of $1 \leq p < \infty$, the spaces $B^\alpha_\tau(L_\tau(\mathbb{R}^d))$, $1/\tau = \alpha/d + 1/p$, occur, as we know, in nonlinear approximation. If we choose the wavelets normalized in L_p, then the characterization (7.28) becomes simply

$$|f|_{B^\alpha_\tau(L_\tau(\mathbb{R}^d))} \asymp \left(\sum_{I \in D} c_{I,p}(f)^\tau\right)^{1/\tau}. \tag{7.29}$$

7.4. Nonlinear wavelet approximation

In this and the next subsections, we shall consider n-term approximation by wavelet sums. The results we present hold equally well in the univariate and the multivariate case. However, the notation is somewhat simpler in the univariate case. Therefore, to spare the reader, we shall initially treat only this case. At the end of the section we shall formulate the results for multivariate functions.

The idea of how to utilize wavelets in nonlinear approximation is quite intuitive. If the target function is smooth on a region we can use a coarse resolution (approximation) on that region. This amounts to putting terms in the approximation corresponding to low frequency-terms from dyadic level k with k small. On regions where the target function is not smooth we use higher resolution. This is accomplished by taking more wavelet functions in the approximation, that is, terms from higher dyadic levels. The questions that arise from these intuitive observations are:

(i) exactly how should we measure smoothness to make such demarcations between smooth and nonsmooth?
(ii) how do we allocate terms in a nonlinear strategy?
(iii) are there precise characterizations of the functions that can be approximated with a given approximation order by nonlinear wavelet approximation?

Fortunately, all of these questions have a simple and definitive solution, which we shall presently describe.

We shall limit ourselves to the case of biorthogonal wavelets and approximation in L_p, $1 < p < \infty$. Again, one can work in much more generality. As will be clear from our exposition, what is essential is only the equivalence of function norms with norms on the sequence of wavelet coefficients. Thus, the results we present hold equally well for approximation in the Hardy space H_p (Cohen, DeVore and Hochmuth 1997) and for more general wavelets.

It will also be convenient to consider approximation on all of $\mathbb{R}^d$ (initially on $\mathbb{R}$). In the following section, we shall discuss briefly how results extend to other domains.

Let φ, $\tilde{\varphi}$ be two refinable functions which are in duality as described in

Section 7.2 and let ψ and $\tilde{\psi}$ be their corresponding wavelets. Then, each function $f \in L_p(\mathbb{R})$ has the wavelet decomposition (7.15). We let Σ_n^w denote the set of all functions

$$S = \sum_{I \in \Lambda} a_I \psi_I, \tag{7.30}$$

where $\Lambda \subset D$ is a set of dyadic intervals of cardinality $\#\Lambda \leq n$. Thus Σ_n^w is the set of all functions which are a linear combination of n wavelet functions. In analogy with our previous studies, we define

$$\sigma_n^w(f)_p := \inf_{S \in \Sigma_n^w} \|f - S\|_{L_p(\mathbb{R})}. \tag{7.31}$$

We can characterize the approximation classes for n-term wavelet approximation by proving Jackson and Bernstein inequalities and then invoking the general theory of Section 4.3. The original proofs of these inequalities were given in DeVore, Jawerth and Popov (1992) but we shall follow Cohen, DeVore and Hochmuth (1997) which introduced some simpler techniques.

Given a finite set Λ of intervals, for each $x \in \mathbb{R}$, we let $I(x)$ be the smallest interval in Λ that contains x. If there is no such interval, then we define $I(x) := \mathbb{R}$ and expressions like $|I(x)|^{-1}$ are interpreted as zero. The following lemma of Temlyakov (1998*a*) is a powerful tool in estimating norms of wavelet sums.

Lemma 1 Let $1 < p < \infty$ and Λ be a finite set. If $f \in L_p(\mathbb{R})$ has the wavelet decomposition

$$f = \sum_{I \in \Lambda} c_{I,p}(f) \psi_{I,p}, \tag{7.32}$$

with $|c_{I,p}(f)| \leq M$, for all $I \in \Lambda$, then

$$\|f\|_{L_p(\mathbb{R})} \leq C_1 M \#\Lambda^{1/p}, \tag{7.33}$$

with C_1 an absolute constant. Similarly, if $|c_{I,p}(f)| \geq M$, for all $I \in \Lambda$, then

$$\|f\|_{L_p(\mathbb{R})} \geq C_2 M \#\Lambda^{1/p}, \tag{7.34}$$

with $C_2 > 0$ an absolute constant.

We shall sketch the proof of (7.33) (which is valid for $0 < p < \infty$) since it gives us a chance to show the role of $I(x)$ and the square function. The proof of (7.34) is similar. We have

$$\begin{aligned} \|f\|_{L_p(\mathbb{R})} &\leq \|S(f)\|_{L_p(\mathbb{R})} = C \Big\| \Big(\sum_{I \in \Lambda} c_{I,p}^2 |I|^{-2/p} \chi_I \Big)^{1/2} \Big\|_{L_p(\mathbb{R})} \\ &\leq CM \Big\| \Big(\sum_{I \in \Lambda} |I|^{-2/p} \chi_I \Big)^{1/2} \Big\|_{L_p(\mathbb{R})} \leq CM \| |I(x)|^{-1/p} \|_{L_p(\mathbb{R})}. \end{aligned}$$

If $J \in \Lambda$, then the set $\tilde{J} := \{x : I(x) = J\}$ is a subset of J. It follows that

$$\|f\|_{L_p(\mathbb{R})}^p \le CM^p \int_{\mathbb{R}^d} |I(x)|^{-1}\,\mathrm{d}x \le CM^p \sum_{J\in\Lambda} \int_{\tilde{J}} |J|^{-1} \le CM^p \#\Lambda,$$

which proves (7.33).

We shall now formulate the Jackson inequality for n-term wavelet approximation. Let r be the number of vanishing moments of ψ. Recall that r also represents the order of polynomials that are locally reproduced in $\mathcal{S}(\varphi)$. Recall also that, for $0 < \tau < \infty$, a sequence (a_n) of real numbers is in the Lorentz space $w\ell_\tau := \ell_{\tau,\infty}$ if and only if

$$\#\{n : |a_n| > \epsilon\} \le M^\tau \epsilon^{-\tau} \tag{7.35}$$

for all $\epsilon > 0$. The norm $\|(a_n)\|_{w\ell_\tau}$ is the smallest value of M such that (7.35) holds. Also,

$$\|(a_n)\|_{w\ell_\tau} \le \|(a_n)\|_{\ell_\tau}.$$

Theorem 5 Let $1 < p < \infty$, and $s > 0$, and let $f \in L_p(\mathbb{R})$ and $c_I := c_{I,p}(f)$, $I \in D$, be such that $(c_I)_{I\in D}$ is in $w\ell_\tau$, $1/\tau = s + 1/p$. Then,

$$\sigma_n(f)_p \le Cn^{-s}\|(c_I)\|_{w\ell_\tau}, \quad n = 1, 2, \ldots, \tag{7.36}$$

with the constant C depending only on p and s.

We sketch the proof. We have

$$\#\{I : |c_I| > \epsilon\} \le M^\tau \epsilon^{-\tau}$$

for all $\epsilon > 0$ with $M := \|(c_I)\|_{w\ell_\tau}$. Let $\Lambda_j := \{I : 2^{-j} < |c_I| \le 2^{-j+1}\}$. Then, for each $k = 1, 2, \ldots$, we have

$$\sum_{j=-\infty}^{k} \#\Lambda_j \le CM^\tau 2^{k\tau} \tag{7.37}$$

with C depending only on τ.

Let $S_j := \sum_{I\in\Lambda_j} c_I\psi_I$ and $T_k := \sum_{j=-\infty}^{k} S_j$. Then $T_k \in \Sigma_N$ with $N = CM^\tau 2^{k\tau}$. We have

$$\|f - T_k\|_{L_p(\mathbb{R})} \le \sum_{j=k+1}^{\infty} \|S_j\|_{L_p(\mathbb{R})}. \tag{7.38}$$

We fix $j > k$ and estimate $\|S_j\|_{L_p(\mathbb{R})}$. Since $|c_I| \le 2^{-j+1}$ for all $I \in \Lambda_j$, we have, from Lemma 1 and (7.37),

$$\|S_j\|_{L_p(\mathbb{R})} \le C2^{-j}\#\Lambda_j^{1/p} \le CM^{\tau/p}2^{j(\tau/p-1)}.$$

We therefore conclude from (7.38) that

$$\|f - T_k\|_{L_p(\mathbb{R})} \le CM^{\tau/p} \sum_{j=k+1}^{\infty} 2^{j(\tau/p-1)} \le CM(M2^k)^{\tau/p-1}$$

because $\tau/p - 1 < 0$. In other words, for $N \asymp M^\tau 2^{k\tau}$, we have

$$\sigma_N(f)_p \le CMN^{1/p-1/\tau} = CMN^{-s}.$$

From the monotonicity of σ_n it follows that the last inequality holds for all $N \ge 1$.

Let us note a couple of things about the theorem. First of all there is no restriction on s. However, for large s, the set of functions satisfying $(c_{I,p}(f)) \in w\ell_\tau$ is not a classical smoothness space. We can use the theorem to obtain Jackson inequalities in terms of Besov spaces by using the characterization of Besov spaces by wavelet coefficients. Recall that this characterization applies to $B^s_\tau(L_\tau(\mathbb{R}))$ provided the following two properties hold:

(i) ψ has r vanishing moments with $r > s$
(ii) ψ is in $B^\rho_q(L_\tau)$ for some q and some $\rho > s$.

That is, ψ must have sufficient vanishing moments and sufficient smoothness. Under these assumptions, we have the following result.

Corollary 1 Let $1 < p < \infty$, let $s > 0$ and let $f \in B^s_\tau(L_\tau(\mathbb{R}))$, $1/\tau = s + 1/p$. If ψ satisfies the above two conditions (i) and (ii), then

$$\sigma_n(f)_p \le C|f|_{B^s_\tau(L_\tau(\mathbb{R}))} n^{-s}, \quad n = 1, 2, \ldots, \tag{7.39}$$

with C depending only on p and s.

We have $c_{I,\tau}(f) = c_{I,p}(f)|I|^{1/\tau-1/p} = c_{I,p}(f)|I|^{s/d}$. Thus, from (7.29) we find

$$|f|_{B^s_\tau(L_\tau(\mathbb{R}))} = \|(c_I)\|_{\ell_\tau} \ge \|(c_I)\|_{w\ell_\tau}.$$

Hence (7.39) follows from Theorem 5.

7.5. The Bernstein inequality for n-term wavelet approximation

The following theorem gives the Bernstein inequality which is the companion to (7.39).

Theorem 6 Let $1 < p < \infty$, and let the assumptions of Theorem 5 be valid. If $f = \sum_{I \in \Lambda} c_{I,p}(f)\psi_{I,p}$ with $\#\Lambda \le n$, we have

$$\|f\|_{B^s_\tau(L_\tau(\mathbb{R}))} \le Cn^s \|f\|_{L_p(\mathbb{R})}. \tag{7.40}$$

We sketch the simple proof of this inequality. We first note that, for each $I \in \Lambda$, we have

$$c_I|I|^{-1/p}\chi_I \le S(f),$$

because the left side is one of the terms appearing in the square function $S(f)$. Hence, with $I(x)$ defined as the smallest interval in Λ that contains x, we have, from (7.29),

$$\begin{aligned}
|f|^{\tau}_{B^s_\tau(L_\tau(\mathbb{R}))} &= \int_{\mathbb{R}} \sum_{I\in\Lambda} |c_I|^{\tau}|I|^{-1}\chi_I = \int_{\mathbb{R}} \sum_{I\in\Lambda} c_I^{\tau}|I|^{-\tau/p}\chi_I|I|^{-1+\tau/p}\chi_I \\
&\le C\int_{\mathbb{R}} S(f)^{\tau}\sum_{I\in\Lambda}|I|^{-1+\tau/p}\chi_r \le C\int_{\mathbb{R}} S(f,x)^{\tau}|I(x)|^{-1+\tau/p}\,dx \\
&\le C\left(\int_{\mathbb{R}} S(f,x)^p\right)^{\tau/p}\left(\int_{\mathbb{R}}|I(x)|^{-1}\right)^{1-\tau/p} dx \\
&\le Cn^{1-\tau/p}\|S(f)\|^{\tau}_{L_p(\mathbb{R})} \le Cn^{1-\tau/p}\|f\|^{\tau}_{L_p(\mathbb{R})}.
\end{aligned}$$

7.6. Approximation spaces for n-term wavelet approximation

The Jackson and Bernstein inequalities of the previous sections are equally valid in $\mathbb{R}^d$. The only distinction is that $n^{\pm s}$ should be replaced by $n^{\pm s/d}$. The proofs are identical to the univariate case except for the more elaborate notation needed in the multivariate formulation.

With the Jackson and Bernstein inequalities in hand, we can apply the general machinery of Section 4.3 to obtain the following characterization of the approximation spaces for n-term wavelet approximation. We formulate the results for the multivariate case.

Let $1 < p < \infty$ and $s > 0$ and let $1/\tau := s/d + 1/p$. If ψ satisfies the vanishing moments and smoothness assumptions needed for the Jackson and Bernstein inequalities, then, for any $0 < \gamma < s$ and any $0 < q \le \infty$,

$$\mathcal{A}^{\gamma/d}_q(L_p(\mathbb{R}^d)) = (L_p(\mathbb{R}^d), B^s_\tau(L_\tau(\mathbb{R}^d)))_{\gamma/s,q}. \tag{7.41}$$

Several remarks are in order about (7.41).

Remark 7.5 We have seen the interpolation spaces on the right side of (7.41) before for free knot spline approximation and $d = 1$.

Remark 7.6 For each γ there is one value of q where the right side is a Besov space; namely, when $1/q = \gamma/d + 1/p$, the right side of (7.41) is the Besov space $B^\gamma_q(L_q(\mathbb{R}^d))$ with equivalent norms.

Remark 7.7 There is a description of the interpolation spaces on the right of (7.41) in terms of wavelet coefficients. Namely, a function is in the space $(L_p(\mathbb{R}^d), B^s_\tau(L_\tau(\mathbb{R}^d)))_{\gamma/s,q}$ if and only if $(c_{I,p}(f))_{I\in D}$ is in the Lorentz space

$\ell_{\mu,q}$ where $1/\mu := \gamma/d + 1/p$ and, in fact, we have

$$|f|_{\mathcal{A}_q^{\gamma/d}(L_p)} \asymp \|(c_{I,p}(f))\|_{\ell_{\mu,q}}.$$

This verifies Remark 7.6 that, in the case that $q = \mu$, then $\mathcal{A}_\mu^{\gamma/d}(L_p(\mathbb{R}^d)) = B_\mu^\gamma(\mathbb{R}^d))$ with equivalent norms.

These results can be proved by a slightly finer analysis of n-term wavelet approximation (see Cohen, DeVore and Hochmuth (1997) and Temlyakov (1998*a*))

There is a further connection between n-term approximation and interpolation that we wish to bring out. Let p, s, and τ have the same meaning as above. For each n, let f_n denote a best n-term approximation to f in $L_p(\mathbb{R}^d)$ (which can be shown to exist – see Temlyakov (1998*a*)). It follows from what we have proved and Theorem 3 of Section 4.3 that, for $n = 1, 2, \ldots$, we have

$$K(f, n^{-s}, L_p(\mathbb{R}^d), B_\tau^s(L_\tau(\mathbb{R}^d))) = \|f - f_n\|_{L_p(\mathbb{R}^d)} + n^{-s}|f_n|_{B_\tau^s(L_\tau(\mathbb{R}^d))}.$$

In other words, f_n realizes this K-functional at $t = n^{-s}$.

In summary, n-term wavelet approximation offers an attractive alternative to free knot spline approximation on several counts. In one space dimension (the only case where free knot spline approximation is completely understood), it provides the same approximation efficiency and yet is more easily numerically implementable (as will be discussed subsequently).

7.7. Wavelet decompositions and n-term approximation on domains in $\mathbb{R}^d$

In numerical considerations, we usually deal with functions defined on a finite domain $\Omega \subset \mathbb{R}^d$. The above results can be generalized to that setting in the following way. We assume that the boundary $\partial\Omega$ of of Ω is Lipschitz (it is possible to work under slightly weaker assumptions). Under this assumption, it follows that any function f in the Besov space $B_q^\alpha(\Omega)$ can be extended to all of $\mathbb{R}^d$ in such a way that the extended function Ef satisfies

$$|Ef|_{B_q^\alpha(L_p(\mathbb{R}^d))} \le C|f|_{B_q^\alpha(L_p(\Omega))}. \tag{7.42}$$

We refer the reader to DeVore and Sharpley (1984, 1993) for a discussion of such extensions. The extended function Ef has a wavelet decomposition (7.23) and the results of the previous section can be applied. The n-term approximation to Ef will provide the same order of approximation to f on Ω and one can delete in the approximant all terms corresponding to wavelets that are not active on Ω (that is, all wavelets whose support does not intersect Ω).

While the above remarks concerning extensions are completely satisfactory for theoretical considerations, they are not always easily implementable in numerical settings. Another approach which is applicable in certain settings is the construction of a wavelet basis for the domain Ω. This is particularly suitable in the case of an interval $\Omega \subset \mathbb{R}$. Biorthogonal wavelet bases can be constructed for an interval (see Cohen, Daubechies and Vial (1993)) and can easily be extended to parallelepipeds in $\mathbb{R}^d$ and even polyhedral domains (see Dahmen (1997) and the references therein).

7.8. Thresholding and other numerical considerations

We have thus far concerned ourselves mainly with the theoretical aspects of n-term wavelet approximation. We shall now discuss how this form of approximation is implemented in practice. We assume that approximation takes place on a domain $\Omega \subset \mathbb{R}^d$ which admits a biorthogonal basis as discussed in the previous section. For simplicity of notation, we assume that $d = 1$. We shall also assume that the wavelet decomposition of the target function f is finite and known to us. This provides a good match with certain applications such as image processing. When the wavelet decomposition is not finite, one usually assumes more about f that allows truncation of the wavelet series while retaining the desired level of numerical accuracy.

In the case of approximation in $L_2(\Omega)$, the best n-term approximation to a target function f is obtained by choosing the n terms in the wavelet series (7.20) of f for which the coefficients are largest. A similar strategy applies in the case of $L_p(\mathbb{R})$ approximation. Now, we write f in its wavelet expansion with respect to L_p normalized wavelets (see (7.23)) and choose the n-terms for which $|c_{I,p}(f)|$ is largest. The results of Section 7.4 show that this approximant will provide the Jackson estimates for n-term wavelet approximation. It is remarkable that this simple strategy also gives a near best approximant f_n to f. Temlyakov (1998*a*) has shown that

$$\|f - f_n\|_{L_p(\Omega)} \le C\sigma_n(f)_p, \quad n = 1, 2, \ldots, \tag{7.43}$$

with a constant C independent of f and n.

In numerical implementation, one would like to avoid the expensive sorting inherent in the above description of n-term approximation. This can be done by employing the following strategy known as *thresholding.* We fix the $L_p(\Omega)$ space in which the approximation error is to be measured. Given a tolerance $\epsilon > 0$, we let $\Lambda_\epsilon(f)$ denote the set of all intervals I for which $|c_{I,p}(f)| > \epsilon$ and define the *hard thresholding* operator

$$T_\epsilon(f) := \sum_{I \in \Lambda_\epsilon(f)} c_I(f)\psi_I = \sum_{|c_I(f)|>\epsilon} c_I(f)\psi_I. \tag{7.44}$$

If the target function f is in weak ℓ_τ, with $1/\tau = s + 1/p$, then it follows from the definition of this space that

$$\#(\Lambda_\epsilon) \le M^\tau \epsilon^{-\tau} \tag{7.45}$$

Table 1. *Thresholding values*

Threshold	Number of coefficients	Error
ϵ	$M^\tau \epsilon^{-\tau}$	$M^{\tau/p}\epsilon^{1-\tau/p}$
$M^{-1/(ps)}\eta^{1/(s\tau)}$	$M^{1/s}\eta^{-1/s}$	η
$MN^{-1/\tau}$	N	MN^{-s}

with M the weak ℓ_τ norm of the coefficients. Moreover, arguing as in the proof of Theorem 5, we obtain

$$\|f - T_\epsilon(f)\|_{L_p(\Omega)} \leq CM^{\tau/p}\epsilon^{1-\tau/p}. \tag{7.46}$$

For example, if $\epsilon = MN^{-1/\tau}$, then $\#(\Lambda_\epsilon(f)) \leq N$ and $\|f - T_\epsilon(f)\|_{L_p(\Omega)} \leq CMN^{-s}$. In other words, thresholding provides the Jackson estimate. In this sense, thresholding provides the same approximation efficiency as n-term approximation.

Table 1 records the relationship between thresholding and n-term approximation. Here, $M = |f|_{\ell_{\tau,\infty}}$, ϵ is a thresholding tolerance, η is a prescribed error, and N is a prescribed number of coefficients.

For example, the second row of this table gives bounds on the thresholding parameter and the number of coefficients needed to achieve an error tolerance $\eta > 0$.

Hard thresholding has a certain instability in that coefficients just below the thresholding tolerance are set to zero and those just above are kept intact. This instability can be remedied by *soft thresholding.* Given $\epsilon > 0$, we define

$$s_\epsilon(x) := \begin{cases} 0, & |x| \leq \epsilon, \\ 2(|x| - \epsilon)\operatorname{sign} x, & \epsilon \leq |x| \leq 2\epsilon, \\ x, & |x| > 2\epsilon. \end{cases} \tag{7.47}$$

Then, the soft thresholding operator

$$T'_\epsilon(f) := \sum_{I \in D} s_\epsilon(c_{I,p}(f))\psi_{I,p} \tag{7.48}$$

has the same approximation properties as T_ϵ.

8. Highly nonlinear approximation

Nonlinear wavelet approximation in the form of n-term approximation or thresholding is simple and effective. However, two natural questions arise. How does the effectiveness of this form of approximation depend on the wavelet basis? Secondly, is there any advantage to be gained by adaptively choosing a basis which depends on the target function f? To be reasonable, we would have to limit our search of wavelet bases to a numerically implementable class. An example of such a class is the collection of wavelet packet bases defined in Section 3.7. We call such a class $\mathcal{L}$ of bases a *library.* We shall limit our discussion to approximation in a Hilbert space $\mathcal{H}$ and libraries of orthonormal bases for $\mathcal{H}$. So our problem of nonlinear approximation would be given a target function $f \in \mathcal{H}$, to choose both a basis $B \in \mathcal{L}$ and an n-term approximation to f from this basis. We call such an approximation problem *highly nonlinear* since it involves another layer of nonlinearity in the basis selection.

A closely related form of approximation is n-term approximation from a *dictionary* $\mathbb{D} \subset \mathcal{H}$ of functions. For us, a dictionary will be an arbitrary subset of $\mathcal{H}$. However, dictionaries have to be limited to be computationally feasible. Perhaps the first example of this type of approximation was considered by E. Schmidt (1907), who considered the approximation of functions $f(x,y)$ of two variables by bilinear forms $\sum_{i=1}^{m} u_i(x)v_i(y)$ in $L_2([0,1]^2)$. This problem is closely connected with properties of the integral operator with kernel $f(x,y)$.

We mention some other important examples of dictionaries. In neural networks, one approximates functions of d-variables by linear combinations of functions from the set

$$\{\sigma(a \cdot x + b) : a \in \mathbb{R}^d,\ b \in \mathbb{R}\},$$

where σ is a fixed univariate function. The functions $\sigma(a \cdot x + b)$ are planar waves; also called ridge functions. Usually, σ is required to have additional properties. For example, the sigmoidal functions, which are used in neural networks, are monotone nondecreasing, tend to 0 as $x \to -\infty$, and tend to 1 as $x \to \infty$.

Another example, from signal processing, uses the Gabor functions

$$g_{a,b}(x) := e^{iax} e^{-bx^2}$$

and approximates a univariate function by linear combinations of the elements from

$$\mathbb{D} := \{g_{a,b}(x - c) : a, b, c \in \mathbb{R}\}.$$

Gabor functions are one example of a dictionary of space(time)-frequency atoms. The parameter a serves to position the function $g_{a,b}$ in frequency and c does the same in space. The shape parameter b localizes $g_{a,b}$.

The common feature of these examples is that the family of functions used in the approximation process is redundant. There are many more functions in the dictionary than needed to approximate any target function f. The hope is that the redundancy will increase the efficiency of approximation.

On the other hand, redundancy may slow down the search for good approximations.

Results on highly nonlinear approximation are quite fragmentary and a cohesive theory still needs to be developed. We shall present some of what is known about this theory, both for its usefulness and in the hope of bringing attention to this interesting area.

8.1. Adaptive basis selection

It will be useful to begin by recalling the results of Sections 2 and 5 on n-term approximation using the elements of an orthonormal basis. Let $B := \{\eta_k\}$ be an orthonormal basis for $\mathcal{H}$ and let $\Sigma_n(B)$ denote the functions in $\mathcal{H}$ which can be written as a linear combination of n of the functions η_k, $k = 0, 1, \ldots,$ and further let

$$\sigma_n(f, B) := \sigma_n(f, B)_{\mathcal{H}} := \inf_{S \in \Sigma_n(B)} \|f - S\|_{\mathcal{H}} \tag{8.1}$$

be the corresponding approximation error.

We have seen that the decrease of the approximation errors $\sigma_n(f, B)$ is completely determined by the rearranged coefficients $\langle f, \eta_k \rangle$. As before, we let $\gamma_k(f, B)$ be the kth largest of the absolute values of these coefficients. For example, we have seen that for any $\alpha > 0$, a function f from $\mathcal{H}$ is in $\mathcal{A}^\alpha_\infty$ (*i.e.*, $\sigma_n(f, B) = O(n^{-\alpha})$, $n \to \infty$), if and only if $(\gamma_n(f, B))$ is in weak ℓ_τ (*i.e.*, in $\ell_{\tau,\infty}$) with $\tau := (\alpha + 1/2)^{-1}$. Moreover,

$$\|(\gamma_n(f, B))\|_{\ell_{\tau,\infty}} \asymp |f|_{\mathcal{A}^\alpha_\infty}, \tag{8.2}$$

with constants of equivalency independent of B.

Suppose now that $\mathcal{L} = \{B\}$ is a library of such orthonormal bases B. We define the approximation error

$$\sigma^{\mathcal{L}}_n(f)_{\mathcal{H}} := \inf_{B \in \mathcal{L}} \sigma_n(f, B)_{\mathcal{H}}. \tag{8.3}$$

The approximation classes $\mathcal{A}^\alpha_q(\mathcal{H}, \mathcal{L})$ are defined in the usual way (see Section 4.1). It is of great interest to characterize the approximation classes in concrete settings since this would give us a clear indication of the advantages of adaptive basis selection. A few results are known in discrete settings (see, for instance, Kashin and Temlyakov (1997)). We shall limit ourselves to the following rather trivial observations.

In view of (8.2), we have the upper estimate

$$\sigma^{\mathcal{L}}_n(f)_{\mathcal{H}} \le C n^{-\alpha} \inf_B \|(\gamma_n(f, B))\|_{\ell_{\tau,\infty}} \tag{8.4}$$

with C an absolute constant. Moreover, for any $\alpha > 0$, we have

$$\cap_B \mathcal{A}^\alpha_\infty(\mathcal{H}, B) \subset \mathcal{A}^\alpha_\infty(\mathcal{H}, \mathcal{L}). \tag{8.5}$$

We can interpret these results in the following way. For each basis B, the condition $(\gamma_n(f)) \in \ell_{\tau,\infty}$, $\tau := (\alpha + 1/2)^{-1}$ can be viewed as a smoothness condition on f relative to the basis B. Thus the infimum on the right side of (8.4) can be thought of as the infimum of smoothness conditions relative to the different bases B. Similarly, we can view the classes $\mathcal{A}^\alpha_\infty(\mathcal{H}, B)$ as smoothness classes with respect to the basis B. The right side of (8.5) is an intersection of smoothness classes. Thus, an advantage of optimal basis selection is that we are allowed to take the basis $B \in \mathcal{L}$ in which f is smoothest.

In general (8.4) and (8.5) are not reversible. One can easily construct two basis B_1, B_2, and a target function f so that, as n varies, we alternate between the choices B_1 and B_2 as the best basis selection for varying n. It is less clear whether this remains the case when the library is chosen to have some structure as in the case of wavelet packets. Thus the jury is still out as to whether (8.4) and (8.5) can sometimes be reversed in concrete situations and thereby obtain a characterization of $\mathcal{A}^\alpha_\infty(\mathcal{H}, \mathcal{L})$.

The above discussion for $q = \infty$ generalizes to any $0 < q \leq \infty$.

8.2. Two examples of wavelet libraries

We would be remiss in not mentioning at least a couple of simple examples of libraries of bases that are useful in applications. The understanding of the approximation properties in such examples would go a long way toward understanding highly nonlinear approximation.

Our first example is to generalize the wavelet packets of Section 3.7. Since the situation is completely analogous to that section, we shall be brief. In place of χ_Ω and the Haar function H, we can take any orthogonal scaling function φ and its orthogonal wavelet ψ. We take for $\mathcal{H}$ the space $L_2(\mathbb{R})$. The function φ satisfies the refinement equation (7.8) with refinement coefficients c_k, $k \in \mathbb{Z}$, and likewise the wavelet ψ satisfies (7.10). Therefore, the operators of (3.57) are replaced by

$$A_0 g := \sum_k c_k g(2 \cdot -k); \qquad A_1 g := \sum_k d_k g(2 \cdot -k). \tag{8.6}$$

Then, $A_0(\varphi) = \varphi$, and $A_1(\varphi) = \psi$.

Starting with $\gamma_0 := \varphi$, we generate the functions γ_b and the spaces Γ_b exactly as in Section 3.7. The interpretation using the binary tree of Figure 4 applies and gives the same interpretation of orthonormal bases for $\mathcal{S}_m(\varphi)$. These bases form the library of wavelet packet bases. For further discussion of wavelet packet libraries and their implementation, we refer the reader to Wickerhauser (1994).

For our second example, we take $\mathcal{H} = L_2(\mathbb{R}^2)$ and again consider a compactly supported, refinable function $\varphi \in L_2(\mathbb{R})$ with orthonormal shifts and its corresponding orthogonal wavelet ψ. We define $\psi^0 := \varphi$, $\psi^1 := \psi$. To each vertex e of the unit square $[0,1]^2$, each $j = (j_1, j_2) \in \mathbb{Z}^2$, $k = (k_1, k_2) \in \mathbb{Z}^2$, we associate the function

$$\psi^e_{j,k}(x_1, x_2) := 2^{(k_1+k_2)/2}\psi^{e_1}(2^{k_1}x_1 - j_1)\psi^{e_2}(2^{k_2}x_2 - j_2). \tag{8.7}$$

Each of these functions has $L_2(\mathbb{R}^2)$ norm one. We let $\mathcal{L}$ denote the library of all complete orthonormal systems which can be made up from the functions in (8.7). In particular $\mathcal{L}$ will include the usual wavelet bases given in (7.19) and the hyperbolic basis (7.21), which is the tensor product of the univariate wavelet basis.

As a special case of the above library consider $\varphi = \chi_{[0,1)}$ and $\psi = H$, with H the Haar function. We approximate functions defined on the unit square $\Omega := [0,1)^2$. The library $\mathcal{L}$ includes bases of the following type. We can take an arbitrary partition $\mathcal{P}$ of Ω into dyadic rectangles R. On each R we can take a standard or hyperbolic wavelet Haar basis. This library of bases is also closely related to the CART algorithm studied by Donoho (1997).

8.3. *Approximation using n-terms from a dictionary*

Suppose that $\mathbb{D}$ is a dictionary of functions from $\mathcal{H}$. It will be convenient to assume (without loss of generality in n-term approximation) that each $g \in \mathbb{D}$ has norm one ($\|g\|_{\mathcal{H}} = 1$) and that $-g \in \mathbb{D}$ whenever $g \in \mathbb{D}$. One particular example of a dictionary is to start with an orthonormal basis B for $\mathcal{H}$ and to take $\mathbb{D} := \{\pm b : b \in \mathcal{H}\}$. We shall say that this is the *dictionary generated by* B. For each $n \in \mathbb{N}$, we let $\Sigma_n := \Sigma_n(\mathbb{D})$ denote the collection of all functions in H which can be expressed as a linear combination of at most n elements of $\mathbb{D}$. Thus each function $S \in \Sigma_n$ can be written in the form

$$S = \sum_{g\in\Lambda} c_g g, \quad \Lambda \subset \mathbb{D}, \quad \#\Lambda \le n, \tag{8.8}$$

with the $c_g \in \mathbb{R}$. It may be possible to write an element from $\Sigma_n(\mathbb{D})$ in the form (8.8) in more than one way.

For a function $f \in H$, we define its approximation error

$$\sigma_n(f) := \sigma_n(f, \mathbb{D})_{\mathcal{H}} := \inf_{S\in\Sigma_n} \|f - S\|_{\mathcal{H}}. \tag{8.9}$$

We shall be interested in estimates for σ_n (from above and below). For this purpose, we introduce the following way of measuring smoothness with respect to the dictionary $\mathbb{D}$.

For a general dictionary $\mathbb{D}$, and for any $\tau > 0$, we define the class of functions

$$\mathcal{K}^o_\tau(\mathbb{D}, M) := \left\{ f = \sum_{g\in\Lambda} c_g g : \Lambda \subset \mathbb{D},\ \#\Lambda < \infty \quad \text{and} \quad \sum_{g\in\Lambda} |c_g|^\tau \le M^\tau \right\},$$

and we define $K_\tau(\mathbb{D}, M)$ as the closure (in $\mathcal{H}$) of $\mathcal{K}_\tau^o(\mathbb{D}, M)$. Furthermore, we define $\mathcal{K}_\tau(\mathbb{D})$ as the union of the classes $\mathcal{K}_\tau(\mathbb{D}, M)$ over all $M > 0$. For $f \in \mathcal{K}_\tau(\mathbb{D})$, we define the semi-norm

$$|f|_{\mathcal{K}_\tau(\mathbb{D})} \tag{8.10}$$

as the infimum of all M such that $f \in \mathcal{K}_\tau(\mathbb{D}, M)$. Notice that, when $\tau = 1$, $\mathcal{K}_1$ is the class of functions which are a convex combination of the functions in $\mathbb{D}$.

The case when $\mathbb{D}$ is generated by a basis B is instructive for the results that follow. In this case, n-term approximation from $\mathbb{D}$ is the same as n-term approximation from B which we have analysed in Sections 2 and 5. We have shown that if $1/\tau = \alpha + 1/2$, then f is in the approximation class $\mathcal{A}_\tau^\alpha(\mathbb{D})$ if and only if

$$\sum_k |\langle f, h_k \rangle|^\tau$$

is finite and this last expression is equivalent to $|f|^\tau_{\mathcal{A}_\tau(\mathcal{B})}$. In particular, this shows that

$$\sigma_n(f, \mathbb{D})_{\mathcal{H}} \le Cn^{-\alpha} |f|_{\mathcal{K}_\tau(\mathbb{D})} \tag{8.11}$$

in the special case that $\mathbb{D}$ is given by an orthonormal basis B.

We are now interested in understanding whether (8.11) holds for more general dictionaries $\mathbb{D}$. The results in the following section will show that (8.11) is valid for a general dictionary provided $\alpha \ge 1/2$. The first result of this type was due to Maurey (see Pisier (1980)) who showed that, in the case $\alpha = 1/2$, (8.11) is valid for any dictionary. An iterative algorithm to generate approximants from $\Sigma_n(\mathbb{D})$ that achieves this estimate (for $\alpha = 1/2$) was given by Jones (1992). For $\alpha > 1/2$, (8.11) is proved in DeVore and Temlyakov (1996). For $\alpha < 1/2$ $(1 \le \tau \le 2)$ there seems to be no obvious analogue of (8.11) for general dictionaries.

8.4. Greedy algorithms

The estimate (8.11) can be proved for a general dictionary by using *greedy algorithms* (also known as *adaptive pursuit*). These algorithms are often used computationally as well. We shall mention three examples of greedy algorithms and analyse their approximation properties. In what follows, $\|\cdot\|$ is the norm in $\mathcal{H}$ and $\langle \cdot, \cdot \rangle$ is the inner product in $\mathcal{H}$.

The first algorithm, known as the *pure greedy algorithm*, can be applied for any dictionary $\mathbb{D}$. Its advantage is its simplicity. It begins with a target function f and successively generates approximants $G_m(f) \in \Sigma_m(\mathbb{D})$, $m = 1, 2, \ldots$. In the case that $\mathbb{D}$ is generated by an orthonormal basis B, $G_m(f)$ is a best m-term approximation to f.

If $f \in H$, we let $g = g(f) \in \mathbb{D}$ denote an element from $\mathbb{D}$ which maximizes $\langle f, g \rangle$:

$$\langle f, g(f) \rangle = \sup_{g \in D} \langle f, g \rangle. \tag{8.12}$$

We shall assume for simplicity that such a maximizer exists; if not, suitable modifications are necessary in the algorithms that follow. We define

$$G(f) := G(f, \mathbb{D}) := \langle f, g(f) \rangle g(f)$$

and

$$R(f) := R(f, D) := f - G(f).$$

Then, $G(f)$ is a best one-term approximation to f from $\mathbb{D}$ and $R(f)$ is the residual of this approximation.

Pure greedy algorithm. Initially, we set $R_0(f) := R_0(f, D) := f$ and $G_0(f) := 0$. Then, for each $m \geq 1$, we inductively define

$$\begin{aligned} G_m(f) &:= G_m(f, \mathbb{D}) := G_{m-1}(f) + G(R_{m-1}(f)), \\ R_m(f) &:= R_m(f, \mathbb{D}) := f - G_m(f) = R(R_{m-1}(f)). \end{aligned}$$

The pure greedy algorithm converges to f for each $f \in \mathcal{H}$ (see Davis, Mallat and Avellaneda (1997)). This algorithm is greedy in the sense that at each iteration it approximates the residual $R_m(f)$ as best possible by a single function from $\mathbb{D}$. If $\mathbb{D}$ is generated by an orthonormal basis, then it is easy to see that $G_m(f)$ is a best m-term approximation to f from $\mathbb{D}$ and

$$\sigma_m(f, \mathcal{B})_{\mathcal{H}} = \|f - G_m(f)\|_{\mathcal{H}} = \|R_m(f)\|_{\mathcal{H}}.$$

However, for general dictionaries, this is not the case, and in fact the approximation properties of this algorithm are somewhat in doubt, as we shall now describe.

For a general dictionary $\mathbb{D}$, the best estimate (proved in DeVore and Temlyakov (1996)) known for the pure greedy algorithm is that for each $f \in K_1(\mathbb{D})$ we have

$$\|f - G_m(f)\|_{\mathcal{H}} \leq |f|_{K_1(\mathbb{D})} m^{-1/6}. \tag{8.13}$$

Moreover, the same authors have given an example of a dictionary $\mathbb{D}$ and a function f which is a linear combination of two elements of $\mathbb{D}$ such that

$$\|f - G_m(f)\|_{\mathcal{H}} \geq C m^{-1/2}, \tag{8.14}$$

with C an absolute constant. In other words, for the simplest of functions f (which are in all of the smoothness classes $K_\tau(\mathbb{D})$), the pure greedy algorithm provides approximation of at most order $O(m^{-1/2})$. Thus, this algorithm cannot provide estimates like (8.11) for $\alpha > 1/2$.

There are modifications of the pure greedy algorithm with more favourable approximation properties. We mention two of these: the *relaxed greedy algorithm* and the *orthogonal greedy algorithm.*

Relaxed greedy algorithm. We define $R_0^r(f) := R_0^r(f,\mathbb{D}) := f$ and $G_0^r(f) := G_0^r(f,D) := 0$. For $m = 1$, we define $G_1^r(f) := G_1^r(f,\mathbb{D}) := G_1(f)$ and $R_1^r(f) := R_1^r(f,\mathbb{D}) := R_1(f)$. As before, for a function $h \in H$, let $g = g(h)$ denote a function from $\mathbb{D}$ which maximizes $\langle h, g\rangle$. Then, for each $m \geq 2$, we inductively define

$$\begin{aligned} G_m^r(f) &:= G_m^r(f,\mathbb{D}) := \left(1-\frac{1}{m}\right) G_{m-1}^r(f) + \frac{1}{m} g(R_{m-1}^r(f)), \\ R_m^r(f) &:= R_m^r(f,\mathbb{D}) := f - G_m^r(f). \end{aligned}$$

Thus, the relaxed greedy algorithm is less greedy than the pure greedy algorithm. It makes only modest use of the greedy approximation to the residual at each step. The number $1/m$ appearing at each step is the relaxation parameter.

Algorithms of this type appear in Jones (1992), who showed that the relaxed greedy algorithm provides the approximation order

$$\|f - G_m^r(f)\| \leq C m^{-1/2}, \quad m = 1, 2, \ldots. \tag{8.15}$$

for any $f \in \mathcal{K}_1(\mathbb{D})$. Unfortunately, this estimate requires the knowledge that $f \in K_1(\mathbb{D})$. In the event that this information is not available – as would be the case in most numerical considerations – the choice of relaxation parameter $1/m$ is not appropriate.

The relaxed greedy algorithm gives a constructive proof that (8.11) holds for a general dictionary $\mathbb{D}$ in the case $\alpha = 1/2$. We shall discuss how to prove (8.11) in the next section. But first we want to put out on the table another variant of the greedy algorithm, called the orthogonal greedy algorithm, which removes some of the objections to the choice of the relaxation parameter in the relaxed greedy algorithm.

To motivate the orthogonal greedy algorithm, let us return for a moment to the pure greedy algorithm. This algorithm chooses functions $g_j := G(R_j(f))$, $j = 1, \ldots, m$, to use in approximating f. One of the deficiencies of the algorithm is that it does not provide the best approximation from the span of $g_1, \ldots, g_m$. We can remove this deficiency as follows.

If H_0 is a finite-dimensional subspace of H, we let P_{H_0} be the orthogonal projector from H onto H_0, that is, $P_{H_0}(f)$ is the best approximation to f from H_0.

Orthogonal greedy algorithm. We define $R_0^o(f) := R_0^o(f,D) := f$ and $G_0^o(f) := G_0^o(f,D) := 0$. Then, for each $m \geq 1$, we inductively define

$$H_m := H_m(f) := \operatorname{span}\{g(R_0^o(f)), \ldots, g(R_{m-1}^o(f))\},$$

$$\begin{aligned} G^o_m(f) &:= G^o_m(f,D) := P_{H_m}(f), \\ R^o_m(f) &:= R^o_m(f,D) := f - G^o_m(f). \end{aligned}$$

Thus, the distinction between the orthogonal greedy algorithm and the pure greedy algorithm is that the former takes the best approximation by linear combinations of the functions $G(R_0(f)), \ldots, G(R_{m-1}(f))$ available at each iteration. The first step of the orthogonal greedy algorithm is the same as the pure greedy algorithm. However, they will generally be different at later steps.

DeVore and Temlyakov (1996) have shown (as will be discussed in more detail in the next section) that the orthogonal greedy algorithm satisfies the estimate

$$\|f - G^o_m(f,D)\|_{\mathcal{H}} \le |f|_{K_1(\mathbb{D})} m^{-1/2}. \tag{8.16}$$

Thus, the orthogonal greedy algorithm gives another constructive proof that (8.11) holds for a general dictionary $\mathbb{D}$. However, one should note that the orthogonal greedy algorithm is computationally more expensive in the computation of the best approximation from H_m.

From (8.16), it is easy to prove the following theorem from DeVore and Temlyakov (1996).

Theorem 7 Let $\mathbb{D}$ be any dictionary, let $\alpha \ge 1/2$ and $1/\tau = \alpha + 1/2$. If $f \in K_\tau(\mathbb{D})$, then

$$\sigma_m(f,\mathbb{D})_{\mathcal{H}} \le C|f|_{K_\tau(D)} m^{-\alpha}, \quad m = 1, 2, \ldots, \tag{8.17}$$

where C depends on τ if τ is small.

We sketch the simple proof. It is enough to prove (8.17) for functions f which are a finite sum $f = \sum_j c_j g_j$, $g_j \in D$, with $\sum_j |c_j|^\tau \le M^\tau$. Without loss of generality we can assume that the c_j are positive and nonincreasing. We let $s_1 := \sum_{j=1}^n c_j g_j$ and $R_1 := f - s_1 = \sum_{j>n} c_j g_j$. Now,

$$c_n^\tau \le \frac{1}{n} \sum_{j=1}^n |c_j|^\tau \le \frac{M^\tau}{n}, \quad n = 1, 2, \ldots.$$

Hence, $c_j \le M n^{-1/\tau}$, $j > n$ and it follows that

$$\sum_{j>n} c_j = \sum_{j>n} c_j^{1-\tau} c_j^\tau \le M^{1-\tau} n^{1-1/\tau} \sum_{j>n} c_j^\tau \le M n^{1-1/\tau}.$$

This gives that R_1 is in $K_1(\mathbb{D}, Mn^{1-1/\tau})$. Using (8.16), there is a function s_2 which is a linear combination of at most n of the $g \in \mathbb{D}$ such that

$$\|f - (s_1 + s_2)\| = \|R_1 - s_2\| \le 2Mn^{1-1/\tau} n^{-1/2} = 2Mn^{-\alpha},$$

and (8.17) follows.

8.5. Further analysis of greedy algorithms

To determine the performance of a greedy algorithm, we try to estimate the decrease in error provided by one step of the pure greedy algorithm. Let $\mathbb{D}$ be an arbitrary dictionary. If $f \in \mathcal{H}$ and

$$\rho(f) := \langle f, g(f)\rangle / \|f\|_{\mathcal{H}}, \tag{8.18}$$

where as before $g(f) \in \mathbb{D}$ satisfies

$$\langle f, g(f)\rangle = \sup_{g \in D} \langle f, g\rangle,$$

then

$$R(f)^2 = \|f - G(f)\|_{\mathcal{H}}^2 = \|f\|_{\mathcal{H}}^2(1 - \rho(f)^2). \tag{8.19}$$

The larger $\rho(f)$ is, the better the decrease of the error in the pure greedy algorithm.

The following theorem from DeVore and Temlyakov (1996) estimates the error in approximation by the orthogonal greedy algorithm.

Theorem 8 Let $\mathbb{D}$ be an arbitrary dictionary in $\mathcal{H}$. Then, for each $f \in \mathcal{K}_1(\mathbb{D}, M)$ we have

$$\|f - G_m^o(f, \mathbb{D})\|_{\mathcal{H}} \le M m^{-1/2}. \tag{8.20}$$

Proof. We can assume that $M = 1$ and that f is in $\mathcal{K}_1^o(\mathbb{D}, 1)$. We let $f_m^o := R_m^o(f)$ be the residual at step m of the orthogonal greedy algorithm. Then, from the definition of this algorithm, we have

$$\|f_{m+1}^o\|_{\mathcal{H}} \le \|f_m^o - G(f_m^o, \mathbb{D})\|_{\mathcal{H}}.$$

Using (8.19), we obtain

$$\|f_{m+1}^o\|_{\mathcal{H}}^2 \le \|f_m^o\|_{\mathcal{H}}^2(1 - \rho(f_m^o)^2). \tag{8.21}$$

Since $f \in \mathcal{K}_1^o(\mathbb{D}, 1)$, we can write $f = \sum_{k=1}^N c_k g_k$ with $c_k > 0$, $k = 1, \ldots, N$, and $\sum_{k=1}^N c_k = 1$. By the definition of the orthogonal greedy algorithm, $G_m^o(f) = P_{H_m} f$, and hence $f_m^o = f - G_m^o(f)$ is orthogonal to $G_m^o(f)$. Using this, we obtain

$$\|f_m^o\|_{\mathcal{H}}^2 = \langle f_m^o, f\rangle = \sum_{k=1}^N c_k \langle f_m^o, g_k\rangle \le \rho(f_m^o)\|f_m^o\|_{\mathcal{H}}.$$

Hence,

$$\rho(f_m^o) \ge \|f_m^o\|_{\mathcal{H}}.$$

Using this inequality in (8.21), we find

$$\|f_{m+1}^o\|_{\mathcal{H}}^2 \le \|f_m^o\|_{\mathcal{H}}^2(1 - \|f_m^o\|_{\mathcal{H}}^2).$$

It is now easy to derive from this that $\|f_m^o\|_{\mathcal{H}}^2 \le 1/m$. $\square$

9. Lower estimates for approximation: n-widths

In this section, we shall try to understand better the limitations of linear and nonlinear approximation. Our analysis thus far has relied on the concept of approximation spaces. For example, we started with a sequence of linear or nonlinear spaces X_n and defined the approximation classes A^α_∞ consisting of all functions that can be approximated with accuracy $O(n^{-\alpha})$ by the elements of X_n. We have stressed the importance of characterizing these approximation spaces in terms of something more classical such as smoothness spaces and in fact we have accomplished this in many settings. In this way, we have seen among other things that the classical nonlinear methods of approximation (like free knot splines or n-term approximation) outperform their counterparts in linear approximation.

To make these points more clearly, let us recall perhaps the simplest setting for the results we have presented. Namely, we consider $L_2(\Omega)$-approximation, $\Omega := [0, 1)$, using the Haar wavelet H. Every function in $L_2(\Omega)$ has a decomposition

$$f = a\chi_{[0,1)} + \sum_{I \in D([0,1))} c_I(f) H_I, \quad c_I(f) := \langle f, H_I \rangle, \tag{9.1}$$

with the H_I normalized in $L_2(\Omega)$ and a the average of f over Ω.

In linear approximation, we take as our approximation to f the partial sum of the series (9.1) consisting of the first n terms with respect to the natural order of dyadic intervals (this is the ordering which gives priority first to size and then to orientation from left to right). For this approximation, we have seen that f is approximated in the norm of $L_2(\Omega)$ with accuracy $O(n^{-\alpha})$, $0 < \alpha < 1/2$, if and only if $f \in \mathrm{Lip}(\alpha, L_2(\Omega))$. The upper limit of $1/2$ for the characterization comes about because the Haar wavelet H is in $\mathrm{Lip}(1/2, L_2(\Omega))$ but in no higher-order Lipschitz space.

In nonlinear approximation, we approximated f by taking the partial sum of (9.1) which consists of the n terms with largest coefficients. It is clear that this form of approximation is at least as efficient as the linear approximation. We have seen that we can characterize the functions approximable with order $O(n^{-\alpha})$ by conditions on the wavelet coefficients that roughly correspond to smoothness of order α in L_τ with $1/\tau = \alpha + 1/2$ (see Remark 7.7 on page 118). In fact, it is easy to see that each function in $\mathrm{Lip}(\alpha, L_\gamma(\Omega))$ with $\gamma > \tau$ is approximated with this order by the nonlinear method.

Is this really convincing proof that nonlinear methods outperform linear methods? Certainly it shows that this nonlinear wavelet method outperforms the linear wavelet method. However, what can prevent some other linear method (not the wavelet method just described) from also containing the $\mathrm{Lip}(\alpha, L_\gamma(\Omega))$ classes in its $\mathcal{A}^\alpha_\infty$? There is a way of deciding whether this is possible by using the concept of n-widths, which we now describe.

There are many definitions of n-widths. For our purpose of measuring the performance of linear methods, the following definition of Kolmogorov is most appropriate. If X is a Banach space and K is a compact subset of X, we define

$$d_n(K) := \inf_{\dim(X_n)=n} \sup_{f\in K} E(f, X_n)_X, \tag{9.2}$$

where the infimum is taken over all n-dimensional linear spaces and of course $E(f, X_n)_X$ is the error in approximating f by the elements of X_n in the norm of X. So d_n measures the performance of the best n-dimensional space on the class K.

To answer our question posed above, we would like to know the n-width of the unit ball U_γ^α of $\mathrm{Lip}(\alpha, L_\gamma(\Omega))$ in $L_2(\Omega)$ (this unit ball is a compact subset of $L_2(\Omega)$ provided $\gamma > \tau = (\alpha + 1/2)^{-1}$). The Kolmogorov n-widths of Besov and Lipschitz balls are known and can be found, for example, in Chapter 14 of Lorentz, von Golitschek and Makovoz (1996). We shall limit our discussion to the results relevant to our comparison of linear and nonlinear approximation.

We fix the space $L_p(\Omega)$, $\Omega = [0, 1)$, where approximation is to take place. While we shall discuss only univariate approximation in this section, all results on n-widths hold equally well in the multivariate case. In Figure 8, we use our usual interpretation of smoothness spaces as points in the upper right quadrant to give information about the n-widths of the unit balls $U_r^\alpha(L_q(\Omega))$ of the Besov spaces $B_r^\alpha(L_q(\Omega))$. The shaded region of that figure corresponds to those Besov spaces whose unit ball has n-width $O(n^{-\alpha})$.

Several remarks will complete our understanding of Figure 8 and what it tells us regarding linear and nonlinear methods.

Remark 9.1 The n-width of $U_r^\alpha(L_q(\Omega))$ is never better than $\mathcal{O}(n^{-\alpha})$. In other words, once we know the smoothness index α of the space, this provides a limit as to how effective linear methods can be.

Remark 9.2 The sets $U_r^\alpha(L_p(\Omega))$ which correspond to the Besov spaces on the linear line (L) always have Kolmogorov n-width $\asymp n^{-\alpha}$. Thus, for these spaces the classical methods of approximation such as polynomials or fixed knot splines provide the best order of approximation for these classes.

Remark 9.3 For approximation in $L_p(\Omega)$, with $2 < p \le \infty$, and for $\alpha > 1/p$ there is always a certain range of q (depicted in Figure 8 by the shaded region) where the Kolmogorov n-width of $U_r^\alpha(L_q\Omega))$ is still $\asymp n^{-\alpha}$. This is a rather surprising result of Kashin (1977). We know that classical methods cannot provide this order of approximation because we have characterized their approximation classes $\mathcal{A}_\infty^\alpha(L_p(\Omega))$ and these classes do not contain general functions from $U_r^\alpha(L_q(\Omega))$ once $q < p$. So there are linear spaces with super approximation properties (which to a limited extent

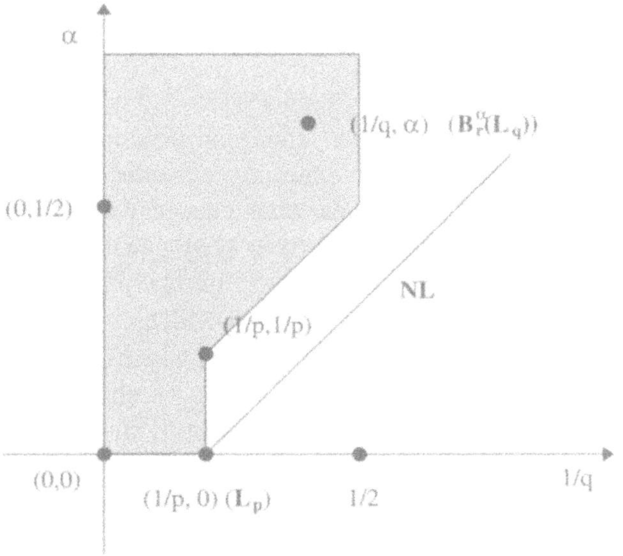

Fig. 8. Shaded region gives $(1/q, \alpha)$ such that $U^\alpha_r(L_q([0,1))$ has n-width of order $O(n^{-\alpha})$ in L_p, $2 \le p \le \infty$

mimic the advantages of nonlinear approximation). What are these spaces? Unfortunately these spaces are not known constructively. They are usually described by probabilistic methods. So, while their existence is known, we cannot put our hands on them and definitely can't use them numerically.

Remark 9.4 The range of q where the super linear spaces come into play always falls well short of the nonlinear line. Thus nonlinear methods always perform better than linear methods, in the sense that their approximation classes are strictly larger.

Remark 9.5 We have not depicted the case $p \le 2$ since in this case there are no Besov balls $U^\alpha_r(L_q(\Omega))$ which have the order $O(n^{-\alpha})$ save for the case $q \ge p$ which we already know from the classical linear theory.

Remark 9.6 Now, here is an important point that is sometimes misunderstood. It is not always safe to say that, for a specific target function, nonlinear methods will perform better than linear methods. Let us forget for a moment the super linear theory since it is not relevant in numerical situations anyway. Given f, there will be a maximal value of α – let's call it α_L – for which f is in $B^\alpha_\infty(L_p(\Omega))$. Then, we know that approximation from classical n-dimensional linear spaces will achieve an approximation rate $O(n^{-\alpha_L})$, but they can do no better. Let us similarly define α_N as the largest value of α for which f is in the space $B^{\alpha_N}_\infty(L_\gamma)$ for some $\gamma > (\alpha + 1/p)^{-1}$; then nonlinear methods such as n-term wavelet approximation will provide an approximation error $O(n^{-\alpha_N})$. If $\alpha_N > \alpha_L$, then certainly nonlinear methods outperform linear methods (at least asymptotically as $n \to \infty$). However, if $\alpha_L = \alpha_N$ then there is no gain in using nonlinear methods to approximate the target function f.

The questions we have posed for linear approximation can likewise be posed for nonlinear methods. For example, consider univariate approximation in $L_p(\Omega)$, $\Omega = [0, 1)$. We know that classical nonlinear methods approximate functions in $B^\alpha_\infty(L_\gamma)$, $\gamma > (\alpha + 1/p)^{-1}$ with accuracy $n^{-\alpha}$. But can it be that other nonlinear methods do better? Questions of this type can be answered by introducing *nonlinear* n-widths.

There are several definitions of nonlinear n-widths, the most prominent of which is the Alexandrov width. However, we shall only be concerned with the manifold n-width, which was introduced by DeVore, Howard and Micchelli (1989), since it fits best with numerical methods. Let X be the space in which we shall measure error (we shall assume that X is equipped with a norm $\|\cdot\|_X$). By a (nonlinear) manifold $\mathcal{M}_n$ of dimension n, we shall mean the image of a continuous mapping $M : \mathbb{R}^n \to X$. (Thus our manifolds are *not* the manifolds of differential topology.) We shall approximate using the elements of $\mathcal{M}_n$. For each compact set $K \subset X$, we define the manifold width

$$\delta_n(K, X) := \inf_{M,a} \sup_{f \in K} \|f - M(a(f))\|_X, \tag{9.3}$$

where the infimum is taken over all manifolds of dimension n and all continuous parameter mappings $a : K \to \mathbb{R}^n$.

We make a couple of remarks which may help explain the nature of the width δ_n.

Remark 9.7 For any compact set, we can select a countable number of points which are dense in K and construct a one-dimensional manifold (a continuous piecewise linear function of $t \in \mathbb{R}$) passing through each of these points. Thus, without the restriction that the approximation arises through a continuous parameter selection a, we would always have $\delta_n(K) = 0$.

Remark 9.8 The function a also guarantees stability of the approximation process. If we perturb f slightly the continuity of a guarantees that the parameters $a(f)$ only change slightly.

The nonlinear widths of each of the Besov balls $U^\alpha_r(L_\tau(\Omega))$ in the space $L_p(\Omega)$ are known. If this ball is a compact subset of $L_p(\Omega)$, then the nonlinear n-width is

$$\delta_n(U^\alpha_r(L_\tau(\Omega)) \asymp n^{-\alpha}, \quad n \to \infty. \tag{9.4}$$

This shows, therefore, that we cannot obtain a better approximation order for these balls than what we obtain via n-term wavelet approximation. However, n-term approximation, as it now stands, is not described as one of the procedures appearing in (9.3). However, this requires only a little massaging. Using certain results from topology, DeVore, Kyriazis, Leviatan and Tikhomirov (1993) have shown nonlinear approximation in terms of soft thresholding of the coefficients can be used to describe an approximation process which provides the upper estimate in (9.3). We shall not go into the details of this construction.

On the basis of the evidence we have thus far provided about linear and nonlinear methods, is it safe to conclude that the nonlinear methods such as n-term wavelet approximation are superior to other nonlinear methods? The answer is definitely not. We only know that if we classify functions according to their Besov smoothness, then for this classification no other nonlinear methods can do better. On the other hand, each nonlinear method will have its approximation classes and these need not be Besov spaces. A case in point where we have seen this is the case of approximation in a Hilbert space by n terms of an orthonormal basis. In this setting, we have seen that the approximation classes depend on the basis and that smoothness of a function for this type of approximation should be viewed as decay of the coefficients with respect to the basis. This will generally not be a Besov space. In other words, there are other ways to measure smoothness in which wavelet performance will not be optimal.

Our discussion thus far has not included lower estimates for optimal basis selection or n-term approximation from a dictionary. We do not know of a concept of widths that properly measures the performance of these highly nonlinear methods of approximation. This is an important open problem in nonlinear approximation because it would shed light on the role of such methods in applications such as image compression (see the section below).

Finally, we want to mention the *VC* dimension of Vapnik and Chervonenkis (see the book of Vapnik (1982)). The VC dimension measures the size of nonlinear sets of functions by looking at the maximum number of sign alternations of its elements. It has an important role in statistical estimation but has not been fully considered in approximation settings. The paper of Mairov and Ratasby (1998) uses VC dimension to define a new n-width and analyses the widths of Besov balls. Their results are similar to those above for nonlinear widths.

10. Applications of nonlinear approximation

Nonlinear methods have found many applications both numerical and analytical. The most prominent of these have been to image processing, statistical estimation, and the numerical and analytic treatment of differential equations. There are several excellent accounts of these matters: see Mallat (1998) for image processing; Donoho and Johnstone (1994), Donoho, Johnstone, Kerkyacharian and Picard (1996) for statistical estimation; Dahmen

(1997) and Dahlke, Dahmen and DeVore (1997) for applications to PDEs. We shall limit ourselves to a broad outline of the use of nonlinear approximation in image processing and PDEs.

10.1. Image processing

We shall discuss the processing of digitized grey-scale images. Signals, colour images, and other variants can be treated similarly but have their own peculiarities. A digitized grey-scale image $\mathcal{I}$ is an array of numbers (called *pixel values*) which represent the grey scale. We assume 8-bit grey-scale images, which means the pixel values range from 0 (black) to 255 (white). We shall also assume (only for the sake of specificity) that the array consists of 1024×1024 pixel values. Given such images, the generic problems of image processing are: compression, noise reduction, feature extraction, and object recognition.

To utilize techniques from mathematical analysis in image processing, it is useful to have a model for images as functions. One such model is to assume that the pixel values are obtained from an underlying intensity function f by averaging over dyadic squares. In our case, the dyadic squares are those in $D_m := D_m(\Omega)$, $\Omega := [0,1)^2$, with $m = 10$, thus resulting in 1024 squares and the same number of pixel values. We denote the pixel values by

$$p_I = \frac{1}{|I|}\int_I f(x)\,\mathrm{d}x, \quad I \in D_m. \tag{10.1}$$

Of course, there is more than one function f with these pixel values. Since the pixel values are integers, another possibility would be to view them as integer quantizations of the averages of f. In other words, other natural models may be proposed. But the main point is to visualize the image as obtained from an intensity function f.

Compression

A grey-scale image $\mathcal{I}$ of the type described is represented by its pixel array, $\mathcal{I} \sim (p_I)_{I\in D_m}$, which is a file of size one megabyte. For the purposes of transmission, storage, or other processing, we would like to represent this image with fewer bits. This can be accomplished in two ways. *Lossless encoding* of the image uses techniques from information theory to encode the image in fewer bits. The encoded image is identical to the original; in other words the process of encoding is reversible. *Lossy compression* replaces the original image by an approximation. This allows for more compression but with the potential loss of fidelity. Lossless encoders will typically result in compression factors of 2–1 which means the original file is reduced by half. Much higher compression factors can be obtained in lossy compression with no perceived degradation of the original image (images compressed by factors of 10–1 are typically indistinguishable from the original).

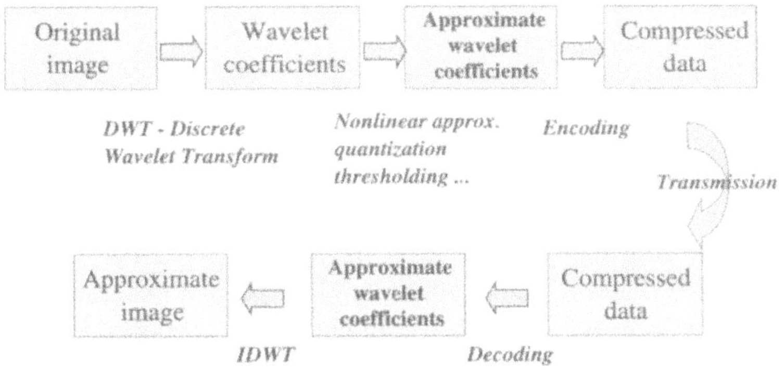

Fig. 9. Schematic of a typical wavelet-based compression algorithm

We can use the techniques of approximation theory and functional analysis for lossy compression. We view the intensity function as our target function and consider methods for approximating it from the pixel values. Wavelet-based methods proceed as follows.

We choose a multivariate scaling function φ and represent the image by the series

$$\mathcal{I} \sim \sum_{I \in D_m} p_I \varphi_I. \tag{10.2}$$

Here p_I, $I \in D_m$, are some appropriate extension of the pixel values. (When using wavelets other than Haar, one has to do some massaging near the boundary, which we shall not discuss.) We use the Fast Wavelet Transform to convert pixel values to wavelet coefficients. This gives the wavelet representation of $\mathcal{I}$:

$$\mathcal{I} \sim P_0 + \sum_{k=0}^{m-1} \sum_{I \in D_k} \sum_{e \in E} a_I^e \psi_I^e, \tag{10.3}$$

where P_0 consists of all the scaling function terms from level 0, and the other notation conforms to our multivariate wavelet notation of Section 7 (see (7.20)).

The problem of image compression is then viewed as nonlinear wavelet approximation and the results of Section 7 can be employed. Figure 9 gives a schematic of typical compression algorithms. We use thresholding to obtain a compressed file $(\tilde{a}_I^e)$ of wavelet coefficients which correspond to a compressed image $\tilde{I}$. The compressed coefficient file is further compressed using a lossless encoder. The encoded compressed file is our compressed representation of the original image. We can reverse the process. From the encoded compressed file of wavelet coefficients, we apply a decoder and then the Inverse Fast Wavelet Transform to obtain the pixel values of the compressed image $\tilde{I}$. The following remarks will help clarify the role of nonlinear approximation in this process.

Remark 10.1 We apply nonlinear wavelet approximation in the form of thresholding (Section 7.8). We choose a value of p (corresponding to the L_p space in which we are to measure error) and retain all coefficients that satisfy $\|a_I^e \psi_I^e\|_{L_p} > \epsilon$. We replace by zero all coefficients for which $\|a_I^e \psi_I^e\|_{L_p} \leq \epsilon$. Soft thresholding can also be used in place of hard thresholding. This gives compressed wavelet coefficients $\bar{a}_I^e$. The larger we choose ϵ the more coefficients $\tilde{a}_I^e$ will be zero. In most applications, p is chosen to be 2. Larger values of p will emphasize edges, smaller values emphasize smoothness.

Remark 10.2 Further compression, in terms of number of bits, can be attained by quantizing the compressed wavelet coefficients. This means that $\bar{a}_I^e$ is replaced by a number $\tilde{a}_I^e$ which requires fewer bits in its binary representation. Quantization can be combined with thresholding by finding $\tilde{a}_I^e$ with the fewest bits which satisfies $\|(a_I^e - \tilde{a}_I^e)\psi_I^e\|_{L_p} \leq \epsilon$.

Remark 10.3 The wavelet coefficient file consisting of the $\tilde{a}_I^e$ is further compressed by using a lossless encoder such as run length encoding or arithmetic encoding. The position of the coefficients must be encoded as well as their value. This can be done by keeping the entire array of coefficients in natural order (which will necessarily have many zero entries) or separately encoding positions.

Remark 10.4 The most efficient wavelet-based compression algorithms, such as the zero tree encoders (see Shapiro (1993) or Xiong, Ramchandran and Orchard (1997)) or bitstream encoder (see Gao and Sharpley (1997)), take advantage of the spatial correlation of the wavelet coefficients. For example, if we represent the coefficients by means of quadtrees with each node of the tree corresponding to one of the dyadic square I appearing in (10.3), then there will be many subtrees consisting only of zero entries, and one tries to encode these efficiently.

Remark 10.5 We can measure the efficiency of compression by the error

$$\sigma_n := \|I - \tilde{I}\|_{L_p}, \tag{10.4}$$

where n is the number of nonzero coefficients in the compressed wavelet file for $\tilde{I}$. Nonlinear approximation theory gives a direct relation between the rate of decrease of σ_n and the smoothness of the intensity function f. For example, consider approximation in L_2. If f is in the Besov class $B_\tau^\alpha(L_\tau)$, $1/\tau = \alpha/2 + 1/2$, then $\sigma_n \leq Cn^{-\alpha/2}$. Indeed, assuming this smoothness for f, one can show that the function in (10.2) inherits this smoothness (see Chambolle, DeVore, Lee and Lucier (1998)) and therefore the claim follows from the results of Sections 7.6–7.7. An inverse theorem provides converse statements that deduce smoothness of the intensity function from the rate of compression. However, for these converse results one must think of varying m, that is, finer and finer pixel representations. The point is that one can associate to each image a smoothness index α which measures its smoothness in the above scale of Besov spaces, and relate this directly with efficiency of wavelet compression (DeVore, Jawerth and Lucier 1992).

Remark 10.6 In image compression, we are not interested in the number of nonzero coefficients of the compressed image *per se*, but rather the number of bits in the encoded coefficient file. This leads one to consider the error

$$\rho_n := \|I - \tilde{I}\|_{L_p}, \tag{10.5}$$

where n is the number of bits in the encoded file of wavelet coefficients for $\tilde{I}$. It has recently been shown by Cohen, Daubechies, Guleryuz and Orchard (1997) that a similar analysis to that developed here for nonlinear wavelet approximation exists for the error ρ_n. For example, they show that if a univariate intensity function f is in the Besov space $B^\alpha_\infty(L_q)$, with $q > \alpha + 1/2$, then with a proper choice of encoder one has $\rho_N \leq N^{-\alpha}$. This matches the error rate σ_n in terms of the number of coefficients. Related results hold in a stochastic setting (see Mallat and Falzon (1997) and Cohen, Daubechies, Guleryuz and Orchard (1997)).

Remark 10.7 Adaptive basis selection for the wavelet packet library has been used successfully in compression. Most applications have been to signal processing (in particular speech signals). There is, however, the interesting application of compressing the FBI fingerprint files. Rather than use a different basis for each file, the current algorithms choose one basis of the wavelet packet library chosen by its performance on a sample collection of fingerprint files.

Noise reduction

Noise reduction is quite similar to compression. If an image is corrupted by noise then the noisy pixel values will be converted to noisy wavelet coefficients. Large wavelet coefficients are thought to carry mostly signal and should be retained; small coefficients are thought to be mostly noise and should be thresholded to zero. Donoho and Johnstone have put forward algorithms for noise reduction (called wavelet shrinkage) which have elements similar to the above theory of compression. We give a brief description of certain aspects of this theory as it relates to nonlinear approximation. We refer the reader to Donoho, Johnstone, Kerkyacharian and Picard (1996), and the papers referenced therein, for a more complete description of the properties of wavelet shrinkage.

Wavelet-based noise reduction algorithms are applied even when the noise characteristics are unknown. However, the theory has its most complete description in the case that the pixel values are corrupted by Gaussian noise. This means we are given a noisy image $\tilde{\mathcal{I}} = \mathcal{I} + \mathcal{N}$ with noisy pixel values

$$\tilde{p}_I = p_I + \eta_I, \tag{10.6}$$

where the p_I are the original (noise-free) pixel values and the η_I are independent, identically distributed Gaussians with mean 0 and variance σ_0^2. If we choose an orthonormal wavelet basis for $L_2(\Omega)$, $\Omega = [0,1)^2$, then the wavelet coefficients computed from the $\tilde{p}_I$ will take the form

$$\tilde{c}_I^e = c_I^e + \epsilon_I^e, \tag{10.7}$$

where c_I^e are the original wavelet coefficients of $\mathcal{I}$ and ϵ_I^e are independent, identically distributed Gaussians with variance $\sigma_0^2 2^{-2m}$. Wavelet shrinkage with parameter $\lambda > 0$ replaces $\tilde{c}_I^e$ by the shrunk coefficients $s_\lambda(c_I^e)$ where

$$s_\lambda(t) := \begin{cases} (|t| - \lambda)\text{sign } t, & \lambda < t, \\ 0, & |t| \le \lambda, \end{cases} \tag{10.8}$$

Thus, large coefficients (*i.e.*, those larger than λ in absolute value) are shrunk by an amount λ and small coefficients are shrunk to zero. We denote the function with these wavelet coefficients by

$$f_\lambda := P_0 + \sum_{j=0}^{m-1} \sum_{I \in D_j} \sum_{e \in E} s_\lambda(\tilde{c}_I^e)\psi_{I,e}, \tag{10.9}$$

with the term P_0 incorporating the scaling functions from the coarsest level. We seek a value of λ which minimizes the expected error

$$E(\|f - f_\lambda\|^2_{L_2(\Omega)}). \tag{10.10}$$

Donoho and Johnstone propose the parameter choice $\lambda^* = \sqrt{2 \ln 2^m} 2^m \sigma_0$ and show its near optimality in several statistical senses. One of the extremal problems studied by them, as well as by DeVore and Lucier (1992), is the following. We assume that the original image intensity function f comes from the the Besov space $B_\tau^\alpha(L_\tau(\Omega))$, with $\tau = (\alpha/2 + 1/2)^{-1}$. We know that these spaces characterize the approximation space $A_\tau^\alpha(L_2(\Omega))$ for bivariate nonlinear wavelet approximation. It can be shown that the above choice of λ gives the noise reduction

$$E(\|f - f_\lambda\|^2) \le C(\lambda)\|f\|^\tau_{B_\tau^\alpha(L_\tau(\Omega))}[\sigma_0 2^{-m}]^{2-\tau}. \tag{10.11}$$

The choice of $\lambda = \lambda^*$ gives an absolute constant $c(\lambda^*)$. A finer analysis of this error was given by Chambolle, DeVore, Lee and Lucier (1998) and shows that choosing the shrinkage parameter to depend on α will result in an improved error estimate.

Significant improvements in noise reduction (at least in the visual quality of the images) can be obtained by using the technique of cycle spinning, as proposed by Coifman and Donoho (1995). The idea behind their method can be described by the following analysis of discontinuities of a univariate function g. The performance of wavelet-based compression and noise reduction algorithms depends on the position of the discontinuities. If a discontinuity of g occurs at a coarse dyadic rational, say $1/2$, it will affect only a few wavelet coefficients. These coefficients will be the ones that are changed by shrinking. On the other hand, if the discontinuity occurs at a fine level rational binary, say 2^{-m}, then all coefficients will feel this discontinuity and can potentially be affected by shrinkage. This less favourable situation can be circumvented by translating the image, so that the discontinuity appears at a coarse binary rational, and then applying wavelet shrinkage to the translated image. The image is shifted back to the original position to obtain the noise reduced image. Since it is not possible to anticipate the position of the discontinuities, Coifman and Donoho propose averaging over all possible shifts. The result is an algorithm that involves $O(m2^{2m})$ computations.

Feature extraction and object recognition

The time-frequency localization of wavelets allows for the extraction of features such as edges and texture. These can then be utilized for object recognition by matching the extraction to a corresponding template for the object to be extracted. Edges and other discontinuities are identifiable by the large wavelet coefficients. These occur at every dyadic level. Retention of high frequency (*i.e.*, the highest level) coefficients is like an artist's sketch of an image.

Feature extraction has been a prominent application of adaptive basis selection and approximation from a dictionary. A dictionary of waveforms is utilized which is robust enough to allow the feature to be approximated with a few terms. Examples are the Gabor functions mentioned in Section 8. In some cases, an understanding of the physics of wave propagation can allow the designing of dictionaries appropriate for the features to be extracted. A good example of this approach in the context of Synthetic Aperture Radar is given by McClure and Carin (1997). The use of adaptive basis selection for feature extraction is well represented in the book of Wickerhauser (1994). The application of greedy algorithms and approximation from dictionaries is discussed in detail in the book of Mallat (1998). Other techniques based on wavelet decompositions can be found in DeVore, Lucier and Yang (1996) (in digital mammography), and DeVore et al. (1997) (in image registration).

10.2. Analytical and numerical methods for PDEs

To a certain extent, one can view the problem of numerically recovering a solution u to a PDE (or system of PDEs) as a problem of approximating the target function u. However, there is a large distinction in the information available about u in numerical computation versus approximation theory. In approximation theory one views information such as point values of a function or wavelet coefficients as known, and constructs methods of approximation using this information. However, in numerical methods for PDEs, the target function is unknown except through the PDE. Thus, the information the approximation theorist wants and loves so much is not available except through numerical computation. In spite of this divergence of viewpoints, approximation theory can be very useful in numerical computation in suggesting numcrical algorithms and, more importantly, to clarify the performance expected from linear and nonlinear numerical methods.

Adaptive methods are commonly used for numerically resolving PDEs. These methods can be viewed as a form of nonlinear approximation with the target function the unknown solution u to the PDE. Most adaptive numerical methods have not even been shown to converge and certainly have not been *theoretically proven* to have numerical efficiency over linear methods. Nevertheless, they have been very successful in practice and their effficiency has been experimentally established.

Nonlinear approximation can be very useful in understanding when and how adaptive numerical methods should be used. For example, from the analysis put forward in this paper, we know that adaptive piecewise polynomial methods, as well as the n-term wavelet approximation methods, have increased efficiency over linear methods when the target function u has certain types of singularities; specifically, singularities that would destroy the smoothness of u in the Sobolev scale but would not impair its smoothness in the Besov scale for nonlinear approximation.

To be more precise, suppose that u is to be approximated in the $L_p(\Omega)$ norm with Ω a domain in $\mathbb{R}^d$. Let α_L be the largest value of α such that u is in the Besov space $B^\alpha_\infty(L_p(\Omega))$. We know that u can be approximated by linear methods such as piecewise polynomial or linear wavelet approximation with accuracy $O(n^{-\alpha_L/d})$, with n the dimension of the linear space. However, we do not know (unless we prove it) whether our particular numerical method has this efficiency. If we wish to establish the efficiency of our particular linear numerical method, we should seek an estimate of the form

$$\|u - u_n\|_{L_p(\Omega)} \le C|u|_{B^{\alpha_L}_\infty(L_p(\Omega))} n^{-\alpha_L/d}, \tag{10.12}$$

where u_n is the approximate solution provided by our numerical method. In many papers, $W^{\alpha_L}(L_p(\Omega))$ is used in place of $B^{\alpha_L}_\infty(L_p(\Omega)$. The form of such estimates is familiar to the numerical analyst in finite element methods where such estimates are known in various settings (especially in the case $p = 2$ since this can be related to the energy norm).

Note that n is related to the numerical effort needed to compute the approximant. However, the number of computations needed to compute an approximant with this accuracy may exceed Cn. This may be the case, for example, in solving elliptic equations with finite element methods, since the coefficients of the unknown solution must be computed as a solution to a matrix equation of size $n \times n$.

We can do a similar analysis for nonlinear methods. According to the results reviewed in this article, the appropriate scale of Besov spaces to gauge the performance of nonlinear algorithms are the $B_q^\alpha(L_q(\Omega))$ where $1/q = \alpha/d + 1/p$ (see Figure 3 in the case $d = 1$). Let α_N be the largest value of α such that u is in the Besov space $B_q^\alpha(L_q(\Omega))$, $1/q = \alpha/d + 1/p$. If $\alpha_N > \alpha_L$, then nonlinear approximation will be more efficient than linear approximation in approximating u and therefore the use of nonlinear methods is completely justified. However, there still remains the question of how to construct a nonlinear algorithm that approximates u with the efficiency $O(n^{-\alpha_N/d})$. If we have a particular nonlinear numerical method in hand and wish to analyse its efficiency, then the correct form of an error estimate for such a nonlinear algorithm would be

$$\|u - u_n\|_{L_p(\Omega)} \leq C|u|_{B_q^{\alpha_N}(L_q(\Omega))} n^{-\alpha_N/d}, \quad 1/q = \alpha_N/d + 1/p. \tag{10.13}$$

How could we decide beforehand whether nonlinear methods offer a benefit over linear methods? This is the role of regularity theorems for PDEs. A typical regularity theorem infers the smoothness of the solution u to a PDE from information such as the coefficients, inhomogeneous term, initial conditions, or boundary conditions. We shall discuss this in a little more detail in a moment, but for now we want to make the point of what form these regularity theorems should take. The most common regularity theorems are in the form of Sobolev regularity and are compatible with the linear theory of numerical methods. Much less emphasis has been placed on the regularity in the nonlinear scale of Besov spaces but this is exactly what we need for an analysis of adaptive, or other nonlinear, algorithms.

To go a little further in our discussion, we shall consider two model problems, one hyperbolic and the other elliptic, to elucidate the points discussed above.

Conservation laws
Consider the scalar univariate conservation law

$$\begin{cases} u_t + f(u)_x = 0, & x \in \mathbb{R},\ t > 0, \\ u(x,0) = u_0(x), & x \in \mathbb{R}, \end{cases} \tag{10.14}$$

where f is a given flux, u_0 a given initial condition, and u is the sought-after solution. This is a well studied nonlinear transport equation with transport velocity $a(u) = f'(u)$ (see, for instance, the book of Godlewski and Raviart (1991)). We shall assume that the flux is strictly convex, which means the transport velocity is strictly increasing. The important fact for us is that, even when the initial condition u_0 is smooth, the solution $u(\cdot,t)$ will develop spontaneous shock discontinuities at later times t.

The proper setting for the analysis of conservation laws is in L_1 and, in particular, the error of numerical methods should be measured in this space. Thus, concerning the performance of linear numerical methods, the question arises as to the possible values of the smoothness parameter α_L of $u(\cdot,t)$ as measured in L_1. It is known that, if the initial condition u_0 is in $BV = \text{Lip}(1, L_1)$, then the solution u remains in this space for all later time $t > 0$. However, since this solution develops discontinuities, no matter how smooth the initial condition is, the Sobolev embedding theorem precludes u being in any Besov space $B^\alpha_\infty(L_1))$ for any $\alpha > 1$. This means that the largest value we can expect for α_L is $\alpha_L = 1$. Thus, the optimal performance we can expect from linear methods of approximation is $O(n^{-1})$, with n the dimension of the linear spaces used in the approximation. Typical numerical methods utilize spaces of piecewise polynomials on a uniform mesh with mesh length h and the above remarks mean that the maximum efficiency we can expect for numerical methods is $O(h)$, $h \to 0$. In reality, the best proven estimates are $O(\sqrt{h})$ under the assumption that $u_0 \in \text{Lip}(1, L_1)$. This discrepancy between the possible performance of numerical algorithms and the actual performance is not unusual. The solution is known to have sufficient regularity to be approximated, for example, by piecewise constants with uniform mesh h to accuracy $O(h)$, but algorithms which capture this accuracy are unkown.

To understand the possible performance of nonlinear methods such as moving grid methods, we should estimate the smoothness of the solution in the nonlinear Besov scale $B^\alpha_\tau(L_\tau))$, $1/\tau = \alpha + 1$, corresponding to approximation in the L_1-norm. A rather surprising result of DeVore and Lucier (1990) shows that, starting with a smooth initial condition u_0, the solution u will be in each of these Besov spaces for all $\alpha > 0$. In other words, depending on the smoothness of u_0, α_N can be arbitrarily large. This means that nonlinear methods such as moving grid methods could provide arbitrarily high efficiency. In fact, such algorithms, based on piecewise polynomial approximation, can be constructed using the method of characteristics (see Lucier (1986) for the case of piecewise linear approximation).

Unfortunately, the situation concerning numerical methods for multivariate conservation laws is not as clear. While the linear theory goes through almost *verbatim*, the nonlinear theory is left wanting. The proper form of nonlinear approximation in the multivariate case would most likely be by piecewise polynomials on free triangulations. As we have noted earlier, it is an unsolved problem in nonlinear approximation to describe the smoothness conditions that govern the efficiency of this type of approximation. For a further discussion of the multivariate case see DeVore and Lucier (1996).

Because of their unique ability to detect singularities in a function, wavelet methods seem a natural candidate for numerical resolution of solutions to conservation laws. However, it is not yet completely clear how wavelets should be used in numerical solvers. Attempts to use wavelets directly in a time-stepping solver have not been completely effective. Ami Harten (1994) and his collaborators have suggested the use of wavelets to compress the computations in numerical algorithms. For example, he proposes the use of a standard time-stepping solver, such as Godunov, based on cell averages for computing the solution at time step t_{n+1} from the numerically computed solution at time step t_n, but to utilize wavelet compression to reduce the number of flux computations in the solution step.

Elliptic equations

An extensive accounting of the role of linear and nonlinear approximation in the solution of elliptic problems is given in Dahmen (1997) and Dahlke, Dahmen and DeVore (1997). We shall therefore limit ourselves to reiterating a couple of important points about the role of regularity theorems and the form of nonlinear estimates. We consider the model problem

$$\begin{aligned} \Delta u &= f \quad \text{on } \Omega \subset \mathbb{R}^d, \\ u &= 0 \quad \text{on } \partial\Omega \end{aligned} \tag{10.15}$$

of Poisson's equation on a domain $\Omega \subset \mathbb{R}^d$ with zero boundary conditions. We are interested in numerical methods for recovering the solution to (10.15) and, in particular, in the question of whether nonlinear methods such as adaptive solvers are of any benefit. We shall also limit our discussion to estimating error in the L_2-norm, although various results are known for general p.

Consider first the case where $f \in L_2(\Omega)$ and Ω has a smooth boundary. Then, the solution u to (10.15) has smoothness $W^2(L_2(\Omega))$. In our previous notation, this means that $\alpha_L = 2$. In general, the solution will not have higher smoothness in the nonlinear Besov scale $B_q^\alpha(L_q(\Omega))$, $1/q = \alpha/d + 1/2$, for L_2 approximation. Therefore $\alpha_N = 2$ and there is no apparent advantage to nonlinear methods. The solution can be approximated by linear spaces of piecewise polynomials of dimension n to accuracy $O(n^{-2/d})$. This accuracy can actually be achieved by finite element methods using uniformly refined partitions. There is no evidence to suggest any better performance using adaptive methods.

If the boundary $\partial\Omega$ of Ω is not smooth then the solutions (10.15) have singularities due to corners or other discontinuities of $\partial\Omega$ (see, for instance, Kondrat'ev and Oleinik (1983)). Regularity theory in the case of a non-smooth boundary is a prominent area of PDEs. For some of the deepest and most recent results see Jerison and Kenig (1995). For example, on a general Lipschitz domain, we can only expect that the solution u to (10.15) is in the Sobolev space $W^{3/2}(L_2(\Omega))$. Thus, in the notation given earlier in this section, we will only have $\alpha_L = 3/2$.

Because of the appearance of singularities due to the boundary, adaptive numerical techniques are suggested for numerically recovering the solution u. We understand that to justify the use of such methods we should determine the regularity of the solution in the scale of Besov spaces $B^\alpha_q(L_q(\Omega))$, $1/q = \alpha/d + 1/2$. Such regularity has been studied by Dahlke and DeVore (1997). They prove, for example, that, for $d = 2, 3, 4$, we have $u \in B^\alpha_q(L_q)$, $1/q = \alpha/d + 1/2$, for each $\alpha < 2$. In other words, $\alpha_N > \alpha_L$ and the use of adaptive methods is completely justified. There are also more general results which apply for any $d > 1$ and show that we always have $\alpha_N > \alpha_L$.

We reiterate that the above results on regularity of elliptic equations only indicate the possibility of constructing nonlinear methods with higher efficiency. It remains a difficult problem to construct adaptive methods and prove that they exhibit the increased accuracy indicated by the approximation theory. The aim is to construct numerical methods that provide the error estimate (10.13). We refer the reader to Dahmen (1997) for a comprehensive discussion of what is known about adaptive methods for elliptic equations.

Acknowledgements

The author thanks Professors de Boor, Cohen, Oskolkov, Petrushev and Temlyakov for their valuable suggestions concerning this survey.

REFERENCES

R. A. Adams (1975), *Sobolev Spaces*, Academic Press, New York.

I. Babuška and M. Suri (1994), 'The p and h versions of the finite element method: basic principles and properties', *SIAM Review* **36**, 578–632.

G. A. Baker, Jr. (1975), *Essentials of Padé Approximants*, Academic Press, New York.

C. Bennett and R. Sharpley (1988), *Interpolation of Operators*, Academic Press, New York.

J. Bergh and J. Löfström (1976), *Interpolation Spaces: An Introduction*, Springer, Berlin.

J. Bergh and J. Peetre (1974), 'On the spaces V_p $(0 < p \leq \infty)$', *Boll. Unione Mat. Ital.* **10**, 632–648.

M. Birman and M. Solomyak (1967), 'Piecewise polynomial approximation of functions of the class W^α_p', *Mat. Sbornik* **2**, 295–317.

C. de Boor (1973), 'Good approximation by splines with variable knots', in *Spline Functions and Approximation* (A. Meir and A. Sharma, eds), Birkhäuser, Basel, pp. 57–72.

C. de Boor, R. DeVore and A. Ron (1993), 'Approximation from shift invariant spaces', *Trans. Amer. Math. Soc.* **341**, 787–806.

L. Brown and B. Lucier (1994), 'Best approximations in L_1 are best in L_p, $p < 1$, too', *Proc. Amer. Math. Soc.* **120**, 97–100.

Yu. Brudnyi (1974), 'Spline approximation of functions of bounded variation', *Soviet Math. Dokl.* **15**, 518–521.

A. P. Calderón (1964*a*), 'Intermediate spaces and interpolation: the complex method', *Studia Math.* **24**, 113–190.

A. P. Calderón (1964*b*), 'Spaces between L_1 and L_∞ and the theorem of Marcinkieiwicz: the complex method', *Studia Math.* **26**, 273–279.

A. Chambolle, R. DeVore, N.-Y. Lee and B. Lucier (1998), 'Nonlinear wavelet image processing: Variational problems, compression, and noise removal through wavelet shrinkage', *IEEE Trans. Image Processing* **7**, 319–335.

A. Cohen, I. Daubechies and J.-C. Feauveau (1992), 'Biorthogonal bases of compactly supported wavelets', *Comm. Pure Appl. Math.* **43**, 485–560.

A. Cohen, I. Daubechies and P. Vial (1993), 'Wavelets on the interval and fast wavelet transforms', *Appl. Comput. Harm. Anal.* **1**, 54–81.

A. Cohen, I. Daubechies, O. Guleryuz and M. Orchard (1997), 'On the importance of combining wavelet-based non-linear approximation in coding strategies'. Preprint.

A. Cohen, R. DeVore and R. Hochmuth (1997), 'Restricted approximation'. Preprint.

A. Cohen, R. DeVore, P. Petrushev and H. Xu (1998), 'Nonlinear approximation and the space BV($\mathbb{R}^2$)'. Preprint.

R. R. Coifman and D. Donoho (1995) Translation invariant de-noising, in *Wavelets in Statistics* (A. Antoniadis and G. Oppenheim, eds), Springer, New York, pp. 125–150.

R. R. Coifman and M. V. Wickerhauser (1992), 'Entropy based algorithms for best basis selection', *IEEE Trans. Inform. Theory* **32**, 712–718.

S. Dahlke and R. DeVore (1997), 'Besov regularity for elliptic boundary value problems', *Commun. Partial Diff. Eqns.* **22**, 1–16.

S. Dahlke, W. Dahmen and R. DeVore (1997), Nonlinear approximation and adaptive techniques for solving elliptic operator equations, in *Multiscale Wavelet Methods for PDEs* (W. Dahmen, A. Kurdila and P. Oswald, eds), Academic Press, pp. 237–284.

W. Dahmen (1997), Wavelet and multiscale methods for operator equations, in *Acta Numerica*, Vol. 6, Cambridge University Press, pp. 55–228.

I. Daubechies (1988), 'Orthonormal bases of compactly supported wavelets', *Comm. Pure Appl. Math.* **41**, 909–996 .

I. Daubechies (1992), *Ten Lectures on Wavelets*, Vol. 61 of *CBMS-NSF Regional Conference Series in Applied Mathematics*, SIAM, Philadelphia.

G. Davis, S. Mallat and M. Avellaneda (1997), 'Adaptive greedy approximations', *Constr. Approx.* **13**, 57–98.

R. A. DeVore (1987), 'A note on adaptive approximation', *Approx. Theory Appl.* **3**, 74–78.

R. A. DeVore and G. G. Lorentz (1993), *Constructive Approximation*, Vol. 303 of *Grundlehren*, Springer, Heidelberg.

R. A. DeVore and B. Lucier (1990), 'High order regularity for conservation laws', *Indiana Math. J.* **39**, 413–430.

R. A. DeVore and B. Lucier (1992), Wavelets, in *Acta Numerica*, Vol. 1, Cambridge University Press, pp. 1–56.
R. A. DeVore and B. Lucier (1996), 'On the size and smoothness of solutions to nonlinear hyperbolic conservation laws', *SIAM J. Math. Anal.* **27**, 684–707.
R. DeVore and V. Popov (1987), 'Free multivariate splines', *Constr. Approx.* **3**, 239–248.
R. DeVore and V. Popov (1988*a*), 'Interpolation of Besov spaces', *Trans. Amer. Math. Soc.* **305**, 397–414.
R. A. DeVore and V. A. Popov (1988*b*), Interpolation spaces and nonlinear approximation, in *Function Spaces and Applications* (M. Cwikel *et al.*, eds), Vol. 1302 of *Lecture Notes in Mathematics*, Springer, Berlin, pp. 191–205.
R. A. DeVore and K. Scherer (1979), 'Interpolation of operators on Sobolev spaces', *Ann. Math.* **109**, 583–599.
R. A. DeVore and K. Scherer (1980), Variable knot, variable degree approximation to x^β, in *Quantitative Approximation* (R. A. DeVore and K. Scherer, eds), Academic Press, New York, pp. 121–131.
R. A. DeVore and R. C. Sharpley (1984), *Maximal Functions Measuring Smoothness*, Memoirs Vol. 293, American Mathematical Society, Providence, RI.
R. A. DeVore and R. C. Sharpley (1993), 'Besov spaces on domains in $\mathbb{R}^d$', *Trans. Amer. Math. Soc.* **335**, 843–864.
R. A. DeVore and V. Temlyakov (1996), 'Some remarks on greedy algorithms', *Adv. Comput. Math.* **5**, 173–187.
R. DeVore and X. M. Yu (1986), 'Multivariate rational approximation', *Trans. Amer. Math. Soc.* **293**, 161–169.
R. DeVore and X. M. Yu (1990), 'Degree of adaptive approximation', *Math. Comput.* **55**, 625–635.
R. DeVore, R. Howard and C. A. Micchelli (1989), 'Optimal nonlinear approximation', *Manuskripta Math.* **63**, 469–478.
R. DeVore, B. Jawerth and B. Lucier (1992), 'Image compression through transform coding', *IEEE Proc. Inform. Theory* **38**, 719–746.
R. DeVore, B. Jawerth and V. Popov (1992), 'Compression of wavelet decompositions', *Amer. J. Math.* **114**, 737–785.
R. DeVore, G. Kyriazis, D. Leviatan and V. M. Tikhomirov (1993), 'Wavelet compression and nonlinear n-widths', *Adv. Comput. Math.* **1**, 197–214.
R. DeVore, B. Lucier and Z. Yang (1996), Feature extraction in digital mammography, in *Wavelets in Biology and Medicine* (A. Aldroubi and M. Unser, eds), CRC, Boca Raton, FL, pp. 145–156.
R. DeVore, W. Shao, J. Pierce, E. Kaymaz, B. Lerner and W. Campbell (1997), Using nonlinear wavelet compression to enhance image registration, in *Wavelet Applications IV, Proceedings of the SPIE Conf.* **3028**, AeroSense 97 Conference, Orlando, FL, April 22–24, 1997 (H. Szu, ed.), pp. 539–551.
R. A. DeVore, S. Konyagin and V. Temlyakov (1998), 'Hyperbolic wavelet approximation', *Constr. Approx.* **14**, 1–26.
D. Donoho (1997), 'CART and best-ortho-basis: a connection', *Ann. Statistics* **25**, 1870–1911.
D. Donoho and I. Johnstone (1994), 'Ideal spatial adaptation via wavelet shrinkage', *Biometrika* **81**, 425–455.

D. Donoho, I. Johnstone, G. Kerkyacharian and D. Picard (1996), 'Wavelet shrinkage asymptotia?', *J. Royal Statistical Soc., Ser. B.* **57**, 301–369.

Z. Gao and R. Sharpley (1997), 'Data compression and the elementary encoding of wavelet compression'. Preprint.

E. Godlewski and P.-A. Raviart (1991), *Hyperbolic Systems of Conservation Laws*, Mathématiques et Applications, Ellipses.

A. Harten (1994), 'Adaptive multiresolution schemes for shock computations', *J. Comput. Phys.* **115**, 319–338.

D. Jerison and C. E. Kenig (1995), 'The inhomogeneous Dirichlet problem in Lipschitz domains', *J. Functional Analysis* **93**, 161–219.

R. Q. Jia (1983), Approximation by smooth bivariate splines on a three directional mesh, in *Approximation Theory IV* (C. K. Chui, L. L. Schumaker and J. Ward, eds), Academic Press, New York, pp. 539–546.

H. Johnen and K. Scherer (1977), On the equivalence of the K-functional and moduli of continuity and some applications, in *Constructive Theory of Functions of Several Variables*, Vol. 571 of *Lecture Notes in Mathematics*, Springer, Berlin, pp. 119–140.

L. Jones (1992), 'A simple lemma on greedy approximation in Hilbert space and convergence results for projection pursuit regression and neural network training', *Ann. Statistics* **20**, 608–613.

J. P. Kahane (1961), *Teoria Constructiva de Functiones*, Course notes, University of Buenos Aires.

B. Kashin (1977) 'The widths of certain finite dimensional sets and classes of smooth functions', *Izvestia* **41**, 334–351.

B. Kashin and V. Temlyakov (1997), 'On best n-term approximation and the entropy of sets in the space L_1', *Math. Notes* **56**, 1137–1157.

V. A. Kondrat'ev and O. A. Oleinik (1983), 'Boundary value problems for partial differential equations in non-smooth domains', *Russian Math. Surveys* **38**, 1–86.

G. Kyriazis (1996) 'Wavelet coefficients measuring smoothness in $H_p(\mathbb{R}^d)$', *Appl. Comput. Harm. Anal.* **3**, 100–119

B. Lucier (1986), 'Regularity through approximation for scalar conservation laws', *SIAM J. Math. Anal.* **19**, 763–773.

G. G. Lorentz, M. von Golitschek and Ju. Makovoz (1996), *Constructive Approximation: Advanced Problems*, Springer, Berlin.

M. McClure and L. Carin (1997), 'Matched pursuits with a wave-based dictionary'. Preprint.

V. Mairov and J. Ratasby (1998), 'On the degree of approximation using manifolds of finite pseudo-dimension'. Preprint.

S. Mallat (1989), 'Multiresolution and wavelet orthonormal bases in $L_2(\mathbb{R})$', *Trans. Amer. Math. Soc.* **315**, 69–87.

S. Mallat (1998), *A Wavelet Tour of Signal Processing*, Academic Press, New York.

S. Mallat and F. Falzon (1997), 'Analysis of low bit rate image transform coding'. Preprint.

Y. Meyer (1990) *Ondelettes et Opérateurs, Vols 1 and 2*, Hermann, Paris.

D. Newman (1964), 'Rational approximation to $|x|$', *Michigan Math. J.* **11**, 11–14.

E. Novak (1996), 'On the power of adaptation', *J. Complexity* **12**, 199–237.

K. Oskolkov (1979), 'Polygonal approximation of functions of two variables', *Math. USSR Sbornik* **35**, 851–861.

P. Oswald (1980), 'On the degree of nonlinear spline approximation in Besov–Sobolev spaces', *J. Approx. Theory* **61**, 131–157.

J. Peetre (1963) *A Theory of Interpolation of Normed Spaces*, Course notes, University of Brasilia.

A. Pekarski (1986), 'Relations between the rational and piecewise polynomial approximations', *Izvestia BSSR, Ser. Mat.-Fiz. Nauk* **5**, 36–39.

V. Peller (1980), 'Hankel operators of the class S_p, investigations of the rate of rational approximation and other applications', *Mat. Sbornik* **122**, 481–510.

P. P. Petrushev (1986), 'Relations between rational and spline approximations in L_p metrics', *J. Approx. Theory* **50**, 141–159.

P. P. Petrushev (1988), Direct and converse theorems for spline and rational approximation and Besov spaces, in *Function Spaces and Applications* (M. Cwikel *et al.*, eds), Vol. 1302 of *Lecture Notes in Mathematics*, Springer, Berlin, pp. 363–377.

P. Petrushev and V. Popov (1987), *Rational Approximation of Real Functions*, Cambridge University Press, Cambridge.

G. Pisier (1980) 'Remarques sur un résultat non publié de B. Maurey', *Seminaire d'Analyse Fonctionelle 1980–81, École Polytechnique, Centre de Mathématiques, Palaiseau.*

E. Schmidt (1907), 'Zur Theorie der linearen und nichtlinearen Integralgleichungen. I', *Math. Ann.* **63**, 433–476.

I. Schoenberg (1946), 'Contributions to the problem of approximation of equidistant data by analytic functions', *Quart. Appl. Math.* **4**, 45–99.

J. Shapiro (1993), An embedded hierarchial image coder using zerotrees of wavelet coefficients, in *Data Compression Conference* (J. A. Storer and M. Cohn, eds), IEEE Computer Society Press, Los Alamitos, CA, pp. 214–223.

E. Stein (1970), *Singular Integrals and Differentiability Properties of Functions*, Princeton University Press, Princeton.

G. Strang and G. Fix (1973), A Fourier analysis of the finite element variational method, in *Constructive Aspects of Functional Analysis* (G. Geymonat, ed.), C.I.M.E., II, Ciclo, 1971, pp. 793–840.

V. Temlyakov (1998*a*), 'Best m-term approximation and greedy algorithms'. Preprint.

V. Temlyakov (1998*b*), 'Nonlinear m-term approximation with regard to the multivariate Haar system'. Preprint.

J. F. Traub, G. W. Wasilkowski and H. Woźniakowski (1988), *Information-Based Complexity*, Academic Press, Boston.

V. N. Vapnik (1982), *Estimation of Dependences Based on Empirical Data*, Springer, Berlin.

M. V. Wickerhauser (1994), *Adapted Wavelet Analysis from Theory to Software*, Peters.

Z. Xiong, K. Ramchandran and M. T. Orchard (1997), 'Space-frequency quantization for wavelet image coding', *Trans. Image Processing* **6**, 677–693.

Acta Numerica (1998), pp. 151–201

Relative perturbation results for matrix eigenvalues and singular values

Ilse C. F. Ipsen*
Department of Mathematics,
North Carolina State University,
Raleigh, NC 27695–8205, USA
E-mail: ipsen@math.ncsu.edu
http://www4.ncsu.edu/~ipsen/info.html

It used to be good enough to bound absolute errors of matrix eigenvalues and singular values. Not any more. Now it is fashionable to bound relative errors. We present a collection of relative perturbation results which have emerged during the past ten years.

No need to throw away all those absolute error bounds, though. Deep down, the derivation of many relative bounds can be based on absolute bounds. This means that relative bounds are not always better. They may just be better sometimes – and exactly when depends on the perturbation.

CONTENTS

* Work supported in part by grant CCR-9400921 from the National Science Foundation.

1. Introduction

Are you concerned about accuracy of matrix eigenvalues or singular values, especially the ones close to zero? If so, this paper is for you!

We present error bounds for eigenvalues and singular values that can be much tighter than the traditional bounds, especially when these values have small magnitude. Our goal is to give some intuition for what the bounds mean and why they hold.

Suppose you have to compute an eigenvalue of a complex square matrix A. Numerical software usually produces a number $\hat{\lambda}$ that is not the desired eigenvalue. So you ask yourself how far away is $\hat{\lambda}$ from an eigenvalue of A? If $\hat{\lambda}$ was produced by a reliable (*i.e.*, backward stable) numerical method, there is a round-off error analysis to assure you that $\hat{\lambda}$ is an eigenvalue of a nearby matrix $A + E$, where E is small in some sense. Then you can use perturbation theory to estimate the error in $\hat{\lambda}$. For instance, when A is diagonalizable, the Bauer–Fike theorem bounds the absolute distance between $\hat{\lambda}$ and a closest eigenvalue λ of A by

$$|\lambda - \hat{\lambda}| \leq \kappa(X)\, \|E\|, \tag{1.1}$$

where $\kappa(X) \equiv \|X\|\, \|X^{-1}\|$ is the condition number of an eigenvector matrix X of A.

The quantity $|\lambda - \hat{\lambda}|$ represents an absolute error. Traditional perturbation theory assesses the quality of a perturbed eigenvalue by bounding absolute errors. However, there are practical situations where small eigenvalues have physical meaning and should be determined to high relative accuracy. Such situations include computing modes of vibration in a finite element context, and computing energy levels in quantum mechanical systems (Demmel, Gu, Eisenstat, Slapničar, Veselić and Drmač 1997, Section 1). Absolute error bounds cannot cope with relative accuracy, especially when confronted with small eigenvalues or singular values. The following section explains why.

1.1. Why absolute bounds don't do the job

If we want relative accuracy, we need relative error bounds. The simplest way to generate a relative error bound is to divide an absolute error bound by an eigenvalue. For instance, dividing the absolute error bound (1.1) by a nonzero eigenvalue λ produces the relative error bound

$$\frac{|\lambda - \hat{\lambda}|}{|\lambda|} \leq \frac{\|E\|}{|\lambda|}.$$

Unlike the absolute bound, though, the relative bound depends on λ. This has several disadvantages. First, each eigenvalue has a different relative bound. Second, the relative bound is smaller for eigenvalues λ that are large in magnitude than for those that are small in magnitude. Third, the relative bound can be pessimistic for eigenvalues of small magnitude, as the following example illustrates.

Example 1.1 The mere act of storing a diagonal matrix

$$A = \begin{pmatrix} \lambda_1 & & \\ & \ddots & \\ & & \lambda_n \end{pmatrix}$$

in floating point arithmetic produces a perturbed matrix

$$A + E = \begin{pmatrix} \lambda_1(1+\epsilon_1) & & \\ & \ddots & \\ & & \lambda_n(1+\epsilon_n) \end{pmatrix},$$

where $|\epsilon_i| \leq \epsilon$ and $\epsilon > 0$ reflects the machine accuracy. According to the absolute perturbation bound (1.1), the error in an eigenvalue $\hat{\lambda}$ of $A + E$ is bounded by

$$\min_i |\lambda_i - \hat{\lambda}| \leq \|E\| = \max_k |\lambda_k \epsilon_k| \leq \epsilon \max_k |\lambda_k|.$$

This bound is realistic for eigenvalues of largest magnitude: if $\hat{\lambda}$ is closest to an eigenvalue $\lambda_{\max}$ of largest magnitude among all eigenvalues of A, then

$$\frac{|\lambda_{\max} - \hat{\lambda}|}{|\lambda_{\max}|} \leq \epsilon.$$

Since the relative error in all eigenvalues does not exceed ϵ, the bound is tight in this case.

However, the bound is too pessimistic for eigenvalues of smallest magnitude: if $\hat{\lambda}$ is closest to an eigenvalue $\lambda_{\min}$ of smallest magnitude among all eigenvalues of A, then

$$\frac{|\lambda_{\min} - \hat{\lambda}|}{|\lambda_{\min}|} \leq \epsilon \frac{|\lambda_{\max}|}{|\lambda_{\min}|}.$$

The bound is much larger than ϵ when the magnitude of the eigenvalues varies widely. Since the relative error does not exceed ϵ, the bound is not tight. □

There are algorithms whose relative error bounds do not depend on the eigenvalues. These algorithms compute all eigenvalues or singular values to high relative accuracy, even those of small magnitude: the dqds algorithm for singular values of bidiagonal matrices (Fernando and Parlett 1994, Parlett 1995), for instance, as well as Jacobi methods for eigenvalues of symmetric positive-definite matrices and for singular values (see Demmel (1997), Section 5.4.3, and Mathias (1995)). Absolute perturbation bounds cannot account for this phenomenon.

Absolute error bounds are well suited for describing the accuracy of fixed point arithmetic. But fixed point arithmetic has been replaced by floating point arithmetic, especially on general purpose machines where many eigenvalue and singular value computations are carried out nowadays. The accuracy of floating point arithmetic is best described by relative errors. In the absence of underflow and overflow, a number α is represented as a floating point number

$$\hat{\alpha} \equiv \alpha(1 + \epsilon_\alpha), \qquad \text{where} \quad |\epsilon_\alpha| \leq \epsilon,$$

and $\epsilon > 0$ reflects the machine accuracy. In IEEE arithmetic, for instance, $\epsilon \approx 10^{-7}$ in single precision and $\epsilon \approx 10^{-16}$ in double precision. Therefore the accuracy of floating point arithmetic can be described by relative error bounds of the form

$$|\hat{\alpha} - \alpha| \leq |\alpha|\, \epsilon \qquad \text{or} \qquad |\hat{\alpha} - \alpha| \leq |\hat{\alpha}|\, \epsilon.$$

Absolute error bounds cannot model this situation.

And even if you never require high relative accuracy from your small eigenvalues or singular values, you can still profit from it. It turns out that intermediate quantities computed to high relative accuracy can sometimes speed up subsequent computations. For instance, computing eigenvalues of a real, symmetric, tridiagonal matrix to high relative accuracy can accelerate eigenvector computations because the time-consuming process of orthogonalizing eigenvectors can be shortened or even avoided (Dhillon, Fann and Parlett 1997, Dhillon 1997).

Now that we have established the need for 'genuine' relative error bounds beyond any shadow of a doubt, it's time to find out what kind of relative bounds are out there.

1.2. Overview

Relative error bounds have been derived in the context of two different perturbation models.

- *Additive* perturbations (Sections 2, 3, 4) represent the perturbed matrix as $A + E$.
- *Multiplicative* perturbations (Sections 5, 6, 7) represent the perturbed matrix as $D_1 A D_2$, where D_1 and D_2 are nonsingular matrices.

The traditional absolute error bounds are derived in the context of additive perturbations.

We group the bounds for eigenvalues (Sections 2, 5) and for singular values (Sections 3, 6) according to a loose order of increasing specialization, as follows.

- *Bauer–Fike type.* Two-norm bounds on the distance between a perturbed eigenvalue and a closest exact eigenvalue.
- *Hoffman–Wielandt type.* Frobenius norm bounds on the sum of squares of all distances between perturbed eigenvalues and corresponding exact eigenvalues, where perturbed and exact eigenvalues are paired up in a one-to-one fashion. Similar for singular values.
- *Weyl type.* Two norm bounds on the largest distance between a perturbed eigenvalue and the corresponding exact eigenvalue, where the ith largest perturbed eigenvalue is paired up with the ith largest exact eigenvalue. Similar for singular values.

There are several different ways to normalize an absolute error $|\lambda - \hat{\lambda}|$ and turn it into a relative error. We present bounds for the relative error measures

$$\frac{|\lambda - \hat{\lambda}|}{|\lambda|}, \qquad \frac{|\lambda - \hat{\lambda}|}{\sqrt{|\lambda|\,|\hat{\lambda}|}}, \qquad \frac{|\lambda - \hat{\lambda}|}{\sqrt[p]{|\lambda|^p + |\hat{\lambda}|^p}},$$

where $1 \leq p \leq \infty$ is an integer. For instance, the traditional relative error $|\lambda - \hat{\lambda}|/|\lambda|$ can be larger or smaller than the second error measure, while it is never smaller than the third. Detailed relationships among the different measures are discussed by Li (1994*a*, Section 2). Since the measures are essentially proportional to each other we disregard any differences among them.

Sections 4 and 7 discuss applications of additive and multiplicative perturbations.

1.3. Notation

We use two norms: the two-norm

$$\|A\| = \max_{x \neq 0} \frac{\|Ax\|}{\|x\|}, \qquad \text{where} \quad \|x\| = \sqrt{x^* x},$$

and the superscript $*$ denotes the conjugate transpose; and the Frobenius norm

$$\|A\|_F = \sqrt{\sum_{i,j} |a_{ij}|^2},$$

where a_{ij} are the elements of the matrix A. The identity matrix of order n is

$$I = \begin{pmatrix} 1 & & \\ & \ddots & \\ & & 1 \end{pmatrix} = \begin{pmatrix} e_1 & \ldots & e_n \end{pmatrix}$$

with columns e_i.

For a matrix A we denote by range(A) its column space, A^{-1} its inverse and $A^\dagger$ its Moore–Penrose inverse. The absolute value matrix $|A|$ has elements $|a_{ij}|$. A matrix inequality of the form $|A| \le |B|$ is meant element-wise, that is, $|a_{ij}| \le |b_{ij}|$ for all i and j.

2. Additive perturbations for eigenvalues

Let A be a complex square matrix. We want to bound the absolute and relative errors in the eigenvalues of the perturbed matrix $A+E$. In the process we show that relative error bounds are as natural as absolute error bounds, and that many relative bounds are implied by absolute bounds.

2.1. Bauer–Fike-type bounds for diagonalizable matrices

The Bauer–Fike theorem bounds the distance between an eigenvalue of $A+E$ and a closest eigenvalue of A. The matrix A must be diagonalizable, while $A+E$ does not have to be.

Let $A = X\Lambda X^{-1}$ be an eigendecomposition of A, where

$$\Lambda = \begin{pmatrix} \lambda_1 & & \\ & \ddots & \\ & & \lambda_n \end{pmatrix},$$

and λ_i are the eigenvalues of A. Let $\hat{\lambda}$ be an eigenvalue of $A+E$.

The Bauer–Fike theorem for the two-norm (Bauer and Fike 1960, Theorem IIIa) bounds the absolute error,

$$\min_i |\lambda_i - \hat{\lambda}| \le \kappa(X)\, \|E\|. \tag{2.1}$$

The relative version of the Bauer–Fike theorem below requires in addition that A be nonsingular.

Theorem 2.1 If A is diagonalizable and nonsingular, then

$$\min_i \frac{|\lambda_i - \hat{\lambda}|}{|\lambda_i|} \le \kappa(X)\, \|A^{-1}E\|,$$

where $\kappa(X) \equiv \|X\|\, \|X^{-1}\|$.

Proof. (See Eisenstat and Ipsen (1997), Corollary 2.2.) The idea is to 'divide' by the eigenvalues of A and apply the absolute error bound (2.1).

Write $(A+E)\hat{x} = \hat{\lambda}\hat{x}$ as

$$(\hat{\lambda}A^{-1} - A^{-1}E)\hat{x} = \hat{x}.$$

This means that 1 is an eigenvalue of $\hat{\lambda}A^{-1} - A^{-1}E$. The matrix $\hat{\lambda}A^{-1}$ has the same eigenvector matrix as A and its eigenvalues are $\hat{\lambda}/\lambda_i$. Apply the

Bauer–Fike theorem (2.1) to $\hat{\lambda}A^{-1}$ and to the perturbed matrix $\hat{\lambda}A^{-1} - A^{-1}E$. □

If we interpret the amplifier $\kappa(X)$ as a condition number for the eigenvalues of A, then absolute and relative error bounds have the same condition number. In the eyes of the Bauer–Fike theorem, this means an eigenvalue is as sensitive in the absolute sense as it is in the relative sense. A comparison of the absolute bound (2.1) and the relative bound in Theorem 2.1 shows that the absolute error is bounded in terms of the absolute perturbation E, while the relative error is bounded in terms of the relative perturbation $A^{-1}E$.

But $A^{-1}E$ is not the only way to express a relative perturbation. Why keep A^{-1} on the left of E? Why not move it to the right, or distribute it on both sides of E? Splitting $A = A_1A_2$ and sandwiching E between the two factors, like $A_1^{-1}EA_2^{-1}$, results in undreamt-of possibilities for relative perturbations.

Theorem 2.2 Let A be diagonalizable and nonsingular. If $A = A_1A_2$ where A_1 and A_2 commute, then

$$\min_i \frac{|\lambda_i - \hat{\lambda}|}{|\lambda_i|} \leq \kappa(X)\, \|A_1^{-1}EA_2^{-1}\|.$$

Proof. (See Eisenstat and Ipsen (1997), Corollary 2.4.) The idea is to apply Theorem 2.1 to the similarity transformations

$$A_2AA_2^{-1} \qquad \text{and} \qquad A_2(A+E)A_2^{-1}.$$

Fortunately, similarity transformations preserve eigenvalues. And the commutativity of A_1 and A_2 prevents the similarity from changing A,

$$A_2\, A\, A_2^{-1} = A_2\,(A_1A_2)\,A_2^{-1} = A_2A_1 = A_1A_2 = A.$$

Therefore we retain the condition number of the original eigenvector matrix X. □

When $A_1 = A$ and $A_2 = I$, Theorem 2.2 reduces to the relative bound in Theorem 2.1. Setting $A_1 = I$ and $A_2 = A$ gives (Eisenstat and Ipsen 1997, Corollary 2.5)

$$\min_i \frac{|\lambda_i - \hat{\lambda}|}{|\lambda_i|} \leq \kappa(X)\, \|EA^{-1}\|.$$

This bound includes Theorem 3.17 of Veselić and Slapničar (1993) as a special case. Another popular choice for A_1 and A_2 is a square root $A^{1/2}$ of A. In this case Theorem 2.2 gives (Eisenstat and Ipsen 1997, Corollary 2.6)

$$\min_i \frac{|\lambda_i - \hat{\lambda}|}{|\lambda_i|} \leq \kappa(X)\, \|A^{-1/2}EA^{-1/2}\|.$$

2.2. Bauer–Fike-type bounds for normal matrices

Normal matrices have unitary eigenvector matrices but, in contrast to Hermitian or real symmetric matrices, their eigenvalues are not necessarily real. Normal matrices include Hermitian, skew-Hermitian, real symmetric, real skew-symmetric, diagonal, unitary and real orthogonal matrices.

Since the condition number of a unitary eigenvector matrix equals one, the Bauer–Fike theorem applied to a normal matrix A simplifies. The absolute error bound (2.1) becomes

$$\min_i |\lambda_i - \hat{\lambda}| \leq \|E\|,$$

while the corresponding relative bound requires again that A be nonsingular.

Theorem 2.3 Let A be normal and nonsingular. If $A = A_1 A_2$, where A_1 and A_2 commute, then

$$\min_i \frac{|\lambda_i - \hat{\lambda}|}{|\lambda_i|} \leq \|A_1^{-1} E A_2^{-1}\|.$$

Proof. This follows immediately from Theorem 2.2. □

Therefore eigenvalues of normal matrices are well conditioned, in the absolute as well as in many relative senses. The relative bound in Theorem 2.3 is tight for diagonal matrices A and component-wise perturbations E, like those in Example 1.1.

For the relative bound to remain in effect, A_1 and A_2 have to commute. Our choices for 'commuting factorizations' have been, so far:

$$(A_1, A_2) = (A, I), \qquad (A_1, A_2) = (I, A), \qquad (A_1, A_2) = (A^{1/2}, A^{1/2}).$$

But since A is normal there is another commuting factorization: the polar factorization. Every square matrix A has a polar factorization $A = HU$, where $H \equiv (AA^*)^{1/2}$ is Hermitian positive-semidefinite and U is unitary (Horn and Johnson 1985, Theorem 7.3.2). The matrix H is always unique, while U is only unique when A is nonsingular. In particular, when A is Hermitian positive-definite, $H = A$ and U is the identity. We use the fact that polar factors of normal nonsingular matrices commute in the following sense (Eisenstat and Ipsen 1997, Lemma 3.2):

$$HU = UH = H^{1/2} U H^{1/2}.$$

Theorem 2.4 If A is normal and nonsingular, with Hermitian positive-definite polar factor H, then

$$\min_i \frac{|\lambda_i - \hat{\lambda}|}{|\lambda_i|} \leq \|H^{-1/2} E H^{-1/2}\|.$$

Proof. (See Eisenstat and Ipsen (1997), Theorem 3.3.) As $A = H^{1/2}UH^{1/2}$, we can set $A_1 = H^{1/2}U$ and $A_2 = H^{1/2}$ in Theorem 2.3 to get

$$\|A_1^{-1}EA_2^{-1}\| = \|U^*H^{-1/2}EH^{-1/2}\| = \|H^{-1/2}EH^{-1/2}\|.$$

□

Therefore the eigenvalues of a normal matrix have the same relative error bound as the eigenvalues of its positive-definite polar factor. This suggests that the eigenvalues of a normal matrix are as well conditioned as the eigenvalues of its positive-definite polar factor. More generally we conclude that eigenvalues of normal matrices are no more sensitive than eigenvalues of Hermitian positive-definite matrices.

2.3. Hoffman–Wielandt-type bounds for diagonalizable matrices

The Hoffman–Wielandt theorem establishes a one-to-one pairing between all eigenvalues of A and $A+E$ and bounds the sum of all pairwise distances in the Frobenius norm. This requires not only A but also $A+E$ to be diagonalizable.

Let A and $A+E$ be diagonalizable matrices with eigendecompositions $A = X\Lambda X^{-1}$ and $A+E = \hat{X}\hat{\Lambda}\hat{X}^{-1}$, respectively. The eigenvalues are

$$\Lambda = \begin{pmatrix} \lambda_1 & & \\ & \ddots & \\ & & \lambda_n \end{pmatrix}, \qquad \hat{\Lambda} = \begin{pmatrix} \hat{\lambda}_1 & & \\ & \ddots & \\ & & \hat{\lambda}_n \end{pmatrix}.$$

The extension of the Hoffman–Wielandt theorem from normal to diagonalizable matrices (Elsner and Friedland 1995, Theorem 3.1) bounds the absolute error,

$$\sqrt{\sum_{i=1}^{n} |\lambda_i - \hat{\lambda}_{\tau(i)}|^2} \leq \kappa(\hat{X})\,\kappa(X)\,\|E\|_F, \tag{2.2}$$

for some permutation τ. The condition numbers $\kappa(X)$ and $\kappa(\hat{X})$ are expressed in the two-norm to make the bound tighter, since the two-norm never exceeds the Frobenius norm.

We can obtain a relative Hoffman–Wielandt-type bound from a stronger version of (2.2) that deals with eigenvalues of matrix products. To this end write the perturbed matrix as $AC+E$, where C must have the same eigenvector matrix as $AC+E$. The bound (2.2) is the special case where $C = I$. The eigendecomposition of C is

$$C = \hat{X}\Gamma\hat{X}^{-1}, \qquad \text{where} \quad \Gamma = \begin{pmatrix} \gamma_1 & & \\ & \ddots & \\ & & \gamma_n \end{pmatrix}.$$

The eigendecompositions of A and the perturbed matrix remain the same:

$$A = X\Lambda X^{-1}, \qquad AC + E = \hat{X}\hat{\Lambda}\hat{X}^{-1}.$$

The stronger Hoffman–Wielandt-type bound given below bounds the sum of squares of absolute errors in the products of the eigenvalues of A and C.

Lemma 2.5 If A, C and $AC + E$ are diagonalizable, then there exists a permutation τ such that

$$\sqrt{\sum_{i=1}^{n} |\lambda_i \gamma_{\tau(i)} - \hat{\lambda}_{\tau(i)}|^2} \leq \kappa(\hat{X})\, \kappa(X)\, \|E\|_F.$$

Proof. See Eisenstat and Ipsen (1997), Theorem 5.1. □

Now we are ready for the relative bound. The stronger absolute bound in Lemma 2.5 implies a relative version of the original Hoffman–Wielandt-type bound (2.2), provided A is nonsingular.

Theorem 2.6 Let A and $A + E$ be diagonalizable. If A is nonsingular then there exists a permutation τ such that

$$\sqrt{\sum_{i=1}^{n} \left(\frac{|\lambda_i - \hat{\lambda}_{\tau(i)}|}{|\lambda_i|} \right)^2} \leq \kappa(\hat{X})\, \kappa(X)\, \|A^{-1}E\|_F.$$

Proof. (See Eisenstat and Ipsen (1997), Corollary 5.2.) Since $A^{-1}(A+E) - A^{-1}E = I$ we can set

$$\bar{A} \equiv A^{-1}, \qquad C \equiv A + E, \qquad \bar{E} \equiv -A^{-1}E.$$

Then $\bar{A}$ is diagonalizable with eigenvector matrix X and eigenvalues λ_i^{-1}; C is diagonalizable with eigenvector matrix $\hat{X}$ and eigenvalues $\hat{\lambda}_i$; and $\bar{A}C + \bar{E} = \hat{X} I \hat{X}^{-1}$ is diagonalizable, where the eigenvalues are 1 and one can choose $\hat{X}$ as an eigenvector matrix. Applying Lemma 2.5 to $\bar{A}$, C and $\bar{E}$ gives

$$\sum_{i=1}^{n} |\lambda_i^{-1} \hat{\lambda}_{\tau(i)} - 1|^2 \leq \kappa(\hat{X})^2 \kappa(X)^2 \|A^{-1}E\|_F^2.$$

□

2.4. Hoffman–Wielandt-type bounds for Hermitian matrices

When A and $A + E$ are Hermitian, the permutation in the Hoffman–Wielandt theorem is the identity, provided exact and perturbed eigenvalues are numbered as

$$\lambda_n \leq \cdots \leq \lambda_1, \qquad \hat{\lambda}_n \leq \cdots \leq \hat{\lambda}_1.$$

The Hoffman–Wielandt theorem for Hermitian matrices (see Bhatia (1997), Exercise III.6.15, and Löwner (1934)) bounds the absolute error by

$$\sqrt{\sum_{i=1}^{n} |\lambda_i - \hat{\lambda}_i|^2} \leq \|E\|_F. \tag{2.3}$$

The relative bound below requires in addition that A and $A+E$ be positive-definite.

Theorem 2.7 If A and $A+E$ are Hermitian positive-definite, then

$$\sqrt{\sum_{i=1}^{n} \frac{|\lambda_i - \hat{\lambda}_i|^2}{\lambda_i \hat{\lambda}_i}} \leq \|A^{-1/2}EA^{-1/2}\,(I + A^{-1/2}EA^{-1/2})^{-1/2}\|_F.$$

Proof. See Li (1994*a*), Theorem 3.2, and Li and Mathias (1997), Proposition 3.4′. □

As a consequence, a small $\|A^{-1/2}EA^{-1/2}\|_F$ guarantees a small eigenvalue error. If one does not mind dealing with majorization theory, one can derive bounds that are stronger than Theorem 2.7 and hold for any unitarily invariant norm (Li and Mathias 1997, Proposition 3.4, (3.19)).

2.5. Weyl-type bounds

Weyl's perturbation theorem (Bhatia 1997, Corollary III.2.6) bounds the worst distance between the ith eigenvalues of Hermitian matrices A and $A+E$ in the two-norm

$$\max_{1\leq i\leq n} |\lambda_i - \hat{\lambda}_i| \leq \|E\|. \tag{2.4}$$

The absolute bound (2.4) implies a relative bound, provided that A is positive-definite. There is no restriction on E other than being Hermitian.

Theorem 2.8 Let A and $A+E$ be Hermitian. If A is also positive-definite, then

$$\max_{1\leq i\leq n} \frac{|\lambda_i - \hat{\lambda}_i|}{|\lambda_i|} \leq \|A^{-1/2}EA^{-1/2}\|.$$

Proof. (See Eisenstat and Ipsen (1997), Corollary 4.2, and Mathias (1997*b*), Theorem 2.3.) We reproduce the proof from Eisenstat and Ipsen (1997) because it explains how the absolute bound (2.4) implies the relative bound. Fix an index i. Let $\hat{x}$ be an eigenvector of $A+E$ associated with $\hat{\lambda}_i$, that is,

$$(A+E)\hat{x} = \hat{\lambda}_i \hat{x}.$$

Multiplying $(\hat{\lambda}_i I - E)\hat{x} = A\hat{x}$ by $A^{-1/2}$ on both sides gives

$$(\bar{A} + \bar{E})\, z = z,$$

where

$$\bar{A} \equiv \hat{\lambda}_i A^{-1}, \qquad \bar{E} \equiv -A^{-1/2} E A^{-1/2}, \qquad z \equiv A^{1/2}\hat{x}.$$

Hence 1 is an eigenvalue of $\bar{A} + \bar{E}$.

We show that it is actually the $(n-i+1)$st eigenvalue and argue as in the proof of Theorem 2.1 of Eisenstat and Ipsen (1995). Since $\hat{\lambda}_i$ is the ith eigenvalue of $A+E$, 0 must be the ith eigenvalue of

$$(A+E) - \hat{\lambda}_i I = A^{1/2}\left(I - \bar{A} - \bar{E}\right) A^{1/2}.$$

But this is a congruence transformation because square roots of positive-definite matrices are Hermitian. Congruence transformations preserve the inertia. Hence 0 is the ith eigenvalue of $I - \bar{A} - \bar{E}$, and 1 is the $(n-i+1)$st eigenvalue of $\bar{A} + \bar{E}$.

Applying Weyl's theorem (2.4) to $\bar{A}$ and $\bar{A} + \bar{E}$ gives

$$\max_{1 \le j \le n} \left| \frac{\hat{\lambda}_i}{\lambda_{n-j+1}} - \mu_j \right| \le \|\bar{E}\| = \|A^{-1/2} E A^{-1/2}\|,$$

where μ_j are the eigenvalues of $\bar{A} + \bar{E}$. When $j = n-i+1$, then $\mu_j = 1$ and we get the desired bound. □

The following example illustrates what the relative bound in Theorem 2.8 looks like when E is a component-wise relative perturbation.

Example 2.1 (See Mathias (1997*b*), pp. 6, 7.) Let's first subject a single diagonal element of a Hermitian positive-definite matrix A to a component-wise relative perturbation. Say, a_{jj} is perturbed to $a_{jj}(1+\epsilon)$. The perturbed matrix is $A+E$, where $E = \epsilon e_j e_j^T$ and e_j is the jth column of the identity matrix. Then

$$\begin{aligned} \|A^{-1/2} E A^{-1/2}\| &= |\epsilon| \, \|A^{-1/2} e_j e_j^T A^{-1/2}\| = |\epsilon| \, |e_j^T A^{-1/2} A^{-1/2} e_j| \\ &= |\epsilon| \, (A^{-1})_{jj}, \end{aligned}$$

where $(A^{-1})_{jj}$ is the jth diagonal element of A^{-1} (which is positive since A is positive-definite). The relative error bound in Theorem 2.8 is

$$\max_{1 \le i \le n} \frac{|\lambda_i - \hat{\lambda}_i|}{|\lambda_i|} \le |\epsilon| \, (A^{-1})_{jj}.$$

This means that a small relative error in a diagonal element of a Hermitian positive-definite matrix causes only a small relative error in the eigenvalues if the corresponding diagonal element of the inverse is not much larger than one.

Next we'll subject a pair of off-diagonal elements to a component-wise relative perturbation. Say, a_{jk} and a_{kj} are perturbed to $a_{jk}(1+\epsilon)$ and $a_{kj}(1+\epsilon)$, respectively. The perturbed matrix is $A+E$, where $E = \epsilon(e_j e_k^T +$

$e_j e_k^T$). In this case we get

$$\|A^{-1/2}EA^{-1/2}\| \leq 2|\epsilon| \sqrt{(A^{-1})_{jj}(A^{-1})_{kk}}.$$

The relative error bound in Theorem 2.8 becomes

$$\max_{1\leq i\leq n} \frac{|\lambda_i - \hat{\lambda}_i|}{|\lambda_i|} \leq 2|\epsilon| \sqrt{(A^{-1})_{jj}(A^{-1})_{kk}}.$$

This means that a small relative error in a pair of off-diagonal elements of a Hermitian positive-definite matrix causes only a small relative error in the eigenvalues if the product of the corresponding diagonal elements in the inverse is not much larger than one. □

The bound below, for a different error measure, is similar to the Frobenius norm bound in Theorem 2.7.

Theorem 2.9 If A and $A+E$ are Hermitian positive-definite then

$$\max_{1\leq i\leq n} \frac{|\lambda_i - \hat{\lambda}_i|}{\sqrt{\lambda_i \hat{\lambda}_i}} \leq \|A^{-1/2}EA^{-1/2}\,(I + A^{-1/2}EA^{-1/2})^{-1/2}\|.$$

Proof. See Li (1994*a*), Theorem 3.2, and Li and Mathias (1997), Proposition 3.4′. □

2.6. Weyl-type bounds for more restrictive perturbations

It is possible to get a Weyl-type bound for eigenvalues of Hermitian matrices without officially asking for positive-definiteness. The price to be paid, however, is a severe restriction on E to prevent perturbed eigenvalues from switching sign.

Theorem 2.10 Let A and $A+E$ be Hermitian. If $0 < \epsilon_l \leq \epsilon_u$ and

$$\epsilon_l\, x^*Ax \leq x^*Ex \leq \epsilon_u\, x^*Ax \qquad \text{for all } x,$$

then

$$\epsilon_l\, \lambda_i \leq \hat{\lambda}_i - \lambda_i \leq \epsilon_u\, \lambda_i, \quad 1 \leq i \leq n.$$

Proof. (This is a consequence of Barlow and Demmel (1990), Lemma 1.) The Minimax principle for eigenvalues of Hermitian matrices (Bhatia 1997, Corollary III.1.2) implies

$$\lambda_i = \max_{\dim(S)=i} \min_{x\in S} \frac{x^*Ax}{x^*x} = \min_{x\in S_0} \frac{x^*Ax}{x^*x},$$

for some subspace S_0 of dimension i. Then

$$\hat{\lambda}_i = \max_{\dim(S)=i} \min_{x\in S} \frac{x^*(A+E)x}{x^*x} \geq \min_{x\in S_0} \frac{x^*(A+E)x}{x^*x} = \frac{x_0^*(A+E)x_0}{x_0^*x_0}$$

for some $x_0 \in S_0$. The above expression for λ_i and the assumption imply

$$\frac{x_0^*(A+E)x_0}{x_0^*x_0} \geq \min_{x \in S_0} \frac{x^*Ax}{x^*x} + \epsilon_l \frac{x_0^*Ax_0}{x_0^*x_0} \geq \lambda_i + \epsilon_l \lambda_i,$$

where we have also used the fact that $\epsilon_l > 0$. Hence $\epsilon_l \lambda_i \leq \hat{\lambda}_i - \lambda_i$. The upper bound is proved similarly using the characterization

$$\lambda_i = \min_{\dim(S)=n-i+1} \max_{x \in S} \frac{x^*Ax}{x^*x}.$$

□

Therefore the relative error in the eigenvalues lies in the same interval as the relative perturbation. The relative error bound in Theorem 2.10 implies

$$(1+\epsilon_l)\,\lambda_i \leq \hat{\lambda}_i \leq (1+\epsilon_u)\,\lambda_i,$$

where ϵ_l and ϵ_u are positive. Hence $\hat{\lambda}_i$ has the same sign as λ_i, and $|\hat{\lambda}_i| > |\lambda_i|$. Thus the restriction on the perturbation is strong enough that it not only forces A and $A+E$ to have the same inertia, but it also pushes the perturbed eigenvalues farther from zero than the exact eigenvalues.

The restriction on the perturbation in the following bound is slightly weaker. It uses the polar factor technology from Theorem 2.4.

Theorem 2.11 Let A and $A+E$ be Hermitian. If H is the positive-semidefinite polar factor of A and if, for some $0 < \epsilon < 1$,

$$|x^*Ex| \leq \epsilon\, x^*Hx \qquad \text{for all } x,$$

then

$$|\lambda_i - \hat{\lambda}_i| \leq \epsilon\, |\lambda_i|.$$

Proof. This is a consequence of Veselić and Slapničar (1993), Theorem 2.1. The assumption implies

$$x^*(A-\epsilon H)x \leq x^*(A+E)x \leq x^*(A+\epsilon H)x.$$

If $A = X\Lambda X^*$ is an eigendecomposition of A, then, because A is Hermitian, the polar factor is $H = X|\Lambda|X^*$, where $|\Lambda|$ is the matrix whose elements are the absolute values of Λ. Hence A and H have the same eigenvectors. A min-max argument as in the proof of Theorem 2.10 establishes

$$\lambda_i - \epsilon|\lambda_i| \leq \hat{\lambda}_i \leq \lambda_i + \epsilon|\lambda_i|.$$

□

Therefore the relative error in the eigenvalues is small if the relative perturbation with regard to the polar factor is small. Since $0 < \epsilon < 1$ the relative error bound implies that $\hat{\lambda}_i$ has the same sign as λ_i. Hence the assumptions in Theorem 2.11 ensure that $A+E$ has the same inertia as A.

Theorem 2.11 applies to component-wise relative perturbations. Matrix inequalities below of the form $|E| \le \epsilon|A|$ are to be interpreted element-wise.

Corollary 2.12 Let A and $A+E$ be Hermitian, and $|E| \le \epsilon\,|A|$ for some $\epsilon > 0$. If H is the positive-definite polar factor of A and, for some $\eta > 0$,

$$\epsilon\,|x|^*|A||x| \le \eta\, x^* H x \qquad \text{for all } x,$$

then

$$|\lambda_i - \hat{\lambda}_i| \le \eta\,|\lambda_i|.$$

Proof. This is a consequence of Veselić and Slapničar (1993), Theorem 2.11. We merely need to verify that the assumptions of Theorem 2.11 hold,

$$|x^*Ex| \le |x|^*|E||x| \le \epsilon\,|x|^*|A||x| \le \eta\, x^*Hx.$$

□

2.7. Congruence transformations for positive-definite matrices

All the bounds we have presented so far for positive-definite matrices contain the term $A^{-1/2}EA^{-1/2}$. Since A is Hermitian positive-definite, $A^{1/2}$ is Hermitian, which makes $A^{-1/2}EA^{-1/2}$ Hermitian. This in turn implies that the two-norm and Frobenius norm of $A^{-1/2}EA^{-1/2}$ are invariant under congruence transformations. We say that two square matrices A and M are *congruent* if $A = D^*MD$ for some nonsingular matrix D. If A is Hermitian positive-definite, so is M because congruence transformations preserve the inertia.

We start out by showing that the bound in Theorem 2.8 is invariant under congruence transformations.

Corollary 2.13 Let A and be Hermitian positive-definite and $A+E$ Hermitian. If $A = D^*MD$ and $A+E = D^*(M+F)D$, where D is nonsingular, then

$$\max_{1\le i\le n} \frac{|\lambda_i - \hat{\lambda}_i|}{|\lambda_i|} \le \|M^{-1/2}FM^{-1/2}\|.$$

Proof. (See Mathias (1997*b*), Theorem 2.4.) Start with the bound in Theorem 2.8,

$$\max_{1\le i\le n} \frac{|\lambda_i - \hat{\lambda}_i|}{|\lambda_i|} \le \|A^{-1/2}EA^{-1/2}\|.$$

Positive-definiteness is essential here. Since A is Hermitian positive-definite, it has a Hermitian square root $A^{1/2}$. Hence $A^{-1/2}EA^{-1/2}$ is Hermitian. This implies that the norm is an eigenvalue,

$$\|A^{-1/2}EA^{-1/2}\| = \max_{1\le j\le n} |\lambda_j(A^{-1/2}EA^{-1/2})|.$$

Now comes the trick. Since eigenvalues are preserved under similarity transformations, we can reorder the matrices in a circular fashion until all grading matrices have cancelled each other out, that is,

$$\begin{aligned}\lambda_j(A^{-1/2}EA^{-1/2}) &= \lambda_j(A^{-1}E) = \lambda_j(D^{-1}\,M^{-1}F\,D) = \lambda_j(M^{-1}F)\\ &= \lambda_j(M^{-1/2}FM^{-1/2}).\end{aligned}$$

At last we recover the norm

$$\max_{1\le j\le n} |\lambda_j(M^{-1/2}FM^{-1/2})| = \|M^{-1/2}FM^{-1/2}\|.$$

□

Corollary 2.13 extends Theorem 2 of Barlow and Demmel (1990) and Theorem 2.3 of Demmel and Veselić (1992) to a larger class of matrices. It suggests that the eigenvalues of $A+E$ have the same error bound as the eigenvalues of $M+F$. We can interpret this to mean that the eigenvalues of a Hermitian positive-definite matrix behave as well as the eigenvalues of any matrix congruent to A. The example below illustrates this.

Example 2.2 (See Demmel and Veselić (1992), p. 1211.) The matrix

$$A = \begin{pmatrix} 10^{40} & 10^{29} & 10^{19}\\ 10^{29} & 10^{20} & 10^{9}\\ 10^{19} & 10^{9} & 1 \end{pmatrix}$$

is symmetric positive-definite with eigenvalues (to six decimal places)

$$1.00000\cdot 10^{40}, \qquad 9.90000\cdot 10^{19}, \qquad 9.81818\cdot 10^{-1}.$$

If we write $A = DMD$, where

$$M = \begin{pmatrix} 1 & .1 & .1\\ .1 & 1 & .1\\ .1 & .1 & 1 \end{pmatrix}, \qquad D = \begin{pmatrix} 10^{20} & & \\ & 10^{10} & \\ & & 1 \end{pmatrix},$$

then the eigenvalues of M are (to six decimal places)

$$9.00000\cdot 10^{-1}, \qquad 9.00000\cdot 10^{-1}, \qquad 1.20000.$$

Corollary 2.13 implies that the widely varying eigenvalues of A, and in particular the very small ones, are as impervious to changes in M as the uniformly sized eigenvalues of M.

As long as 30 years ago, structural engineers considered congruence transformations like the one above where D is diagonal and all diagonal elements of M are equal to one (Rosanoff, Glouderman and Levy 1968, pp. 1041,1050). They observed that such an equilibration 'reduce[s] the ratio of extreme eigenvalues' (Rosanoff et al. 1968, p. 1045), and that 'equilibration is of major importance in measurement of matrix conditioning' (Rosanoff et al. 1968, p. 1059). □

From the circular reordering argument in the proof of Corollary 2.13 it also follows that the other bounds for positive-definite matrices are invariant under congruences. One bound is Theorem 2.9.

Corollary 2.14 Let A and $A+E$ be Hermitian positive-definite. If $A = D^*MD$ and $A+E = D^*(M+F)D$, where D is nonsingular, then

$$\max_{1\le i\le n} \frac{|\lambda_i - \hat{\lambda}_i|}{\sqrt{\lambda_i \hat{\lambda}_i}} \le \|M^{-1/2}FM^{-1/2}\,(I + M^{-1/2}FM^{-1/2})^{-1/2}\|.$$

Proof. See Li (1994*a*), Theorem 3.2, and Li and Mathias (1997), Proposition 3.4′. □

The other bound that is also invariant under congruences is the Frobenius norm bound Theorem 2.7.

Corollary 2.15 Let A and $A+E$ be Hermitian positive-definite. If $A = D^*MD$ and $A+E = D^*(M+F)D$, where D is nonsingular, then

$$\sqrt{\sum_{i=1}^{n} \frac{|\lambda_i - \hat{\lambda}_i|^2}{\lambda_i \hat{\lambda}_i}} \le \|M^{-1/2}FM^{-1/2}\,(I + M^{-1/2}FM^{-1/2})^{-1/2}\|_F.$$

Proof. (See Li (1994*a*), Theorem 3.2, and Li and Mathias (1997), Proposition 3.4′.) Since the Frobenius norm sums up squares of eigenvalues, the bound from Theorem 2.7 can be written as

$$\|A^{-1/2}EA^{-1/2}\,(I + A^{-1/2}EA^{-1/2})^{-1/2}\|_F^2 = \sum_{i=1}^{n} \frac{\mu_i^2}{1+\mu_i},$$

where μ_i are the eigenvalues of the Hermitian matrix $A^{-1/2}EA^{-1/2}$. The circular reordering argument from the proof of Corollary 2.13 implies that μ_i are also the eigenvalues of $M^{-1/2}FM^{-1/2}$. □

One may wonder what's so interesting about congruence transformations. One can use congruence transformations to pull the grading out of a matrix (see Barlow and Demmel (1990), Section 2, Demmel and Veselić (1992), Sections 1, 2.1, and Mathias (1995). Consider the matrix A in Example 2.2. It has elements of widely varying magnitude that decrease from top to bottom. The diagonal matrix D removes the grading and produces a matrix M, where $M \equiv D^{-1}AD^{-1}$, all of whose elements have about the same order of magnitude and all of whose eigenvalues are of about the same size.

More generally we say that a Hermitian positive-definite matrix A is *graded*, or *scaled*, if $A = DMD^*$ and the eigenvalues of M vary much less in magnitude than the eigenvalues of A (Mathias 1995, Section 1).

2.8. Congruence transformations for indefinite matrices

Since the application of congruence transformations is not restricted to Hermitian positive-definite matrices, we may as well try to find out whether indefinite matrices are invariant under congruences. It turns out that the resulting error bounds are weaker than the ones for positive-definite matrices because they require stronger assumptions.

If we are a little sneaky (by extracting the congruence from the polar factor rather than the matrix proper) then the bound for normal matrices in Theorem 2.4 becomes invariant under congruences.

Corollary 2.16 Let A be normal and nonsingular, with Hermitian positive-definite polar factor H. If D is nonsingular and

$$E = DE_1D^*, \qquad H = DM_1D^*,$$

then

$$\min_i \frac{|\lambda_i - \hat{\lambda}|}{|\lambda_i|} \leq \|M_1^{-1/2} E_1 M_1^{-1/2}\|.$$

Proof. See Eisenstat and Ipsen (1997), Corollary 3.4. □

This means the error bound for eigenvalues of a normal matrix is the same as the error bound for eigenvalues of the best scaled version of its positive-definite polar factor.

Let's return to Weyl-type bounds, but now under the condition that the congruence transformation is real diagonal. Theorem 2.11 leads to a bound that is essentially scaling invariant. It is similar to the one above, in the sense that the scaling matrix is extracted from its positive-definite polar factor. However, now the perturbations are restricted to be component-wise relative.

Corollary 2.17 Let $A = DMD$ be nonsingular Hermitian, where D is diagonal with positive diagonal elements, and let $H = DM_1D$ be the positive-definite polar factor of A. If $A + E$ is Hermitian and $|E| \leq \epsilon\, |A|$ for some $\epsilon > 0$, then

$$\max_{1 \leq i \leq n} \frac{|\lambda_i - \hat{\lambda}_i|}{|\lambda_i|} \leq \epsilon \|\,|M|\,\|\,\|M_1^{-1}\|.$$

Proof. This is a consequence of Veselić and Slapničar (1993), Theorem 2.13. We use variational inequalities to show that the assumptions of Corollary 2.12 are fulfilled. Since D is positive-definite,

$$|x|^*|A|\,|x| = |x|^* D\,|M|\,D|x| \leq \|\,|M|\,\|\, x^* D^2 x \qquad \text{for all } x.$$

Variational inequalities imply

$$x^* D^2 x \leq \|M_1^{-1}\|\, x^* H x.$$

Therefore

$$\epsilon\, |x|^*|A|\, |x| \le \eta\, x^* H x \qquad \text{for all } x,$$

where $\eta \equiv \epsilon\, \|\, |M|\, \|\, \|M_1^{-1}\|$. Now apply Corollary 2.12. □

The amplifying factor $\|\, |M|\, \|\, \|M_1^{-1}\|$ in the bound is almost like the condition number of the absolute value matrix $|M|$, or the condition number of the scaled polar factor M_1. Therefore the relative error in the eigenvalues of a Hermitian matrix is small if the polar factor and the absolute value of the matrix are well scaled.

The following bound is similar in the sense that it applies to a column scaling of a Hermitian matrix $A = MD$. In contrast to Corollary 2.17, however, the scaled matrix M is in general no longer Hermitian, and the inverse M^{-1} now appears in the bound rather than the inverse M_1^{-1} of the scaled polar factor.

Corollary 2.18 Let $A = MD$ be nonsingular Hermitian and D diagonal with positive diagonal elements. If $A + E$ is Hermitian, and $|E| \le \epsilon\, |A|$ for some $\epsilon > 0$, then

$$\max_{1\le i\le n} \frac{|\lambda_i - \hat{\lambda}_i|}{|\lambda_i|} \le \epsilon \|\, |M|\, \|\, \|M^{-1}\|.$$

Proof. (See Veselić and Slapničar (1993), Theorem 3.16.) First we take care of the scaling matrix D. The technology of previous proofs requires that D appear on both sides of the matrix. That's why we consider $A^2 = A^*A = DM^*MD$. The component-wise perturbation implies

$$|x^*E^2x| \le \epsilon^2\, |x|^*|A|^2\, |x|.$$

Proceeding as in the proof of Corollary 2.17 gives

$$|x|^*|A|^2\, |x| \le \|\, |M|\, \|^2\, \|M^{-1}\|^2\, x^*A^2x \qquad \text{for all } x.$$

Hence

$$|x^*E^2x| \le \eta^2\, x^*A^2x,$$

where $\eta \equiv \epsilon \|\, |M|\, \|\, \|M_1^{-1}\|$. Now that we have got rid of D, we need to undo the squares. In order to take the positive square root without losing the monotonicity, we need positive-definite matrices under the squares. Polar factors do the job.

If H and H_E are the Hermitian positive-definite polar factors of A and E, respectively, then

$$x^*E^2x = x^*H_E^2x, \qquad x^*A^2x = x^*H^2x.$$

Therefore

$$x^*H_E^2x \le \eta^2\, x^*H^2x \qquad \text{for all } x.$$

Now comes the trick. Because H_E and H are Hermitian positive-definite we can apply the fact that the square root is operator-monotone (Bhatia 1997, Proposition V.I.8) and conclude

$$x^* H_E x \leq \eta\, x^* H x.$$

Since $|x^* E x| \leq x^* H_E x$, Theorem 2.11 applies. □

The next bound, the last one in this section, applies to general perturbations. Compared to other bounds it severely constrains the size of the perturbation by forcing it to be smaller than any eigenvalue of any principal submatrix.

Theorem 2.19 Let $A = DMD$ and $A + E = D(M + F)D$ be real, symmetric and D a real nonsingular diagonal matrix. Among all eigenvalues of principal submatrices of M, let μ be the smallest in magnitude. If $\|F\| < |\mu|$ then

$$-\|F\| \frac{2|\mu| - \|F\|}{|\mu|^2} \leq \frac{\hat{\lambda}_i - \lambda_i}{\lambda_i} \leq \|F\| \frac{2|\mu| - \|F\|}{(|\mu| - \|F\|)^2}, \quad 1 \leq i \leq n.$$

Proof. See Gu and Eisenstat (1993), Corollary 5. □

2.9. Ritz values

Ritz values are 'optimal' approximations to the eigenvalues of Hermitian matrices.

Let A be a Hermitian matrix of order n and Q a matrix with m orthonormal columns. Then $W \equiv Q^* A Q$ is a matrix of order m whose eigenvalues

$$\hat{\lambda}_1 \geq \cdots \geq \hat{\lambda}_m,$$

are called *Ritz values* of A (Parlett 1980, Section 11.3). The corresponding residual is $R \equiv AQ - QW$. Ritz values are optimal in the following sense. Given Q, the norm of R can only increase if we replace W by another matrix, that is (Parlett 1980, Theorem 11-4-5),

$$\|R\| = \|AQ - QW\| \leq \|AQ - QC\|$$

for all matrices C of order m.

Moreover, one can always find m eigenvalues of A that are within absolute distance $\|R\|$ of the Ritz values (Parlett 1980, Theorem 11-5-1),

$$\max_{1 \leq j \leq m} |\lambda_{\tau(j)} - \hat{\lambda}_j| \leq \|R\|$$

for some permutation τ.

Unfortunately the corresponding relative error bounds are not as simple. They are expressed in terms of angles between the subspaces range(Q), range(AQ), and range($A^{-1}Q$). Let $0 \leq \theta_1 \leq \pi/2$ be the maximal principal angle between range(Q) and range(AQ), and $0 \leq \theta_2 \leq \pi/2$ be the maximal principal angle between range(AQ) and range($A^{-1}Q$).

Theorem 2.20 If A is nonsingular Hermitian, then there exists a permutation τ such that

$$\max_{1 \leq i \leq m} \frac{|\lambda_{\tau(i)} - \hat{\lambda}_i|}{|\lambda_{\tau(i)}|} \leq \sin\theta_1 + \tan\theta_2.$$

Proof. (See Drmač (1996a), Theorem 3, Proposition 5.) In order to exhibit the connection to previous results we sketch the idea for the proof. First express the Ritz values as an additive perturbation. To this end define the Hermitian perturbation

$$E \equiv -(RQ^* + QR^*).$$

Then Q is an invariant subspace of $A + E$,

$$(A + E)Q = QW,$$

and the eigenvalues of W are eigenvalues of $A + E$.

Now proceed as in the proof of Corollary 2.18 and look at the squares,

$$x^*E^2x = x^*A^*\, A^{-*}E^*EA^{-1}\, Ax \leq \|EA^{-1}\|^2 x^*A^2x \qquad \text{for all } x.$$

Undo the squares using polar factors and the operator-monotonicity of the square root, and apply Theorem 2.11. Hence the eigenvalues $\mu_1 \geq \cdots \geq \mu_n$ of $A + E$ satisfy

$$\max_{1 \leq i \leq n} \frac{|\lambda_i - \mu_i|}{|\lambda_i|} \leq \|EA^{-1}\|.$$

Let $\hat{\lambda}_1 \geq \cdots \geq \hat{\lambda}_m$ be those μ_i that are also eigenvalues of W, and let τ be a permutation that numbers the eigenvalues of A corresponding to μ_i first. Then

$$\max_{1 \leq i \leq m} \frac{|\lambda_{\tau(i)} - \hat{\lambda}_i|}{|\lambda_{\tau(i)}|} \leq \|EA^{-1}\|.$$

We still have to worry about $\|EA^{-1}\|$. Write

$$-EA^{-1} = (I - QQ^*)AQQ^*A^{-1} + QQ^*(I - AQQ^*A^{-1}).$$

Here QQ^* is the orthogonal projector onto range(Q), while AQQ^*A^{-1} is the oblique projector onto range(AQ) along range(Q^*A^{-1}). This expression for EA^{-1} appears in Drmač (1996a), Theorem 3. It can be bounded above by $\sin\theta_1 + \tan\theta_2$. □

Therefore the relative error in the Ritz values of $W = Q^*AQ$ is small if both subspace angles θ_1 and θ_2 are small. Things simplify when the matrix A is also positive-definite because there is only one angle to deal with.

Let A be Hermitian positive-definite with Cholesky factorization $A = LL^*$. Let $0 \le \theta \le \pi/2$ be the maximal principal angle between range(L^*Q) and range$(L^{-1}Q)$.

Theorem 2.21 If A is Hermitian positive-definite and if $\sin\theta < 1$ then there exists a permutation τ so that

$$\max_{1\le i\le m} \frac{|\lambda_{\tau(i)} - \hat{\lambda}_i|}{|\lambda_{\tau(i)}|} \le \frac{\sin\theta}{1-\sin\theta}.$$

Proof. See Drmač (1996*a*), Theorem 6. □

Theorem 2.21 can be extended to semi-definite matrices (Drmač and Hari 1997).

3. Additive perturbations for singular values

Let B be a complex matrix. We want to estimate the absolute and the relative errors in the singular values of the perturbed matrix $B + F$. For definiteness we assume that B is tall and skinny, that is, B is $m \times n$ with $m \ge n$ (if this is not the case just consider B^*).

Perturbation bounds for singular values are usually derived by first converting the singular value problem to an Hermitian eigenvalue problem.

3.1. Converting singular values to eigenvalues

The singular value decomposition of a $m \times n$ matrix B, $m \ge n$, is

$$B = U \begin{pmatrix} \Sigma \\ 0 \end{pmatrix} V^*,$$

where the left singular vector matrix U and the right singular vector matrix V are unitary matrices of order m and n, respectively. The nonnegative diagonal matrix Σ of order n contains the singular values σ_i of B,

$$\Sigma = \begin{pmatrix} \sigma_1 & & \\ & \ddots & \\ & & \sigma_n \end{pmatrix},$$

where

$$\sigma_1 \ge \cdots \ge \sigma_n \ge 0.$$

There are two popular ways to convert a singular value problem to an eigenvalue problem.

- The eigenvalues of

$$A \equiv \begin{matrix} \\ m \\ n \end{matrix} \begin{matrix} m & n \\ \begin{pmatrix} 0 & B \\ B^* & 0 \end{pmatrix} \end{matrix}$$

are

$$\sigma_1, \ldots, \sigma_n, -\sigma_1, \ldots, -\sigma_n, \underbrace{0, \ldots, 0}_{m-n}.$$

Therefore the singular values of B are the n largest eigenvalues of A (Horn and Johnson 1985, Theorem 7.3.7).

- The eigenvalues of B^*B are

$$\sigma_1^2, \ldots, \sigma_n^2.$$

Therefore the singular values of B are the positive square roots of the eigenvalues of B^*B (Horn and Johnson 1985, Lemma 7.3.1).

Since singular values are eigenvalues of a Hermitian matrix, they are well conditioned in the absolute sense.

3.2. Hoffman–Wielandt-type bounds

We bound the sum of squares of all distances between the ith exact and perturbed singular values in terms of the Frobenius norm.

The singular values of B and $B+F$ are, respectively,

$$\sigma_1 \geq \cdots \geq \sigma_n \geq 0, \qquad \hat{\sigma}_1 \geq \cdots \geq \hat{\sigma}_n \geq 0.$$

Converting the singular value problem to an eigenvalue problem *à la* Section 3.1 and applying the Hoffman–Wielandt theorem for Hermitian matrices (2.3) lead immediately to the absolute error bound

$$\sqrt{\sum_{i=1}^{n} |\sigma_i - \hat{\sigma}_i|^2} \leq \|F\|_F.$$

The relative bound below requires in addition that both matrices be nonsingular.

Theorem 3.1 Let B and $B+F$ be nonsingular. If $\|FB^{-1}\| < 1$, then

$$\sqrt{\sum_{i=1}^{n} \frac{|\sigma_i - \hat{\sigma}_i|^2}{\sigma_i \hat{\sigma}_i}} \leq \frac{1}{2} \, \|(I+FB^{-1})^* - (I+FB^{-1})^{-1}\|_F.$$

Therefore the error in the singular values is small if $I + FB^{-1}$ is close to being unitary (or orthogonal). This is the case when $B+F = (I+FB^{-1})\,B$ is more or less a unitary transformation away from B.

Proof. See Li (1994*a*), Theorem 4.3. □

3.3. Weyl-type bounds

We bound the worst-case distance between the ith exact and perturbed singular values in terms of the two-norm.

The absolute error bound is an immediate consequence of Weyl's Perturbation Theorem (2.4)

$$|\sigma_i - \hat{\sigma}_i| \leq \|F\|, \quad 1 \leq i \leq n.$$

The corresponding relative bound below restricts the range of F but not its size. Here $B^\dagger$ is the Moore–Penrose inverse of B.

Theorem 3.2 If range$(B+F) \subset$ range(B), then

$$|\sigma_i - \hat{\sigma}_i| \leq \sigma_i \, \|B^\dagger F\|, \quad 1 \leq i \leq n.$$

If range$((B+F)^*) \subset$ range(B^*), then

$$|\sigma_i - \hat{\sigma}_i| \leq \sigma_i \, \|FB^\dagger\|, \quad 1 \leq i \leq n.$$

Proof. (See Di Lena, Peluso and Piazza (1993), Theorem 1.1.) We prove the first bound; the proof for the second one is similar.

Life would be easy if we could pull B out of F, say if $F = BC$ for some matrix C. Then we could write $B + F = B(I + C)$ and apply inequality (3.3.26) of Horn and Johnson (1991),

$$(1 - \|C\|)\sigma_i \leq \hat{\sigma}_i \leq (1 + \|C\|)\sigma_i,$$

to get the relative bound

$$|\sigma_i - \hat{\sigma}_i| \leq \sigma_i \, \|C\|.$$

It turns out that the range condition is exactly what is needed to pull B out of F. This is because range$(B+F) \subset$ range(B) implies $F = BC_1$ for some C_1. This allows us to write

$$F = BC_1 = BB^\dagger B \, C_1 = BB^\dagger F.$$

Consequently, setting $C \equiv B^\dagger F$ gives the desired result. □

When B has full column rank the second range condition in Theorem 3.2 is automatically satisfied.

Corollary 3.3 If B has full column rank then

$$|\sigma_i - \hat{\sigma}_i| \leq \sigma_i \, \|FB^\dagger\|, \quad 1 \leq i \leq n.$$

Proof. (See Di Lena et al. (1993), Remark 1.1.) If B has full column rank n then its rows span n-space. Hence range$((B+F)^*) \subset$ range(B^*) for any F, and the second relative bound in Theorem 3.2 holds. □

Therefore singular values of full-rank matrices are well conditioned in the absolute as well as relative sense. This may sound implausible at first, in particular when B has full rank while $B+F$ is rank-deficient. In this case all singular values of B are nonzero while at least one singular value of $B+F$ is zero. Hence a zero singular value of $B+F$ must have relative error equal to one. How can the singular values of B be well conditioned? The answer is that in this case the relative perturbation $\|B^\dagger F\|$ is large. The following example illustrates that $\|B^\dagger F\|$ is large when B and $B+F$ differ in rank.

Example 3.1 Let

$$B = \begin{pmatrix} 2 & 0 \\ 0 & \eta \\ 0 & 0 \end{pmatrix}, \qquad F = \begin{pmatrix} 0 & 0 \\ 0 & -\eta \\ 0 & 0 \end{pmatrix},$$

where $\eta \neq 0$. The rank of B is two, while the rank of $B+F$ is one.

The relative error in the singular value $\hat{\sigma} = 0$ of $B+F$ is equal to one because $|\eta - 0|/|\eta| = 1$. Since

$$B^\dagger = \begin{pmatrix} \frac{1}{2} & 0 & 0 \\ 0 & \frac{1}{\eta} & 0 \end{pmatrix}, \qquad B^\dagger F = \begin{pmatrix} 0 & 0 \\ 0 & -1 \end{pmatrix},$$

we get $\|B^\dagger F\| = 1$. Corollary 3.3 gives

$$\max_i \frac{|\sigma_i - \hat{\sigma}|}{\sigma_i} \leq 1.$$

Therefore Corollary 3.3 is tight for the zero singular values of $B+F$. □

Corollary 3.3 extends Demmel and Veselić (1992, Lemma 2.12) to matrices that do not necessarily have full rank (Di Lena et al. 1993, Remark 1.2). When B is nonsingular Corollary 3.3 implies that both range conditions in Theorem 3.2 hold automatically.

Corollary 3.4 If B is nonsingular, then

$$\max_{1 \leq i \leq n} \frac{|\sigma_i - \hat{\sigma}_i|}{\sigma_i} \leq \min\left\{\|B^{-1}F\|, \|FB^{-1}\|\right\}.$$

The following bound, for a different error measure, is similar to the Frobenius norm bound Theorem 3.1. It requires that both B and $B+F$ be nonsingular.

Theorem 3.5 Let B and $B+F$ be nonsingular. Then

$$\max_{1 \leq i \leq n} \frac{|\sigma_i - \hat{\sigma}_i|}{\sqrt{\sigma_i \hat{\sigma}_i}} \leq \frac{1}{2} \|(I + FB^{-1})^* - (I + FB^{-1})^{-1}\|.$$

Proof. See Li (1994*a*), Theorem 4.3. □

As in Theorem 3.1, the error in the singular values is small if $I + FB^{-1}$ is close to being unitary (or orthogonal). This is the case when $B + F = (I + FB^{-1})\, B$ is more or less a unitary transformation away from B.

3.4. Congruence transformations

We start with one-sided grading of matrices B with full-column rank. That is, $B = CD$ or $B = DC$, where D is nonsingular. In the event D is diagonal, CD represents a column scaling while DC represents a row scaling.

All relative singular value bounds presented so far are invariant under one-sided grading from the appropriate side. That's because the bounds contain terms of the form $B^\dagger F$ or $FB^\dagger$. Consider $B^\dagger F$, for instance. Grading from the right is sandwiched in the middle, between $B^\dagger$ and F, and therefore cancels out.

Let's first look at the Hoffman–Wielandt-type bound for graded matrices, which follows directly from Theorem 3.1.

Corollary 3.6 Let $B = CD$ and $B + F = (C + G)D$ be nonsingular. If $\|GC^{-1}\| < 1$, then

$$\sqrt{\sum_{i=1}^{n} \frac{|\sigma_i - \hat{\sigma}_i|^2}{\sigma_i \hat{\sigma}_i}} \leq \frac{1}{2}\, \|(I + GC^{-1})^* - (I + GC^{-1})^{-1}\|_F.$$

Proof. (See Li (1994*a*), Theorem 4.3.) The grading is sandwiched in the middle of the relative perturbation FB^{-1}, and cancels out,

$$FB^{-1} = (GD)\,(CD)^{-1} = G\,DD^{-1}\,C^{-1} = GC^{-1}.$$

□

Moving right along to the two-norm, we see that Corollary 3.3 is invariant under grading from the right.

Corollary 3.7 If $B = CD$ has full column rank, and if $B+F = (C+G)D$, then

$$|\sigma_i - \hat{\sigma}_i| \leq \sigma_i\, \|GC^\dagger\|, \quad 1 \leq i \leq n.$$

Proof. (See Di Lena et al. (1993), Remark 1.2.) Full column rank is needed to extract the grading matrix from the inverse,

$$B^\dagger = (CD)^\dagger = D^\dagger C^\dagger = D^{-1} C^\dagger.$$

□

Therefore the relative error in the singular values of $B+F$ is small if there is a grading matrix D that causes the relative perturbation of the graded matrices $\|GC^\dagger\|$ to be small. For instance, suppose $B = CD$ has columns whose norms vary widely while the columns of C are almost orthonormal.

If the perturbed matrix is scaled in the same way then the error bound in Corollary 3.7 ignores the scaling and acts as if it saw the well-behaved matrices C and $C+G$. Corollary 3.7 extends Theorem 2.14 of Demmel and Veselić (1992) to a larger class of matrices.

The other two-norm bound, Theorem 3.5, is also invariant under grading.

Corollary 3.8 Let $B = CD$ and $B + F = (C + G)D$ be nonsingular. If $\|GC^{-1}\| < 1$, then

$$\max_{1 \le i \le n} \frac{|\sigma_i - \hat{\sigma}_i|}{\sqrt{\sigma_i \hat{\sigma}_i}} \le \frac{1}{2} \|(I + GC^{-1})^* - (I + GC^{-1})^{-1}\|.$$

Proof. See Li (1994*a*), Theorem 4.3. □

Finally we present the only bound that is invariant under grading from both sides. It requires that the grading matrices be real diagonal, and it restricts the size of the perturbation more severely than the other bounds.

Theorem 3.9 Let $B = D_l C D_r$ and $B + F = D_l(C + G)D_r$ be real symmetric, where D_l and D_r are real nonsingular diagonal matrices. Among the singular values of all square submatrices of B, let θ be the smallest one. If $\|G\| < \theta$, then

$$-\|G\| \frac{2\theta - \|G\|}{\theta^2} \le \frac{\hat{\sigma}_i - \sigma_i}{\sigma_i} \le \|G\| \frac{2\theta - \|G\|}{(\theta - \|G\|)^2}, \quad 1 \le i \le n.$$

Proof. See Gu and Eisenstat (1993), Corollary 10. □

4. Some applications of additive perturbations

We discuss Jacobi's method for computing singular values and eigenvalues, and deflation of triangular and bidiagonal matrices.

4.1. Jacobi's method for singular values

Jacobi's method is generally viewed as a method that computes eigenvalues and singular values to optimal accuracy. It was Jacobi's method that first attracted attention to invariance of eigenvalue and singular values error bounds under congruence (Demmel and Veselić 1992, Mathias 1995, Rosanoff et al. 1968). We give a very intuitive plausibility argument, shoving many subtleties under the rug, to explain the high accuracy and invariance under grading of Jacobi's method. Our discussion runs along the lines of Demmel (1997), Section 5.4.3, and Mathias (1995), Sections 2, 3. Other detailed accounts can be found in Demmel and Veselić (1992) and Drmač (1996*b*). An attempt at a geometric interpretion of Jacobi's high accuracy is made in Rosanoff et al. (1968), pp. 1045–6.

A one-sided Jacobi method computes the singular values of a tall and skinny matrix by applying a sequence of orthogonal transformations on the right side of the matrix. The duty of each orthogonal transformation is to orthogonalize two columns of the matrix. The method stops once all columns are sufficiently orthogonal to each other. At this point the singular values are approximated by the column norms, that is, the Euclidean lengths of the columns.

For simplicity, assume that B is a real nonsingular matrix of order n. Let D be a row scaling of B, that is, $B = DC$, where D is diagonal. We show that the one-sided Jacobi method ignores the row scaling. When Jacobi applies an orthogonal transformation Q to B, the outcome in floating point arithmetic is $BQ + F$. Corollary 3.4 implies that the singular values $\hat{\sigma}_i$ of $BQ + F$ satisfy

$$\max_{1 \leq i \leq n} \frac{|\sigma_i - \hat{\sigma}_i|}{\sigma_i} \leq \|C^{-1}G\| \leq \|C^{-1}\| \, \|G\|,$$

where $F = DG$.

Let's bound the squared error $\|G\|$. Round-off error analysis tells us that the error in the ith row is

$$\|e_i^T F\| \leq \epsilon \beta_i \, \|e_i^T B\| + O(\epsilon^2),$$

where β_i depends on the matrix size n and $\epsilon > 0$ reflects the machine accuracy. Now the crucial observation is that the orthogonal transformations happen on one side of the matrix and the scaling on the other side. Because Q operates on columns it does not mix up different rows and therefore preserves the row scaling. This means we can pull the ith diagonal element of D out of $e_i^T F$,

$$\|e_i^T F\| \leq \epsilon \beta_i \, \|e_i^T B\| = \epsilon \beta_i \, \|e_i^T (DC)\| = |d_{ii}| \, \epsilon \beta_i \, \|e_i^T C\|.$$

This gives a bound for the ith row of G,

$$\|e_i^T G\| = |d_{ii}|^{-1} \, \|e_i^T F\| \leq \epsilon \beta_i \, \|e_i^T C\| + O(\epsilon^2).$$

The total error is therefore bounded by

$$\|G\| \leq \epsilon \beta \, \|C\| + O(\epsilon^2),$$

where β depends on n. Therefore the error bound for the singular values of $BQ + F$ is independent of the row scaling

$$\max_{1 \leq i \leq n} \frac{|\sigma_i - \hat{\sigma}_i|}{\sigma_i} \leq \epsilon \beta \, \kappa(C) + O(\epsilon^2).$$

This means Jacobi's method produces singular values of B but acts as if it saw C instead. That's good, particularly if D manages to pull out all the grading. Then all singular values of C have about the same magnitude and

$\kappa(C)$ is close to one. Therefore the above bound $\kappa(C)\,\beta\epsilon$ tends to be on the order of machine accuracy ϵ, implying that the relative error in the singular values is on the order of machine accuracy.

The argument is more complicated when the orthogonal transformations are applied on the same side as the scaling matrix. Fortunately the resulting error bounds do not tend to be much weaker (Mathias 1995, Section 4).

4.2. Jacobi's method for eigenvalues

A two-sided Jacobi method computes eigenvalues of a real symmetric positive-definite matrix by applying a sequence of orthogonal similarity transformations to the matrix. An orthogonal similarity transformation operates on two rows, i and j, and two columns, i and j, to zero out elements (i,j) and (j,i). The method stops once all off-diagonal elements are sufficiently small. At this point the eigenvalues are approximated by the diagonal elements.

Let A be real symmetric positive-definite of order n, and $A = DMD$, where D is a nonsingular diagonal matrix. The Jacobi method computes the eigenvalues of a matrix $A+E$. According to Corollary 2.13, the error bound for the eigenvalues $\hat{\lambda}_i$ of $A+E$ is

$$\max_{1\le i\le n} \frac{|\lambda_i - \hat{\lambda}_i|}{|\lambda_i|} \le \|M^{-1/2}FM^{-1/2}\| \le \|M^{-1}\|\,\|F\|, \tag{4.1}$$

where $E = DFD$. One can show that the error is bounded by

$$\|F\| \le \alpha\epsilon\,\|M\| + O(\epsilon^2),$$

where α depends on n and $\epsilon > 0$ reflects the machine accuracy. Therefore

$$\max_{1\le i\le n} \frac{|\lambda_i - \hat{\lambda}_i|}{|\lambda_i|} \le \alpha\epsilon\,\kappa(M) + O(\epsilon^2).$$

This means that the relative error in the eigenvalues is small, provided the amplifier $\kappa(M)$ is small.

The amplifier $\kappa(M)$ can be minimized via an appropriate choice of the scaling matrix D. If

$$D = \begin{pmatrix} \sqrt{a_{11}} & & \\ & \ddots & \\ & & \sqrt{a_{nn}} \end{pmatrix},$$

then all diagonal elements of M are equal to one. Therefore (van der Sluis 1969, Theorem 4.1)

$$\kappa(M) \le n \min_S \kappa(SAS),$$

where the minimum ranges over all nonsingular diagonal matrices S. This means that a diagonal scaling that makes all diagonal elements the same gives the minimal condition number among all diagonal scalings (up to a factor of matrix size n).

We claimed above that the error $\|F\|$ in (4.1) is small. Let's examine in more detail why. The error F comes about because of floating point arithmetic and because of the fact that eigenvalues are approximated by diagonal elements when the off-diagonal elements are small but not necessarily zero. Let's ignore the round-off error, and ask why ignoring small off-diagonal elements results in a small $\|F\|$. The trick here is to be clever about what it means to be 'small enough'.

Suppose the Jacobi method has produced a matrix A whose off-diagonal elements are small compared to the corresponding diagonal elements,

$$|a_{ij}| \leq \epsilon\sqrt{a_{ii}a_{jj}}.$$

This implies $|m_{ij}| \leq \epsilon$ where m_{ij} are the elements of the graded matrix $M = D^{-1}AD^{-1}$ and D is the above grading matrix with $\sqrt{a_{ii}}$ on the diagonal. Since the diagonal elements of M are equal to one, we can write $M = I + F$, where F contains all the off-diagonal elements of M and

$$\|F\| \leq (n-1)\epsilon.$$

Therefore $\|F\|$ is small, and (4.1) implies that the error in the eigenvalues is bounded by

$$\max_{1\leq i\leq n} \frac{|\lambda_i - \hat{\lambda}_i|}{|\lambda_i|} \leq \|M^{-1}\|\,(n-1)\epsilon.$$

Furthermore, one can bound $\|M^{-1}\|$ in terms of ϵ,

$$\|M^{-1}\| = \|(I+F)^{-1}\| \leq \frac{1}{1-\|F\|} \leq \frac{1}{1-(n-1)\epsilon}.$$

Replacing this in the error bound gives

$$\max_{1\leq i\leq n} \frac{|\lambda_i - \hat{\lambda}_i|}{|\lambda_i|} \leq \frac{(n-1)\epsilon}{1-(n-1)\epsilon}.$$

Therefore, ignoring small off-diagonal elements produces a small relative error. A detailed discussion of relative bounds for eigenvalues of scaled, almost diagonal matrices can be found in Hari and Drmač (1997) and Section 1 of Matejaš and Hari (1998).

The preceding arguments illustrate that Jacobi's method views a matrix in the best possible light, that is, in its optimally scaled version. Therefore eigenvalues produced by Jacobi's method tend to have relative accuracy close to machine precision. This is as accurate as it gets. In this sense Jacobi's method is considered optimally accurate.

One way to implement a two-sided Jacobi method is to apply a one-sided method to a Cholesky factor (Barlow and Demmel 1990, Mathias 1996, Veselić and Hari 1989). Let A be a Hermitian positive-definite matrix with Cholesky decomposition $A = L^*L$. The squares of the singular values of L are the eigenvalues of A. The singular values of L can be computed by the one-sided Jacobi method from Section 4.1. The preliminary Cholesky factorization does not harm the accuracy of the eigenvalues. Here is why. The computed Cholesky factor is the exact Cholesky factor of a matrix $A + E$, where (Mathias 1995, Lemma 2.6)

$$|e_{ij}| \leq \gamma\epsilon\sqrt{a_{ii}a_{jj}},$$

and γ depends on n. These perturbations have the same form as the ones above, hence lead to a small relative error. A similar argument shows that the squares of the diagonal elements of L are often good approximations to the eigenvalues of A (Mathias 1996).

4.3. Deflation of block triangular matrices

When a matrix is tall and skinny, or short and fat, one can save operations by first converting it to a skinny, short matrix before computing singular values. This can be accomplished by applying a QR decomposition and then computing the singular values of the resulting triangular matrix (Chan 1982). If done properly, the relative accuracy of the singular values is preserved (Mathias 1995, Theorem 3.2).

Suppose we compute the singular values of a triangular matrix by reducing the matrix to diagonal form, say by a Jacobi or QR method. Partition the triangular matrix as

$$B = \begin{pmatrix} B_{11} & B_{12} \\ & B_{22} \end{pmatrix}.$$

If the off-diagonal block B_{12} were zero then the problem of finding the singular values of B could be split into the two smaller, independent subproblems of finding the singular values of B_{11} and of B_{22}. However, if the off-diagonal block B_{12} is not zero, we want to know when it can be thrown away without causing too much harm to the singular values of B. The process of discarding information in a matrix to reduce the problem complexity is called 'deflation'.

The deflated matrix and the perturbation are, respectively,

$$B + F = \begin{pmatrix} B_{11} & \\ & B_{22} \end{pmatrix}, \qquad F = \begin{pmatrix} 0 & -B_{12} \\ 0 & 0 \end{pmatrix}.$$

Corollary 3.4 implies the following relative bound for the singular values $\hat{\sigma}_i$ of the deflated matrix $B + F$.

Corollary 4.1 If B is nonsingular, then

$$\max_{1\leq i\leq n} \frac{|\sigma_i - \hat{\sigma}_i|}{\sigma_i} \leq \min\left\{\|B_{11}^{-1}B_{12}\|,\ \|B_{12}B_{22}^{-1}\|\right\}.$$

Proof. See Di Lena et al. (1993), Theorem 2.1. □

This means the singular values of the deflated matrix have small relative error when the off-diagonal block is small compared to one of the diagonal blocks.

Now let's suppose a preliminary ordering of the singular values has already taken place. Say, the large singular values have floated to the top of the matrix B while the smaller ones have sunk to the bottom. The bound below is useful when the singular values of the top diagonal block are well separated from the singular values of the bottom block.

Theorem 4.2 If

$$\sigma_{\min}(B_{11}) \geq \alpha > \beta \geq \sigma_{\max}(B_{22}),$$

then

$$\max_{1\leq i\leq n} |\sigma_i - \hat{\sigma}_i| \leq \sigma_i \frac{\|B_{12}\|^2}{\alpha^2 - \beta^2}.$$

Proof. See Di Lena et al. (1993), Theorem 2.2. □

This means the singular values of the deflated matrix have small relative error if the off-diagonal block is small compared to the separation between the singular values of the two diagonal blocks. Theorem 4.2 is an extension of Demmel and Kahan (1990), Theorem 5. Other bounds that profit from a strong singular value separation appear in Chandrasekaran and Ipsen (1995), Theorem 5.2.1, Eisenstat and Ipsen (1995), Section 5, and Mathias and Stewart (1993), Theorem 3.1. Bounds for almost diagonal matrices that can take advantage of scaling are derived in Matejaš and Hari (1998).

4.4. Deflation of bidiagonal matrices

Triangular matrices are often further reduced to bidiagonal form before singular values are computed. A bidiagonal matrix is of the form

$$B = \begin{pmatrix} \alpha_1 & \beta_1 & & \\ & \ddots & \ddots & \\ & & \alpha_{n-1} & \beta_{n-1} \\ & & & \alpha_n \end{pmatrix}.$$

Bidiagonal matrices can also arise when one computes the vibrational frequencies of a linear mass-spring system (Demmel et al. 1997, Section 12.1).

There are several algorithms for computing singular values of a bidiagonal matrix to high relative accuracy (Demmel and Kahan 1990, Fernando and Parlett 1994). Because such algorithms apply a sequence of transformations to reduce B to diagonal form, they need to decide when an off-diagonal element β_j is small enough to be neglected without severely harming the singular values.

Suppose we are contemplating the removal of a single off-diagonal element. Here $B+F$ is equal to B, except for the off-diagonal element in row j and column $j+1$, which is equal to zero. Then $F = -\beta_j e_j e_{j+1}^T$ and $\|F\| = |\beta_j|$. Corollary 3.4 implies the following bound.

Corollary 4.3 If B is nonsingular bidiagonal, then

$$\max_{1\le i\le n} \frac{|\sigma_i - \hat{\sigma}_i|}{\sigma_i} \le |\beta_j| \min\left\{\|B^{-1}e_j\|, \|B^{-*}e_{j+1}\|\right\}.$$

Proof. See Di Lena et al. (1993), Section 3. □

This means that if we remove a small element from row j and column $j+1$ of a bidiagonal matrix, then the relative error in the singular values of the deflated matrix is small if column j or row $j+1$ of B^{-1} are small in norm. Similar bounds, but for a different error measure, appear in Deift, Demmel, Li and Tomei (1991), Theorem 4.7, and Demmel and Kahan (1990), Theorem 4.

Corollary 4.3 justifies the use of Convergence Criterion 1 (see Demmel and Kahan (1990), Section 2, and Deift et al. (1991), Section 4) in the zero-shift Golub–Kahan algorithm for computing singular values of bidiagonal matrices. The practical usefulness of this bound also derives from the fact that it can be computed via the simple recursion below.

Corollary 4.4 If B is nonsingular bidiagonal, then

$$\max_{1\le i\le n} \frac{|\sigma_i - \hat{\sigma}_i|}{\sigma_i} \le \min\{\sqrt{r_j}, \sqrt{c_j}\},$$

where

$$r_1 = \frac{\beta_1^2}{\alpha_1^2}, \qquad r_j = \frac{\beta_j^2}{\alpha_j^2}(1 + r_{j-1}), \quad 2 \le j \le n,$$

and

$$c_{n-1} = \frac{\beta_{n-1}^2}{\alpha_n^2}, \qquad c_j = \frac{\beta_j^2}{\alpha_{j+1}^2}(1 + c_{j+1}), \quad n-2 \ge j \ge 1.$$

Proof. See Di Lena et al. (1993), Theorems 3.1, 3.2. □

When the shift in the Golub–Kahan algorithm or the qd algorithm is nonzero, it can be incorporated into the perturbation bounds (see Eisenstat and Ipsen (1995), Theorem 5.7, and Fernando and Parlett (1994)).

5. Multiplicative perturbations for eigenvalues

We shift gears and represent the perturbed matrix from now on as D_1AD_2 where D_1 and D_2 are nonsingular. When $D_2 = D_1^{-1}$ this is just a similarity transformation, which means that A and D_1AD_2 have the same eigenvalues. When $D_2 = D_1^*$ this is a congruence transformation, which means that A and D_1AD_2 have the same inertia when A is Hermitian. Since the nonsingularity of D_1 and D_2 forces A and D_1AD_2 to have the same rank, multiplicative perturbations are less powerful than additive perturbations.

Then why are multiplicative perturbations useful? It turns out that it is sometimes easier to express a component-wise relative perturbation of a sparse matrix as a multiplicative perturbation than as an additive perturbation. The following example illustrates how natural multiplicative perturbations can be, especially for bidiagonal and tridiagonal matrices.

Example 5.1 (See Barlow and Demmel (1990), p. 770, Eisenstat and Ipsen (1995), Corollary 4.1.) Consider the real, symmetric tridiagonal matrix

$$A = \begin{pmatrix} 0 & \alpha_1 & & & & \\ \alpha_1 & 0 & \alpha_2 & & & \\ & \alpha_2 & 0 & \alpha_3 & & \\ & & \alpha_3 & 0 & \alpha_4 & \\ & & & \alpha_4 & 0 & \alpha_5 \\ & & & & \alpha_5 & 0 \end{pmatrix}.$$

Such a matrix occurs, for instance, when one converts the singular value problem of a bidiagonal matrix to an eigenvalue problem (see Section 3.1). A component-wise relative perturbation of a single off-diagonal pair in A produces the perturbed matrix

$$\hat{A} = \begin{pmatrix} 0 & \alpha_1 & & & & \\ \alpha_1 & 0 & \alpha_2 & & & \\ & \alpha_2 & 0 & \beta\alpha_3 & & \\ & & \beta\alpha_3 & 0 & \alpha_4 & \\ & & & \alpha_4 & 0 & \alpha_5 \\ & & & & \alpha_5 & 0 \end{pmatrix},$$

where $\beta \neq 0$. For instance, β could be of the form $\beta = 1 + \epsilon$, where $|\epsilon|$ does not exceed machine epsilon. The perturbed matrix $\hat{A}$ can be represented as a multiplicative perturbation $\hat{A} = D^TAD$, where

$$D = \begin{pmatrix} \sqrt{\beta} & & & & & \\ & 1/\sqrt{\beta} & & & & \\ & & \sqrt{\beta} & & & \\ & & & \sqrt{\beta} & & \\ & & & & 1/\sqrt{\beta} & \\ & & & & & \sqrt{\beta} \end{pmatrix}.$$

In this case a component-wise relative perturbation of an off-diagonal pair can be represented as a multiplicative perturbation. □

The simple-minded approach of disguising a multiplicative perturbation as an additive perturbation, that is,

$$D_1AD_2 = A + E, \qquad \text{where} \quad E = D_1AD_2 - A,$$

produces a perturbation matrix E that may not be small or meaningful.

There are different techniques for deriving multiplicative perturbation bounds, and some of them are compared by Li and Mathias (1997, Section 4.2). Here we start from absolute perturbation bounds and show that they imply many of relative bounds.

5.1. Bauer–Fike-type bounds

Again we start with a diagonalizable matrix, and we bound the distance of a perturbed eigenvalue to a closest exact eigenvalue in terms of the two-norm.

Let $A = X\Lambda X^{-1}$ be an eigendecomposition of A, where

$$\Lambda = \begin{pmatrix} \lambda_1 & & \\ & \ddots & \\ & & \lambda_n \end{pmatrix},$$

and λ_i are the eigenvalues of A. Let $\hat{\lambda}$ be an eigenvalue of the perturbed matrix D_1AD_2 and $\hat{x} \neq 0$ a corresponding unit eigenvector,

$$(D_1AD_2)\,\hat{x} = \hat{\lambda}\hat{x}, \qquad \|\hat{x}\| = 1,$$

with residual

$$r \equiv A\hat{x} - \hat{\lambda}\hat{x}.$$

This time we use the Bauer–Fike theorem with residual bound (Bauer and Fike 1960, Theorem IIIa),

$$\min_{1\le i\le n} |\lambda_i - \hat{\lambda}| \le \kappa(X)\, \|r\|. \tag{5.1}$$

The relative error bound below for the eigenvalue $\hat{\lambda}$ of the perturbed matrix D_1AD_2 measures the error relative to the perturbed eigenvalue rather than an exact eigenvalue.

Theorem 5.1 If A is diagonalizable, then

$$\min_{1\le i\le n} |\lambda_i - \hat{\lambda}| \le |\hat{\lambda}|\, \kappa(X)\, \|I - D_1^{-1}D_2^{-1}\|.$$

Proof. (See Eisenstat and Ipsen (1996), Theorem 6.1.) The idea is to concoct a residual that contains the factor $\hat{\lambda}$ and then to use the absolute bound (5.1).

From $(D_1AD_2)\hat{x} = \hat{\lambda}\hat{x}$ follows

$$A\,z = \hat{\lambda}\,D_1^{-1}D_2^{-1}\,z, \qquad \text{where} \quad z \equiv D_2\hat{x}/\|D_2\hat{x}\|.$$

The residual for $\hat{\lambda}$ and z is

$$f \equiv Az - \hat{\lambda}z = \hat{\lambda}\left(D_1^{-1}D_2^{-1} - I\right)z,$$

and it contains $\hat{\lambda}$ as a factor. Now apply the absolute bound (5.1) to f. □

The perturbed matrix D_1AD_2 is not required to be diagonalizable. As in the case of additive perturbations, $\kappa(X)$ can be interpreted as a condition number. The factor $\|I - D_1^{-1}D_2^{-1}\|$ represents a relative deviation from similarity, because

$$I - D_1^{-1}D_2^{-1} = \left(D_2 - D_1^{-1}\right)D_2^{-1}$$

represents a difference relative to D_2.

There are two cases in which the bound in Theorem 5.1 is guaranteed to be zero and hence tight. First, when $D_1 = D_2^{-1}$, because similar matrices have the same eigenvalues. Second, when $\hat{\lambda} = 0$, because A and D_1AD_2 are singular and both have a zero eigenvalue.

5.2. Hoffman–Wielandt-type bounds for diagonalizable matrices

Based on a one-to-one correspondence between exact and perturbed eigenvalues, we bound the sum (of squares) of all distances between exact and perturbed eigenvalues in terms of the Frobenius norm. In contrast to the previous section, the perturbed matrix must now also be diagonalizable.

Let A and D_1AD_2 be diagonalizable with respective eigendecompositions

$$A = X\Lambda X^{-1}, \qquad D_1AD_2 = \hat{X}\hat{\Lambda}\hat{X}^{-1}.$$

The eigenvalues are

$$\Lambda = \begin{pmatrix} \lambda_1 & & \\ & \ddots & \\ & & \lambda_n \end{pmatrix}, \qquad \hat{\Lambda} = \begin{pmatrix} \hat{\lambda}_1 & & \\ & \ddots & \\ & & \hat{\lambda}_n \end{pmatrix}.$$

Theorem 5.2 If A and D_1AD_2 are nonsingular and diagonalizable, then there exists a permutation τ such that

$$\sqrt{\sum_{i=1}^{n}\left(\frac{|\lambda_i - \hat{\lambda}_{\tau(i)}|}{|\lambda_i|}\right)^2} \le \kappa(X)\kappa(\hat{X})\,\|D_2\|\|D_1 - D_2^{-1}\|_F.$$

There also exists a permutation σ such that

$$\sqrt{\sum_{i=1}^{n} \left(\frac{|\lambda_i - \hat{\lambda}_{\sigma(i)}|}{|\lambda_i|} \right)^2} \leq \kappa(X)\kappa(\hat{X}) \, \|D_1\| \|D_1^{-1} - D_2\|_F.$$

Proof. The first bound is Theorem 2.1 of Li (1997), while the second one is Theorem 2.1′ of Li (1997). □

Therefore the relative error bound can only be small if D_1 and D_2 are close to being a similarity transformation. A similar bound in Proposition 3.5 of Li and Mathias (1997) applies to a one-sided perturbation and matrices with positive eigenvalues and holds in any unitarily invariant norm.

5.3. Hoffman–Wielandt-type bounds for Hermitian matrices

Hoffman–Wielandt-type bounds for Hermitian matrices require that the perturbed matrix also be Hermitian. This means that the perturbed matrix had better be of the form DAD^*, where D is nonsingular. Since the perturbed matrix is congruent to A, it has the same inertia as A. Number the eigenvalues of A and DAD^* so that

$$\lambda_n \leq \cdots \leq \lambda_1, \qquad \hat{\lambda}_n \leq \cdots \leq \hat{\lambda}_1.$$

Theorem 5.3 If A and DAD^* are Hermitian and nonsingular, then

$$\sqrt{\sum_{i=1}^{n} \frac{|\lambda_i - \hat{\lambda}_i|^2}{|\lambda_i \hat{\lambda}_i|}} \leq \|D^* - D^{-1}\|_F.$$

Proof. See Li and Mathias (1997), Corollary 3.2′. A proof for the special case of positive-definite matrices appears in Li (1994*a*), Theorem 3.1. □

Therefore, the relative error in the eigenvalues of DAD^* is small if D is close to a unitary (or orthogonal) matrix. The bound (Li 1997, Theorem 2.2) is weaker than Theorem 5.3 (Li and Mathias 1997, Section 4.1). Majorization theory can deliver bounds that are stronger than Theorem 5.3 and hold for any unitarily invariant norm (Li and Mathias 1997, Proposition 3.2, (3.8)).

5.4. Ostrowski-type bounds

In 1959 Ostrowski presented the first relative perturbation bounds for eigenvalues. He created a multiplicative perturbation DAD^* of a Hermitian matrix A, where D is nonsingular; and he bounded the ratio of exact and perturbed eigenvalues in terms of the smallest and largest eigenvalues of DD^* (Ostrowski (1959), Horn and Johnson (1985), Theorem 4.5.9),

$$\lambda_{\min}(DD^*) \, \lambda_i \leq \hat{\lambda}_i \leq \lambda_i \, \lambda_{\max}(DD^*). \tag{5.2}$$

Ostrowski's theorem can also be phrased in terms of absolute values of eigenvalues.

Theorem 5.4 If A and DAD^* are Hermitian, then

$$\frac{|\lambda_i|}{\|(D^*D)^{-1}\|} \leq |\hat{\lambda}_i| \leq |\lambda_i| \, \|D^*D\|.$$

Proof. See Eisenstat and Ipsen (1995), Theorem 2.1. □

This bound is tight, for instance, when D is a multiple of an orthogonal matrix. The following example illustrates what the bound looks like in the case of tridiagonal matrices.

Example 5.2 (See Eisenstat and Ipsen (1995), Corollary 4.1.) Return to the symmetric tridiagonal matrix with zero diagonal and its single-element perturbation in Example 5.1. In this case the bound in Theorem 5.4 amounts to

$$\frac{1}{\eta}|\lambda_i| \leq |\hat{\lambda}_i| \leq \eta \, |\lambda_i|$$

(Kahan (1966), pp. 49ff., Demmel and Kahan (1990), Theorem 2), where $\eta \equiv \max\{|\beta|, 1/|\beta|\}$. Therefore the ratio between perturbed and exact eigenvalues is close to one if the perturbation $|\beta|$ is close to one.

This bound can be extended to the perturbation of any number of off-diagonal pairs of a real symmetric tridiagonal matrix with zero diagonal (Demmel and Kahan 1990, Corollary 1). □

Ostrowski's theorem (5.2) can also be extended to products of eigenvalue ratios (Li and Mathias 1997, Theorem 2.3).

5.5. Weyl-type bounds

Ostrowski's theorem leads to a relative Weyl-type bound for multiplicative perturbations.

Theorem 5.5 If A and DAD^* are Hermitian, then

$$|\lambda_i - \hat{\lambda}_i| \leq |\lambda_i| \, \|DD^* - I\|, \quad 1 \leq i \leq n.$$

Proof. (See Eisenstat and Ipsen (1995), Theorem 2.1.) The proof is similar to that of Theorem 2.8 for additive perturbations. Fix an index i. Since 0 is the ith eigenvalue of $A - \lambda_i I$, Sylvester's Law of Inertia (Horn and Johnson 1985, Theorem 4.5.8) implies that 0 is the ith eigenvalue of $D(A - \lambda_i I)D^*$. Write

$$D(A - \lambda_i I)D^* = DAD^* - \lambda_i DD^* = \bar{A} + \bar{E},$$

where

$$\bar{A} \equiv DAD^* - \lambda_i I, \qquad \bar{E} \equiv \lambda_i (I - DD^*).$$

Applying Weyl's absolute bound (2.4) to $\bar{A}$ and $\bar{A}+\bar{E}$ gives

$$\max_{1\leq j\leq n} |\lambda_j(\bar{A}) - \lambda_j(\bar{A}+\bar{E})| \leq \|\bar{E}\| = |\lambda_i| \, \|DD^* - I\|.$$

In particular, for $j = i$,

$$|0 - (\hat{\lambda}_i - \lambda_i)| \leq |\lambda_i| \, \|DD^* - I\|.$$

□

The bound above holds even for zero eigenvalues. The factor $\|DD^* - I\|$ represents the deviation of the congruence transformation from similarity. This means that if the perturbed matrix is a congruence transformation of the original matrix, then the relative error in the perturbed eigenvalues is small if the congruence transformation is close to similarity.

Example 5.3 We can apply Theorem 5.5 to the symmetric tridiagonal matrix with zero diagonal in Example 5.1. If we assume that the multiplicative perturbation α is of the form $\alpha = 1 + \epsilon$, then

$$|\lambda_i - \hat{\lambda}_i| \leq |\lambda_i| \, |\epsilon|.$$

Therefore a small relative error in a pair of off-diagonal elements causes only a small relative error in the eigenvalues.

The bound below is similar in spirit to Theorem 5.5 but applies to a different error measure.

Theorem 5.6 If A and DAD^* are nonsingular Hermitian, then

$$\max_{1\leq i\leq n} \frac{|\lambda_i - \hat{\lambda}_i|}{\sqrt{\lambda_i \hat{\lambda}_i}} \leq \|D^{-1} - D^*\|.$$

Proof. See Li and Mathias (1997), Proposition 3.2′. □

This bound was first derived for the special case of positive-definite matrices (Li 1994*a*, Theorem 3.1).

6. Multiplicative perturbations for singular values

The perturbed matrix is represented as $D_1 B D_2$, where D_1 and D_2 are nonsingular diagonal matrices. Such a perturbation can occur, for instance, when a one-sided Jacobi method is applied to B. In this case the computed singular values are exact singular values of a matrix $D_1 B D_2$ where D_1 and D_2 are close to the identity (Demmel 1997, Section 5.4.3).

Again let B be a tall and skinny matrix, that is, B is $m \times n$ with $m \geq n$. The singular values of B are

$$\sigma_1 \geq \cdots \geq \sigma_n \geq 0.$$

The perturbed matrix is represented as $D_1 B D_2$, where D_1 and D_2 are nonsingular. The singular values of $D_1 B D_2$ are

$$\hat{\sigma}_1 \geq \cdots \geq \hat{\sigma}_n \geq 0.$$

When D_1 is diagonal it represents a row scaling, while a diagonal D_2 represents a column scaling.

6.1. Ostrowski-type bounds

Let's first determine by which factor the singular values of B change when multiplicative perturbations D_1 and D_2 are applied.

Theorem 6.1 We have

$$\frac{\sigma_i}{\|D_1^{-1}\| \, \|D_2^{-1}\|} \leq \hat{\sigma}_i \leq \sigma_i \, \|D_1\| \, \|D_2\|.$$

Proof. (See Eisenstat and Ipsen (1995), Theorem 3.1.) Convert the problem to an eigenvalue problem as in Section 3.1 and apply the eigenvalue result Theorem 5.4. □

This means that if D_1 and D_2 are almost unitary then the norms in Theorem 6.1 are almost one, and a perturbed singular value differs from the corresponding exact singular value by a factor close to one.

Theorem 6.1 can reproduce perturbation bounds for component-wise perturbations of bidiagonal matrices from Barlow and Demmel (1990), Theorem 1, Deift et al. (1991), Theorem 2.12, and Demmel and Kahan (1990), Corollary 2. The example below illustrates how.

Example 6.1 (See Eisenstat and Ipsen (1995), Corollary 4.2.) Consider the bidiagonal matrix

$$B = \begin{pmatrix} \alpha_1 & \beta_1 & & \\ & \alpha_2 & \beta_2 & \\ & & \alpha_3 & \beta_3 \\ & & & \alpha_4 \end{pmatrix}$$

and its component-wise perturbation

$$\hat{B} = \begin{pmatrix} \gamma_1\alpha_1 & \gamma_2\beta_1 & & \\ & \gamma_3\alpha_2 & \gamma_4\beta_2 & \\ & & \gamma_5\alpha_3 & \gamma_6\beta_3 \\ & & & \gamma_7\alpha_4 \end{pmatrix},$$

where $\gamma_j \neq 0$. For instance, if $\gamma_j = 1+\epsilon_j$ for small ϵ_j then $\hat{B}$ is a component-wise relative perturbation of B.

Write $\hat{B} = D_1 B D_2$, where D_1 takes care of the odd-numbered, diagonal perturbations, while D_2 takes care of the even-numbered, off-diagonal perturbations, like so:

$$D_1 = \begin{pmatrix} \gamma_1 & & & \\ & \gamma_1\gamma_3/\gamma_2 & & \\ & & \gamma_1\gamma_3\gamma_5/\gamma_4 & \\ & & & \gamma_1\gamma_3\gamma_5\gamma_7/\gamma_6 \end{pmatrix},$$

$$D_2 = \begin{pmatrix} 1 & & & \\ & \gamma_2/\gamma_1 & & \\ & & \gamma_2\gamma_4/\gamma_3 & \\ & & & \gamma_2\gamma_4\gamma_6/\gamma_5 \end{pmatrix}.$$

The perturbation D_1 operates on rows and D_2 operates on columns. The resulting interference is apparent in the increasing products and the denominators, where each D_i undoes part of the other's action in its own territory.

Application of Theorem 6.1 yields

$$\frac{1}{\eta}\,\sigma_i \leq \hat{\sigma}_i \leq \eta\,\sigma_i,$$

where $\eta \equiv \prod_{j=1}^{7} \max\{|\gamma_j|, 1/|\gamma_j|\}$. This means that if each factor γ_j is close to one then the ratio of perturbed to exact singular values is close to one. □

These bounds are actually realistic. There are algorithms that deliver singular values of bidiagonal matrices to high relative accuracy: the dqds algorithm (Fernando and Parlett 1994) and, to a large extent, a fine-tuned zero-shift version of the Golub–Kahan algorithm (Demmel and Kahan 1990, Deift et al. 1991).

6.2. Hoffman–Wielandt-type bounds

Now let's bound the sum of squares of all relative errors.

Theorem 6.2 If B and $D_1 B D_2$ have full column rank, then

$$\sqrt{\sum_{i=1}^{n} \frac{|\sigma_i - \hat{\sigma}_i|^2}{\sigma_i \hat{\sigma}_i}} \leq \frac{1}{2}\left(\|D_1^* - D_1^{-1}\|_F + \|D_2^* - D_2^{-1}\|_F\right).$$

Proof. See Li and Mathias (1997), Proposition 3.3′. □

The terms $\|D_1^* - D_1^{-1}\|_F$ and $\|D_2^* - D_2^{-1}\|_F$ indicate how far D_1 and D_2, respectively, are from being unitary (or orthogonal). The relative error is small if D_1 and D_2 are close to unitary. A weaker version of Theorem 6.2 appears in Theorem 4.1 of Li (1994*a*). As with Theorem 5.3, majorization theory yields a bound stronger than Theorem 6.2 that holds in any unitarily invariant norm (Li and Mathias 1997, Proposition 3.3, (3.12)).

The bound given below is similar in spirit but applies to a different error measure.

Theorem 6.3 If B and D_1BD_2 have full column rank, then

$$\sqrt{\sum_{i=1}^{n}\left(\frac{|\sigma_i-\hat{\sigma}_i|}{\sqrt[p]{\sigma_i^p+\hat{\sigma}_i^p}}\right)^2} \leq \frac{1}{2^{1+1/p}}\left(\|D_1^*-D_1^{-1}\|_F+\|D_2^*-D_2^{-1}\|_F\right),$$

where $1 \leq p \leq \infty$ is an integer.

Proof. (See Li (1994*a*), Theorem 4.2.) This follows from Theorem 6.2 because

$$\left(\frac{2}{\sigma_i^p+\hat{\sigma}_i^p}\right)^{1/p} \leq \frac{1}{\sigma_i\hat{\sigma}_i}.$$

□

6.3. Weyl-type bounds

At last we bound individual relative errors.

Theorem 6.4 $|\sigma_i-\hat{\sigma}_i| \leq \sigma_i \max\left\{\|I-D_1^{-1}D_1^{-*}\|, \|I-D_2^{-*}D_2^{-1}\|\right\}.$

Proof. (See Eisenstat and Ipsen (1995), Theorem 3.3.) Convert the singular value problem to a large eigenvalue problem as in Section 3.1 and apply Theorem 5.5 to the eigenvalue problem. □

The factors $\|I-D_1^{-1}D_1^{-*}\|$ and $\|I-D_2^{-*}D_2^{-1}\|$ represent relative deviations of D_1 and D_2, respectively, from being unitary (or orthogonal). Hence the relative error in the singular values of D_1BD_2 is small if D_1 and D_2 are close to unitary.

The following inequality is like the Frobenius norm bound in Theorem 6.2.

Theorem 6.5 If B and D_1BD_2 have full column rank, then

$$\max_{1\leq i\leq n}\frac{|\sigma_i-\hat{\sigma}_i|}{\sqrt{\sigma_i\hat{\sigma}_i}} \leq \frac{1}{2}\left(\|D_1^*-D_1^{-1}\|+\|D_2^*-D_2^{-1}\|\right).$$

Proof. See Li and Mathias (1997), Proposition 3.3′. □

A weaker bound appears in Theorem 4.1 of Li (1994*a*).

The following bound is a counterpart of the Frobenius norm bound in Theorem 6.3.

Theorem 6.6 If B and D_1BD_2 have full column rank, then

$$\max_{1\leq i\leq n}\frac{|\sigma_i-\hat{\sigma}_i|}{\sqrt[p]{|\sigma_i|^p+|\hat{\sigma}_i|^p}} \leq \frac{1}{2^{1+1/p}}\left(\|D_1^*-D_1^{-1}\|+\|D_2^*-D_2^{-1}\|\right),$$

where $1 \leq p \leq \infty$ is an integer.

Proof. (See Li (1994*a*), Theorem 4.2.) This follows from Theorem 6.5 in the same way that Theorem 6.3 follows from Theorem 6.2. □

7. Some applications of multiplicative perturbations

We discuss component-wise perturbations of generalized bidiagonal matrices, deflation of triangular matrices, and rank-revealing decompositions.

7.1. Component-wise perturbations of generalized bidiagonal matrices

Example 6.1 illustrates that small relative changes in any bidiagonal matrix cause only small relative changes in its singular values, regardless of the value of the nonzero matrix elements. Are there other matrices with this pleasant property? The answer is not really. The only other matrices with this property are those whose sparsity structure is 'essentially' bidiagonal (Demmel and Gragg 1993). At first glance this result looks pretty negative. It appears to suggest that we can forget about high relative accuracy for matrices other than bidiagonals. But then again, not all perturbations are component-wise relative perturbations. Just think about the perturbations caused by the deflation of triangular matrices in Sections 4.3 and 4.4. There is still plenty of room for singular values of all kinds of matrices to have relative accuracy, but mostly not with regard to component-wise relative perturbations.[1]

So, what are 'essentially' bidiagonal matrices? We define an undirected bipartite graph G of a matrix B as follows. Each row of B is represented by a node r_i, and each column by a node c_i. There is an edge between r_i and c_j if and only if element (i,j) of B is nonzero. The matrix B is called *biacyclic* if its graph G is acyclic (Demmel and Gragg 1993, Section 1). Examples of biacyclic matrices, in addition to bidiagonal matrices, include the following 'half arrow' matrices (Demmel and Gragg 1993, Section 5):

$$\begin{pmatrix} * & & & & * \\ & * & & & * \\ & & * & & * \\ & & & * & * \\ & & & & * \end{pmatrix}, \qquad \begin{pmatrix} * & & & * & & \\ & * & & & * & \\ & & * & & & * \\ & & & * & & * \\ & & & & * & * \\ & & & & & * \end{pmatrix}.$$

The nice thing about subjecting biacyclic matrices to component-wise relative perturbations is that we get an Ostrowski-type bound. This means that changing an element of a biacyclic matrix by a factor does not change

[1] However, if we are willing to restrict the values of the matrix elements and impose signs on the nonzero entries of a matrix so as to forestall cancellation in the computation of certain quantities, then one can also obtain high relative accuracy (Demmel et al. 1997).

the singular values by more than this factor, regardless of the value of the nonzero matrix elements. No other matrices possess this property.

Theorem 7.1 Let B be a matrix of order n. The following two conditions are equivalent:

- B is biacyclic
- if $\hat{B}$ is equal to B except for element (k, l), which is multiplied by $\gamma \neq 0$, then the singular values $\hat{\sigma}_i$ of $\hat{B}$ satisfy

$$\min\left\{|\gamma|, |\gamma^{-1}|\right\} \sigma_i \leq \hat{\sigma}_i \leq \sigma_i \max\left\{|\gamma|, |\gamma^{-1}|\right\}, \quad 1 \leq i \leq n.$$

Proof. (See Demmel and Gragg (1993), Theorem 1.) The result also follows directly from Theorem 6.1 (Eisenstat and Ipsen 1995, Corollary 4.3). □

Theorem 7.1 implies that a small relative perturbation in a matrix element causes small relative changes in the singular values, regardless of the values of the nonzero matrix elements, if and only if the matrix has an acyclic graph. To see this, consider $\gamma = 1 + \epsilon$ for some $\epsilon > 0$. Theorem 7.1 implies the relative error bound

$$|\sigma_i - \hat{\sigma}_i| \leq \sigma_i \epsilon, \quad 1 \leq i \leq n.$$

Theorem 7.1 can be extended to the case where all elements of a biacyclic matrix are multiplied by nonzero factors. This gives a bound similar to the one in Example 6.1 (see Demmel and Gragg (1993), p. 206, and Eisenstat and Ipsen (1995), Corollary 4.3).

7.2. Deflation of block triangular matrices, again

First we prove an auxiliary bound for a special multiplicative perturbation which is useful for modelling deflation in triangular matrices. Let

$$B = \begin{pmatrix} B_{11} & B_{12} \\ & B_{22} \end{pmatrix}$$

be a block triangular matrix of order n, and let the perturbed matrix be DB or BD, where

$$D = \begin{pmatrix} I & X \\ & I \end{pmatrix}$$

is partitioned commensurately with B.

Theorem 7.2 The singular values $\hat{\sigma}_i$ of DB, or BD, satisfy

$$|\sigma_i - \hat{\sigma}_i| \leq \sigma_i \, \|X\|, \quad 1 \leq i \leq n.$$

If in addition B is nonsingular, then

$$\max_{1 \leq i \leq n} \frac{|\sigma_i - \hat{\sigma}_i|}{\sqrt{\sigma_i \hat{\sigma}_i}} \leq \frac{1}{2} \|X\|.$$

Proof. The first inequality (Eisenstat and Ipsen 1995, Lemma 5.1) follows from Theorem 6.1,

$$\frac{\sigma_i}{\|D^{-1}\|} \leq \hat{\sigma}_i \leq \sigma_i \|D\|,$$

and from

$$\|D\| = \|D^{-1}\| \leq 1 + \|X\|.$$

The second inequality (Li 1994*a*, Corollary 4.1) follows from Theorem 6.5. □

Now look at a perturbed matrix that is a deflated version of the block triangular matrix B,

$$\hat{B} = \begin{pmatrix} B_{11} & \\ & B_{22} \end{pmatrix}.$$

The following bound is the same as the one in Corollary 4.1, which was derived in the context of additive pertubations.

Theorem 7.3 If B_{11} or B_{22} are nonsingular, then the singular values of B and $\hat{B}$ satisfy

$$|\sigma_i - \hat{\sigma}_i| \leq \sigma_i \min\left\{\|B_{11}^{-1} B_{12}\|, \|B_{12} B_{22}^{-1}\|\right\}.$$

Proof. (See Eisenstat and Ipsen (1995), Theorem 5.2.) This follows directly from Theorem 7.2. □

Let's see what happens for bidiagonal matrices. We write the bidiagonal matrix and its perturbation so as to highlight the action,

$$B = \begin{pmatrix} B_{11} & \beta_j e_j e_1^T \\ & B_{22} \end{pmatrix}, \qquad \hat{B} = \begin{pmatrix} B_{11} & \\ & B_{22} \end{pmatrix}.$$

Both matrices are bidiagonal, and $\hat{B}$ is equal to B, except for the off-diagonal element in row j and column $j+1$, which is equal to zero. Application of Theorem 7.3 to B and $\hat{B}$ produces a bound similar to the one in Corollary 4.3.

Theorem 7.4 If B and $\hat{B}$ are nonsingular, then

$$\frac{1}{1+\eta} \leq \frac{\hat{\sigma}_i}{\sigma_i} \leq 1 + \eta,$$

where

$$\eta \equiv |\beta_j| \min\left\{\|B_{11}^{-1} e_j\|, \|B_{22}^{-*} e_1\|\right\}.$$

Proof. See Eisenstat and Ipsen (1995), Theorem 5.5. □

This bound, like Corollary 4.3 in the context of additive perturbations, can be used to justify Convergence Criterion 1 in the Golub–Kahan algorithm with zero shift (Eisenstat and Ipsen 1995, Corollary 5.6).

7.3. Rank-revealing decompositions

A rank-revealing decomposition is a cheap imitation of a singular value decomposition. It can serve as an intermediate step in the high-accuracy computation of a singular value decomposition. Passing through a rank-revealing decomposition on the way to a singular value decomposition allows one to represent all errors in terms of multiplicative perturbations (Demmel et al. 1997, Section 3).

Consider an $m \times n$ matrix B, $m \geq n$, of rank r. Decompose $B = XDY^*$ where X is $m \times r$, Y is $n \times r$, and D is diagonal of order r and $r \leq n$. This means D is nonsingular, and X and Y have full column rank. The decomposition $B = XDY^*$ is a *rank-revealing decomposition of* A if X and Y are well conditioned, that is, if

$$\kappa(X) \equiv \|X\| \, \|X^\dagger\| \qquad \text{and} \qquad \kappa(Y) \equiv \|Y\| \, \|Y^\dagger\|$$

are close to one (Demmel et al. 1997, Definition 2.1).

A singular value decomposition qualifies as the luxury edition of a rank revealing decomposition because X and Y are orthogonal, hence perfectly conditioned. Gaussian elimination with complete pivoting may be a more affordable model. Here $X = P_l L$ and $Y = UP_u$, where P_l and P_u are permutation matrices, L is unit lower triangular and U is unit upper triangular. Because all entries of L and U are bounded by one in absolute value, X and Y tend to be well conditioned.

Suppose the computed version of our rank revealing decomposition is $\hat{B} = \hat{X}\hat{D}\hat{Y}^*$. If the elements of the diagonal matrix $\hat{D}$ have high relative accuracy, and $\hat{X}$ and $\hat{Y}$ have high norm-wise accuracy, then the singular values $\hat{\sigma}_i$ of $\hat{B}$ have small relative error.

Theorem 7.5 Let

$$\hat{D} = D + \Delta, \qquad \hat{X} = X + E, \qquad \hat{Y} = Y + F.$$

If, for some $0 \leq \epsilon < 1$,

$$\frac{|\Delta_{ii}|}{|D_{ii}|} \leq \epsilon, \qquad \frac{\|E\|}{\|X\|} \leq \epsilon, \qquad \frac{\|F\|}{\|Y\|} \leq \epsilon$$

then

$$|\sigma_i - \hat{\sigma}_i| \leq \sigma_i \, (2\eta + \eta^2),$$

where

$$\eta \equiv \epsilon(2 + \epsilon) \max\{\kappa(X), \kappa(Y)\}.$$

Proof. (See Demmel et al. (1997), Theorem 2.1.) The idea is to express Δ, E and F as multiplicative perturbations. Since X has full column rank, it has a left-inverse $X^\dagger$ and we can write

$$\hat{B} = (X + E)\hat{D}\hat{Y}^* = (I + EX^\dagger)X\hat{D}\hat{Y}^* = D_1\, X\hat{D}\hat{Y}^*,$$

where $D_1 \equiv I + EX^\dagger$ is a multiplicative perturbation and

$$\|D_1\| \leq 1 + \|E\|\, \|X^\dagger\| \leq 1 + \epsilon\kappa(X).$$

Similarly one can show $\hat{B} = D_1 B D_2$, where

$$\|D_2\| \leq 1 + \kappa(Y)\,(2\epsilon + \epsilon^2).$$

Application of Theorem 6.1 gives the desired bound. □

Therefore, if X and Y are well conditioned the relative error in the singular values of $\hat{B}$ is proportional to the accuracy ϵ of the rank-revealing decomposition. Note that the error bound depends on $\kappa(X)$ and $\kappa(Y)$ but not on $\kappa(D)$. That's because of the stricter requirement for the perturbation Δ of D, which must be a component-wise relative perturbation.

8. The end

We have seen that many absolute perturbation bounds imply relative bounds. Examples include the bounds by Bauer and Fike, Hoffman and Wielandt, and Weyl. So there is no question of existence. Relative error bounds always exist, for any matrix and for any perturbation.

Like absolute bounds, relative bounds become stronger when the matrices have structure. A Weyl-type bound for Hermitian positive-definite matrices, for instance, is stronger than a Bauer–Fike-type bound for diagonalizable matrices. In contrast to absolute bounds, though, relative bounds can impose more stringent conditions on the matrices to achieve the corresponding bound. For example, most relative bounds for additive perturbations require that the original matrix be nonsingular.

Therefore relative error bounds are not necessarily stronger than absolute error bounds. They just rely for their accuracy on different perturbations. Consider eigenvalues of normal matrices, for instance. A small absolute perturbation E guarantees a small absolute error, while a small relative perturbation, such as

$$\|A^{-1/2}EA^{-1/2}\| \qquad \text{or} \qquad \|I - D_1^{-1}D_2^{-1}\|,$$

guarantees a small relative error. This means that before requesting high relative accuracy you'd better be sure to have a small relative perturbation.

Several theses have been written on the subject of relative error bounds in the context of Jacobi methods for computing singular values (Drmač 1994), eigendecompositions of Hermitian matrices (Slapničar 1992), and eigenvalues of skew-symmetric matrices (Pietzsch 1993), as well as fast algorithms for computing eigendecompositions of real symmetric tridiagonal matrices (Dhillon 1997).

We have omitted the following issues in our discussion of relative error bounds:

- generalized eigenvalue problems (see Barlow and Demmel (1990), Hari and Drmač (1997), Li (1994*a*), Veselić and Slapničar (1993))
- sensitivity of eigenvalues and singular values to perturbations in the factors of a matrix (see Dhillon (1997), Demmel et al. (1997), Parlett (1997), Veselić and Slapničar (1993))
- relative errors in the form of derivatives when the matrix elements depend smoothly on a parameter (see Deift et al. (1991), Section 2, Parlett (1997), Theorem 1).

It is also possible to derive relative perturbation bounds for invariant subspaces and singular vector spaces. These are generally bounds on the angle between an exact and perturbed invariant subspace in terms of a relative eigenvalue separation as opposed to an absolute eigenvalue separation. Many of the papers cited here also discuss bounds for subspaces. Papers solely dealing with subspaces include, among others, Eisenstat and Ipsen (1994), Li (1994*b*), Mathias (1997*a*), Mathias and Veselić (1995), Slapničar and Veselić (1995), and Truhar and Slapničar (1997).

Acknowledgement
I thank Stan Eisenstat and Vjeran Hari for carefully reading the manuscript and for many helpful suggestions.

REFERENCES

J. Barlow and J. Demmel (1990), 'Computing accurate eigensystems of scaled diagonally dominant matrices', *SIAM J. Numer. Anal.* **27**, 762–91.

F. Bauer and C. Fike (1960), 'Norms and exclusion theorems', *Numer. Math.* **2**, 137–41.

R. Bhatia (1997), *Matrix Analysis*, Springer, New York.

T. Chan (1982), 'An improved algorithm for computing the singular value decomposition', *ACM Trans. Math. Software* **8**, 72–83.

S. Chandrasekaran and I. Ipsen (1995), 'Analysis of a QR algorithm for computing singular values', *SIAM J. Matrix Anal. Appl.* **16**, 520–35.

P. Deift, J. Demmel, L. Li and C. Tomei (1991), 'The bidiagonal singular value decomposition and Hamiltonian mechanics', *SIAM J. Numer. Anal.* **28**, 1463–1516.

J. Demmel (1997), *Applied Numerical Linear Algebra*, SIAM, Philadelphia.

J. Demmel and W. Gragg (1993), 'On computing accurate singular values and eigenvalues of matrices with acyclic graphs', *Linear Algebra Appl.* **185**, 203–17.

J. Demmel and W. Kahan (1990), 'Accurate singular values of bidiagonal matrices', *SIAM J. Sci. Statist. Comput.* **11**, 873–912.

J. Demmel and K. Veselić (1992), 'Jacobi's method is more accurate than QR', *SIAM J. Matrix Anal. Appl.* **13**, 1204–45.

J. Demmel, M. Gu, S. Eisenstat, I. Slapničar, K. Veselić and Z. Drmač (1997), Computing the singular value decomposition with high relative accuracy, Technical report, Computer Science Division, University of California, Berkeley, CA.

I. Dhillon (1997), A new $O(n^2)$ algorithm for the symmetric tridiagonal eigenvalue/eigenvector problem, PhD thesis, University of California, Berkeley, California.

I. Dhillon, G. Fann and B. Parlett (1997), Application of a new algorithm for the symmetric eigenproblem to computational quantum chemistry, in *Proceedings of the Eighth SIAM Conference on Parallel Processing for Scientific Computing*, SIAM, Philadelphia.

G. Di Lena, R. Peluso and G. Piazza (1993), 'Results on the relative perturbation of the singular values of a matrix', *BIT* **33**, 647–53.

Z. Drmač (1994), Computing the singular and the generalized singular values, PhD thesis, Fachbereich Mathematik, Fernuniversität Gesamthochschule Hagen, Germany.

Z. Drmač (1996*a*), 'On relative residual bounds for the eigenvalues of a Hermitian matrix', *Linear Algebra Appl.* **244**, 155–64.

Z. Drmač (1996*b*), 'On the condition behaviour in the Jacobi method', *SIAM J. Matrix Anal. Appl.* **17**, 509–14.

Z. Drmač and V. Hari (1997), 'Relative residual bounds for the eigenvalues of a Hermitian semidefinite matrix', *SIAM J. Matrix Anal. Appl.* **18**, 21–9.

S. Eisenstat and I. Ipsen (1994), Relative perturbation bounds for eigenspaces and singular vector subspaces, in *Applied Linear Algebra*, SIAM, Philadelphia, pp. 62–5.

S. Eisenstat and I. Ipsen (1995), 'Relative perturbation techniques for singular value problems', *SIAM J. Numer. Anal.* **32**, 1972–88.

S. Eisenstat and I. Ipsen (1996), Relative perturbation results for eigenvalues and eigenvectors of diagonalisable matrices, Technical Report CRSC-TR96-6, Center for Research in Scientific Computation, Department of Mathematics, North Carolina State University.

S. Eisenstat and I. Ipsen (1997), Three absolute perturbation bounds for matrix eigenvalues imply relative bounds, Technical Report CRSC-TR97-16, Center for Research in Scientific Computation, Department of Mathematics, North Carolina State University. Under review for *SIAM J. Matrix Anal. Appl.*

L. Elsner and S. Friedland (1995), 'Singular values, doubly stochastic matrices, and applications', *Linear Algebra Appl.* **220**, 161–9.

K. Fernando and B. Parlett (1994), 'Accurate singular values and differential qd algorithms', *Numer. Math.* **67**, 191–229.

M. Gu and S. Eisenstat (1993), Relative perturbation theory for eigenproblems, Research Report YALEU/DCS/RR-934, Department of Computer Science, Yale University.

V. Hari and Z. Drmač (1997), 'On scaled almost diagonal Hermitian matrix pairs', *SIAM J. Matrix Anal. Appl.* **18**, 1000–1012.

R. Horn and C. Johnson (1985), *Matrix Analysis*, Cambridge University Press.

R. Horn and C. Johnson (1991), *Topics in Matrix Analysis*, Cambridge University Press.

W. Kahan (1966), Accurate eigenvalues of a symmetric tri-diagonal matrix, Technical Report CS41, Computer Science Department, Stanford University. Revised June 1968.

C. Li and R. Mathias (1997), On the Lidskii–Mirsky–Wielandt theorem, Technical report, Department of Mathematics, College of William and Mary, Williamsburg, VA.

R. Li (1994*a*), Relative perturbation theory: (I) eigenvalue variations, LAPACK working note 84, Computer Science Department, University of Tennessee, Knoxville. Revised May 1997.

R. Li (1994*b*), Relative perturbation theory: (II) eigenspace variations, LAPACK working note 85, Computer Science Department, University of Tennessee, Knoxville. Revised May 1997.

R. Li (1997), 'Relative perturbation theory: (III) more bounds on eigenvalue variation', *Linear Algebra Appl.* **266**, 337–45.

K. Löwner (1934), 'Über monotone Matrix Funktionen', *Math. Z.* **38**, 177–216.

J. Matejaš and V. Hari (1998), 'Scaled almost diagonal matrices with multiple singular values', *Z. Angew. Math. Mech.* **78**, 121–31.

R. Mathias (1995), 'Accurate eigensystem computations by Jacobi methods', *SIAM J. Matrix Anal. Appl.* **16**, 977–1003.

R. Mathias (1996), 'Fast accurate eigenvalue methods for graded positive-definite matrices', *Numer. Math.* **74**, 85–103.

R. Mathias (1997*a*), 'A bound for matrix square root with application to eigenvector perturbation', *SIAM J. Matrix Anal. Appl.* **18**, 861–7.

R. Mathias (1997*b*), 'Spectral perturbation bounds for positive definite matrices', *SIAM J. Matrix Anal. Appl.* **18**, 959–80.

R. Mathias and G. Stewart (1993), 'A block QR algorithm and the singular value decomposition', *Linear Algebra Appl.* **182**, 91–100.

R. Mathias and K. Veselić (1995), A relative perturbation bound for positive-definite matrices, revised December 1996.

A. Ostrowski (1959), 'A quantitative formulation of Sylvester's law of inertia', *Proc. Nat. Acad. Sci.* **45**, 740–4.

B. Parlett (1980), *The Symmetric Eigenvalue Problem*, Prentice Hall.

B. Parlett (1995), 'The new qd algorithms', in *Acta Numerica*, Vol. 4, Cambridge University Press, pp. 459–91.

B. Parlett (1997), Spectral sensitivity of products of bidiagonals. Unpublished manuscript.

E. Pietzsch (1993), Genaue Eigenwertberechnung nichtsingulärer schiefsymmetrischer Matrizen, PhD thesis, Fachbereich Mathematik, Fernuniversität Gesamthochschule Hagen, Germany.

R. Rosanoff, J. Glouderman and S. Levy (1968), Numerical conditions of stiffness matrix formulations for frame structures, in *Proceedings of the Conference on Matrix Methods in Structural Mechanics*, AFFDL-TR-68-150, Wright-Patterson Air Force Base, Ohio, pp. 1029–60.

I. Slapničar (1992), Accurate symmetric eigenreduction by a Jacobi method, PhD thesis, Fernuniversität Gesamthochschule Hagen, Germany.

I. Slapničar and K. Veselić (1995), 'Perturbations of the eigenprojections of a factorised Hermitian matrix', *Linear Algebra Appl.* **218**, 273–80.

N. Truhar and I. Slapničar (1997), Relative perturbation bounds for invariant subspaces of indefinite Hermitian matrices. Unpublished manuscript.

A. van der Sluis (1969), 'Condition, equilibration, and pivoting in linear algebraic systems', *Numer. Math.* **15**, 74–86.

K. Veselić and V. Hari (1989), A note on a one-sided Jacobi algorithm, *Numer. Math.* **56**, 627–33.

K. Veselić and I. Slapničar (1993), 'Floating-point perturbations of Hermitian matrices', *Linear Algebra Appl.* **195**, 81–116.

Acta Numerica (1998), pp. 203–285

Stability for time-dependent differential equations

Heinz-Otto Kreiss*
Department of Mathematics,
UCLA, Los Angeles, CA 90095, USA
E-mail: `kreiss@math.ucla.edu`

Jens Lorenz†
Department of Mathematics and Statistics,
UNM, Albuquerque, NM 87131, USA
E-mail: `lorenz@math.unm.edu`

In this paper we review results on asymptotic stability of stationary states of PDEs. After scaling, our normal form is $u_t = Pu + \varepsilon f(u, u_x, \ldots) + F(x,t)$, where the (vector-valued) function $u(x,t)$ depends on the space variable x and time t. The differential operator P is linear, $F(x,t)$ is a smooth forcing, which decays to zero for $t \to \infty$, and $\varepsilon f(u, \ldots)$ is a nonlinear perturbation. We will discuss conditions that ensure $u \to 0$ for $t \to \infty$ when $|\varepsilon|$ is sufficiently small. If this holds, we call the problem asymptotically stable.

While there are many approaches to show asymptotic stability, we mainly concentrate on the resolvent technique. However, comparisons with the Lyapunov technique will also be given. The emphasis on the resolvent technique is motivated by the recent interest in pseudospectra.

* Supported by Office of Naval Research N 00014 92 J 1890.
† Supported by NSF Grant DMS-9404124 and DOE Grant DE-FG03-95ER25235.

CONTENTS

1. Introduction

A basic result in the stability theory of ODEs can be formulated as follows. If $y_0 \in \mathbb{R}^n$ is a fixed point of the ODE system

$$y_t \equiv \frac{dy}{dt} = \Phi(y), \tag{1.1}$$

where $\Phi : \mathbb{R}^n \to \mathbb{R}^n$ is a C^1 vector field, then y_0 is asymptotically stable,[1] if all eigenvalues of the Jacobian $A = \Phi_y(y_0)$ have negative real parts. By definition, asymptotic stability of y_0 means the following.

(1) For all $\mu > 0$, there is $\varepsilon > 0$ so that[2] $|y_0 - y_1| < \varepsilon$ implies $|y_0 - y(t; y_1)| < \mu$ for all $t \geq 0$. Here $y(t; y_1)$ is the solution of (1.1) with $y = y_1$ at $t = 0$.
(2) There exists $\varepsilon_0 > 0$ such that $|y_0 - y(t; y_1)| \to 0$ as $t \to \infty$ whenever $|y_0 - y_1| < \varepsilon_0$.

Without going into details here (they are given in Section 2), the result can be made plausible. One introduces a new variable $u(t)$ by

$$y(t; y_1) = y_0 + \varepsilon u(t), \quad |y_0 - y_1| = \varepsilon,$$

[1] in the sense of Lyapunov (Lyapunov 1956)
[2] With $\langle u, v\rangle = \sum_j u_j v_j$ and $|u|^2 = \langle u, u\rangle$ we denote the Euclidean inner product and norm. The corresponding matrix norm is $|A| = \max\{|Au| : |u| = 1\}$.

for which one obtains

$$\begin{aligned} \varepsilon u_t &= \Phi(y_0 + \varepsilon u) \\ &= \varepsilon A u + \varepsilon^2 f(u), \quad |f(u)| \leq C|u|^2. \end{aligned}$$

Therefore,

$$u_t = Au + \varepsilon f(u). \tag{1.2}$$

By assumption, all eigenvalues of A have negative real parts, which implies exponential decay of the solutions of the homogeneous equation $u_t = Au$. With proper arguments, the decay estimate can be extended to the nonlinear system (1.2) if $|\varepsilon|$ is sufficiently small, and asymptotic stability follows.

There are two kinds of difficulty in generalizing this basic stability result from ODEs to PDEs, one regarding the linear problem corresponding to $u_t = Au$, the other the small nonlinearity. To be more specific:

(1) In the PDE case, the linear operator corresponding to the matrix A might have a continuous spectrum and it might not be sufficient to look only at eigenvalues. Instead of exponential decay one might obtain only algebraic decay for solutions of the linear problem.
(2) In the PDE case, different norms enter the picture. For the linear problem, one might obtain decay of solutions in one norm, but not in another. Therefore, it is possible that a small nonlinear term $\varepsilon f(u)$ leads to asymptotic stability, whereas a term $\varepsilon f(u, u_x)$ does not.

Despite these difficulties, we will take the ODE case as a guideline. It will be convenient to generalize (1.2) to a system of the form

$$u_t = Au + \varepsilon f(t, u) + F(t) \tag{1.3}$$

where $f(t, 0) = 0$. For PDEs our corresponding normal form is

$$u_t = Pu + \varepsilon f(x, t, u, Du, \ldots, D^r u) + F(x, t). \tag{1.4}$$

Here x varies in a spatial domain Ω and $D^j u$ denotes the array of all spatial derivatives of $u = u(x, t)$ of order j.

The concept of asymptotic stability used in this paper is similar to Lyapunov's, but slightly more restrictive. For the ODE (1.3) our concept is as follows. We first consider the linear problem, obtained for $\varepsilon = 0$, with homogeneous initial condition

$$u = 0 \quad \text{at} \quad t = 0.$$

The forcing $F(t)$ will drive the system away from $u = 0$, and we ask if $u(t)$ will approach zero as $t \to \infty$ if $\lim_{t\to\infty} F(t) = 0$ (or if $F(t) \to 0$ as $t \to \infty$ at a certain rate). If this is so, we call (1.3) *linearly asymptotically stable.* If the same holds whenever $|\varepsilon|$ is sufficiently small, then we call (1.3) *nonlinearly asymptotically stable* (or simply *stable*).

For the ODE case, we will show in Section 2 that linear and also nonlinear asymptotic stability of (1.3) are both equivalent to the eigenvalue condition[3]

$$\operatorname{Re}\lambda < 0 \quad \text{for all} \quad \lambda \in \sigma(A). \tag{1.5}$$

In contrast, this eigenvalue condition is sufficient, but not necessary for asymptotic stability in the sense of Lyapunov of the zero solution of (1.3) with $F \equiv 0$. For example, for the equation $u_t = -u^3$ the zero solution is asymptotically stable in the sense of Lyapunov.

In short, the stability concept that we use here is slightly more restrictive than Lyapunov's, but also more robust. This makes it easier to generalize to PDEs, which is the main interest of the paper. In all cases, our sufficient conditions on asymptotic stability also provide estimates of the solution $u(x,t)$ for $0 \leq t \leq T$ by $F(x,t)$ for $0 \leq t \leq T$, in suitable norms, and the constant in the estimate will be independent of T and of $|\varepsilon| \leq \varepsilon_0$. Therefore, asymptotic stability in Lyapunov's sense, where only the initial data are perturbed, can always be shown under our assumptions. See also Section 8.1.

(The introduction of the inhomogeneous term $F(t)$ can also be motivated by numerical considerations. When one interpolates the numerical values, one obtains a function u which satisfies a perturbed differential equation. The size of the perturbation is measured by $F(t)$.)

A more detailed outline of the paper follows. (Sections 2 to 5 are partially based on Kreiss and Lui (1997).) In Section 2 we use the Lyapunov technique and the resolvent technique (or Laplace-transform technique) to show asymptotic stability in the ODE setting if (1.5) holds. The solution-operator technique, which plays a major role for nonlinear wave equations, will be illustrated briefly in Section 2.3, but we refer to Racke (1992) for a comprehensive treatment.

In Sections 3 to 5 we assume that the linear operator P in (1.4) has constant coefficients and that the boundary conditions are periodic. Then, for the linear problem $u_t = Pu+F$, one can use Fourier expansions in space. The following two points will be emphasized.

(1) Assuming the Cauchy problem for $u_t = Pu + F$ to be well posed, the eigenvalues of the symbols $\hat{P}(i\omega)$ tell what kind of resolvent estimate holds, *i.e.*, how many derivatives one gains when estimating u by F.

(2) If one gains q derivatives in the resolvent estimate, then the nonlinearity f in (1.4) may depend on all space derivatives of u of order $\leq q$.

[3] With $\sigma(A)$ we denote the set of all eigenvalues of A.

For parabolic problems, these results are satisfactory. However, simple hyperbolic equations like

$$u_t = u_x - u + \varepsilon u u_x + F(x,t)$$

cannot be treated in this way, because one does not gain a derivative in the resolvent estimate. As we will show in Section 5 on hyperbolic problems, the Lyapunov technique can be applied to overcome the difficulty.

To give a problem where the resolvent technique is more adequate, consider the parabolic equation

$$u_t = Pu \equiv u_{xx} + (a(x)u)_x, \quad 0 \leq x \leq 1, \quad t \geq 0, \tag{1.6}$$

with boundary conditions

$$u(0,t) = u(1,t) = 0, \quad t \geq 0, \tag{1.7}$$

and initial condition

$$u(x,0) = u_0(x), \quad 0 \leq x \leq 1.$$

The ODE operator

$$Pu = u_{xx} + (a(x)u)_x$$

has the adjoint $P^*u = u_{xx} - a(x)u_x$, which makes it easy to show by the maximum principle that all eigenvalues of P are negative. (Here we use the boundary conditions (1.7), for example.) Then, by our results in Section 6, the resolvent technique applies and asymptotic stability follows, even for fully nonlinear perturbations $\varepsilon f(x,t,u,u_x,u_{xx})$. If one wanted to obtain this result by the Lyapunov technique, one would have to construct an inner product $(\cdot,\cdot)_{\mathcal{H}}$ such that

$$(u, Pu)_{\mathcal{H}} \leq -c(u,u)_{\mathcal{H}}, \quad c > 0.$$

However, it is not clear how to construct such an inner product, whose norm also has to be strong enough to bound u_{xx}.

It is natural to ask if one can combine the resolvent technique and the Lyapunov technique, using the strengths of each. This is indeed possible as demonstrated by Kreiss, Kreiss and Lorenz (1998*a*, 1998*b*). One area of applications is the study of mixed parabolic–hyperbolic systems. In the present paper we do not pursue this, however.

Sections 7 to 11 deal with PDEs on unbounded spatial domains Ω. In fact, we only consider the cases $\Omega = \mathbb{R}^d$ and $\Omega = \mathcal{H}^d$, a half-space. As will be shown in Sections 7 and 9, the unboundedness of the domain does not lead to difficulties as long as the linear operator P has a strictly negative zero-order term. A more interesting situation occurs when the zero-order term of P vanishes (or is semi-negative).

Using the resolvent technique, we will discuss this for parabolic problems on all space in Section 8 and will derive a weak resolvent estimate. The form of the linear estimate dictates more specific assumptions on the nonlinearity. The results apply to viscous conservation laws.

In Section 10 we consider the scalar parabolic model problem

$$u_t = \Delta u + a_1 u_{x_1} + \cdots + a_d u_{x_d} + G(x,t), \quad x_1 \geq 0, \tag{1.8}$$

on a half-space, again in the critical situation when the zero-order term vanishes, and show a weak resolvent estimate. The derivation is elementary but requires some rather technical analysis of ordinary BVPs on the half-line. It will become clear that the underlying hyperbolic part in (1.8) is important. In particular, the conditions for the weak resolvent estimate depend on the sign of the characteristic speed a_1 in relation to the boundary $x_1 = 0$.

Finally, in Section 11 we sketch stability results for parabolic systems on the real line, which are applicable to travelling waves.

The main emphasis in this paper is the derivation of stability results by the resolvent technique. Trefethen's work on pseudospectra (see Trefethen (1997) and the references given there) was a major motivation for this emphasis. Once a stability problem is cast into the normal form (1.4), one might want to answer the following questions.

(1) Applying Laplace transformation to the linear problem

$$u_t = Pu + F(x,t), \quad u(x,0) = 0, \tag{1.9}$$

one obtains the resolvent equation

$$(sI - P)\tilde{u} = \tilde{F}, \quad \operatorname{Re} s \geq 0. \tag{1.10}$$

What kind of estimates of $\tilde{u}$ by $\tilde{F}$ can one obtain? How do the estimates depend on s and how do they translate into estimates for physical variables?

(2) Given certain estimates for the linear problem (1.9), what kind of nonlinear problems (1.4) can one treat for small $|\varepsilon|$?

(3) How are the resolvent estimates for (1.10) related to properties of the spectrum or the pseudospectrum of P?

(4) If P depends on parameters – like the Reynolds number – how do the constants in the resolvent estimate scale as functions of the parameters and how does this affect the size $|\varepsilon|$ of the nonlinear terms for which one retains stability?

In this paper we only address the first two questions, though the other two are clearly of great interest. We remark that Romanov (1973) has considered plane-parallel Couette flow and has obtained an upper bound $-c\nu$ (with $c > 0$) for the real parts of the spectral values of a corresponding linearized operator P. Kreiss, Lundbladh and Henningson (1994b) have made attempts to obtain a bound for the resolvent $(sI - P)^{-1}$ and to address the fourth question for these flows. For the ODE case, we will sketch some simple related observations in Section 2.4.

2. Ordinary differential equations

Consider an initial value problem

$$u_t = Au + \varepsilon f(t,u) + F(t), \quad t \geq 0; \quad u(0) = u_0. \tag{2.1}$$

Here $A \in \mathbb{C}^{n\times n}$ is a constant matrix, the functions

$$f : \mathbb{C}^n \times [0,\infty) \to \mathbb{C}^n, \quad F : [0,\infty) \to \mathbb{C}^n$$

are assumed to be C^∞, for simplicity, and $u_0 \in \mathbb{C}^n$ is a given initial vector. We will always assume that $F(t)$ is a bounded function and set

$$|F|_\infty = \sup_{t\geq 0} |F(t)|.$$

The function $f(t,u)$ is assumed to vanish at $u = 0$. More precisely, we assume that

$$\text{for all } c_1 > 0 \text{ there exists } C_1 > 0 \text{ with } \quad |f(u,t)| \leq C_1|u| \quad \text{if } |u| \leq c_1. \tag{2.2}$$

Our concept of asymptotic stability is as follows.

Definition 2.1 Problem (2.1) with $\varepsilon = 0$ is called linearly asymptotically stable if

$$\lim_{t\to\infty} F(t) = 0 \quad \text{implies} \quad \lim_{t\to\infty} u(t) = 0. \tag{2.3}$$

Furthermore, (2.1) is called nonlinearly asymptotically stable if (2.3) holds for all sufficiently small $|\varepsilon|$.

If A has an eigenvalue λ with $\operatorname{Re}\lambda \geq 0$ and $A\phi = \lambda\phi, \phi \neq 0$, then $u(t) = e^{\lambda t}\phi$ solves $u_t = Au$, but $u(t)$ does not tend to zero as $t \to \infty$. Therefore, the following condition is necessary for asymptotic stability of (2.1).

Eigenvalue condition.

$$\operatorname{Re}\lambda < 0 \quad \text{for all} \quad \lambda \in \sigma(A). \tag{2.4}$$

We will now show that (2.4) *characterizes* asymptotic stability.

Theorem 2.1 Under assumption (2.2) the eigenvalue condition (2.4) is necessary and sufficient for linear (and also for nonlinear) asymptotic stability of (2.1).

For illustration, we will prove Theorem 2.1 in two different ways, namely the Lyapunov technique (or energy estimate) and the resolvent technique.[4]

[4] The assumption for $F(t)$ is slightly different in the result proved by the resolvent technique: see Theorem 2.3.

In the ODE case, both techniques lead to the same characterization of asymptotically stable problems (Theorem 2.1). In the PDE case, the two techniques have different strengths and weaknesses, however, as we will show in the subsequent sections.

2.1. The Lyapunov technique

The linear problem

Let us first show linear asymptotic stability under assumption (2.4). The easiest case occurs if A is normal, where we can use the following result.[5]

Lemma 2.1 Let A be normal and let

$$\operatorname{Re}\lambda \leq -\delta < 0 \quad \text{for all} \quad \lambda \in \sigma(A). \tag{2.5}$$

Then we have

$$A + A^* \leq -2\delta I. \tag{2.6}$$

Proof. There is a unitary matrix U such that

$$U^*AU = \Lambda = \operatorname{diag}(\lambda_j), \quad \operatorname{Re}\lambda_j \leq -\delta.$$

Therefore,

$$A + A^* = U(\Lambda + \bar{\Lambda})U^* \leq -2\delta I.$$

□

Now assume that A is normal and let $u_t = Au + F(t)$. Then we obtain

$$\begin{aligned}
\frac{\mathrm{d}}{\mathrm{d}t}|u|^2 &= \langle u, u_t\rangle + \langle u_t, u\rangle \\
&= \langle u, (A + A^*)u\rangle + \langle u, F\rangle + \langle F, u\rangle \\
&\leq -2\delta|u|^2 + 2|u||F| \\
&\leq -\delta|u|^2 + \frac{1}{\delta}|F|^2.
\end{aligned}$$

This implies

$$\begin{aligned}
|u(t)|^2 &\leq \mathrm{e}^{-\delta t}|u_0|^2 + \frac{1}{\delta}\int_0^t \mathrm{e}^{-\delta(t-\tau)}|F(\tau)|^2\,\mathrm{d}\tau \\
&\leq \mathrm{e}^{-\delta t}|u_0|^2 + \frac{1}{\delta^2}\max_{0\leq\tau\leq t}|F(\tau)|^2;
\end{aligned} \tag{2.7}$$

thus

$$|u(t)|^2 \leq |u_0|^2 + \frac{1}{\delta^2}|F|_\infty^2 =: c_0^2, \quad t \geq 0. \tag{2.8}$$

[5] If $H_1, H_2 \in \mathbb{C}^{n\times n}$ are Hermitian matrices, then we write $H_1 \leq H_2$ if and only if $u^*H_1u \leq u^*H_2u$ for all $u \in \mathbb{C}^n$. Similarly, we write $H_1 < H_2$ if and only if $u^*H_1u < u^*H_2u$ for all $u \in \mathbb{C}^n, u \neq 0$.

Clearly, we may start the estimate at any t_0 and obtain

$$|u(t)|^2 \leq e^{-\delta(t-t_0)}|u(t_0)|^2 + \frac{1}{\delta^2} \max_{t_0 \leq \tau \leq t} |F(\tau)|^2. \tag{2.9}$$

Now recall the assumption $\lim_{t\to\infty} F(t) = 0$. If $\eta > 0$ is given, there exists $t_0(\eta)$ with

$$\frac{1}{\delta^2} \sup_{t_0(\eta) \leq \tau < \infty} |F(\tau)|^2 \leq \eta^2.$$

Therefore, (2.9) and (2.8) imply

$$|u(t)|^2 \leq 2\eta^2 \quad \text{for} \quad t \geq t_1(\eta),$$

and we have shown $\lim_{t\to\infty} u(t) = 0$.

If A is not normal, the crucial inequality $A + A^* \leq -2\delta I$ does not follow from (2.5), in general. However, by changing the inner product, we can basically still argue as before. The following notation will be used.

Notation
If $H > 0$ is a positive definite Hermitian matrix, then a scalar product and norm are determined by

$$\langle u, v\rangle_H = u^* H v, \quad |u|_H^2 = u^* H u, \quad u, v \in \mathbb{C}^n. \tag{2.10}$$

The corresponding matrix norm is

$$|B|_H = \max_{|u|_H = 1} |Bu|_H.$$

We note that the inequalities

$$\frac{1}{c} I \leq H \leq cI, \quad c \geq 1, \tag{2.11}$$

are equivalent to the norm estimates

$$\frac{1}{c}\, |u|^2 \leq |u|_H^2 \leq c|u|^2 \quad \text{for all} \quad u \in \mathbb{C}^n, \tag{2.12}$$

and therefore (2.11) implies

$$\frac{1}{c}\, |B|_H \leq |B| \leq c|B|_H, \quad B \in \mathbb{C}^{n\times n}. \tag{2.13}$$

Lemma 2.1 has the following generalization to arbitrary matrices $A \in \mathbb{C}^{n\times n}$.

Theorem 2.2 If

$$\operatorname{Re} \lambda \leq -\delta < -\delta_1 < 0 \quad \text{for all} \quad \lambda \in \sigma(A), \tag{2.14}$$

then there exists a positive definite Hermitian matrix H with

$$HA + A^* H \leq -2\delta_1 H < 0. \tag{2.15}$$

Proof. By Schur's theorem there exists a unitary matrix U so that

$$U^*AU = \Lambda + R = \mathrm{diag}(\lambda_j) + R, \quad \mathrm{Re}\,\lambda_j \le -\delta < 0,$$

where R is strictly upper triangular. We set

$$D = \mathrm{diag}(1, \varepsilon, \ldots, \varepsilon^{n-1}), \quad \varepsilon > 0,$$

and consider

$$D^{-1}U^*AUD = \Lambda + D^{-1}RD.$$

The entries in $D^{-1}RD$ are $\mathcal{O}(\varepsilon)$. Setting $S = (UD)^{-1}$ one obtains

$$\begin{aligned} SAS^{-1} + (SAS^{-1})^* &= \Lambda + \bar{\Lambda} + \mathcal{O}(\varepsilon) \\ &\le -2\delta_1 I \end{aligned}$$

for sufficiently small ε. Therefore, if one defines $H = S^*S$, one obtains the desired inequality

$$\begin{aligned} HA + A^*H &= S^*(\Lambda + \bar{\Lambda} + \mathcal{O}(\varepsilon))S \\ &\le -2\delta_1 H \end{aligned}$$

for sufficiently small $\varepsilon > 0$. □

Using $|\cdot|_H$ one can estimate the solutions $u(t)$ of $u_t = Au + F(t)$ as before,

$$\begin{aligned} \frac{\mathrm{d}}{\mathrm{d}t}|u|_H^2 &= \langle u, u_t\rangle_H + \langle u_t, u\rangle_H \\ &= \langle u, (HA + A^*H)u\rangle + \langle u, F\rangle_H + \langle F, u\rangle_H \\ &\le -2\delta_1|u|_H^2 + 2|u|_H|F|_H \\ &\le -\delta_1|u|_H^2 + \frac{1}{\delta_1}|F|_H^2. \end{aligned}$$

Proceeding in the same way as above, we find that $u(t) \to 0$ as $t \to \infty$. Here the equivalence of the norms $|\cdot|$ and $|\cdot|_H$ is used. The equivalence is valid, of course, since we work in a finite dimensional setting. Let us note, however, that the inclusion (2.11) leads to an equivalence constant; see (2.12). This observation will be useful for PDEs, where we deal with infinite families of matrices.

The nonlinear problem

Now consider the nonlinear problem (2.1). For simplicity, let A be normal so that we can use the Euclidean norm for our estimates. In the general case, the same arguments apply with $|\cdot|_H$.

Recall the estimate (2.8) for $\varepsilon = 0$, where $c_0 > 0$ without loss of generality. By continuity, for any ε there exists $T_\varepsilon > 0$ such that

$$|u(t)|^2 \le 3c_0^2 =: c_1^2 \quad \text{for} \quad 0 \le t \le T_\varepsilon. \tag{2.16}$$

Using (2.2) we obtain

$$\begin{aligned}\frac{\mathrm{d}}{\mathrm{d}t}|u|^2 &\le -2\delta|u|^2 + 2|\varepsilon||u||f(t,u)| + 2|u||F| \\ &\le -\delta|u|^2 + \frac{1}{\delta}|F|^2 + 2|\varepsilon|C_1|u|^2 \end{aligned} \tag{2.17}$$

in $0 \le t \le T_\varepsilon$. Henceforth, assume that $|\varepsilon|$ is so small that

$$2|\varepsilon|C_1 \le \frac{\delta}{2}. \tag{2.18}$$

Then we find

$$\frac{\mathrm{d}}{\mathrm{d}t}|u|^2 \le -\frac{\delta}{2}|u|^2 + \frac{1}{\delta}|F|^2;$$

thus

$$\begin{aligned}|u(t)|^2 &\le \mathrm{e}^{-\delta t/2}|u_0|^2 + \frac{1}{\delta}\int_0^t \mathrm{e}^{-\delta(t-\tau)/2}|F(\tau)|^2\,\mathrm{d}\tau \\ &\le |u_0|^2 + \frac{2}{\delta^2}\max_{0\le\tau\le t}|F(\tau)|^2 \\ &\le 2c_0^2. \end{aligned} \tag{2.19}$$

Therefore, always assuming (2.18), we have shown that $|u(t)|^2 \le 3c_0^2$ in $0 \le t \le T_\varepsilon$ implies $|u(t)|^2 \le 2c_0^2$ in $0 \le t \le T_\varepsilon$. A simple continuation and contradiction argument yields that $u(t)$ exists for all $t \ge 0$ and

$$|u(t)| \le 2c_0^2 \quad \text{for all} \quad t \ge 0.$$

With the same arguments as before, we find

$$|u(t)|^2 \le \mathrm{e}^{-\delta(t-t_0)/2}2c_0^2 + \frac{2}{\delta^2}\sup_{t_0\le\tau<\infty}|F(\tau)|^2,$$

and therefore (2.3) holds. This proves Theorem 2.1.

2.2. The resolvent technique

Consider again the ODE (2.1). To illustrate the resolvent technique, we will prove the following result.

Theorem 2.3 Assume (2.2), (2.4) and let $F \in L_2$. Then $\lim_{t\to\infty} u(t) = 0$ if $|\varepsilon|$ is sufficiently small.

It will be convenient to have homogeneous initial data, which can always be achieved by applying the simple transformation[6]

$$u(t) = \mathrm{e}^{-t}u_0 + v(t).$$

[6] For generalizations, it is important to note that we do not make use of the exponential decay of e^{At}.

The function $v(t)$ satisfies $v(0) = 0$ and

$$v_t = Av + \varepsilon\Big(f(t, \mathrm{e}^{-t}u_0 + v) - f(t, \mathrm{e}^{-t}u_0)\Big) + G(t)$$

with

$$G(t) = F(t) + \mathrm{e}^{-t}(Au_0 + u_0) + \varepsilon f(t, \mathrm{e}^{-t}u_0), \quad \lim_{t\to\infty} G(t) = 0.$$

If we set

$$\phi(\xi) = f(t, \mathrm{e}^{-t}u_0 + \xi v), \quad 0 \le \xi \le 1,$$

we can write

$$\begin{aligned} f(t, \mathrm{e}^{-t}u_0 + v) - f(t, \mathrm{e}^{-t}u_0) &= \phi(1) - \phi(0) = \int_0^1 \phi'(\xi)\,\mathrm{d}\xi \\ &= \left(\int_0^1 f_u(t, \mathrm{e}^{-t}u_0 + \xi v)\,\mathrm{d}\xi\right) v \\ &=: g(t, v). \end{aligned} \tag{2.20}$$

Thus, for v we obtain a transformed equation

$$v_t = Av + \varepsilon g(t, v) + G(t), \quad v(0) = 0,$$

of similar form with homogeneous initial data.

For notational convenience, we will assume that the initial data in (2.1) are already homogeneous.

Properties of the Laplace transform

Let us first recall some elementary properties of the Laplace transform. If $g(t)$ is a continuous function of $t \ge 0$ with values in $\mathbb{C}^n$, which satisfies a growth restriction

$$|g(t)| \le K\mathrm{e}^{\alpha t}, \quad t \ge 0,$$

then its Laplace transform is the analytic function

$$\tilde{g}(s) = \int_0^\infty \mathrm{e}^{-st} g(t)\,\mathrm{d}t, \quad \mathrm{Re}\, s > \alpha.$$

The inverse transform reads

$$g_0(t) = \frac{1}{2\pi i}\int_{\eta - \mathrm{i}\infty}^{\eta + \mathrm{i}\infty} \mathrm{e}^{st}\tilde{g}(s)\,\mathrm{d}s, \quad \eta > \alpha,$$

where

$$g_0(t) = g(t) \quad \text{for } t > 0, \quad g_0(0) = \frac{1}{2}g(0), \quad g_0(t) = 0 \quad \text{for } t < 0.$$

Here the path of integration is $\Gamma_\eta = \{s = \eta + \mathrm{i}\xi, -\infty < \xi < \infty\}$.

Parseval's relation reads

$$\int_0^\infty \mathrm{e}^{-2\eta t}|g(t)|^2\,\mathrm{d}t = \frac{1}{2\pi}\int_{-\infty}^{\infty} |\tilde{g}(\eta + \mathrm{i}\xi)|^2\,\mathrm{d}\xi, \quad \eta > \alpha. \tag{2.21}$$

If $g \in C^1$ and

$$|g(t)| + |g_t(t)| \le K\mathrm{e}^{\alpha t}, \quad t \ge 0,$$

then

$$\tilde{g}_t(s) = s\tilde{g}(s) - g(0), \quad \operatorname{Re} s > \alpha,$$

as follows directly from the definition through integration by parts.

Estimates in the linear case

Now consider (2.1) with $\varepsilon = 0, u_0 = 0$, and first assume that $F(t) = 0$ for $t > T$. Then $\tilde{F}(s)$ and $\tilde{u}(s)$ are well defined for $\operatorname{Re} s \ge 0$. Laplace transformation gives us

$$s\tilde{u}(s) = A\tilde{u}(s) + \tilde{F}(s), \quad \operatorname{Re} s \ge 0; \tag{2.22}$$

thus

$$\tilde{u}(s) = (sI - A)^{-1}\tilde{F}(s), \quad \operatorname{Re} s \ge 0. \tag{2.23}$$

For any $A \in \mathbb{C}^{n\times n}$ the matrix-valued function

$$(sI - A)^{-1}, \quad s \in \mathbb{C} \setminus \sigma(A),$$

is called the resolvent of A. We list some of its elementary properties.

Lemma 2.2

(a) For any $A \in \mathbb{C}^{n\times n}$ we have

$$|(sI - A)^{-1}| \le \frac{1}{|s| - |A|} \quad \text{if } |s| > |A|.$$

(b) If $\operatorname{Re}\lambda < 0$ for all $\lambda \in \sigma(A)$, then

$$R := \sup_{\operatorname{Re} s \ge 0} |(sI - A)^{-1}| \tag{2.24}$$

is finite. By definition, R is the resolvent constant of A.

(c) If A is normal and

$$\max_{\lambda\in\sigma(A)} \operatorname{Re}\lambda =: -\delta < 0,$$

then $R = \frac{1}{\delta}$.

Proof.

(a) Let $|s| > |A|$ and let

$$sv = Av + b, \quad b, v \in \mathbb{C}^n.$$

Then one obtains $|s||v| \le |A||v| + |b|$; thus $|v| \le (|s| - |A|)^{-1}|b|$.

(b) Define $\Omega = \{s \in \mathbb{C} \; : \; \mathrm{Re}\, s \ge 0, \; |s| \le |A| + 1\}$. By compactness,

$$\max_{s \in \Omega} |(sI - A)^{-1}| =: R_1$$

is finite. Together with (a) one obtains $R \le \max\{R_1, 1\}$.

(c) There is a unitary matrix U such that $U^*AU = \Lambda = \mathrm{diag}(\lambda_j)$ is diagonal. If

$$s = \eta + \mathrm{i}\xi, \quad \eta \ge 0, \quad \lambda_j = \alpha_j + \mathrm{i}\beta_j, \quad \alpha_j \le -\delta,$$

then one obtains

$$\begin{aligned} |(sI - A)^{-1}|^2 &= |(sI - \Lambda)^{-1}|^2 \\ &= \max_j \frac{1}{|s - \lambda_j|^2} \\ &= \max_j \frac{1}{(\eta - \alpha_j)^2 + (\xi - \beta_j)^2} \\ &\le \frac{1}{\delta^2}. \end{aligned}$$

In the last estimate equality holds for $s = \mathrm{i}\beta_j$ if $\mathrm{Re}\, \alpha_j = -\delta$. This proves the lemma. □

Assuming the eigenvalue condition (2.4), we obtain from (2.23)

$$|\tilde{u}(s)| \le R\, |\tilde{F}(s)|, \quad \mathrm{Re}\, s \ge 0.$$

Then Parseval's relation (see (2.21) and set $\eta = 0$) implies that

$$\int_0^\infty |u(t)|^2 \, \mathrm{d}t \le R^2 \int_0^\infty |F(t)|^2 \, \mathrm{d}t. \tag{2.25}$$

So far, we have assumed that $F(t)$ vanishes for large t. However, if $F \in L_2$ is arbitrary, we can approximate F by a sequence F_n with $F_n(t) = F(t)$ for $t \le n$ and $F_n(t) = 0$ for $t \ge n + 1$. A simple limit argument shows that the estimate (2.25) still holds.

Remark 2.1 Using the definition of the resolvent constant R and Parseval's relation, it is not difficult to show that R is the *best* constant for which (2.25) holds for all $F \in L_2$.

The estimate (2.25) is not strong enough to yield a nonlinear stability result, because one cannot bound point values $u(t)$ in terms of the L_2-integral.

An estimate of point values is generally necessary, however, to control nonlinearities $f(t, u)$. We can strengthen the estimate (2.25) as follows.

Theorem 2.4 Consider

$$u_t = Au + F(t), \quad t \geq 0, \quad u(0) = 0,$$

and assume the eigenvalue condition (2.4). There is a constant K_1, depending only on A, so that

$$\int_0^\infty \left(|u(t)|^2 + |u_t(t)|^2 \right) \mathrm{d}t \leq K_1 \int_0^\infty |F(t)|^2 \, \mathrm{d}t. \tag{2.26}$$

Proof. As before, we first assume $F(t) = 0$ for large t. Since $u(0) = 0$, Laplace transformation yields

$$\tilde{u}_t(s) = s\tilde{u}(s) = A\tilde{u}(s) + \tilde{F}(s);$$

thus

$$\begin{aligned} |\tilde{u}_t(s)| &\leq |A||\tilde{u}(s)| + |\tilde{F}(s)| \\ &\leq (|A|R + 1)|\tilde{F}(s)|. \end{aligned}$$

Using the same arguments as above, the desired estimate follows. □

Since values of $F(t)$ for $t > T$ do not affect the solution $u(t)$ for $t \leq T$, it is not difficult to show that the estimate (2.26) can be restricted to any finite time interval, that is,

$$\int_0^T \left(|u(t)|^2 + |u_t(t)|^2 \right) \mathrm{d}t \leq K_1 \int_0^T |F(t)|^2 \, \mathrm{d}t. \tag{2.27}$$

The left-hand side of (2.27) can now be used to bound u in maximum norm. This follows easily from $u(0) = 0$ and

$$\begin{aligned} \frac{\mathrm{d}}{\mathrm{d}t}|u|^2 &= \langle u, u_t \rangle + \langle u_t, u \rangle \\ &\leq 2|u||u_t| \leq |u|^2 + |u_t|^2. \end{aligned}$$

The resulting estimate is a simple example of a Sobolev inequality, which we state next.

Lemma 2.3 Let $u(t), a \leq t \leq b$, denote a C^1-function. Then we have

$$\max_{a \leq t \leq b} |u(t)|^2 \leq \left(1 + \frac{1}{b-a} \right) \int_a^b |u|^2 \, \mathrm{d}t \; + \; \int_a^b |u_t|^2 \, \mathrm{d}t. \tag{2.28}$$

If $u(t^*) = 0$ for some $a \leq t^* \leq b$, then

$$\max_{a \leq t \leq b} |u(t)|^2 \leq \int_a^b \left(|u|^2 + |u_t|^2 \right) \mathrm{d}t. \tag{2.29}$$

Applying (2.28), for $b \to \infty$, one obtains that

$$\sup_{t \geq t_0} |u(t)|^2 \leq \int_{t_0}^{\infty} \left(|u|^2 + |u_t|^2 \right) dt. \tag{2.30}$$

The right-hand side tends to zero for $t_0 \to \infty$ since $\int_0^\infty (|u|^2 + |u_t|^2)\, dt$ is finite by Theorem 2.4. Thus we have shown the claim of Theorem 2.3 for $\varepsilon = 0$.

Nonlinear stability

Since our estimates are strong enough to control u in maximum norm, it is not difficult to extend the result to small $|\varepsilon|$. The arguments are as follows.

For any ε there exists $T_\varepsilon > 0$ with

$$\int_0^{T_\varepsilon} \left(|u|^2 + |u_t|^2 \right) dt \leq 4K_1 \int_0^\infty |F|^2\, dt =: c_1^2. \tag{2.31}$$

By (2.29),

$$|u(t)| \leq c_1, \quad 0 \leq t \leq T_\varepsilon,$$

thus

$$|f(t, u(t))| \leq C_1 |u(t)|, \quad 0 \leq t \leq T_\varepsilon,$$

and therefore

$$\int_0^{T_\varepsilon} |f(t, u(t))|^2\, dt \leq C_1^2 \int_0^{T_\varepsilon} |u|^2\, dt. \tag{2.32}$$

Application of (2.27), with $F(t)$ replaced by $F(t) + \varepsilon f(t, u(t))$, yields

$$\int_0^{T_\varepsilon} \left(|u|^2 + |u_t|^2 \right) dt \leq 2K_1 \int_0^{T_\varepsilon} \left(|F|^2 \; + \; |\varepsilon|^2 C_1^2 |u|^2 \right) \; dt.$$

Assuming that $|\varepsilon|$ is so small that

$$2K_1 |\varepsilon|^2 C_1^2 \leq \frac{1}{3},$$

we obtain

$$\int_0^{T_\varepsilon} \left(|u|^2 + |u_t|^2 \right) dt \leq 3K_1 \int_0^\infty |F|^2\, dt = \frac{3}{4} c_1^2. \tag{2.33}$$

To summarize, from (2.31) we could conclude (2.33) under the above smallness assumption for $|\varepsilon|$. A simple continuation and contradiction argument yields existence of $u(t)$ for all $t \geq 0$, and

$$\int_0^\infty \left(|u|^2 + |u_t|^2 \right) dt \leq \frac{3}{4} c_1^2 < \infty.$$

As before, $\lim_{t \to \infty} u(t) = 0$ follows.

Outline of generalizations to PDEs

It is worthwhile to re-emphasize the main points of the above arguments and to indicate generalizations. There are essentially three steps.

(1) An estimate for a linear problem $u_t = Pu + F(t)$ of the type

$$\|u\|_U \le K_1 \|F\|_V. \tag{2.34}$$

Such an estimate will generalize (2.26).

(2) A Sobolev inequality like

$$|u|_\infty \le K_2 \|u\|_U. \tag{2.35}$$

Such an estimate will generalize (2.29). In the PDE case, if the nonlinearity f depends also on u_x, *etc.*, then $\|u\|_U$ has to be strong enough to bound $|u_x|_\infty$, *etc.*, too.

(3) An estimate of the nonlinear term of the following type. If $|u|_\infty \le \kappa$ then

$$\|f(t, u(t))\|_V \le C_\kappa \|u\|_U. \tag{2.36}$$

Such an estimate will generalize (2.32).

Now assume we have established (2.34), (2.35), (2.36). Then, in order to bound the solution of the nonlinear problem

$$u_t = Pu + \varepsilon f(t, u) + F(t)$$

for small $|\varepsilon|$, we formally proceed as follows. We consider $\varepsilon f(t, u(t))$ as part of the forcing so that (2.34) yields

$$\|u\|_U \le K_1(\|F\|_V + |\varepsilon|\ \|f\|_V). \tag{2.37}$$

Assuming, tentatively, that

$$\|u\|_U \le 2K_1 \|F\|_V, \tag{2.38}$$

one obtains from (2.35)

$$|u|_\infty \le 2K_1 K_2 \|F\|_V =: \kappa.$$

Then (2.36) implies

$$\|f(t, u(t))\|_V \le C_\kappa \|u\|_U$$

and (2.37) yields

$$\|u\|_U \le K_1(\|F\|_V + |\varepsilon|\ C_\kappa\ \|u\|_U).$$

If ε satisfies the restriction

$$|\varepsilon| K_1 C_\kappa \le \frac{1}{3},$$

then the previous estimate gives us the desired bound

$$\|u\|_U \leq \frac{3}{2}K_1\|F\|_V,$$

which is consistent with the tentative assumption (2.38). As in the proof of Theorem 2.4, these formal arguments can be made rigorous by restricting the relevant estimates to finite intervals $0 \leq t \leq T$. Here it is assumed that the norm $\|u\|_U$ is strong enough to guarantee continuation of a local solution.

2.3. The solution-operator technique

There are cases where it is best to argue directly with the solution operator of the linear homogeneous equation. For illustration, we consider the ODE initial value problem

$$u_t = -\frac{\gamma}{t+1}\, u + \varepsilon u^\rho + F(t), \quad u(0) = u_0, \tag{2.39}$$

with constants $\gamma > 0, \rho \geq 1$, and a continuous forcing $F(t)$ which satisfies an estimate

$$|F(t)| \leq \frac{K}{(t+1)^\beta}, \quad \beta > 0. \tag{2.40}$$

We ask for conditions on β and ρ which imply that, for sufficiently small $|\varepsilon|$, $u(t)$ converges to zero for $t \to \infty$ as fast as the solution of the homogeneous equation

$$u_t = -\frac{\gamma}{t+1}\, u, \quad u(0) = u_0. \tag{2.41}$$

The solution of (2.41) is

$$u_h(t) = (t+1)^{-\gamma}u_0,$$

and the solution of (2.39) with $\varepsilon = 0$ is

$$\begin{aligned} u(t) &= (t+1)^{-\gamma}u_0 + \int_0^t \left(\frac{\xi+1}{t+1}\right)^\gamma F(\xi)\, d\xi \\ &= : u_h(t) + J(t). \end{aligned}$$

Using (2.40) we can bound the integral term as follows:

$$|J(t)| \leq K(t+1)^{-\gamma}\int_0^t (\xi+1)^{\gamma-\beta}\, d\xi,$$

which shows that the solution $u(t)$ decays like $(t+1)^{-\gamma}$ if $\gamma - \beta < -1$.

Now consider the nonlinear problem (2.39). Writing εu^ρ as forcing, one obtains

$$\begin{aligned}\phi(t) := (t+1)^\gamma |u(t)| &\le |u_0| + K' + |\varepsilon| \int_0^t (\xi+1)^\gamma |u(\xi)|^\rho \, d\xi \\ &\le |u_0| + K' + |\varepsilon| \left(\max_{0\le\xi\le t} \phi(x) \right) \int_0^t (\xi+1)^{\gamma-\gamma\rho} \, d\xi.\end{aligned}$$

The integral is finite if $\gamma - \gamma\rho < -1$. Thus, if one makes the assumptions

$$\beta > 1+\gamma \quad \text{and} \quad \rho > 1 + \frac{1}{\gamma},$$

then the function $\phi(t)$ remains bounded if $|\varepsilon|$ is sufficiently small and, therefore, $|u(t)| \le C(t+1)^{-\gamma}$.

For certain classes of PDEs the solution operator technique is very powerful. This is true, in particular, if an explicit solution for the linear part of the equation is available, which yields accurate estimates of the solution operator. We refer to the book by Racke (1992) for applications to nonlinear wave equations and other systems of PDEs on all space. Earlier references with similar ideas are Kawashima (1987), Klainerman and Ponce (1983), Matsumura and Nishida (1979), Shatah (1982), and Strauss (1981).

2.4. Remarks on the size of perturbations

Remark 2.2 An important and interesting problem is to quantify the size of the perturbation that one is allowed to apply to a stable system without losing stability. This has been emphasized in the recent work of Trefethen (1997). In general, the question is difficult, and any answer will depend on the norms that are used. Some insight can be obtained as follows. Assuming that $u_t = Au + F(t)$ is linearly stable, we add a perturbation $\varepsilon f(u) = \varepsilon Bu$, which is also linear. (See Remark 2.3 below for nonlinear perturbations.) Laplace transformation yields

$$s\tilde{u} = A\tilde{u} + \varepsilon B\tilde{u} + \tilde{F}; \tag{2.42}$$

thus

$$\tilde{u} = \varepsilon(sI - A)^{-1}B\tilde{u} + (sI - A)^{-1}\tilde{F}.$$

Recall that $R = \sup_{\mathrm{Re}\, s\ge 0} |(sI-A)^{-1}|$ is the resolvent constant of A. If

$$|\varepsilon B|R \le \frac{1}{2}, \tag{2.43}$$

say, then $|\tilde{u}| \le 2R|\tilde{F}|$, and we obtain essentially the same estimate of u by F as in the unperturbed case. Thus, if (2.43) holds, stability is retained. On the other hand, if $|\varepsilon B|R \ge 1$, then the system (2.42) may be singular, allowing for instability, and we conclude that (2.43) is a realistic condition for retaining stability. This shows the importance of the resolvent constant.

How is the resolvent constant R related to the eigenvalues of A? Trefethen defines the spectral abscissa of A by

$$\alpha(A) = \max_{\lambda \in \sigma(A)} \operatorname{Re} \lambda.$$

Then, if A is normal and $\alpha(A) = -\delta < 0$, we have $R = \frac{1}{\delta}$, as noted in Lemma 2.2(c). In this case, condition (2.43) reads

$$|\varepsilon B| \leq \frac{1}{2}\delta. \tag{2.44}$$

If A is not normal, then $R \gg \frac{1}{\delta}$ is possible, as emphasized by Trefethen (1997). Clearly, if $R \gg \frac{1}{\delta}$, the requirement (2.43) is much more restrictive than (2.44). For this reason it is important to obtain good estimates for the resolvent constant R. The techniques for proving the Kreiss matrix theorem (Kreiss and Lorenz 1989) can be applied, in principle, but the treatment of concrete examples may be formidable. For the alternative approach of computing the pseudospectrum of A numerically, we refer to Trefethen (1997). (If $\sigma_\varepsilon(A) = \{z \in \mathbb{C} : |(zI - A)^{-1}| \geq 1/\varepsilon\}$ denotes the ε-pseudospectrum of A, then $R = 1/\varepsilon_0$, where ε_0 $\{\varepsilon > 0 : \sigma_\varepsilon(A)$ lies in the left half-plane$\}$.)

Remark 2.3 Consider the linear problem $u_t = Au + F(t), u(0) = 0$. If $F(0) = 0$ (which can always be enforced by the transformation $u = te^{-t}F(0) + v$), we obtain from (2.25) and $u_{tt} = Au_t + F_t$ the estimate

$$\int_0^\infty \left(|u|^2 + |u_t|^2\right) \mathrm{d}t \leq R^2 \int_0^\infty \left(|F|^2 + |F_t|^2\right) \mathrm{d}t. \tag{2.45}$$

The left-hand side bounds $\sup_t |u(t)|^2$, and we can use (2.45) instead of (2.26) to show nonlinear stability. Then our arguments, given above, show that the size of R is again crucial for determining the size of $|\varepsilon|$ which retains stability for the nonlinear perturbed problem.

3. Parabolic systems: periodic boundary conditions

Consider a parabolic equation

$$u_t = Pu + \varepsilon f\left(x, t, u, Du, D^2u\right) + F(x,t), \quad x \in \mathbb{R}^d, \quad t \geq 0, \tag{3.1}$$

where P is a constant coefficient operator,

$$Pu = \Delta u + \sum_{j=1}^{d} A_j D_j u + Bu. \tag{3.2}$$

Here

$$D_j = \frac{\partial}{\partial x_j}, \qquad \Delta = D_1^2 + \cdots + D_d^2, \tag{3.3}$$

and $A_j, B \in \mathbb{C}^{n\times n}$ are constant matrices. The function $u(x,t)$ takes values in $\mathbb{C}^n$. For simplicity, f and F are assumed to be C^∞ functions of their arguments; the nonlinearity f may depend on x, t, u and

$$Du = (D_1 u, \ldots, D_d u), \qquad D^2 u = (D_i D_j u)_{1\le i,j\le d} \ .$$

A main assumption is that $F(x,t)$ and $f(x,t,u,Du,D^2u)$ are 2π-periodic in each x_j, and we seek a solution $u(x,t)$ with the same spatial periodicity property. In other words, the space variable x lives in the d-torus $\mathbb{T}^d = (\mathbb{R}/(2\pi\mathbb{Z}))^d$.

We want to discuss asymptotic stability using the Laplace transform (or resolvent) technique and will assume, without loss of generality, a homogeneous initial condition

$$u(x,0) = 0, \quad x \in \mathbb{R}^d. \tag{3.4}$$

Setting

$$v = (u, Du, D^2u),$$

we will assume that $f(x,t,v)$ vanishes at $v = 0$. More precisely, with an integer p specified below, we require the following.

Assumption 3.1 For all $c_1 > 0$ there exists $C_1 > 0$ with

$$\begin{aligned} |D_x^\beta f(x,t,v)| &\le C_1|v| \quad \text{if } |v| \le c_1, \;\; |\beta| \le p, \\ |D_x^\beta D_v^\gamma f(x,t,v)| &\le C_1 \quad \text{if } |v| \le c_1, \;\; |\beta| + |\gamma| \le p, \;\; |\gamma| \ge 1. \end{aligned} \tag{3.5}$$

We will show nonlinear asymptotic stability of (3.1) in the sense of the following theorem.[7]

Theorem 3.1 Let Assumption 3.1 hold with $p = d + 5$, and let

$$\int_0^\infty \|F(\cdot,t)\|_{H^p}^2 \, dt < \infty \tag{3.6}$$

be finite. If $|\varepsilon|$ is small enough, then $\max_x |u(x,t)| \to 0$ as $t \to \infty$.

Remark 3.1 The value $p = d + 5$ is not optimal; that is, it suffices to require Assumption 3.1 and (3.6) with a smaller value of p. For most applications this is uninteresting, since the assumptions hold for all p if they hold for some small p. See also Remark 3.2 on page 228.

[7] We use the standard notation $\|u\|_{H^p}^2 = \sum_{|\alpha|\le p} \|D^\alpha u\|^2$ where $\|u\|^2 = \int_{\mathbb{T}^d} |u(x)|^2 \, dx$, $D^\alpha = D_1^{\alpha_1} \ldots D_d^{\alpha_d}$ and $|\alpha| = \sum_j \alpha_j$.

The linear problem

To begin with, consider the linear problem $u_t = Pu + F$. Applying Fourier expansion in space, we obtain

$$\hat{u}_t(\omega, t) = \hat{P}(\mathrm{i}\omega)\hat{u}(\omega, t) + \hat{F}(\omega, t), \quad \omega \in \mathbb{Z}^d. \tag{3.7}$$

Here

$$\hat{u}(\omega, t) = (2\pi)^{-d/2} \int_{\mathbb{T}^d} \mathrm{e}^{-\mathrm{i}\omega\cdot x} u(x, t)\,\mathrm{d}x, \quad \omega \cdot x = \sum_j \omega_j x_j,$$

and

$$\hat{P}(\mathrm{i}\omega) = -|\omega|^2 I + \mathrm{i} \sum_{j=1}^{d} \omega_j A_j + B \tag{3.8}$$

is the symbol of P. Denoting the Laplace transform of $\hat{u}(\omega, t)$ by $\tilde{u}(\omega, s)$, we obtain from (3.7)

$$s\tilde{u} = \hat{P}(\mathrm{i}\omega)\tilde{u} + \tilde{F}. \tag{3.9}$$

As we will show, good estimates of u in terms of F can be obtained if the matrices $\hat{P}(\mathrm{i}\omega)$ satisfy the eigenvalue condition of the ODE case uniformly in ω. Accordingly, we make the following assumption, which we will discuss at the end of the section.

Assumption 3.2 (Eigenvalue Assumption) There exists $\delta > 0$ such that

$$\operatorname{Re} \lambda \le -\delta < 0 \quad \text{for all} \quad \lambda \in \sigma(\hat{P}(\mathrm{i}\omega)), \quad \omega \in \mathbb{Z}^d.$$

If Assumption 3.2 is satisfied, the matrices $sI - \hat{P}(\mathrm{i}\omega)$ are nonsingular for $\operatorname{Re} s \ge 0$, and one has the following uniform estimate for the resolvents of $\hat{P}(\mathrm{i}\omega)$.

Lemma 3.1 If Assumption 3.2 holds, then there is a constant K_1 with

$$|(sI - \hat{P}(\mathrm{i}\omega))^{-1}| \le \frac{K_1}{|\omega|^2 + 1} \quad \text{for all} \ \operatorname{Re} s \ge 0, \quad \omega \in \mathbb{Z}^d. \tag{3.10}$$

Proof. First consider large $|\omega|$. We write

$$\begin{aligned} sI - \hat{P}(\mathrm{i}\omega) &= \left(s + |\omega|^2\right) I - \hat{Q}(\omega) \qquad (\text{with } |\hat{Q}| \le C(|\omega| + 1)) \\ &= \left(s + |\omega|^2\right)\left(I - \mathcal{O}\left(\frac{1}{|\omega|}\right)\right). \end{aligned} \tag{3.11}$$

Thus, there is $C > 0$ such that (3.10) holds for $|\omega| \ge C$.

There are only finitely many $\omega \in \mathbb{Z}^d$ with $|\omega| \le C$. For these ω-vectors, we apply the resolvent estimate of Lemma 2.2b. □

Using (3.9) and (3.10) we can estimate $\tilde{u}$ in terms of $\tilde{F}$,

$$\left(|\omega|^2 + 1\right)|\tilde{u}(\omega, s)| \le K_1|\tilde{F}(\omega, s)|,$$

and Parseval's relation yields the basic inequality

$$\int_0^\infty \|u(\cdot, t)\|_{H^2}^2 \, dt \le K_2 \int_0^\infty \|F(\cdot, t)\|^2 \, dt. \tag{3.12}$$

If we first apply D^α to the differential equation $u_t = Pu + F$ and sum the resulting inequalities (3.12) over $|\alpha| \le p$, we find

$$\int_0^\infty \|u(\cdot, t)\|_{H^{p+2}}^2 \, dt \le K_2 \int_0^\infty \|F(\cdot, t)\|_{H^p}^2 \, dt, \quad p = 0, 1, 2, \ldots. \tag{3.13}$$

Using the differential equation $u_t = Pu + F$, we can also estimate u_t and its space derivatives. For example,

$$\|u_t\| \le C(\|u\|_{H^2} + \|F\|)$$

since P is of second order. Furthermore, since values of $F(x, t)$ for $t > T$ do not affect the solution $u(x, t)$ for $t \le T$, we can restrict the resulting estimates to any finite time interval. Let us summarize these results.

Theorem 3.2 Let $u_t = Pu + F$ satisfy the Eigenvalue Assumption (Assumption 3.2). There is a constant K, independent of F and T, with

$$\int_0^T \left(\|u\|_{H^{p+2}}^2 + \|u_t\|_{H^p}^2\right) dt \le K \int_0^T \|F\|_{H^p}^2 \, dt, \quad p = 0, 1, 2, \ldots. \tag{3.14}$$

To estimate u (and some of its derivatives) in maximum norm, we will make use of the following Sobolev inequality.

Theorem 3.3 Let $v \in H^p(\mathbb{T}^d)$. If $p > \frac{d}{2}$ then $v \in C(\mathbb{T}^d)$ and

$$|v|_\infty \le C_{p,d}\|v\|_{H^p}. \tag{3.15}$$

The constant $C_{p,d}$ does not depend on v.

By (2.29) we have, for any x,

$$\max_{0 \le t \le T} |u(x, t)|^2 \le \int_0^T \left(|u(x, t)|^2 + |u_t(x, t)|^2\right) dt.$$

Taking the maximum over $x \in \mathbb{T}^d$ and using the above Sobolev inequality, we find that

$$\max_{0 \le t \le T} |u(\cdot, t)|_\infty^2 \le C \int_0^T \left(\|u\|_{H^p}^2 + \|u_t\|_{H^p}^2\right) dt \quad \text{if } p > \frac{d}{2}. \tag{3.16}$$

Again, we can apply the same estimate to derivatives $D^\alpha u$ and obtain

$$\max_{0 \le t \le T} |D^\alpha u(\cdot, t)|_\infty^2 \le C \int_0^T \left(\|u\|_{H^p}^2 + \|u_t\|_{H^p}^2\right) dt \quad \text{if } p > |\alpha| + \frac{d}{2}. \tag{3.17}$$

In terms of $v = (u, Du, D^2u)$, we have the following bound:

$$\max_{0\le t\le T} |D^\alpha v(\cdot,t)|_\infty^2 \le C \int_0^T \left(\|u\|_{H^p}^2 + \|u_t\|_{H^p}^2\right) dt \quad \text{if } p > |\alpha|+2+\frac{d}{2}. \quad (3.18)$$

(Clearly, estimates (3.17) and (3.18) are valid for any sufficiently regular function $u(x,t)$ with $u = 0$ at $t = 0$; the PDE has not been used.)

After these preparations, let us prove Theorem 3.1.

The nonlinear problem

Consider (3.1) for any ε. There exists $T_\varepsilon > 0$ with

$$\int_0^{T_\varepsilon} \left(\|u\|_{H^{p+2}}^2 + \|u_t\|_{H^p}^2\right) dt \le 4K \int_0^\infty \|F\|_{H^p}^2\, dt. \quad (3.19)$$

(The consideration of (3.19) is motivated by the fact that (3.19) is valid for $\varepsilon = 0$ with $4K$ replaced by K; see Theorem 3.2.) We want to prove that (3.19) holds for $T_\varepsilon = \infty$ if $|\varepsilon|$ is sufficiently small. As before, we set

$$v = (u, Du, D^2u)$$

and use the linear estimate of Theorem 3.2 with $F(x,t)$ replaced by $F(x,t)+\varepsilon f(x,t,v(x,t))$, to obtain

$$\int_0^{T_\varepsilon} \left(\|u\|_{H^{p+2}}^2 + \|u_t\|_{H^p}^2\right) dt$$

$$\le\ 2K \int_0^\infty \|F\|_{H^p}^2\, dt + 2K|\varepsilon|^2 \int_0^{T_\varepsilon} \|f(\cdot,t,v(\cdot,t))\|_{H^p}^2\, dt. \quad (3.20)$$

It remains to prove that we can bound the integral $\int_0^{T_\varepsilon} \|f\|_{H^p}^2\, dt$ in terms of the left-hand side of (3.20). Basically, this turns out to be possible because the left-hand side of (3.20) dominates the maximum norm of sufficiently many derivatives of v if p is large. As we will see, our choice $p = d + 5$ suffices.

The main technical difficulty is treated in the following theorem. In its proof, the simple estimate

$$\|\phi\psi\| \le |\phi|_\infty \|\psi\|$$

for the L_2-norm of the product of two functions is used. The definition of κ in the theorem is motivated by (3.18).

Theorem 3.4 (estimate based on chain rule) Let $v : \mathbb{T}^d \to \mathbb{C}^m$ and $f : \mathbb{C}^m \to \mathbb{C}^n$ denote functions of class C^p, where $p = d + 5$. Assume

$$|v(x)| \le B, \quad x \in \mathbb{T}^d,$$

and let C_B denote a bound for the derivatives of $f(v)$ in the ball $|v| \le B$,

$$|D_v^\gamma f(v)| \le C_B \quad \text{if } |v| \le B,\ \ |\gamma| \le p.$$

We set

$$\kappa := \max\{|D^\alpha v|_\infty \ : \ |\alpha| + 2 + \frac{d}{2} < p\}.$$

(The maximum is taken over all multi-indices α with $|\alpha|+2+\frac{d}{2} < p$.) Then the composite function $f(v(x)), x \in \mathbb{T}^d$, satisfies

$$\|D^\alpha(f \circ v)\| \le CC_B \left(1 + \kappa^{p-1}\right) \|v\|_{H^p} \tag{3.21}$$

for $1 \le |\alpha| \le p$. The constant C is independent of v and f.

Proof. By the chain rule

$$D^\alpha(f \circ v)(x) = \sum_\sigma c_\sigma \phi_\sigma(v(x)) D^{\sigma_1} v \dots D^{\sigma_k} v,$$

where $\sigma_1, \dots, \sigma_k$ are multi-indices with

$$\sigma_1 + \cdots + \sigma_k = \alpha,$$

$\phi_\sigma(v)$ is a derivative of $f(v)$ of order $\le p$, and c_σ are numerical coefficients. Therefore,

$$\|D^\alpha(f \circ v)\| \le C_1 C_B \sum_\sigma \|D^{\sigma_1} v \dots D^{\sigma_k} v\|.$$

A factor $D^{\sigma_j} v$ can be bounded in maximum norm by κ if

$$p > |\sigma_j| + 2 + \frac{d}{2}.$$

Suppose there are two factors, $D^{\sigma_1} v$ and $D^{\sigma_2} v$, say, whose maximum norm *cannot* be bounded by κ. Then

$$p \le |\sigma_1| + 2 + \frac{d}{2} \quad \text{and} \quad p \le |\sigma_2| + 2 + \frac{d}{2};$$

thus

$$2p \le |\sigma_1| + |\sigma_2| + 4 + d \le p + d + 4.$$

However, this contradicts our choice $p = d + 5$. Therefore, each product

$$D^{\sigma_1} v \dots D^{\sigma_k} v$$

contains at most one factor which cannot be estimated in maximum norm by κ. We conclude

$$\|D^{\sigma_1} v \dots D^{\sigma_k} v\| \le C \left(1 + \kappa^{p-1}\right) \|v\|_{H^p}, \quad 1 \le k \le p.$$

This proves the theorem. □

Remark 3.2 In our bound of $\|D^{\sigma_1}v \dots D^{\sigma_k}v\|$, we have only used the simple inequality $\|\phi\psi\| \le |\phi|_\infty \|\psi\|$ and Sobolev's inequality. The condition on p can be relaxed if one instead uses Hölder's inequality and a Gagliardo–Nirenberg inequality. See, for example, Hagstrom and Lorenz (1995) or Racke (1992).

Note that the estimate (3.21) does not hold, in general, for $\alpha = 0$, as the example $f \equiv 1$ shows. In our application we assume, however, that f vanishes for $v = 0$. A corresponding result is formulated next.

Theorem 3.5 Let $v : \mathbb{T}^d \to \mathbb{C}^m$ and $f : \mathbb{T}^d \times \mathbb{C}^m \to \mathbb{C}^n$ denote functions of class C^p, where $p = d + 5$. We consider the composite function

$$f(x, v(x)), \quad x \in \mathbb{T}^d.$$

Assuming that

$$|v(x)| \le B, \quad x \in \mathbb{T}^d,$$

we require the following estimates for the derivatives of f in the ball $|v| \le B$:

$$\begin{aligned} |D_x^\beta f(x,v)| &\le C_B|v| \quad \text{if } |v| \le B, \quad |\beta| \le p; \\ |D_x^\beta D_v^\gamma f(x,v)| &\le C_B \quad \text{if } |v| \le B, \quad |\beta| + |\gamma| \le p, \ |\gamma| \ge 1. \end{aligned}$$

Setting

$$\kappa := \max\left\{|D^\alpha v|_\infty \ : \ |\alpha| + 2 + \frac{d}{2} < p\right\},$$

we have

$$\|f(\cdot, v(\cdot))\|_{H^p} \le CC_B\Big(1 + \kappa^{p-1}\Big)\|v\|_{H^p},$$

where C is independent of v and f.

Proof. For $|\alpha| \le p$, the derivative $D^\alpha f(x, v(x))$ is a sum of terms

$$D_x^\beta D_v^\gamma f(x, v(x)) D^{\sigma_1} v \dots D^{\sigma_k} v \tag{3.22}$$

where

$$|\beta| + |\gamma| \le p, \quad |\sigma_1| + \dots + |\sigma_k| \le p.$$

If $|\gamma| \ge 1$, the estimate of (3.22) proceeds as in the proof of the previous theorem. If $\gamma = 0$, the factor $D^{\sigma_1}v \dots D^{\sigma_k}v$ is empty, and we use the estimate $|D_x^\beta f(x,v)| \le C_B|v|$ to obtain

$$\|D_x^\beta f\| \le C_B\|v\|.$$

The claim follows. □

It is now easy to complete the proof of nonlinear asymptotic stability stated in Theorem 3.1. The term $\int \|f\|_{H^p}^2\, dt$ on the right-hand side of (3.20) can be bounded as follows:

$$\int_0^{T_\varepsilon} \|f\|_{H^p}^2\, dt \le C_1 \int_0^{T_\varepsilon} \|v\|_{H^p}^2\, dt \le C_2 \int_0^{T_\varepsilon} \|u\|_{H^{p+2}}^2\, dt.$$

Therefore, if

$$2K|\varepsilon|^2 C_2 \le \frac{1}{3},$$

then one obtains from (3.20)

$$\int_0^{T_\varepsilon} \left(\|u\|_{H^{p+2}}^2 + \|u_t\|_{H^p}^2 \right) dt \le 3K \int_0^\infty \|F\|_{H^p}^2\, dt. \tag{3.23}$$

Since (3.19) implies (3.23), we can conclude that (3.23) is valid for $T_\varepsilon = \infty$. Convergence $\max_x |u(x,t)| \to 0$ as $t \to \infty$ (and even for $v = (u, Du, D^2u)$ instead of u) follows from

$$\int_{t_0}^\infty \left(\|u\|_{H^{p+2}}^2 + \|u_t\|_{H^p}^2 \right) dt \quad \to 0 \quad \text{as} \quad t_0 \to \infty.$$

Discussion of the Eigenvalue Assumption

Since $\hat{P}(\mathrm{i}\omega) = -|\omega|^2 I + \mathcal{O}(|\omega|)$ for large $|\omega|$, it follows that the Eigenvalue Assumption (Assumption 3.2) is always satisfied for large $|\omega|$. If the Eigenvalue Assumption is violated, there exists $\omega \in \mathbb{Z}^d$ and $\phi \in \mathbb{C}^n$ with

$$\hat{P}(\mathrm{i}\omega)\phi = \lambda\phi, \quad \mathrm{Re}\,\lambda \ge 0, \quad \phi \ne 0.$$

If we set

$$u(x,t) = \frac{1}{\lambda+1}(\mathrm{e}^{\lambda t} - \mathrm{e}^{-t})\phi, \quad F(x,t) = \mathrm{e}^{-t}\phi,$$

then $u_t = Pu + F$, but u does not tend to zero as $t \to \infty$. This shows that the Eigenvalue Assumption is necessary for linear asymptotic stability.

A simple sufficient condition for the Eigenvalue Assumption is

$$A_j = A_j^*, \quad j = 1, \dots, d; \quad B + B^* \le -2\delta I < 0.$$

In this case, $\hat{P}(\mathrm{i}\omega) + \hat{P}^*(\mathrm{i}\omega) \le B + B^* \le -2\delta I$, which implies $\mathrm{Re}\,\lambda \le -\delta$ for all eigenvalues λ of $\hat{P}(\mathrm{i}\omega)$.

Obviously, $\hat{P}(\mathrm{i}\omega) = B$ for $\omega = 0$. This shows that the Eigenvalue Assumption can only be satisfied if all eigenvalues of the zero-order term B of P have negative real parts. There are cases of interest, however, where $B = 0$ or B has a nontrivial null-space. Then it may still be true that $\mathrm{Re}\,\lambda \le -\delta < 0$ for all $\lambda \in \sigma(\hat{P}(\mathrm{i}\omega))$ if $\omega \in \mathbb{Z}^d$ and $\omega \ne 0$. Under such a restricted eigenvalue assumption one can prove a restricted form of asymptotic stability by taking a suitable projection. See, for example, Hagstrom and Lorenz (1995) for applications.

4. General PDEs: periodic boundary conditions

In this section we consider PDEs of the form

$$u_t = Pu + \varepsilon f(x,t,u) + F(x,t), \quad x \in \mathbb{R}^d, \quad t \geq 0, \tag{4.1}$$

with initial condition

$$u(x,0) = 0, \quad x \in \mathbb{R}^d. \tag{4.2}$$

Here P is a linear constant coefficient operator, that is,

$$P = \sum_{|\nu| \leq m} A_\nu D^\nu, \quad A_\nu \in \mathbb{C}^{n\times n}, \tag{4.3}$$

$$D^\nu = D_1^{\nu_1} \dots D_d^{\nu_d}, \quad D_j = \frac{\partial}{\partial x_j}, \quad |\nu| = \nu_1 + \dots + \nu_d. \tag{4.4}$$

As in the previous section, we will assume that $F(x,t)$ and $f(x,t,u)$ are C^∞ functions, which are 2π-periodic in each variable x_j. We seek a solution $u(x,t)$ with the same spatial periodicity. We will also discuss (4.1) with more general nonlinearities

$$f(x,t,u,Du,\dots,D^r u),$$

where $D^j u$ denotes the array of all spatial derivatives of u of order j. As before, it will be assumed that the nonlinear term $f(x,t,v)$ vanishes for $v = 0$ with

$$v = (u, Du, \dots, D^r u).$$

Our aim in this section is twofold.

(1) We want to explain the significance of the eigenvalue assumption

$$\operatorname{Re}\lambda \leq -\delta < 0 \quad \text{for all } \lambda \in \sigma(\hat{P}(i\omega)), \quad \omega \in \mathbb{Z}^d, \tag{4.5}$$

for the general class of operators (4.3). In fact, as we will show, if the Cauchy problem for $u_t = Pu$ is well posed in L_2, the eigenvalue assumption (4.5) always implies a resolvent estimate leading to nonlinear asymptotic stability of (4.1).

(2) The number r of derivatives of u, which are allowed in the nonlinearity f, depends on the resolvent estimate in a simple way. On the Fourier–Laplace side, one needs a bound

$$(|\omega|^q + 1)|\tilde{u}(\omega,s)| \leq K_1|\tilde{F}(\omega,s)| \quad \text{for all } \operatorname{Re}s \geq 0, \quad \omega \in \mathbb{Z}^d, \tag{4.6}$$

with $q \geq r$. If the Cauchy problem for $u_t = Pu$ is well posed, then

(4.6) is equivalent to the eigenvalue condition

$$\mathrm{Re}\,\lambda \le -(|\omega|^q + 1)\delta < 0 \quad \text{for all} \;\; \lambda \in \sigma(\hat{P}(\mathrm{i}\omega)), \;\; \omega \in \mathbb{Z}^d. \qquad (4.7)$$

The simple eigenvalue condition (4.5) leads to (4.7) with $q = 0$. If q is even and $u_t = Pu$ is a parabolic system of order q, then – by the definition of parabolicity – condition (4.7) is always satisfied for large $|\omega|$. In other words, for parabolic systems of order q, the conditions (4.5) and (4.7) are equivalent and can be checked, in principle, by computing the eigenvalues of finitely many matrices $\hat{P}(\mathrm{i}\omega)$, $\omega \in \mathbb{Z}^d$.

Well-posedness

Let us briefly review the concept of well-posedness in L_2 and first consider the linear problem $u_t = Pu$ under an initial condition $u(x,0) = u_0(x)$. We make a Fourier expansion of the initial data,

$$u_0(x) = \sum_{\omega \in \mathbb{Z}^d} \mathrm{e}^{\mathrm{i}\omega\cdot x}\hat{u}_0(\omega), \qquad (4.8)$$

and, tentatively, of the solution,

$$u(x,t) = \sum_{\omega \in \mathbb{Z}^d} \mathrm{e}^{\mathrm{i}\omega\cdot x}\hat{u}(\omega,t), \qquad (4.9)$$

for each $t \ge 0$. Introducing the symbol of P,

$$\hat{P}(\kappa) = \sum_{|\nu|\le m} \kappa_1^{\nu_1}\dots\kappa_d^{\nu_d} A_\nu, \quad \kappa \in \mathbb{C}^d, \qquad (4.10)$$

and observing that

$$P(\mathrm{e}^{\mathrm{i}\omega\cdot x}\phi) = \hat{P}(\mathrm{i}\omega)\phi, \quad \phi \in \mathbb{C}^n, \qquad (4.11)$$

we obtain formally

$$\hat{u}_t(\omega,t) = \hat{P}(\mathrm{i}\omega)\hat{u}(\omega,t), \quad \hat{u}(\omega,0) = \hat{u}_0(\omega); \qquad (4.12)$$

thus

$$\hat{u}(\omega,t) = \mathrm{e}^{\hat{P}(\mathrm{i}\omega)t}\hat{u}_0(\omega). \qquad (4.13)$$

The formal process is justified for $t \ge 0$ if the matrix exponentials in (4.13) have a limited exponential growth rate, which is uniform for all $\omega \in \mathbb{Z}^d$. Well-posedness can be defined accordingly; for details, see Kreiss and Lorenz (1989), for example.

Definition 4.1 The 2π-periodic Cauchy problem for $u_t = Pu$ is well posed (in L_2) if there are real constants K and α with

$$|\mathrm{e}^{\hat{P}(\mathrm{i}\omega)t}| \le K\mathrm{e}^{\alpha t} \quad \text{for all} \;\; \omega \in \mathbb{Z}^d, \; t \ge 0. \qquad (4.14)$$

The L_2-inner product and norm are defined by

$$(u,v) = \int_{\mathbb{T}^d} u^*(x)v(x), \quad \|u\|^2 = (u,u), \quad u,v \in L_2(\mathbb{T}^d, \mathbb{C}^n).$$

Then, if the 2π-periodic Cauchy problem is well posed and $u_0 \in C^\infty$, the formula (4.9) gives us the solution $u(x,t)$, which is C^∞ and satisfies

$$\|u(\cdot,t)\| \le K e^{\alpha t}\|u_0\|, \quad t \ge 0.$$

As usual, boundedness of the assignment $u_0 \to u(\cdot,t)$ in L_2 implies that we can obtain a generalized solution for all initial data $u_0(x)$ in L_2.

Basic resolvent estimate

The following result says that the eigenvalue assumption (4.5) implies a resolvent estimate, whenever the Cauchy problem is well posed.

Theorem 4.1 Assume that $P = \sum_{|\nu|\le m} A_\nu D^\nu$ satisfies the following two conditions.

(1) The 2π-periodic Cauchy problem for $u_t = Pu$ is well posed, that is, there are constants α and K with

$$|e^{\hat{P}(i\omega)t}| \le K e^{\alpha t} \quad \text{for all } \omega \in \mathbb{Z}^d, \quad t \ge 0.$$

(2) There is $\delta > 0$ with

$$\operatorname{Re}\lambda \le -\delta < 0 \quad \text{for all } \lambda \in \sigma(\hat{P}(i\omega)), \quad \omega \in \mathbb{Z}^d.$$

Then there is a constant K with

$$|(sI - \hat{P}(i\omega))^{-1}| \le K \quad \text{for all } \operatorname{Re}s \ge 0, \quad \omega \in \mathbb{Z}^d. \tag{4.15}$$

A main tool for the proof is the Kreiss matrix theorem, which we formulate next. (See Kreiss and Lorenz (1989).)

Theorem 4.2 Let $\mathcal{F}$ denote any set of matrices $A \in \mathbb{C}^{n\times n}$, where n is fixed. Then the following conditions are equivalent.

(1) There is a constant K_1 with

$$|e^{At}| \le K_1 \quad \text{for all } A \in \mathcal{F}, \ t \ge 0. \tag{4.16}$$

(2) For all $A \in \mathcal{F}$ and all $s \in \mathbb{C}$ with $\operatorname{Re}s > 0$ the matrix $sI - A$ is nonsingular, and there is a constant K_2 with

$$|(sI - A)^{-1}| \le \frac{K_2}{\operatorname{Re}s} \quad \text{for all } A \in \mathcal{F}, \ \operatorname{Re}s > 0. \tag{4.17}$$

(3) There are constants K_{31}, K_{32} and, for all $A \in \mathcal{F}$, there is a transformation $S = S(A) \in \mathbb{C}^{n\times n}$ with $|S| + |S^{-1}| \le K_{31}$ so that

$$SAS^{-1} = \begin{pmatrix} \lambda_1 & b_{12} & \cdots & \cdots & b_{1n} \\ 0 & \lambda_2 & b_{23} & \cdots & b_{2n} \\ \vdots & & \ddots & & \vdots \\ \vdots & & & \lambda_{n-1} & b_{n-1,n} \\ 0 & \cdots & \cdots & 0 & \lambda_n \end{pmatrix} \tag{4.18}$$

is upper-triangular, the diagonal is ordered,

$$0 \ge \operatorname{Re}\lambda_1 \ge \cdots \ge \operatorname{Re}\lambda_n,$$

and the upper-diagonal elements satisfy

$$|b_{jk}| \le K_{32}|\operatorname{Re}\lambda_j|, \quad 1 \le j < k \le n.$$

(4) There is a positive constant K_4 and, for each $A \in \mathcal{F}$, there is a Hermitian matrix $H = H(A) \in \mathbb{C}^{n\times n}$ with

$$\frac{1}{K_4} I \le H \le K_4 I\ , \quad HA + A^*H \le 0. \tag{4.19}$$

Proof of Theorem 4.1. There are constants $\alpha \ge 0, K > 0$ with

$$|e^{(\hat{P}(i\omega)-\alpha I)t}| \le K \quad \text{for all } \omega \in \mathbb{Z}^d, \ \ t \ge 0.$$

By the Kreiss matrix theorem – applied to the family $\hat{P}(i\omega) - \alpha I$, $\omega \in \mathbb{Z}^d$, – there is a bounded transformation $S = S(\omega)$ with

$$S(\hat{P}(i\omega) - \alpha I)S^{-1} = \begin{pmatrix} \lambda_1 - \alpha & b_{12} & \cdots & \cdots & b_{1n} \\ 0 & \lambda_2 - \alpha & b_{23} & \cdots & b_{2n} \\ \vdots & & \ddots & & \vdots \\ \vdots & & & \lambda_{n-1} - \alpha & b_{n-1,n} \\ 0 & \cdots & \cdots & 0 & \lambda_n - \alpha \end{pmatrix}$$

and

$$|b_{jk}| \le K_{32}\ |\operatorname{Re}\lambda_j - \alpha|\ , \quad j < k.$$

Using the assumption $\operatorname{Re}\lambda_j \le -\delta < 0 \le \alpha$, we obtain that

$$\begin{aligned} |\operatorname{Re}\lambda_j - \alpha| &= |\operatorname{Re}\lambda_j| \left(1 + \frac{\alpha}{|\operatorname{Re}\lambda_j|}\right) \\ &\le 2\left(1 + \frac{\alpha}{\delta}\right) \left|\operatorname{Re}\frac{\lambda_j}{2}\right| \\ &\le 2\left(1 + \frac{\alpha}{\delta}\right) \left|\operatorname{Re}\lambda_j + \frac{\delta}{2}\right|. \end{aligned} \tag{4.20}$$

Therefore,

$$|b_{jk}| \le K'_{32} \left|\operatorname{Re}\lambda_j + \frac{\delta}{2}\right|, \quad j < k. \tag{4.21}$$

Considering $S(\hat{P}(\mathrm{i}\omega)+\frac{\delta}{2}I)S^{-1}$, we obtain from the estimates (4.21) that the Kreiss matrix theorem also applies to the family $\hat{P}(\mathrm{i}\omega)+\frac{\delta}{2}I$, $\omega \in \mathbb{Z}^d$. Now we use the second characterization in Theorem 4.2 and find

$$\left|\left(\left(s-\frac{\delta}{2}\right)I - \hat{P}(\mathrm{i}\omega)\right)^{-1}\right| \le \frac{K_2}{\operatorname{Re} s}, \quad \operatorname{Re} s > 0.$$

In particular, this implies the estimate

$$\left|\left(sI - \hat{P}(\mathrm{i}\omega)\right)^{-1}\right| \le \frac{K_2}{\delta/2}, \quad \operatorname{Re} s \ge 0,$$

and the theorem is proved. □

Asymptotic stability of (4.1)
Consider the linear problem

$$u_t = Pu + F(x,t), \quad u(x,0) = 0,$$

and let the assumptions of Theorem 4.1 hold. Fourier expansion and Laplace transformation lead to the family of linear algebraic equations

$$(sI - \hat{P}(\mathrm{i}\omega))\tilde{u}(\omega,s) = \tilde{F}(\omega,s), \quad \operatorname{Re} s \ge 0, \quad \omega \in \mathbb{Z}^d, \tag{4.22}$$

and (4.15) yields

$$|\tilde{u}(\omega,s)| \le K|\tilde{F}(\omega,s)|.$$

By Parseval's relation this translates into the estimate

$$\int_0^\infty \|u(\cdot,t)\|^2 \,\mathrm{d}t \le K_1 \int_0^\infty \|F(\cdot,t)\|^2 \,\mathrm{d}t.$$

Applying this basic inequality to $D^\alpha u$ and using $u_t = Pu + F$ to estimate time derivatives, one obtains

$$\int_0^\infty \left(\|u\|^2_{H^p} + \|u_t\|^2_{H^{p-m}}\right) \mathrm{d}t \le K_2 \int_0^\infty \|F\|^2_{H^p} \,\mathrm{d}t, \quad p = m, m+1, \ldots. \tag{4.23}$$

(Recall that m is the order of P.) If $p - m > \frac{d}{2}$ and the right-hand side of (4.23) is finite, we obtain a bound for $\sup_t |u(\cdot,t)|^2$. With the same arguments as in Section 3, this shows linear asymptotic stability.

Theorem 4.3 Let p denote the smallest integer with $p - m > \frac{d}{2}$, and assume

$$\int_0^\infty \|F(\cdot,t)\|^2_{H^p} \,\mathrm{d}t < \infty.$$

If P satisfies the conditions of Theorem 4.1, then $|u(\cdot,t)|_\infty \to 0$ as $t \to \infty$.

The extension to the nonlinear problem (4.1), where $f = f(x,t,u)$ does not depend on the derivatives of u, proceeds as before. Formally, we obtain from (4.23)

$$L^2 := \int_0^T \left(\|u\|_{H^p}^2 + \|u_t\|_{H^{p-m}}^2 \right) dt$$
$$\leq \; 2K_2 \int_0^\infty \|F\|_{H^p}^2 \, dt \; + \; 2K_2|\varepsilon|^2 \int_0^T \|f\|_{H^p}^2 \, dt. \qquad (4.24)$$

Here $f = f(x,t,u(x,t))$. We have (compare (3.17))

$$\max_{0\le t\le T} |D^\alpha u(\cdot,t)|_\infty^2 \le CL^2 \quad \text{if} \;\; p > |\alpha| + m + \frac{d}{2}. \qquad (4.25)$$

We estimate $\|f\|_{H^p}$ by applying the chain rule and have to consider

$$\|D^{\sigma_1}u \ldots D^{\sigma_k}u\|, \quad |\sigma_1| + \cdots + |\sigma_k| \le p.$$

(See the proofs of Theorems 3.4 and 3.5.) If there are two factors, $D^{\sigma_1}u$ and $D^{\sigma_2}u$, say, which are not dominated in maximum norm by L, then

$$p \le |\sigma_1| + m + \frac{d}{2} \quad \text{and} \quad p \le |\sigma_2| + m + \frac{d}{2};$$

thus

$$2p \le |\sigma_1| + |\sigma_2| + 2m + d \le p + 2m + d.$$

Therefore, if we choose $p = 2m + d + 1$, there can be at most one factor $D^{\sigma_j}u$ that cannot be dominated in maximum norm by L, and one obtains

$$\|f(\cdot,t,u(\cdot,t))\|_{H^p} \le C\|u(\cdot,t)\|_{H^p}.$$

By the same arguments as in Section 3 we have proved nonlinear asymptotic stability.

Theorem 4.4 Let $p = 2m + d + 1$. If $\int_0^\infty \|F\|_{H^p}^2 \, dt$ is finite and $f(x,t,u)$ satisfies Assumption 3.1 (with v replaced by u), then $|u(\cdot,t)|_\infty \to 0$ as $t \to \infty$ for sufficiently small $|\varepsilon|$.

Resolvent estimate gaining derivatives
It is not difficult to generalize Theorem 4.1 as follows.

Theorem 4.5 Assume that the 2π-periodic Cauchy problem for $u_t = Pu$ is well posed and that the symbols $\hat{P}(i\omega)$ satisfy the following eigenvalue condition:

$$\operatorname{Re}\lambda \le -(|\omega|^q + 1)\delta < 0 \quad \text{for all} \;\; \lambda \in \sigma(\hat{P}(i\omega)), \;\; \omega \in \mathbb{Z}^d,$$

where q is a nonnegative integer. Then there is a constant K with

$$|(sI - \hat{P}(i\omega))^{-1}| \leq \frac{K}{|\omega|^q + 1} \quad \text{for all} \quad \operatorname{Re} s \geq 0, \quad \omega \in \mathbb{Z}^d. \tag{4.26}$$

Proof. The proof, based on the Kreiss matrix theorem, proceeds in exactly the same way as the proof of Theorem 4.1. Just note that, instead of (4.20), we have here

$$|\operatorname{Re} \lambda_j - \alpha| \leq 2\left(1 + \frac{\alpha}{\delta}\right) \left|\operatorname{Re} \lambda_j + \frac{\delta}{2}(|\omega|^q + 1)\right|.$$

□

For the linear problem $u_t = Pu + F$, (4.26) translates into the estimates

$$\int_0^\infty \|u\|^2_{H^q} \, dt \leq K_1 \int_0^\infty \|F\|^2 \, dt \tag{4.27}$$

and

$$\int_0^\infty \left(\|u\|^2_{H^{p+q}} + \|u_t\|^2_{H^{p+q-m}}\right) dt \leq K_2 \int_0^\infty \|F\|^2_{H^p} \, dt, \quad p \geq m - q. \tag{4.28}$$

The left-hand side of (4.28) bounds $\sup_t |u(\cdot, t)|^2_\infty$ if $p + q - m > \frac{d}{2}$, and one obtains linear asymptotic stability.

Theorem 4.6 Let P satisfy the assumptions of Theorem 4.5 and assume $\int_0^\infty \|F\|^2_p \, dt < \infty$, where p is the smallest integer with $p + q - m > \frac{d}{2}$. Then $\lim_{t \to \infty} \max_x |u(x, t)| = 0$ if $|\varepsilon|$ is sufficiently small.

Nonlinear asymptotic stability when f depends on Du, etc.
Let us assume again that P satisfies the conditions of Theorem 4.5. Thus, in the linear estimate we gain q derivatives; see (4.27) and (4.28). Recall that m is the order of P and $m \geq q$. (This follows from (4.26) for $|\omega| \to \infty$.) Let the nonlinearity f depend on (x, t, v) where

$$v = (u, Du, \ldots, D^r u).$$

We want to explain why one obtains nonlinear stability if

$$r \leq q,$$

but cannot allow $r > q$, in general. Here we assume, as before, that $f(x, t, v)$ vanishes for $v = 0$. More precisely, we require Assumption 3.1 for sufficiently large p.

In order to control $f(x, t, v)$, we choose p so large that the left-hand side of (4.28) dominates $|v|^2_\infty$. Consequently, since v contains $D^r u$, we let p be so large that

$$p + q - m > r + \frac{d}{2}. \tag{4.29}$$

(Further restrictions on p will appear below.) Then we have (see (4.28))

$$\int_0^T \left(\|u\|_{H^{p+q}}^2 + \|u_t\|_{H^{p+q-m}}^2 \right) dt$$
$$\leq 2K_2 \int_0^\infty \|F\|_{H^p}^2 \, dt + 2K_2 |\varepsilon|^2 \int_0^T \|f\|_{H^p}^2 \, dt, \tag{4.30}$$

with

$$f = f(x, t, v(x,t)).$$

Denote the left-hand side of (4.30) by L^2; thus

$$\max_{0 \leq t \leq T} |D^\alpha v(\cdot, t)|_\infty^2 \leq CL^2 \quad \text{if } p + q - m > |\alpha| + r + \frac{d}{2}.$$

We estimate $\|f\|_{H^p}$ by applying the chain rule (see the proofs of Theorems 3.4 and 3.5), which leads to the consideration of

$$\|D^{\sigma_1} v \ldots D^{\sigma_k} v\| \quad \text{with } |\sigma_1| + \cdots + |\sigma_k| \leq p.$$

If there are two factors, $D^{\sigma_1} v$ and $D^{\sigma_2} v$, say, which cannot be estimated in maximum norm by CL, then

$$p + q - m \leq |\sigma_1| + r + \frac{d}{2} \quad \text{and} \quad p + q - m \leq |\sigma_2| + r + \frac{d}{2};$$

thus

$$2p + 2q - 2m \leq p + 2r + d.$$

Therefore, we choose p so large that (4.29) holds and

$$p > 2(r - q) + d + 2m.$$

Under these conditions on p we have

$$\|f(\cdot, t, v(\cdot, t))\|_{H^p} \leq C \|v\|_{H^p}, \quad C = C(L). \tag{4.31}$$

Thus far, no restriction on the relation between r (the number of derivatives of u in f) and q (the number of derivatives gained in the resolvent estimate) has occured. Clearly, $\|v\|_{H^p} \approx \|u\|_{H^{p+r}}$, and therefore,

$$\int_0^T \|f\|_{H^p}^2 \, dt \leq C \int_0^T \|u\|_{H^{p+r}}^2 \, dt \tag{4.32}$$

by (4.31). If $r \leq q$ and $|\varepsilon|$ is small enough, we obtain the desired bound

$$\int_0^T \left(\|u\|_{H^{p+q}}^2 + \|u_t\|_{H^{p+q-m}}^2 \right) dt \leq 3K_2 \int_0^\infty \|F\|_{H^p}^2 \, dt \tag{4.33}$$

from (4.30), and nonlinear stability follows. On the other hand, if $r > q$ and we substitute (4.32) on the right-hand side of (4.30), the new right-hand side contains higher derivatives of u than the left; then we cannot obtain a bound for u.

Let us summarize our result of nonlinear asymptotic stability.

Theorem 4.7 Consider

$$u_t = Pu + \varepsilon f(x,t,v) + F(x,t), \quad u(x,0) = 0,$$

with

$$v = (u, Du, \ldots, D^r u).$$

Let P satisfy the conditions of Theorem 4.5 with $q \geq r$. Furthermore, let Assumption 3.1 hold for f and let $\int_0^\infty \|F\|_{H^p}^2 \, dt < \infty$, where $p = 2m+d+1$. Under these assumptions, $\lim_{t\to\infty} |v(\cdot,t)|_\infty = 0$ if $|\varepsilon|$ is sufficiently small.

Discussion

As noted above, the eigenvalue condition (4.26) is reasonable if $u_t = Pu$ is a parabolic system of order q. In this case Theorem 4.7 states that the nonlinearity may depend on all space derivatives of u of order $\leq q$.

Now consider the hyperbolic equation

$$u_t = u_x - u + F(x,t).$$

We have $\hat{P}(i\omega) = i\omega - 1$, and the simple eigenvalue condition (4.5) is satisfied with $\delta = -1$. However, the resolvent estimate (4.26) is only fulfilled with $q = 0$, as the choice $s = i\omega$ in (4.26) shows. Therefore, by Theorem 4.7, we may add a nonlinear term $\varepsilon f(x,t,u)$, but dependency of f on u_x is not allowed.

In the next section we will treat hyperbolic problems in more generality using the Lyapunov technique. It will become clear that certain nonlinearities $\varepsilon f(x,t,u,u_x)$ still lead to asymptotic stability, though the resolvent technique fails.

5. Hyperbolic problems: periodic boundary conditions

Consider a first-order system

$$u_t = Pu + \varepsilon \sum_{j=1}^d B_j(u) D_j u + F(x,t), \quad x \in \mathbb{R}^d, \quad t \geq 0, \tag{5.1}$$

with initial condition

$$u(x,0) = u_0(x), \quad x \in \mathbb{R}^d, \tag{5.2}$$

where P is a constant coefficient operator,

$$Pu = \sum_{j=1}^d A_j D_j u + Bu.$$

Our main assumption is symmetry, that is,

$$A_j = A_j^*, \quad B_j(u) = B_j^*(u), \quad j = 1, \ldots, d, \tag{5.3}$$

and negativity of the zero-order term,

$$B + B^* \le -2\delta I < 0. \tag{5.4}$$

Because of (5.3), system (5.1) is called *symmetric hyperbolic*. The functions $B_j(u), F(x,t)$, and $u_0(x)$ are assumed to be of class C^∞, for simplicity, and $F(x,t), u_0(x)$, and $u(x,t)$ are 2π-periodic in each x_j. In addition, it will be convenient here to assume that all quantities are real.

As remarked at the end of the previous section, (5.1) cannot be treated by resolvent estimates, but, as we will see, by the Lyapunov technique.

The basic energy estimate

First consider (5.1) for $\varepsilon = 0$. As in Section 2.1, we consider the 'change in energy' of the solution:

$$\begin{aligned}\frac{\mathrm{d}}{\mathrm{d}t}\|u(\cdot,t)\|^2 &= \frac{\mathrm{d}}{\mathrm{d}t}(u,u)\\ &= 2(u,u_t)\\ &= 2\left(u, \sum_j A_j D_j u\right) + 2(u,Bu) + 2(u,F).\end{aligned} \tag{5.5}$$

Using the symmetry of A_j, integration by parts, and the periodic boundary conditions, one obtains

$$\begin{aligned}(u, A_j D_j u) &= (A_j u, D_j u)\\ &= -(A_j D_j u, u);\end{aligned}$$

thus

$$(u, A_j D_j u) = 0.$$

Furthermore,

$$(u, Bu) = (B^* u, u) = (u, B^* u);$$

thus

$$(u, Bu) = \frac{1}{2}(u, (B + B^*)u) \le -\delta\|u\|^2.$$

Equation (5.5) yields

$$\begin{aligned}\frac{\mathrm{d}}{\mathrm{d}t}\|u\|^2 &\le -2\delta\|u\|^2 + 2\|u\|\,\|F\|\\ &\le -\delta\|u\|^2 + \frac{1}{\delta}\|F\|^2,\end{aligned}$$

and we obtain the basic energy estimate

$$\begin{aligned} \|u(\cdot,t)\|^2 \;&\leq\; \mathrm{e}^{-\delta t}\|u_0\|^2 \;+\; \frac{1}{\delta}\int_0^t \mathrm{e}^{-\delta(t-\tau)}\|F(\cdot,\tau)\|^2\,\mathrm{d}\tau \\ &\leq\; \mathrm{e}^{-\delta t}\|u_0\|^2 \;+\; \frac{1}{\delta^2}\max_{0\leq\tau\leq t}\|F(\cdot,\tau)\|^2. \end{aligned} \tag{5.6}$$

This estimate is completely analogous to (2.7).

Clearly, from $u_t = Pu + F$ we find $D^\alpha u = PD^\alpha u + D^\alpha F$, and summing the resulting estimates over all α with $|\alpha| \leq p$, we obtain

$$\|u(\cdot,t)\|^2_{H^p} \leq \mathrm{e}^{-\delta t}\|u_0\|^2_{H^p} \;+\; \frac{1}{\delta^2}\max_{0\leq\tau\leq t}\|F(\cdot,\tau)\|^2_{H^p}, \quad p = 0,1,\ldots. \tag{5.7}$$

By the Sobolev inequality stated in Theorem 3.3, we can bound $|u(\cdot,t)|_\infty$ by $\|u(\cdot,t)\|_{H^p}$ if $p > \frac{d}{2}$. Therefore, arguing exactly as in the ODE case in Section 2.1, we have proved the following result of linear asymptotic stability.

Theorem 5.1 Let $P = \sum_j A_j D_j + B$ satisfy the assumptions $A_j = A_j^*$ and $B + B^* \leq -2\delta I < 0$. Furthermore, assume $\lim_{t\to\infty}\|F(\cdot,t)\|_{H^p} = 0$ where p is the smallest integer with $p > \frac{d}{2}$. Then we have $\lim_{t\to\infty}|u(\cdot,t)|_\infty = 0$.

For the linear problem, we could also have used the resolvent technique if $\int_0^\infty \|F\|^2_{H^p}\,\mathrm{d}t$ were finite.

The nonlinear problem

For the solution of (5.1) we consider again the 'change in energy',

$$\begin{aligned} \frac{\mathrm{d}}{\mathrm{d}t}\|u\|^2 \;&=\; 2(u,Pu) + 2(u,F) + 2\varepsilon\sum_j (u, B_j(u)D_j u) \\ &\leq\; -\delta\|u\|^2 \;+\; \frac{1}{\delta}\|F\|^2 \;+\; 2|\varepsilon|\sum_j |(u, B_j(u)D_j u)|. \end{aligned} \tag{5.8}$$

Using the symmetry of $B_j(u)$, integration by parts, and the periodic boundary conditions, we obtain

$$\begin{aligned} (u, B_j(u)D_j u) \;&=\; (B_j(u)u, D_j u) \\ &=\; -(B_j(u)D_j u, u) - (B_j'(u)(D_j u)u, u); \end{aligned}$$

thus

$$(u, B_j(u)D_j u) = -\frac{1}{2}(B_j'(u)(D_j u)u, u). \tag{5.9}$$

Here

$$|B_j'(u(\cdot,t))|_\infty \leq C_1(1 + |u|_\infty)$$

and one finds

$$|(u, B_j(u)D_j u)| \leq C_1(1 + |u|_\infty)|D_j u|_\infty\|u\|^2. \tag{5.10}$$

We can substitute this estimate into (5.8), but the resulting inequality does not lead to a bound for $\|u\|$, because $|u|_\infty$ and $|D_j u|_\infty$ cannot be bounded in terms of $\|u\|$.

To obtain an inequality that closes, we consider $\frac{\mathrm{d}}{\mathrm{d}t}\|u\|^2_{H^p}$ for sufficiently large p. As we will see below, the choice $p = d+2$, which we now make, is sufficient. For $|\alpha| \le p$, we apply D^α to the differential equation (5.1) and obtain

$$D^\alpha u_t = PD^\alpha u + D^\alpha F + \varepsilon \sum_j D^\alpha(B_j D_j u).$$

Therefore,

$$\frac{\mathrm{d}}{\mathrm{d}t}\|D^\alpha u\|^2 \le -\delta\|D^\alpha u\|^2 + \frac{1}{\delta}\|D^\alpha F\|^2 + 2|\varepsilon| \sum_j |(D^\alpha u, D^\alpha(B_j D_j u))|. \quad (5.11)$$

By Leibnitz's rule,

$$D^\alpha(B_j D_j u) = \sum_{\beta+\gamma=\alpha} c_{\alpha\beta}(D^\beta B_j)(D^\gamma D_j u)$$

with numerical coefficients $c_{\alpha\beta}$. The most 'dangerous' term occurs for $\beta = 0, \gamma = \alpha$. On the right-hand side of (5.11), this term contributes

$$(D^\alpha u, B_j D^\alpha D_j u),$$

and, if $|\alpha| = p$, then $p+1$ derivatives are applied to u. However, using the symmetry of B_j and integration by parts, we can remove one derivative and find, as in (5.10),

$$|(D^\alpha u, B_j D^\alpha D_j u)| \le C_1(1 + |u|_\infty)|D_j u|_\infty \|D^\alpha u\|^2. \quad (5.12)$$

Now let

$$|\beta| \ge 1, \ \beta + \gamma = \alpha, \quad \text{thus} \quad |\gamma| \le p-1,$$

and consider

$$\Big|(D^\alpha u, (D^\beta B_j)(D^\gamma D_j u))\Big| \le \|D^\alpha u\| \|(D^\beta B_j)(D^\gamma D_j u)\|. \quad (5.13)$$

Just as in the proof of Theorem 3.4, we apply the chain rule to write $D^\beta(B_j(u(x,t)))$ as a sum. Then one finds that $(D^\beta B_j)(D^\gamma D_j u)$ is a sum of terms

$$(D_u^\nu B_j) D^{\sigma_1} u \ldots D^{\sigma_k} u, \quad 2 \le k \le p+1,$$

where

$$|\sigma_1| + \cdots + |\sigma_k| \le p+1 \quad \text{and} \quad |\sigma_j| \le p, \ \ j = 1, \ldots, k.$$

Here $D_u^\nu B_j$ is a derivative of $B_j(u)$, and thus

$$\Big|(D_u^\nu B_j)(u(\cdot,t))\Big|_\infty \le C(1 + |u|_\infty). \quad (5.14)$$

It remains to bound

$$\|D^{\sigma_1}u \dots D^{\sigma_k}u\|.$$

By Sobolev's inequality,

$$|D^{\sigma_j}u|_\infty \le C\|u\|_{H^p} \quad \text{if} \quad p > |\sigma| + \frac{d}{2}.$$

Suppose there exist two factors, $D^{\sigma_1}u$ and $D^{\sigma_2}u$, say, which *cannot* be bounded in maximum norm by $\|u\|_{H^p}$. Then we would have

$$p \le |\sigma_1| + \frac{d}{2} \quad \text{and} \quad p \le |\sigma_2| + \frac{d}{2},$$

and thus

$$2p \le |\sigma_1| + |\sigma_2| + d \le p + 1 + d,$$

in contradiction to our choice $p = d+2$. We conclude that each factor $D^{\sigma_j}u$, with at most one exception, can be bounded in maximum norm by $\|u\|_{H^p}$. This implies

$$\|D^{\sigma_1}u \dots D^{\sigma_k}u\| \le C\left(1 + \|u\|_{H^p}^{p-1}\right)\|u\|_{H^p}^2, \tag{5.15}$$

since $2 \le k \le p+1$. To summarize, the inequalities (5.12), (5.13), (5.14), and (5.15) yield

$$|(D^\alpha u, D^\alpha(B_jD_ju))| \le C\left(1 + \|u\|_{H^p}^{p+1}\right)\|u\|_{H^p}^2.$$

We substitute this bound into the right-hand side of (5.11) and sum over all α with $|\alpha| \le p$ to find

$$\frac{\mathrm{d}}{\mathrm{d}t}\|u\|_{H^p}^2 \le -\delta\|u\|_{H^p}^2 + \frac{1}{\delta}\|F\|_{H^p}^2 + 2|\varepsilon|C\left(1 + \|u\|_{H^p}^{p+1}\right)\|u\|_{H^p}^2. \tag{5.16}$$

The constant C is independent of ε and t; it depends on the size of B_j and its derivatives.

Using the differential inequality (5.16) and elementary ODE arguments, one obtains the following result of nonlinear asymptotic stability.

Theorem 5.2 Consider (5.1) under the assumptions (5.3) and (5.4). If $\lim_{t\to\infty}\|F(\cdot,t)\|_{H^p} = 0$ for $p = d+2$, then $\lim_{t\to\infty}\|u(\cdot,t)\|_{H^p} = 0$ for sufficiently small $|\varepsilon|$. In particular, $|u(\cdot,t)|_\infty \to 0$ as $t \to \infty$.

Generalizations

It is straightforward to generalize Theorem 5.2 to the case

$$B_j = B_j(x,t,u)$$

as long as one has symmetry $A_j = A_j^*, B_j = B_j^*, j = 1, \dots, d$. Also, without difficulty, a zero-order term $\varepsilon f(x,t,u)$ with $f(x,t,0)=0$ can be included in (5.1); more precisely, Assumption 3.1 is required.

The arguments become much more involved if the symmetry assumption is dropped. Let us explain the difficulty. We start with the linear case, and assume that

$$u_t = Pu \quad \text{with } Pu = \sum_j A_j D_j u + Bu,$$

is strongly hyperbolic; that is, for all $\omega \in \mathbb{R}^d$ the eigenvalues of $\sum_j \omega_j A_j$ are real and semi-simple, and there is a transformation $S = S(\omega)$ with

$$|S(\omega)| + |S^{-1}(\omega)| \le \text{const}$$

so that

$$S\left(\sum_j \omega_j A_j\right) S^{-1}$$

is diagonal. If one assumes, in addition, the eigenvalue condition

$$\operatorname{Re}\lambda \le -\delta < 0 \quad \text{for all } \lambda \in \sigma(\hat{P}(\mathrm{i}\omega)), \quad \omega \in \mathbb{Z}^d, \tag{5.17}$$

then one can use the characterization (4) in the Kreiss matrix theorem (Theorem 4.2) and construct matrices $H = H(\omega)$ with the following properties:

$$0 < \frac{1}{C} I \le H(\omega) = H^*(\omega) \le CI;$$
$$H(\omega)\hat{P}(\omega) + \hat{P}^*(\omega)H(\omega) \le -\delta H(\omega).$$

Using the matrices $H(\omega)$, which form a so-called symmetrizer, one defines a new inner product on $L_2 = L_2(\mathbb{T}^d, \mathbb{C}^n)$ by

$$(u,v)_{\mathcal{H}} = \sum_{\omega\in\mathbb{Z}^d} \hat{u}^*(\omega)H(\omega)\hat{v}(\omega).$$

The $\mathcal{H}$-inner product is equivalent to the L_2-inner product, and the operator P becomes negative in the sense that

$$2(u,Pu)_{\mathcal{H}} \le -\delta\|u\|^2_{\mathcal{H}}.$$

(This is the main point of the construction.) For solutions of the linear equation $u_t = Pu + F$, one then obtains without difficulty

$$\frac{\mathrm{d}}{\mathrm{d}t}\|u\|^2_{\mathcal{H}} \le -\delta\|u\|^2_{\mathcal{H}} + 2\|u\|_{\mathcal{H}}\|F\|_{\mathcal{H}}$$

and can derive a satisfactory energy estimate.

However, to treat a nonlinear equation

$$u_t = Pu + \varepsilon\sum_j B_j(u)D_j u + F(x,t),$$

the construction is not fine enough, even if $B_j(u) = B_j^*(u)$. The difficulty is that the rule

$$(D^\alpha u, B_j D^\alpha D_j u) = (B_j D^\alpha u, D^\alpha D_j u) \tag{5.18}$$

is not valid if the L_2-inner product is replaced by the $\mathcal{H}$-inner product, and a rule like (5.18) – together with integration by parts – is needed to remove a derivative from the 'dangerous' term $D^\alpha D_j u$. For this reason it is necessary, in general, to refine the construction of $H(\omega)$ by terms of order ε and to construct a symmetrizer adjusted to the solution of the nonlinear problem. Details of the construction, which uses elementary properties of pseudodifferential operators, are carried out in Kreiss, Kreiss and Lorenz (1998*b*) and Kreiss, Ortiz and Reula (1998*c*). It is assumed that either the unperturbed system is strictly hyperbolic or that the eigenvalues of the full symbol have constant multiplicities.

6. Parabolic problems in bounded domains

Model problem and basic estimate

Consider the parabolic equation

$$u_t = Pu + \varepsilon f(x, t, u, u_x, u_{xx}) + F(x,t), \quad 0 \le x \le 1, \quad t \ge 0, \tag{6.1}$$

where P is the second-order operator

$$Pu = u_{xx} + a(x)u_x + b(x)u.$$

We require the initial and boundary conditions

$$u(x,0) = 0, \quad 0 \le x \le 1; \quad u(0,t) = u_x(1,t) = 0, \quad t \ge 0. \tag{6.2}$$

The given scalar functions $a(x), b(x), F(x,t)$ and $f(x,t,u,u_x,u_{xx})$ are assumed to be of class C^∞, for simplicity, and compatibility of the data with the boundary conditions is assumed. The functions f, f_t, f_{tt}, and f_{ttt} are required to satisfy Assumption 3.1 with $v = (u, u_x, u_{xx})$ for all sufficiently large p.

For $\varepsilon = 0$, Laplace transformation leads to a family of ordinary BVPs for $\tilde{u} = \tilde{u}(x,s)$, namely

$$s\tilde{u} = \tilde{u}_{xx} + a(x)\tilde{u}_x + b(x)\tilde{u} + \tilde{F}(x,s), \quad 0 \le x \le 1, \tag{6.3}$$

$$\tilde{u}(0,s) = \tilde{u}_x(1,s) = 0. \tag{6.4}$$

A basic observation is that one can always obtain good estimates of $\tilde{u}$ by $\tilde{F}$, gaining two derivatives, if $\operatorname{Re} s \ge 0$ and $|s|$ is sufficiently large.

Lemma 6.1 There are constants C and K, depending only on $|a|_\infty + |b|_\infty$, so that the solution of (6.3), (6.4) satisfies

$$|s|^2\|\tilde{u}\|^2 + |s|\|\tilde{u}_x\|^2 + \|\tilde{u}_{xx}\|^2 \le K\|\tilde{F}\|^2 \tag{6.5}$$

if $\operatorname{Re} s \ge 0$ and $|s| \ge C$.

Proof. Take the L_2-inner product of (6.3) with $\tilde{u}(x,s)$ and use integration by parts to obtain

$$s\|\tilde{u}\|^2 + \|\tilde{u}_x\|^2 = (\tilde{u}, a\tilde{u}_x) + (\tilde{u}, b\tilde{u}) + (\tilde{u}, \tilde{F}) =: R. \tag{6.6}$$

The absolute value of the right-hand side is bounded by

$$\begin{aligned}|R| &\leq |a|_\infty \|\tilde{u}\|\|\tilde{u}_x\| + |b|_\infty \|\tilde{u}\|^2 + \|\tilde{u}\|\|\tilde{F}\| \\ &\leq \frac{1}{2}\|\tilde{u}_x\|^2 + K_1\|\tilde{u}\|^2 + \|\tilde{u}\|\|\tilde{F}\|.\end{aligned}$$

Taking the real part of (6.6), we find

$$\operatorname{Re} s\|\tilde{u}\|^2 + \frac{1}{2}\|\tilde{u}_x\|^2 \leq K_1\|\tilde{u}\|^2 + \|\tilde{u}\|\|\tilde{F}\|. \tag{6.7}$$

Case 1: $\operatorname{Re} s \geq |\operatorname{Im} s|$; thus $|s| \leq \sqrt{2}\operatorname{Re} s$.
If $K_1 \leq \frac{|s|}{2\sqrt{2}}$, we obtain from (6.7)

$$\frac{|s|}{2\sqrt{2}}\|\tilde{u}\|^2 + \frac{1}{2}\|\tilde{u}_x\|^2 \leq \|\tilde{u}\|\|\tilde{F}\| \leq \frac{|s|}{4\sqrt{2}}\|\tilde{u}\|^2 + \frac{\sqrt{2}}{|s|}\|\tilde{F}\|^2;$$

thus

$$|s|^2\|\tilde{u}\|^2 + |s|\|\tilde{u}_x\|^2 \leq 8\|\tilde{F}\|^2. \tag{6.8}$$

Case 2: $0 \leq \operatorname{Re} s \leq |\operatorname{Im} s|$, thus $|s| \leq \sqrt{2}|\operatorname{Im} s|$.
First, from (6.7) and $\operatorname{Re} s \geq 0$ we find

$$\frac{1}{2}\|\tilde{u}_x\|^2 \leq K_1\|\tilde{u}\|^2 + \|\tilde{u}\|\|\tilde{F}\|. \tag{6.9}$$

Also, taking the imaginary part of (6.6), we have

$$|\operatorname{Im} s|\|\tilde{u}\|^2 \leq K_2\Big(\|\tilde{u}\|^2 + \|\tilde{u}_x\|^2\Big) + \|\tilde{u}\|\|\tilde{F}\|,$$

and, together with (6.9), we obtain

$$|\operatorname{Im} s|\|\tilde{u}\|^2 \leq K_3\Big(\|\tilde{u}\|^2 + \|\tilde{u}\|\|\tilde{F}\|\Big).$$

Recalling that $|s| \leq \sqrt{2}|\operatorname{Im} s|$, we obtain, as before,

$$|s|^2\|\tilde{u}\|^2 \leq K_4\|\tilde{F}\|^2 \quad \text{for} \quad |s| \geq C,$$

if C is sufficiently large. Together with (6.9), we have shown that

$$|s|^2\|\tilde{u}\|^2 + |s|\|\tilde{u}_x\|^2 \leq K_5\|\tilde{F}\|^2 \quad \text{for} \quad |s| \geq C. \tag{6.10}$$

Since we have proved such an estimate already in Case 1 (see (6.8)), it is clear that (6.10) is generally valid for $\operatorname{Re} s \geq 0, |s| \geq C$. Finally, using the differential equation (6.3), we can estimate $\|\tilde{u}_{xx}\|$ by $K_5(|s|\|\tilde{u}\| + \|\tilde{u}_x\| + \|\tilde{F}\|)$, and the lemma is proved. □

A simple implication of Lemma 6.1 is the unique solvability of the BVP

$$s\phi = P\phi + g(x), \qquad \phi(0) = \phi_x(1) = 0,$$

where

$$P\phi = \phi_{xx} + a\phi_x + b\phi,$$

provided that $\operatorname{Re} s \geq 0, |s| \geq C$. Here $g(x)$ is any inhomogeneous term. In particular, it follows that the eigenvalue problem

$$P\phi = \lambda\phi, \qquad \phi(0) = \phi_x(1) = 0 \tag{6.11}$$

does not have an eigenvalue λ with $\operatorname{Re}\lambda \geq 0, |\lambda| \geq C$.

To obtain asymptotic stability, we formulate the following eigenvalue condition.

Assumption 6.1 The eigenvalue problem (6.11) has no eigenvalue λ with $\operatorname{Re}\lambda \geq 0$.

Remark 6.1 Lemma 6.1 excludes *large* eigenvalues λ with nonnegative real part but, depending on $a(x)$ and $b(x)$, eigenvalues λ with $\operatorname{Re}\lambda \geq 0$ and $|\lambda| \leq C$ can exist, of course. Since one only has to examine a compact λ-region, Assumption 6.1 can be tested with standard numerical procedures. Also, sufficient conditions for Assumption 6.1 are well known from the maximum principle. For example, if $a(x)$ and $b(x)$ are real and $b(x) \leq 0$ for all x or $b(x) - a_x(x) \leq 0$ for all x, then Assumption 6.1 holds.

Assumption 6.1 together with Lemma 6.1 gives us a strong resolvent estimate.

Theorem 6.1 Consider

$$P\phi = \phi_{xx} + a(x)\phi_x + b(x)\phi, \quad 0 \leq x \leq 1.$$

Then Assumption 6.1 holds if and only if there is a constant K such that

$$s\tilde{u} = P\tilde{u} + \tilde{F}, \quad \tilde{u}(0,s) = \tilde{u}_x(1,s), \quad \operatorname{Re} s \geq 0, \tag{6.12}$$

implies

$$\|\tilde{u}\|_{H^2} \leq K\|\tilde{F}\|. \tag{6.13}$$

Proof. First, assuming that the estimate (6.13) holds for all $\operatorname{Re} s \geq 0$, there can be no eigenvalue λ with $\operatorname{Re}\lambda \geq 0$; that is, Assumption 6.1 holds.

Second, assume that the eigenvalue problem (6.11) has no eigenvalue λ with $\operatorname{Re}\lambda \geq 0$. Then, for every fixed s with $\operatorname{Re} s \geq 0$, the BVP (6.12) has a unique solution, and the estimate (6.13) holds with $K = K(s)$. Furthermore, if s varies in a compact region, the constants $K(s)$ can be chosen uniformly bounded, as one can prove by a contradiction argument and Arcela's theorem. Therefore, Lemma 6.1 completes the proof. □

By Parseval's relation (with $\operatorname{Re} s = 0$), (6.13) translates into

$$\int_0^\infty \|u(\cdot,t)\|_{H^2}^2 \, dt \le K_1 \int_0^\infty \|F(\cdot,t)\|^2 \, dt. \tag{6.14}$$

Estimates for derivatives and linear stability

In the case of periodic boundary conditions, we could apply an estimate like (6.14) directly to $D^\alpha u$ since $D^\alpha u_t = PD^\alpha u + D^\alpha F$. However, in the present case, the boundary conditions $u(0,t) = u_x(1,t) = 0$ have been used to derive (6.14), and $D^\alpha u$ does not satisfy these boundary conditions, in general. Instead, we differentiate with respect to t to obtain

$$u_{tt} = Pu_t + F_t.$$

Let us assume that

$$u_t(x,0) = 0, \quad 0 \le x \le 1. \tag{6.15}$$

We will show below that this is no restriction. Then we can apply (6.14) to u_t and find

$$\int_0^\infty \|u_t\|_{H^2}^2 \, dt \le K_1 \int_0^\infty \|F_t\|^2 \, dt. \tag{6.16}$$

Further, if $u_{tt}(x,0) \equiv 0$, then we can repeat the process and find

$$\int_0^\infty \|u_{tt}\|_{H^2}^2 \, dt \le K_1 \int_0^\infty \|F_{tt}\|^2 \, dt. \tag{6.17}$$

Since

$$u_{txx} = u_{xxxx} + (au_x)_{xx} + (bu)_{xx} + F_{xx},$$

it is not difficult to show that (6.14) and (6.16) imply

$$\int_0^\infty \left(\|u\|_{H^4}^2 + \|u_t\|_{H^2}^2\right) dt \le K_2 \int_0^\infty \left(\|F\|_{H^2}^2 + \|F_t\|^2\right) dt. \tag{6.18}$$

The Sobolev inequality, which we stated in Theorem 3.3 for periodic functions, remains valid without periodicity and, therefore, the estimate (3.17) applies here. Consequently, the left-hand side of (6.18) dominates

$$\sup_t \left(|u(\cdot,t)|_\infty^2 + |u_x(\cdot,t)|_\infty^2\right).$$

Using the same arguments as in Section 3, we obtain $\lim_{t\to\infty} |u(\cdot,t)|_\infty = 0$ if the right hand side of (6.18) is finite.

It remains to show that the assumption (6.15) is not restrictive. To this end, consider (6.1), (6.2) and make the change of variables

$$u(x,t) = te^{-t}\phi(x) + v(x,t),$$

where $\phi(x)$ will be determined. At $t = 0$ we have

$$F(x,0) = u_t(x,0) = \phi(x) + v_t(x,0).$$

Therefore, if we choose $\phi(x) = F(x,0)$ then we obtain $v_t(x,0) = 0$ for the new variable.

Nonlinear stability

For illustration, let us assume first that the nonlinearity f in (6.1) has the form $f = f(u_x)$. Proceeding as in Section 3, we consider (see (6.18))

$$\begin{aligned} &\int_0^T \left(\|u\|_{H^4}^2 + \|u_t\|_{H^2}^2\right) dt \qquad (6.19) \\ &\leq 2K_2 \int_0^\infty \left(\|F\|_{H^2}^2 + \|F_t\|^2\right) dt + 2K_2|\varepsilon|^2 \int_0^T \left(\|f\|_{H^2}^2 + \|(f)_t\|^2\right) dt. \end{aligned}$$

Here

$$f = f(u_x(x,t)), \qquad (f)_t = f'(u_x(x,t))u_{xt}(x,t).$$

As noted above, the left-hand side of (6.19) dominates $\max_{0\leq t\leq T} |u_x(\cdot,t)|_\infty^2$ and, therefore, the nonlinearity is controlled in maximum norm. Furthermore,

$$\int_0^T \|u_{xt}\|^2 \, dt$$

is also bounded by the left-hand side of (6.19). Using the same arguments as in Section 3, we find that $\lim_{t\to\infty} |u(\cdot,t)|_\infty = 0$ if the right-hand side of (6.19) is finite and $|\varepsilon|$ is sufficiently small.

In the general case $f = f(x,t,u,u_x,u_{xx})$ we proceed similarly. After proper initialization, we have the following generalization of (6.18), restricted to a finite time interval:

$$\begin{aligned} L^2 &:= \int_0^T \left(\|u\|_{H^8}^2 + \|u_t\|_{H^6}^2 + \|u_{tt}\|_{H^4}^2 + \|u_{ttt}\|_{H^2}^2\right) dt \\ &\leq K_3 \int_0^T \left(\|F\|_{H^6}^2 + \|F_t\|_{H^4}^2 + \|F_{tt}\|_{H^2}^2 + \|F_{ttt}\|^2\right) dt, \end{aligned}$$

The left-hand side bounds $\sup_t |D^\alpha u|_\infty^2$ for $\alpha \leq 5$. (Here $D = \partial/\partial x$.) To treat the nonlinear problem, we need to bound

$$\int_0^T \left(\|f\|_{H^6}^2 + \|(f)_t\|_{H^4}^2 + \|(f)_{tt}\|_{H^2}^2 + \|(f)_{ttt}\|^2\right) dt$$

in terms of L^2. Applying the chain rule and expressing any t-derivative on u by two x-derivatives (using the differential equation), one needs to consider

$$\|D^{\sigma_1}u \ldots D^{\sigma_k}u\|, \quad \sigma_1 + \cdots + \sigma_k \leq 8.$$

Since derivatives up to order 5 are controlled in maximum norm by L, one finds that

$$\|D^{\sigma_1}u\dots D^{\sigma_k}u\|^2 \le C\|u\|^2_{H^8}.$$

The remaining arguments are as in Section 3.

Generalizations
It is not difficult to generalize the key result, Lemma 6.1, to the Laplace transforms of parabolic systems

$$u_t = (A(x)u_x)_x + B(x)u_x + C(x)u + F(x,t) \equiv Pu + F(x,t)$$

under initial and boundary conditions

$$u(x,0) = 0, \quad 0 \le x \le 1; \quad R_0u(0,t) = R_1u(1,t) = 0, \quad t \ge 0.$$

Here $A(x), B(x)$, and $C(x)$ are smooth matrix functions, and parabolicity requires

$$A(x) = A^*(x) \ge \alpha I > 0.$$

We also assume that the boundary conditions $R_0\tilde{u} = R_1\tilde{u} = 0$ imply

$$\langle \tilde{u}, A\tilde{u}_x\rangle\Big|_0^1 = 0.$$

(This boundary term appears when $(\tilde{u}, (A\tilde{u}_x)_x)$ is integrated by parts.) In particular, one can use a Dirichlet or Neumann condition. Under these assumptions, one obtains that a strong resolvent estimate

$$\|\tilde{u}\|_{H^2} \le K\|\tilde{F}\| \quad \text{for all } \operatorname{Re} s \ge 0$$

holds if and only if the eigenvalue problem

$$P\phi = \lambda\phi, \qquad R_0\phi = R_1\phi = 0,$$

has no eigenvalue λ with $\operatorname{Re}\lambda \ge 0$. The arguments are the same as in the proof of Theorem 6.1. Again, the eigenvalue condition can be tested numerically since the existence of large eigenvalues λ with $\operatorname{Re}\lambda \ge 0$ is excluded by analytical arguments.

7. PDEs on all space with negative zero-order term

The purpose of this section is to extend the results of Sections 3, 4 and 5 from the case of periodic boundary conditions to problems on all space. The extension is easy, because the constant coefficient operator P is assumed to have a 'negative' zero-order term. This allows us to obtain a resolvent estimate that is valid uniformly up to $\operatorname{Re} s = 0$. For hyperbolic problems, the zero-order term leads to exponential decay. Equations without such a zero-order term are treated in Section 8.

7.1. Problems on all space with strong resolvent estimate

Consider a Cauchy problem

$$u_t = Pu + \varepsilon f(x,t,u,Du,\ldots,D^r u) + F(x,t), \quad x \in \mathbb{R}^d, t \geq 0 \tag{7.1}$$

with homogeneous initial condition

$$u(x,0) = 0, \quad x \in \mathbb{R}^d. \tag{7.2}$$

As in Section 4, the operator P has constant coefficients

$$P = \sum_{|\nu| \leq m} A_\nu D^\nu, \quad A_\nu \in \mathbb{C}^{n\times n}. \tag{7.3}$$

The functions $F(x,t)$ and $f(x,t,v)$ with

$$v = (u, Du, \ldots, D^r u)$$

are assumed to be of class C^∞, for simplicity. Furthermore, let $f(x,t,0) = 0$; more precisely, we require Assumption 3.1 with a sufficiently large p. For the function $F(x,t)$ we assume[8]

$$\|D^\alpha F(\cdot,t)\| < \infty, \quad \text{for all } \alpha, \quad t \geq 0,$$

and

$$\int_0^\infty \|F(\cdot,t)\|^2_{H^p}\, \mathrm{d}t < \infty$$

for a sufficiently large p.

We will always assume that the Cauchy problem $u_t = Pu, u(x,0) = u_0(x)$, is well posed in L_2, that is, there are constants K and α with

$$\left| \mathrm{e}^{\hat{P}(\mathrm{i}\omega)t} \right| \leq K\mathrm{e}^{\alpha t} \quad \text{for all } \omega \in \mathbb{R}^d, \quad t \geq 0.$$

(See, for example, Kreiss and Lorenz (1989) for a discussion of well-posedness in L_2.)

For $\varepsilon = 0$, Fourier transformation in x and Laplace transformation in t yield the family of linear algebraic equations

$$s\tilde{u}(\omega,s) = \hat{P}(\mathrm{i}\omega)\tilde{u}(\omega,s) + \hat{F}(\omega,s). \tag{7.4}$$

Here the Fourier–Laplace transform of u is

$$\tilde{u}(\omega,s) = (2\pi)^{-d/2} \int_0^\infty \int_{\mathbb{R}^d} \mathrm{e}^{-st-\mathrm{i}\omega\cdot x} u(x,t)\, \mathrm{d}x\, \mathrm{d}t.$$

As in Section 4, an eigenvalue condition for the symbols $\hat{P}(\mathrm{i}\omega)$ leads to a strong resolvent estimate.

[8] The L_2-inner product and norm are now defined by $(u,v) = \int_{\mathbb{R}^d} u^*(x)v(v)\,\mathrm{d}x$, $\|u\|^2 = (u,u)$, *i.e.*, the domain of integration is $\mathbb{R}^d$ instead of $\mathbb{T}^d$.

Theorem 7.1 Assume that the Cauchy problem for $u_t = Pu$ is well posed and that there are constants $q \in \{0, 1, \ldots\}$ and $\delta > 0$ with

$$\operatorname{Re}\lambda \le -\Big(|\omega|^q + 1\Big)\delta < 0 \quad \text{for all} \quad \lambda \in \sigma(\hat{P}(\mathrm{i}\omega)), \quad \omega \in \mathbb{R}^d.$$

Then there is a constant K with

$$\left|\Big(sI - \hat{P}(\mathrm{i}\omega)\Big)^{-1}\right| \le \frac{K}{|\omega|^q + 1} \quad \text{for all} \quad \omega \in \mathbb{R}^d, \quad \operatorname{Re} s \ge 0. \tag{7.5}$$

Proof. The proof, based on the Kreiss matrix theorem, is the same as the proof of Theorem 4.5. The essential argument is given in the proof of Theorem 4.1. One only has to replace $\omega \in \mathbb{Z}^d$ by $\omega \in \mathbb{R}^d$. □

Given that P satisfies the assumptions of Theorem 7.1, one obtains from (7.4)

$$|\tilde{u}(\omega, s)| \le \frac{K}{|\omega|^q + 1} |\tilde{F}(\omega, s)| \quad \text{for all} \quad \omega \in \mathbb{R}^d, \quad \operatorname{Re} s \ge 0. \tag{7.6}$$

Then Parseval's relation (with $\operatorname{Re} s = 0$) yields

$$\int_0^\infty \|u(\cdot, t)\|_{H^q}^2 \, \mathrm{d}t \le K_1 \int_0^\infty \|F(\cdot, t)\|^2 \, \mathrm{d}t.$$

We can apply this estimate to $D^\alpha u$ and can also obtain bounds for u_t and $D^\alpha u_t$ using the differential equation $u_t = Pu + F$. Therefore,

$$\int_0^\infty \Big(\|u(\cdot, t)\|_{H^{p+q}}^2 + \|u_t(\cdot, t)\|_{H^{p+q-m}}^2\Big) \, \mathrm{d}t \le K_2 \int_0^\infty \|F(\cdot, t)\|_{H^p}^2 \, \mathrm{d}t, \quad p \ge m - q. \tag{7.7}$$

(Here m is the order of P.) The Sobolev inequality, which we formulated in Theorem 3.3 for periodic functions, is also valid for $u \in H^p(\mathbb{R}^d)$. Consequently, the left-hand side of (7.7) dominates

$$\sup_t |D^\alpha u(\cdot, t)|_\infty^2 \quad \text{if} \quad p + q - m > |\alpha| + \frac{d}{2}.$$

In exactly the same way as we have proved Theorem 4.7, we obtain the following result.

Theorem 7.2 Consider the problem (7.1), (7.2) and assume that P satisfies the conditions of Theorem 7.1 with $q \ge r$. Furthermore, let Assumption 3.1 hold for f and let $\int_0^\infty \|F\|_{H^p}^2 \, \mathrm{d}t < \infty$, where $p = 2m + d + 1$. Under these assumptions $\lim_{t\to\infty} |v(\cdot, t)|_\infty = 0$ if $|\varepsilon|$ is sufficiently small. (Here $v = (u, Du, \ldots, D^r u)$.)

7.2. Hyperbolic problems on all space

Consider a first-order system

$$u_t = Pu + \varepsilon \sum_{j=1}^{d} B_j(u) D_j u + F(x,t), \quad x \in \mathbb{R}^d, \quad t \geq 0, \tag{7.8}$$

with initial condition

$$u(x,0) = u_0(x), \quad x \in \mathbb{R}^d. \tag{7.9}$$

Here P has constant coefficients,

$$Pu = \sum_{j=1}^{d} A_j D_j u + Bu.$$

We assume symmetry,

$$A_j = A_j^*, \quad B_j(u) = B_j^*(u), \quad j = 1, \ldots, d, \tag{7.10}$$

and negativity of the zero-order term,

$$B + B^* \leq -2\delta I < 0. \tag{7.11}$$

The functions $B_j(u), F(x,t)$, and $u_0(x)$ are assumed to be of class C^∞ and, for convenience, all quantities are assumed to be real. Furthermore, let

$$\|u_0\|_{H^p} < \infty, \quad \|F(\cdot,t)\|_{H^p} < \infty \quad \text{for all } p = 0, 1, \ldots \quad \text{and all } t \geq 0. \tag{7.12}$$

Under these assumptions, one knows local (in time) existence of a C^∞ solutions $u(x,t)$, and this solution satisfies

$$\|u(\cdot,t)\|_{H^p} < \infty, \quad p = 0, 1, \ldots. \tag{7.13}$$

in its interval of existence. (See, for example, Kreiss and Lorenz (1989).)

To discuss stability, we first let $\varepsilon = 0$. Consider the 'change in energy',

$$\begin{aligned} \frac{1}{2}\frac{\mathrm{d}}{\mathrm{d}t}\|u\|^2 &= (u, u_t) \\ &= \sum_j (u, A_j D_j u) + (u, Bu) + (u, F). \end{aligned}$$

Using the symmetry of A_j and integration by parts, one finds that

$$(u, A_j D_j u) = (A_j u, D_j u) = -(A_j D_j u, u),$$

and therefore $(u, A_j D_j u) = 0$. When one integrates by parts, boundary terms appear. However, these terms are zero, since the solution u decays to zero for $|x| \to \infty$. This follows from (7.13). For this reason, all arguments used in the spatially periodic case in Section 5 can be used here in the same way. Instead of Theorem 5.2 one obtains the following result.

Theorem 7.3 Consider the symmetric hyperbolic system (7.1) with initial condition (7.2) under the assumptions described above. Also, for $p = d+2$ we assume $\int_0^\infty \|F\|_{H^p}^2 \, dt < \infty$. Then we have $\lim_{t\to\infty} |u(\cdot,t)|_\infty = 0$ if $|\varepsilon|$ is sufficiently small.

The generalizations outlined at the end of Section 5 can also be made in the all-space case. The only difference is that one has to work with Fourier integrals instead of Fourier sums. In particular, if $H(\omega)$ denotes a bounded symmetrizer satisfying

$$H(\omega)\hat{P}(\mathrm{i}\omega) + \hat{P}^*(\mathrm{i}\omega)H(\omega) \le -\delta H(\omega),$$

then the $\mathcal{H}$-inner product becomes

$$(u,v)_{\mathcal{H}} = \int_{\mathbb{R}^d} \hat{u}^*(\omega)H(\omega)v(\omega)\,\mathrm{d}\omega.$$

With respect to this inner product, the linear operator P is negative:

$$(u,Pu)_{\mathcal{H}} + (Pu,u)_{\mathcal{H}} \le -\delta\|u\|_{\mathcal{H}}^2.$$

Perturbed hyperbolic systems

$$u_t = Pu + \varepsilon \sum_j B_j(x,t,u)D_j u + F(x,t),$$

which are either strictly hyperbolic or whose full symbol

$$\sum_j \omega_j A_j + \varepsilon \sum_j \omega_j B(_j(x,t,u), \quad |\omega| = 1,$$

has eigenvalues with constant multiplicities, can again be treated by constructing a norm adjusted to the solution.

8. Parabolic problems on all space with weak resolvent estimate

In this section we consider viscous conservation laws of the form

$$u_t = Pu + \varepsilon_1 P_1 u + \varepsilon_2 \sum_{j=1}^d D_j f_j(u) + \sum_{j=1}^d D_j F_j(x,t), \quad x \in \mathbb{R}^d, \quad t \ge 0, \tag{8.1}$$

with homogeneous initial conditions

$$u(x,0) = 0, \quad x \in \mathbb{R}^d.$$

Here, the operator P has constant coefficients, that is,

$$Pu = \Delta + \sum_{j=1}^d A_j D_j u, \quad A_j \in \mathbb{R}^{n\times n}. \tag{8.2}$$

The term $\varepsilon_1 P_1 u$ describes linear perturbations with variable coefficients, so that,

$$P_1 u = \sum_{j=1}^{d} D_j(B_j(x,t)u), \tag{8.3}$$

and the nonlinear functions $f_j(u)$ vanish quadratically at $u = 0$. Note that all terms on the right-hand side of (8.1) are derivative terms, that is, (8.1) has conservation form. In particular, the constant coefficient operator P does not have a negative zero-order term, and therefore the results of Section 7 do not apply here. The aim of the section is to show that the resolvent technique can still be used, but one needs more specific assumptions about the form of the perturbation terms. Let us list our assumptions for the terms $B_j(x,t), f_j(u)$, and $F_j(x,t)$ appearing in (8.1).

Assumption 8.1

(1) $F_j(x,t), B_j(x,t)$, and $f_j(u)$ are of class C^∞ and

$$f_j(0) = 0, \quad Df_j(0) = 0;$$

(2)
$$\int_0^\infty \int_{\mathbb{R}^d} |F(x,t)|\, dx\, dt < \infty;$$

(3)
$$\int_0^\infty \|F(\cdot,t)\|_{H^p}^2\, dt < \infty, \quad p = 0, 1, \ldots;$$

(4)
$$\int_0^\infty \|B(\cdot,t)\|^2\, dt < \infty;$$

(5)
$$\sup_{x,t} |D^\alpha B(x,t)| < \infty \quad \text{for all } \alpha.$$

Our standard form (8.1) together with these assumptions and homogeneous initial conditions might seem very restrictive, but one can often enforce the requirements by simple transformations. Let us illustrate this.

8.1. *Transformation to standard form*

Consider a system

$$u_t = Pu + \sum_{j=1}^{d} D_j f_j(u) \tag{8.4}$$

of viscous conservation laws, where P has the form (8.2) and where the flux functions $f_j : \mathbb{R}^n \to \mathbb{R}^n$ are of class C^∞ and vanish quadratically at $u = 0$. Clearly, $u \equiv 0$ is a solution of (8.4), and we are interested in its asymptotic stability in the sense of Lyapunov; that is, we consider (8.4) with small initial data

$$u(x,0) = \varepsilon U_0(x), \quad x \in \mathbb{R}^d. \tag{8.5}$$

Conditions on $U_0(x)$ will be derived below. For simplicity, let us assume that the flux functions $f_j(u)$ are quadratic; that is, there are symmetric bilinear functions $Q_j : \mathbb{R}^n \times \mathbb{R}^n \to \mathbb{R}^n$ with

$$f_j(u) = Q_j(u,u).$$

(If $H_{jk} \in \mathbb{R}^{n\times n}$ is the Hessian of the kth component of f_j at $u = 0$, then $(Q_j(u,v))_k = \frac{1}{2}v^t H_{jk}u$.) If we write $u(x,t) = \varepsilon v(x,t)$, then (8.4), (8.5) becomes

$$v_t = Pv + \varepsilon \sum_{j=1}^{d} D_j f_j(v), \quad v(x,0) = U_0(x). \tag{8.6}$$

To derive estimates by Laplace transformation, it will be convenient to initialize first. To this end, we introduce a new variable $w(x,t)$ by

$$v(x,t) = \mathrm{e}^{-t}U_0(x) + w(x,t),$$

for which we obtain

$$w(x,0) = 0. \tag{8.7}$$

Setting

$$\bar{v}(x,t) = \mathrm{e}^{-t}U_0(x)$$

we find that the function $w(x,t)$ satisfies

$$w_t = Pw + \varepsilon \sum_{j=1}^{d} D_j f_j(\bar{v} + w) + P\bar{v} - v_t,$$

where

$$f_j(\bar{v} + w) = f_j(\bar{v}) + f_j(w) + 2Q_j(\bar{v}, w).$$

Thus we can write

$$w_t = Pw + \varepsilon \sum_{j=1}^{d} D_j(B_j(x,t)w) + \varepsilon \sum_{j=1}^{d} D_j f_j(w) + G(x,t)$$

with

$$G = P\bar{v} - \bar{v}_t + \varepsilon \sum_{j=1}^{d} D_j f_j(\bar{v})$$

and matrices $B_j(x,t)$ determined by

$$B_j(x,t)w = 2Q_j(\bar{v}(x,t), w).$$

If we now assume that the initial function $U_0(x)$ has the form

$$U_0(x) = \sum_{j=1}^{d} D_j U_{0j}(x),$$

where $U_{0j} \in C^\infty$ and

$$D^\alpha U_{0j} \in L_1 \cap L_\infty \quad \text{for all} \quad \alpha,$$

then our construction shows that the inhomogeneous term $G(x,t)$ can be written as

$$G(x,t) = \sum_{j=1}^{d} D_j F_j(x,t),$$

where $F_j \in C^\infty$ and $D^\alpha F_j \in L_1 \cap L_\infty$ for all α. Therefore, $w(x,t)$ solves equation (8.1). Also, it is not difficult to show that Assumption 8.1 is satisfied.

8.2. Estimates for the unperturbed problem

Consider the linear equation

$$u_t = Pu + \sum_{j=1}^{d} D_j F_j(x,t), \quad x \in \mathbb{R}^d, \quad t \geq 0, \tag{8.8}$$

with initial condition

$$u(x,0) = 0, \quad x \in \mathbb{R}^d. \tag{8.9}$$

Here $P = \Delta + \sum_j A_j D_j$, and the $F_j(x,t)$ satisfy the relevant conditions of Assumption 8.1. We also require the following.

Assumption 8.2 The system $u_t = \sum_j A_j D_j u$ is strongly hyperbolic; that is, for all $\omega \in \mathbb{R}^d$ the eigenvalues of $\sum_j \omega_j A_j$ are real and semi-simple, and there is a transformation $S = S(\omega)$ with

$$|S(\omega)| + |S^{-1}(\omega)| \leq \text{const}$$

so that

$$S\left(\sum_j \omega_j A_j\right) S^{-1} =: \Lambda(\omega)$$

is diagonal.

Fourier–Laplace transformation of (8.8) yields

$$s\tilde{u}(\omega, s) = \hat{P}(\mathrm{i}\omega)\tilde{u}(\omega, s) + \mathrm{i}\sum_j \omega_j \tilde{F}_j(\omega, s), \tag{8.10}$$

with

$$\hat{P}(\mathrm{i}\omega) = -|\omega|^2 + \mathrm{i}\sum_j \omega_j A_j.$$

The following technical lemma contains a crucial estimate of the resolvent of $\hat{P}(\mathrm{i}\omega)$.

Lemma 8.1 There is a constant C_1, independent of $\omega \in \mathbb{R}^d$ and $\eta \geq 0$, with

$$\int_{-\infty}^{\infty} \left| \left((\eta + \mathrm{i}\xi)I - \hat{P}(\mathrm{i}\omega) \right)^{-1} \right|^2 \mathrm{d}\xi \leq C_1 |\omega|^{-2}. \tag{8.11}$$

Proof. Let $s = \eta + \mathrm{i}\xi, \eta \geq 0$. Using the transformation $S = S(\omega)$ of Assumption 8.2, we have

$$S(sI - \hat{P})S^{-1} = \left(s + |\omega|^2\right) I - \mathrm{i}\Lambda,$$

where $\Lambda = \mathrm{diag}(\lambda_k), \lambda_k \in \mathbb{R}$. Therefore,

$$\left|(sI - \hat{P})^{-1}\right|^2 \leq C \sum_{k=1}^{n} \frac{1}{(\eta + |\omega|^2)^2 + (\xi - \lambda_k)^2}. \tag{8.12}$$

Clearly, for $\eta \geq 0$ and $\omega \neq 0$,

$$\int_{-\infty}^{\infty} \frac{\mathrm{d}\xi}{(\eta + |\omega|^2)^2 + (\xi - \lambda_k)^2} \leq \int_{-\infty}^{\infty} \frac{\mathrm{d}\xi}{|\omega|^4 + \xi^2} = \pi |\omega|^{-2}.$$

This proves the lemma. □

We use the abbreviation

$$M(F,T) = \left(\int_0^T \int_{\mathbb{R}^d} |F(x,t)| \, \mathrm{d}x \, \mathrm{d}t \right)^2$$

for the square of the L_1-norm of F over space and the time interval $0 \leq t \leq T$. Recall that $M(F, \infty)$ is finite by Assumption 8.1. From the definition of the Fourier–Laplace transform, we obtain directly

$$|\tilde{F}(\omega, s)|^2 \leq M(F, \infty) \quad \text{for all} \quad \omega \in \mathbb{R}^d, \quad \mathrm{Re}\, s \geq 0. \tag{8.13}$$

Therefore, using (8.10),

$$\begin{aligned} |\tilde{u}(\omega, s)|^2 &\leq \left|(sI - \hat{P})^{-1}\right|^2 |\omega|^2 \left|\tilde{F}(\omega, s)\right|^2 \\ &\leq M(F, \infty) |\omega|^2 \left|(sI - \hat{P})^{-1}\right|^2. \end{aligned} \tag{8.14}$$

We now apply Parseval's relation (see (2.21)) with $\eta = 0$ to obtain

$$\begin{aligned} \int_0^{\infty} |\hat{u}(\omega, t)|^2 \, \mathrm{d}t &= \frac{1}{2\pi} \int_{-\infty}^{\infty} |\tilde{u}(\omega, \mathrm{i}\xi)|^2 \, \mathrm{d}\xi \\ &\leq C_2 M(F, \infty), \quad \omega \in \mathbb{R}^d. \end{aligned} \tag{8.15}$$

In the last estimate we have used (8.14) and (8.11). The bound (8.15) will give us the crucial estimate for the small-ω projection of the solution.

Definition 8.1 Let $u = u(x), x \in \mathbb{R}^d$, denote an L_2-function with Fourier transform $\hat{u}(\omega)$. Let $\hat{u}(\omega) = \hat{u}^I(\omega) + \hat{u}^{II}(\omega)$, where

$$\hat{u}^I(\omega) = \hat{u}(\omega) \quad \text{for } |\omega| \le 1, \qquad \hat{u}^I(\omega) = 0 \quad \text{for } |\omega| > 1,$$

and let $u^I(x)$ and $u^{II}(x)$ denote the corresponding inverse Fourier transforms. We call u^I and u^{II} the small-ω and the large-ω projections of u, respectively.

A similar notation, $u^{I,II}(x,t)$ and $\tilde{u}^{I,II}(\omega, s)$, will be used for functions $u(x,t)$ and their Fourier–Laplace transforms. Note that the time variable t and the dual variable s are irrelevant for the projections. Also, if $u(x,t)$ is a smooth function with derivatives in L_2, then differentiation and projection commute, because differentiation corresponds to multiplication on the Fourier side, which clearly commutes with cut-off. In particular, one obtains

$$\left|(D^\alpha u^I)^{\sim}(\omega, s)\right| \le |\omega|^{|\alpha|} |\tilde{u}^I(\omega, s)| \le |\tilde{u}^I(\omega, s)|. \tag{8.16}$$

Theorem 8.1 Let $u(x,t)$ solve (8.8), (8.9), and recall Assumptions 8.1 and 8.2. For any $p = 0, 1, \ldots$, there exists C_p, independent of F and T, with

$$\int_0^T \left(\|u\|^2_{H^{p+1}} + \|u_t\|^2_{H^{p-1}}\right) \mathrm{d}t \le C_p \left(M(F,T) + \int_0^T \|F\|^2_{H^p} \,\mathrm{d}t\right). \tag{8.17}$$

Proof.

(1) We first estimate u^I. By Parseval's relation we have

$$\|u^I(\cdot, t)\|^2 = \int_{|\omega| \le 1} |\hat{u}(\omega, t)|^2 \,\mathrm{d}\omega.$$

Integrating this equation in time and observing (8.15), we find

$$\int_0^\infty \|u^I(\cdot, t)\|^2 \,\mathrm{d}t \le C M(F, \infty).$$

By (8.16) we obtain the same estimate for every derivative $D^\alpha u^I$. Therefore,

$$\int_0^\infty \|u^I\|^2_{H^p} \,\mathrm{d}t \le C_p M(F, \infty), \quad p = 0, 1, \ldots. \tag{8.18}$$

(2) The estimate of the large-ω projection u^{II} proceeds like the estimates in Sections 3 to 5. First note that (8.12) implies

$$(|\omega|^2 + 1) \left|\left(sI - \hat{P}(\mathrm{i}\omega)\right)^{-1}\right| \le C, \quad |\omega| \ge 1, \quad \eta \ge 0.$$

Therefore,

$$\int_0^\infty \|u^{II}\|^2_{H^{p+1}}\,\mathrm{d}t \le C_p \int_0^\infty \|F\|^2_{H^p}\,\mathrm{d}t, \quad p = 0, 1, \ldots. \tag{8.19}$$

(Note that we only gain one derivative, because F appears in differentiated form as forcing.) Together with (8.18) we have derived

$$\int_0^\infty \|u\|^2_{H^{p+1}}\,\mathrm{d}t \le C_p \left(M(F,\infty) + \int_0^\infty \|F\|^2_{H^p}\,\mathrm{d}t \right), \quad p = 0, 1, \ldots. \tag{8.20}$$

To estimate time derivatives, we use the differential equation. Clearly,

$$D^\alpha u_t = P D^\alpha u + \sum_j D_j D^\alpha F_j$$

yields

$$\|u_t\|^2_{H^{p-1}} \le C\Big(\|u\|^2_{H^{p+1}} + \|F\|^2_{H^p} \Big).$$

Together with (8.20), the estimate (8.17) follows for $T = \infty$. Since values of $F(x,t)$ for $t > T$ do not affect the solution $u(x,t)$ for $t \le T$, it follows that (8.17) also holds for all finite T. □

Remark 8.1 Suppose the inhomogeneous term in (8.8) is a general function

$$G \in L_1(\mathbb{R}^d \times [0,\infty)),$$

that is, we do not assume the structure $G = \sum_j D_j F_j$. Then we still have

$$|\tilde{G}(\omega, s)|^2 \le M(G,\infty) < \infty, \quad \omega \in \mathbb{R}^d, \quad \mathrm{Re}\, s \ge 0.$$

Therefore, $\tilde{u} = (sI - \hat{P})^{-1}\tilde{G}$ and Lemma 8.1 yield

$$\begin{aligned} \int_0^\infty |\hat{u}(\omega,t)|^2\,\mathrm{d}t &= \frac{1}{2\pi}\int_{-\infty}^\infty |\tilde{u}(\omega,\xi)|^2\,\mathrm{d}\xi \\ &\le CM(G,\infty)|\omega|^{-2}. \end{aligned} \tag{8.21}$$

In one and two space dimensions,

$$\int_{|\omega|\le 1} |\omega|^{-2}\,\mathrm{d}\omega = \infty,$$

and consequently we cannnot obtain a bound of $\int_0^\infty \|u^I\|^2\,\mathrm{d}t$ from (8.21).

However, if the number of space dimensions is $d \geq 3$, then

$$\int_{|\omega|\leq 1} |\omega|^{-2}\, d\omega$$

is finite, and (8.21) yields

$$\int_0^\infty \|u^I\|^2\, dt \leq C_1 M(G,\infty).$$

For this reason, the structural assumptions on the forcing and the perturbation terms in (8.1) can be relaxed for $d \geq 3$.

By (3.16), the left-hand side of (8.17) dominates $\max_{0\leq t\leq T} |u(\cdot,t)|_\infty^2$ if $p-1 > \frac{d}{2}$. With the same arguments as in Section 3, linear asymptotic stability follows.

Theorem 8.2 Consider (8.8), (8.9) under Assumptions 8.1 and 8.2. Then we have $\lim_{t\to\infty} |u(\cdot,t)| = 0$.

8.3. Stability for the perturbed problem

Consider (8.1) with initial condition $u(x,0) = 0$ and recall Assumptions 8.1 and 8.2. We want to show asymptotic stability, if $\varepsilon_1^2 + \varepsilon_2^2$ is small enough. The basic idea is the same as in Section 3: We use the linear estimate of Theorem 8.1 with F replaced by

$$F(x,t) + \varepsilon_1 \sum_j B_j(x,t)u(x,t) + \varepsilon_2 \sum_j f_j(u(x,t)), \tag{8.22}$$

that is, we treat the perturbation terms as forcings.

To be specific, let $p = d+3$ and let C_p be fixed with (8.17). For all $\varepsilon_1, \varepsilon_2$ there exists $T = T(\varepsilon_1, \varepsilon_2)$ so that

$$\begin{aligned} L^2 &:= \int_0^T \left(\|u\|_{H^{p+1}}^2 + \|u_t\|_{H^{p-1}}^2\right) dt \\ &\leq 4C_p \left(M(F,\infty) + \int_0^\infty \|F\|_{H^p}^2\, dt\right) =: R^2. \end{aligned} \tag{8.23}$$

By (3.17), we have

$$\max_{0\leq t\leq T} |D^\alpha u(\cdot,t)|_\infty^2 \leq L^2 \quad \text{if } p-1 > |\alpha| + \frac{d}{2}.$$

From (8.17) (with F replaced by (8.22)) we find

$$\begin{aligned} &\int_0^T \left(\|u\|_{H^{p+1}}^2 + \|u_t\|_{H^{p-1}}^2\right) dt \\ &\quad \leq C_p \left(M(F + \varepsilon_1 Bu + \varepsilon_2 f(u), T) + \int_0^T \|F + \varepsilon_1 Bu + \varepsilon_2 f(u)\|_{H^p}^2\, dt\right). \end{aligned} \tag{8.24}$$

Here

$$M(F + \varepsilon_1 Bu + \varepsilon_2 f(u), T) \le 2M(F, T) + 4\varepsilon_1^2 M(Bu, T) + 4\varepsilon_2 M(f(u), T)$$

and

$$\|F + \varepsilon_1 Bu + \varepsilon_2 f(u)\|_{H^p}^2 \le 2\|F\|_{H^p}^2 + 4\varepsilon_1^2 \|Bu\|_{H^p}^2 + 4\varepsilon_2^2 \|f(u)\|_{H^p}^2,$$

and it remains to show that the quantities

$$M(Bu, T), \quad M(f(u), T), \quad \int_0^T \|Bu\|_{H^p}^2, \quad \int_0^T \|f(u)\|_{H^p}^2$$

can be estimated by $K_R L^2$, with a constant K_R depending only on the (fixed) right-hand side of (8.23). Firstly,

$$\begin{aligned} M(Bu, T) &\le \left(\int_0^T \int_{\mathbb{R}^d} |B(x,t)||u(x,t)| \, \mathrm{d}x \, \mathrm{d}t \right)^2 \\ &\le \int_0^T \int_{\mathbb{R}^d} |B|^2 \, \mathrm{d}x \, \mathrm{d}t \int_0^T \|u\|^2 \, \mathrm{d}t \\ &\le C_B L^2, \end{aligned}$$

where we have used Assumption 8.1. Secondly, since f vanishes quadratically at $u = 0$ and $|u(x,t)| \le L \le R$, we have $|f(u)| \le C_R |u|^2$. Therefore,

$$\begin{aligned} M(f(u), T) &= \left(\int_0^T \int_{\mathbb{R}^d} |f(u(x,t))| \, \mathrm{d}x \, \mathrm{d}t \right)^2 \\ &\le C_R^2 \left(\int_0^T \int_{\mathbb{R}^d} |u|^2 \, \mathrm{d}x \, \mathrm{d}t \right)^2 \\ &\le C_R^2 L^4 \le C_R' L^2. \end{aligned}$$

Thirdly, by Leibnitz's rule,

$$D^\alpha (B_j u) = \sum_{\beta \le \alpha} c_{\alpha\beta} (D^{\alpha - \beta} B_j)(D^\beta u);$$

thus

$$\|Bu\|_{H^p}^2 \le C_B \|u\|_{H^p}^2$$

if we observe Assumption 8.1. This implies

$$\int_0^T \|Bu\|_{H^p}^2 \, \mathrm{d}t \le C_B L^2.$$

Finally, a bound

$$\int_0^T \|f(u)\|_{H^p}^2\, dt \le C(R)L^2$$

is shown exactly as in Section 3. (See Theorems 3.4 and 3.5.) These arguments prove that there is a constant K_R, independent of T and ε_1 and ε_2, such that

$$\begin{aligned} L^2 &:= \int_0^T \left(\|u\|_{H^{p+1}}^2 + \|u_t\|_{H^{p-1}}^2 \right) dt \\ &\le 2C_p \left(M(F,\infty) + \int_0^\infty \|F\|_{H^p}^2\, dt \right) + K_R L^2 \left(\varepsilon_1^2 + \varepsilon_2^2\right). \qquad (8.25) \end{aligned}$$

If we choose $\varepsilon_1^2 + \varepsilon_2^2$ so small that

$$K_R \left(\varepsilon_1^2 + \varepsilon_2^2\right) \le \frac{1}{3},$$

then (8.25) implies

$$\int_0^T \left(\|u\|_{H^{p+1}}^2 + \|u_t\|_{H^{p-1}}^2 \right) dt \le 3C_p \left(M(F,\infty) + \int_0^\infty \|F\|_{H^p}^2\, dt \right).$$

The remaining arguments are as in Section 3.

Theorem 8.3 Consider (8.1) with initial condition $u(x,0) = 0$ and recall Assumptions 8.1, 8.2. If $\varepsilon_1^2 + \varepsilon_2^2$ is small enough, then $\lim_{t\to\infty} |u(\cdot,t)|_\infty = 0$.

9. Half-space problems with strong resolvent estimate

Let $\mathcal{H}^d$ denote the half-space

$$\mathcal{H}^d = \{x = (x_1, x_2, \ldots, x_d) :\ x \in \mathbb{R}^d, x_1 \ge 0\}$$

with boundary

$$\partial\mathcal{H}^d = \{x \in \mathcal{H}^d :\ x_1 = 0\}.$$

As a model problem, we consider the scalar parabolic equation

$$\begin{aligned} u_t &= \Delta u + \sum_{j=1}^d a_j D_j u + bu + G(x,t) \\ &=: Pu + G(x,t), \quad x \in \mathcal{H}^d, \quad t \ge 0, \qquad (9.1) \end{aligned}$$

with initial condition

$$u(x,0) = 0, \quad x \in \mathcal{H}^d, \qquad (9.2)$$

and Dirichlet boundary condition

$$u(x,t) = 0, \quad x \in \partial\mathcal{H}^d, \quad t \ge 0. \qquad (9.3)$$

In (9.1) the coefficients $a_1, \ldots, a_d$, and b are real constants, and $G(x,t)$ is a smooth function, which decays sufficiently fast for $|x| \to \infty$ and for $t \to \infty$. Also, it will be assumed that $G(x,t)$ and the initial and boundary conditions are compatible at $(x,t) = (0,0)$. It is sufficient, but not necessary, that G is of class C^∞ and has compact support in $\mathcal{H}^d \times (0,\infty)$.

As in the previous sections, we will derive estimates of u in terms of the inhomogeneous term G. Then the form of the estimate will tell which type of nonlinear perturbation can be added. As we will see, the stability of the problem (9.1), (9.2), (9.3) depends crucially on the sign of b. If $b > 0$, the problem is unstable. If $b < 0$, the case treated in this section, one can derive a strong resolvent estimate. The case $b = 0$ is more intricate and is treated in the next section.

We remark that the maximum principle can be applied to the scalar equation (9.1) and also to nonlinear perturbations of it. While this principle is very useful for scalar problems, it does not apply to systems of equations in a natural way. In contrast, the approach we present here can be generalized to systems.

Proceeding formally, we Fourier transform in the tangential variables

$$x_- = (x_2, \ldots, x_d)$$

and Laplace transform in t to derive a family of ordinary BVPs on the half-line $0 \le x_1 < \infty$. Using the notation

$$\tilde{u}(x_1,\omega,s) = (2\pi)^{-(d-1)/2} \int_0^\infty \int_{\mathbb{R}^{d-1}} \mathrm{e}^{-st-\mathrm{i}\omega\cdot x_-} u(x_1,x_-,t)\, dx_-\, dt,$$
$$\omega = (\omega_2, \ldots, \omega_d) \in \mathbb{R}^{d-1}, \quad \operatorname{Re} s \ge 0,$$

we obtain

$$s\tilde{u} = D_1^2\tilde{u} - |\omega|^2\tilde{u} + a_1 D_1\tilde{u} + \mathrm{i}\sum_{j=2}^d \omega_j a_j \tilde{u} + b\tilde{u} + \tilde{G}(x_1,\omega,s),$$

that is,

$$D_1^2\tilde{u} + a_1 D_1\tilde{u} - \sigma\tilde{u} = -\tilde{G}(x_1,\omega,s) \tag{9.4}$$

with

$$\sigma = s + |\omega|^2 - \mathrm{i}\rho - b, \quad \rho = \sum_{j=2}^d \omega_j a_j. \tag{9.5}$$

The boundary condition (9.3) transforms to

$$\tilde{u}(0,\omega,s) = 0, \quad \omega \in \mathbb{R}^{d-1}, \quad \operatorname{Re} s \ge 0. \tag{9.6}$$

In Section 9.1 we present some elementary results on ordinary BVPs of the type (9.4), (9.6).

9.1. Auxiliary results on BVPs on the half-line

To simplify notation, we write x instead of x_1, a instead a_1, u instead of $\tilde{u}$, and $F(x)$ instead of $-\tilde{G}(x_1, \omega, s)$. Then the BVP (9.4), (9.6) reads

$$u_{xx} + au_x - \sigma u = F(x), \quad 0 \le x < \infty; \quad u(0) = 0. \tag{9.7}$$

Before discussing (9.7), we look at even simpler first-order equations.

Lemma 9.1 Let $\operatorname{Re}\lambda < 0$ and consider

$$u_x = \lambda u + F(x), \quad 0 \le x < \infty,$$

where[9] $F \in C \cap L_2$. The general solution

$$u(x) = \mathrm{e}^{\lambda x} u(0) + \int_0^x \mathrm{e}^{\lambda(x-\xi)} F(\xi)\, \mathrm{d}\xi \tag{9.8}$$

satisfies the estimate

$$\|u\|^2 \le C_\lambda \Big(|u(0)|^2 + \|F\|^2 \Big)$$

with a constant C_λ independent of $u(0)$ and F.

Proof. For $u_1(x) = \mathrm{e}^{\lambda x} u(0)$ we have

$$\|u_1\|^2 = |u(0)|^2 \int_0^\infty \mathrm{e}^{2\operatorname{Re}\lambda x}\, \mathrm{d}x = \frac{|u(0)|^2}{2|\operatorname{Re}\lambda|}.$$

Let $u_2(x)$ denote the integral term in the general solution. Applying the Cauchy–Schwartz inequality, we find

$$\begin{aligned} |u_2(x)|^2 &\le \left(\int_0^x \mathrm{e}^{\frac{1}{2}\operatorname{Re}\lambda(x-\xi)} \mathrm{e}^{\frac{1}{2}\operatorname{Re}\lambda(x-\xi)} |F(\xi)|\, \mathrm{d}\xi \right)^2 \\ &\le \frac{1}{|\operatorname{Re}\lambda|} \int_0^x \mathrm{e}^{\operatorname{Re}\lambda(x-\xi)} |F(\xi)|^2\, \mathrm{d}\xi. \end{aligned}$$

Therefore,

$$\begin{aligned} \|u_2\|^2 &\le \frac{1}{|\operatorname{Re}\lambda|} \int_0^\infty \int_\xi^\infty \mathrm{e}^{\operatorname{Re}\lambda(x-\xi)} |F(\xi)|^2\, \mathrm{d}x\, \mathrm{d}\xi \\ &= \frac{1}{|\operatorname{Re}\lambda|^2} \|F\|^2. \end{aligned}$$

The estimate of $\|u\|^2$ follows from $\|u\|^2 \le 2\|u_1\|^2 + 2\|u_2\|^2$. □

[9] The condition of continuity, $F \in C$, can be dropped here and in the following if one works with weak derivatives of u instead of classical derivatives.

Lemma 9.2 Let $\operatorname{Re}\lambda > 0$ and consider

$$u_x = \lambda u + F(x), \quad 0 \le x < \infty,$$

where $F \in C \cap L_2$. There is a unique solution in L_2, namely

$$u(x) = -\int_x^\infty e^{\lambda(x-\xi)} F(\xi)\, d\xi. \tag{9.9}$$

This solution satisfies

$$\|u\| \le \frac{1}{\operatorname{Re}\lambda}\|F\|, \qquad |u(0)|^2 \le \frac{1}{2\operatorname{Re}\lambda}\|F\|^2.$$

Proof. If u is given by (9.9), then the estimate of $\|u\|$ follows as in the proof of the previous lemma. Also, the bound for $|u(0)|$ follows from the Cauchy–Schwartz inequality.

The general solution has an additional term $e^{\lambda x}c$, but it is clear that (9.9) is the only solution in L_2. □

In the discussion of (9.7), the roots $\lambda_{1,2}$ of the characteristic equation

$$\lambda^2 + a\lambda - \sigma = 0 \tag{9.10}$$

will be important. We show the following elementary result.

Lemma 9.3 Let $a \in \mathbb{R}$ and let $\operatorname{Re}\sigma \ge 0, \sigma \ne 0$. Then the roots of (9.10) satisfy

$$\operatorname{Re}\lambda_1 < 0 < \operatorname{Re}\lambda_2.$$

Proof. Clearly, $\lambda_1 + \lambda_2 = -a$, $\lambda_1\lambda_2 = -\sigma$. Suppose a root $\lambda_1 = i\alpha$ is purely imaginary. Then $\alpha \ne 0$ since $\sigma \ne 0$. However, $\lambda_2 = -a - i\alpha$, thus

$$\operatorname{Re}\lambda_1\lambda_2 = \alpha^2 = -\operatorname{Re}\sigma \le 0.$$

This contradiction shows that no root can be purely imaginary.

Now let

$$\lambda_1 = x_1 + i\alpha, \quad \lambda_2 = x_2 - i\alpha, \quad x_j, \alpha \in \mathbb{R}.$$

Then

$$\operatorname{Re}\lambda_1\lambda_2 = x_1x_2 + \alpha^2 \le 0$$

implies $x_1x_2 \le 0$. Our previous argument shows $x_j \ne 0$, and the assertion follows. □

Using the previous three lemmas, it is straightforward to obtain the following result for the BVP (9.7).

Theorem 9.1 Consider (9.7) with

$$a \in \mathbb{R}, \quad \operatorname{Re}\sigma \ge 0, \quad \sigma \ne 0, \quad F \in C \cap L_2.$$

The problem has a unique solution $u \in C^2$ with $u, u_x, u_{xx} \in L_2$, and there is a constant $C_{a,\sigma}$, independent of F, with

$$\|u\|_{H^2} \le C_{a,\sigma}\|F\|. \tag{9.11}$$

Proof. Using the variable $v = u_x + au$, we can write (9.7) as a first-order system,

$$\begin{pmatrix} u \\ v \end{pmatrix}_x = \begin{pmatrix} -a & 1 \\ \sigma & 0 \end{pmatrix}\begin{pmatrix} u \\ v \end{pmatrix} + \begin{pmatrix} 0 \\ F \end{pmatrix}. \tag{9.12}$$

The eigenvalues $\lambda_{1,2}$ of the system matrix, which we call A, are the roots discussed in Lemma 9.3, and there is a transformation Q with

$$Q^{-1}AQ = \Lambda = \operatorname{diag}(\lambda_1, \lambda_2).$$

In fact,

$$Q = \begin{pmatrix} \lambda_1 & \lambda_2 \\ \sigma & \sigma \end{pmatrix}.$$

In the variables $V = (V_1, V_2)^T$ determined by

$$\begin{pmatrix} u \\ v \end{pmatrix} = Q\begin{pmatrix} V_1 \\ V_2 \end{pmatrix}$$

the system (9.12) becomes diagonal,

$$V_x = \Lambda V + H(x), \qquad H = Q^{-1}\begin{pmatrix} 0 \\ F \end{pmatrix}.$$

We can now use Lemmas 9.1 and 9.2 to construct and estimate an L_2-solution. Note that the boundary condition $u(0) = 0$ transforms to

$$\lambda_1 V_1(0) + \lambda_2 V_2(0) = 0. \tag{9.13}$$

By Lemma 9.2 the unique L_2-solution of $V_{2x} = \lambda_2 V_2 + H_2$ satisfies

$$\|V_2\|^2 + |V_2(0)|^2 \le C_1\|H_2\|^2, \quad C_1 = C_1(\lambda_2).$$

Using Lemma 9.1 and (9.13) to estimate V_1, we obtain

$$\|V\| \le C_2\|H\|, \quad C_2 = C_2(\lambda_{1,2}),$$

or, in the original (u, v)-variables,

$$\|u\|^2 + \|v\|^2 \le C_3\|F\|^2, \quad C_3 = C_3(\lambda_{1,2}).$$

Since $v = u_x + au$ and since we can use the differential equation to estimate u_{xx}, the inequality (9.11) follows. Clearly, by subtraction, (9.11) also yields uniqueness. □

For later reference, let us note that the Sobolev inequality (2.30) implies

$$\sup_{x \ge x_0}\left(|u(x)|^2 + |u_x(x)|^2\right) \le 2\int_{x_0}^{\infty}\left(|u(x)|^2 + |u_x(x)|^2 + |u_{xx}(x)|^2\right) dx,$$

and, by (9.11), the right-hand side tends to zero for $x_0 \to \infty$. Therefore, the solution $u(x)$ constructed in Theorem 9.1 satisfies

$$\lim_{x\to\infty} (|u(x)| + |u_x(x)|) = 0. \tag{9.14}$$

9.2. Resolvent estimate for $b < 0$

In this section we assume that the coefficient b in the parabolic equation (9.1) is negative. Recall the family of BVPs (9.4), (9.6) derived by Fourier–Laplace transformation, where the ODE (9.4) depends on the parameter

$$\sigma = s + |\omega|^2 - \mathrm{i}\rho - b, \quad \rho = \sum_{j=2}^{d} \omega_j a_j.$$

If $s = \eta + \mathrm{i}\xi, \eta \ge 0$, then the parameter σ satisfies

$$|\sigma|^2 = \left(\eta + |\omega|^2 - b\right)^2 + (\xi - \rho)^2 \ge \left(|\omega|^2 + |b|\right)^2 \ge b^2 > 0 \tag{9.15}$$

and

$$\operatorname{Re}\sigma \ge -b > 0. \tag{9.16}$$

In particular, Theorem 9.1 applies to each BVP (9.4), (9.6), and one finds

$$\|\tilde{u}(\cdot,\omega,s)\|_{H^2} \le C_{a,\sigma}\|\tilde{G}(\cdot,\omega,s)\| \quad \text{for all } \omega \in \mathbb{R}^{d-1}, \ \operatorname{Re} s \ge 0. \tag{9.17}$$

We now use the BVP (9.4), (9.6) – or, in other notation, the BVP (9.7) – directly to sharpen the estimate and to make the dependency on $|\sigma|$ explicit. This will then allow us to apply Parseval's relation.

Theorem 9.2 Consider the BVP (9.7) with

$$a \in \mathbb{R}, \quad \operatorname{Re}\sigma \ge 0, \quad \sigma \ne 0, \quad F \in C \cap L_2,$$

and let u denote the H^2-solution constructed in Theorem 9.1. There is a constant K, independent of a, σ, and F with

$$\|u_{xx}\|^2 + |\sigma|\|u_x\|^2 + |\sigma|^2\|u\|^2 \le K\left(1 + \frac{a^2}{|\sigma|}\right)^2 \|F\|^2. \tag{9.18}$$

Proof. Multiply (9.7) by $\bar{u}(x)$ and integrate over $0 \le x < \infty$ to obtain

$$(u, u_{xx}) - a(u, u_x) - \sigma\|u\|^2 = (u, F).$$

Integration by parts yields

$$-\|u_x\|^2 - a(u, u_x) - \sigma\|u\|^2 = (u, F). \tag{9.19}$$

(Note that the boundary terms are zero because of (9.14) and $u(0) = 0$.) Taking the absolute value of the real part of (9.19) one finds that

$$\|u_x\|^2 + \operatorname{Re}\sigma\|u\|^2 \le \|u\|\|F\| \tag{9.20}$$

and, therefore,

$$\|u_x\|^2 \le \|u\|\|F\|. \tag{9.21}$$

Case 1: $\operatorname{Re}\sigma \ge |\operatorname{Im}\sigma|$, thus $|\sigma| \le \sqrt{2}\operatorname{Re}\sigma$.
In this case, (9.20) yields

$$|\sigma|\|u\|^2 \le \sqrt{2}\|u\|\|F\|,$$

and thus

$$|\sigma|\|u\| \le \sqrt{2}\|F\|. \tag{9.22}$$

Also, from (9.7),

$$\|u_{xx}\| \le |a|\|u_x\| + |\sigma|\|u\| + \|F\|. \tag{9.23}$$

Combining this with estimates (9.21) and (9.22), one obtains

$$\|u_{xx}\|^2 + |\sigma|\|u_x\|^2 + |\sigma|^2\|u\|^2 \le K\left(1 + \frac{a^2}{|\sigma|}\right)\|F\|^2.$$

Case 2: $0 \le \operatorname{Re}\sigma \le |\operatorname{Im}\sigma|$, thus $|\sigma| \le \sqrt{2}|\operatorname{Im}\sigma|$.
Taking the absolute value of the imaginary part of (9.19), one finds that

$$\begin{aligned} \frac{1}{\sqrt{2}}|\sigma|\|u\|^2 &\le \|u\|\|F\| + |a|\|u\|\|u_x\| \\ &\le \|u\|\|F\| + |a|\|u\|^{3/2}\|F\|^{1/2}. \end{aligned} \tag{9.24}$$

In the last estimate (9.21) has been used. From (9.24) we obtain

$$\begin{aligned} \frac{1}{\sqrt{2}}|\sigma|\|u\| &\le \|F\| + |a|\|u\|^{1/2}\|F\|^{1/2} \\ &\le \|F\| + \frac{1}{4}|\sigma|\|u\| + \frac{a^2}{|\sigma|}\|F\| \end{aligned}$$

and, therefore,

$$|\sigma|\|u\| \le 2\sqrt{2}\left(1 + \frac{a^2}{|\sigma|}\right)\|F\|.$$

Combining this with (9.21) and (9.23), the desired bound (9.18) follows. □

The derivation of a strong resolvent estimate is now straightforward. We apply Theorem 9.2 to each BVP (9.4), (9.6) and recall

$$|\sigma| \ge |\omega|^2 + |b| \ge |b| > 0.$$

Therefore,

$$\begin{aligned} &K_1\|\tilde{G}(\cdot,\omega,s)\|^2 \\ &\quad \ge \|\tilde{u}_{x_1x_1}(\cdot,\omega,s)\|^2 + (|\omega|^2+1)\|\tilde{u}_{x_1}(\cdot,\omega,s)\|^2 + (|\omega|^2+1)^2\|\tilde{u}(\cdot,\omega,s)\|^2. \end{aligned}$$

By Parseval's relation we obtain the following result. (Note that the factors $|\omega|^2 + 1$ provide estimates for derivatives in the tangential directions.)

Theorem 9.3 Consider the half-space problem (9.1), (9.2), (9.3) with $b < 0$. There is a constant C, which is independent of G, such that

$$\int_0^\infty \|u(\cdot,t)\|^2_{H^2(\mathcal{H}^d)}\,dt \le C \int_0^\infty \|G(\cdot,t)\|^2_{L_2(\mathcal{H}^d)}\,dt \tag{9.25}$$

for all $G \in L_2(\mathcal{H}^d \times [0,\infty))$.

Remark 9.1 In our derivation of (9.25) we have used more regularity for G than $G \in L_2$. However, by applying a simple approximation argument, it is clear that the assumption $G \in L_2(\mathcal{H}^d \times [0,\infty))$ suffices.

With the same arguments as in Section 6, one can extend the basic estimate (9.25) if G is sufficiently regular and sufficiently high compatibility conditions are satisfied at $(x,t) = (0,0)$. For example, we have

$$u_{tt} = Pu_t + G_t$$

and, if $G(x,0) \equiv 0$, then $u_t = 0$ at $t = 0$. Therefore, (9.25) yields an estimate of u_t and its second space derivatives. Using the differential equation

$$\Delta u_t = \Delta\Delta u + \ldots,$$

one obtains estimates for the fourth space derivatives of u, *etc.* As explained in Section 6, one obtains nonlinear stability for a general perturbation term $\varepsilon f(x,t,u,Du,D^2u)$ added to (9.1).

10. Half-space problems with weak resolvent estimate

In this section we consider the parabolic initial-boundary value problem (9.1), (9.2), (9.3) with $b = 0$. Let us recall the family of ordinary BVPs on the half-line $0 \le x_1 < \infty$, derived by Fourier–Laplace transformation:

$$D_1^2\tilde{u} + a_1 D_1\tilde{u} - \sigma\tilde{u} = -\tilde{G}(x_1,\omega,s), \quad \tilde{u}(0,\omega,s) = 0. \tag{10.1}$$

The differential equation depends on the parameter σ,

$$\sigma = s + |\omega|^2 - i\rho, \quad \rho = \sum_{j=2}^{d} \omega_j a_j, \quad \omega \in \mathbb{R}^{d-1}, \quad \operatorname{Re} s \ge 0. \tag{10.2}$$

If one tries to extend the results of the previous section, where we had assumed $b < 0$, to the case $b = 0$, one faces the difficulty that the crucial estimate of Theorem 9.2 becomes useless for $|\sigma| \approx 0$. In the previous section we had $|\sigma| \ge |b| > 0$, but if $b = 0$ then σ becomes zero for $s = 0, \omega = 0$.

This problem is not just technical. In fact, to obtain linear stability, one is forced to make more restrictive assumptions on the inhomogeneous term

$G(x,t)$ than smoothness and $G \in L_2(\mathcal{H}^d \times (0,\infty))$. As in Section 8, we will assume

$$G(x,t) = \sum_{j=1}^{d} D_j F_j(x,t), \quad x \in \mathcal{H}^d, \quad t \geq 0, \tag{10.3}$$

where the F_j are C^∞ functions and $F_j \in L_1(\mathcal{H}^d \times (0,\infty))$.

Note that the Fourier–Laplace transform of (10.3) is

$$\tilde{G}(x_1,\omega,s) = D_1\tilde{F}_1(x_1,\omega,s) + \mathrm{i}\sum_{j=2}^{d} \omega_j \tilde{F}_j(x_1,\omega,s). \tag{10.4}$$

Several technical estimates are given next.

10.1. Further auxiliary results on BVPs on the half-line

In this section we use the notation

$$\|u\|^2 = \int_0^\infty |u(x)|^2 \, \mathrm{d}x, \qquad \|u\|_1 = \int_0^\infty |u(x)| \, \mathrm{d}x$$

for the L_2-norm and the L_1-norm. We start with estimates for solutions of the first-order equation $u_x = \lambda u + F(x)$.

Lemma 10.1 Let $\mathrm{Re}\,\lambda < 0$ and consider

$$u_x = \lambda u + F(x), \quad 0 \leq x < \infty, \tag{10.5}$$

where $F \in C \cap L_1$. The general solution (9.8) satisfies the estimate

$$\|u\|^2 \leq \frac{2}{|\mathrm{Re}\,\lambda|}\Big(|u(0)|^2 + \|F\|_1^2\Big).$$

Proof. For the integral term $u_2(x)$ in the general solution we have

$$|u_2(x)| \leq \int_0^x \mathrm{e}^{\mathrm{Re}\,\lambda(x-\xi)} |F(\xi)| \, \mathrm{d}\xi \leq \|F\|_1$$

and, therefore,

$$\begin{aligned}
\|u_2\|^2 &= \int_0^\infty |u_2(x)|^2 \, \mathrm{d}x \\
&\leq \|F\|_1 \int_0^\infty \int_\xi^\infty \mathrm{e}^{\mathrm{Re}\,\lambda(x-\xi)} |F(\xi)| \, \mathrm{d}x \, \mathrm{d}\xi \\
&\leq \frac{1}{|\mathrm{Re}\,\lambda|} \|F\|_1^2.
\end{aligned}$$

□

Lemma 10.2 Consider (10.5) with $\text{Re}\,\lambda > 0, F \in C \cap L_1$. The unique L_2-solution (9.9) satisfies

$$|u(0)| \le \|F\|_1, \qquad \|u\|^2 \le \frac{1}{\text{Re}\,\lambda}\|F\|_1^2.$$

Proof. The bound for $|u(0)|$ is obvious, and the bound for $\|u\|^2$ follows in the same way as in the proof of the previous lemma. □

We will also need elementary estimates for the roots $\lambda_{1,2}$ of the characteristic equation

$$\lambda^2 + a\lambda - \sigma = 0 \tag{10.6}$$

for small $|\sigma|$.

Lemma 10.3 Consider (10.6) for

$$a \in \mathbb{R}, \quad a \neq 0, \quad \text{Re}\,\sigma \ge 0, \quad |\sigma| \le \delta.$$

The roots are

$$\lambda_1 = -a + \mathcal{O}(|\sigma|), \qquad \lambda_2 = \frac{\sigma}{a} - \frac{\sigma^2}{a^3} + \mathcal{O}(|\sigma|^3).$$

If $\delta = \delta(a) > 0$ is small enough, then

$$|\text{Re}\,\lambda_2| \ge \frac{1}{2}\left(\frac{\text{Re}\,\sigma}{|a|} + \frac{|\text{Im}\,\sigma|^2}{|a|^3}\right). \tag{10.7}$$

Proof. We only show (10.7). Setting $x = \text{Re}\,\sigma, y = \text{Im}\,\sigma$, the formula for λ_2 yields

$$\text{Re}\,\lambda_2 = \frac{x}{a} - \frac{x^2 - y^2}{a^3} + \mathcal{O}(|x|^3 + |y|^3).$$

Here x/a and y^2/a^3 are not of opposite sign, and the claim follows. □

As motivated by (10.1) and (10.4), we now consider BVPs

$$u_{xx} + au_x - \sigma u = \Phi(x), \quad 0 \le x < \infty; \quad u(0) = 0,$$

under various assumptions on $\Phi(x)$. Theorem 9.1 will guarantee existence of a unique solution $u \in H^2(0, \infty)$. The estimates in the following lemmas hold for this particular solution.

Lemma 10.4 Consider the BVP

$$u_{xx} + au_x - \sigma u = F_x(x), \quad 0 \le x < \infty; \quad u(0) = 0,$$

where

$$a \in \mathbb{R}, \quad \text{Re}\,\sigma \ge 0, \quad \sigma \neq 0,$$

and $F \in C^1, F, F_x \in L_2$. Then the H^2-solution satisfies

$$\|u_x\|^2 + |\sigma|\|u\|^2 \le K\left(1 + \frac{a^2}{|\sigma|}\right)\|F\|^2 \tag{10.8}$$

with a constant K independent of a, σ, and F.

Proof. Integration by parts yields

$$-\|u_x\|^2 - a(u, u_x) - \sigma\|u\|^2 = -(u_x, F). \tag{10.9}$$

Taking the absolute value of the real part, one finds that

$$\|u_x\|^2 + \operatorname{Re}\sigma\|u\|^2 \le \|u_x\|\|F\| \tag{10.10}$$

and, consequently,

$$\|u_x\| \le \|F\|. \tag{10.11}$$

Case 1: $\operatorname{Re}\sigma \ge |\operatorname{Im}\sigma|$, thus $|\sigma| \le \sqrt{2}\operatorname{Re}\sigma$.
In this case, (10.10) and (10.11) yield

$$|\sigma|\|u\|^2 \le \sqrt{2}\|u_x\|\|F\| \le \sqrt{2}\|F\|^2.$$

Together with (10.11) the estimate (10.8) follows.

Case 2: $0 \le \operatorname{Re}\sigma \le |\operatorname{Im}\sigma|$, thus $|\sigma| \le \sqrt{2}|\operatorname{Im}\sigma|$.
Taking the absolute value of the imaginary part of (10.9) and using (10.11), we find that

$$\begin{aligned} \frac{1}{\sqrt{2}}|\sigma|\|u\|^2 &\le \|u_x\|\|F\| + |a|\|u\|\|u_x\| \\ &\le \|F\|^2 + \frac{1}{4}|\sigma|\|u\|^2 + \frac{a^2}{|\sigma|}\|F\|^2. \end{aligned}$$

Again, together with (10.11) the bound (10.8) follows. □

Lemma 10.5 Consider the BVP

$$u_{xx} + au_x - \sigma u = F(x), \quad 0 \le x < \infty; \quad u(0) = 0,$$

where

$$a \in \mathbb{R}, \quad a \ne 0, \quad \operatorname{Re}\sigma \ge 0, \quad \sigma \ne 0,$$

and $F \in C \cap L_1$. There are positive constants δ and K, which depend on a, but are independent of σ and F, such that

$$\|u_x\|^2 + \|u\|^2 \le K\left(1 + \frac{1}{\tau}\right)\|F\|_1^2, \quad \tau = \frac{\operatorname{Re}\sigma}{|a|} + \frac{|\operatorname{Im}\sigma|^2}{|a|^3}, \tag{10.12}$$

if $|\sigma| \le \delta$.

Proof. Let $v = u_x + au$; thus

$$\begin{pmatrix} u \\ v \end{pmatrix}_x = A\begin{pmatrix} u \\ v \end{pmatrix} + \begin{pmatrix} 0 \\ F \end{pmatrix}, \quad A = \begin{pmatrix} -a & 1 \\ \sigma & 0 \end{pmatrix}.$$

The eigenvalues $\{\lambda_1, \lambda_2\}$ of A have been discussed in Lemma 10.3, namely,

$$\lambda_1 = -a + \mathcal{O}(|\sigma|), \qquad \lambda_2 = \frac{\sigma}{a} - \frac{\sigma^2}{a^3} + \mathcal{O}(|\sigma|^3). \tag{10.13}$$

Also, by Lemma 9.3 we have either $\operatorname{Re}\lambda_1 < 0 < \operatorname{Re}\lambda_2$ or $\operatorname{Re}\lambda_1 > 0 > \operatorname{Re}\lambda_2$. Our notation is set by (10.13).

We transform A to diagonal form,

$$Q^{-1}AQ = \operatorname{diag}(\lambda_1, \lambda_2),$$

using the transformation

$$\begin{aligned} Q &= \begin{pmatrix} \lambda_1 & a\lambda_2/\sigma \\ \sigma & a \end{pmatrix} = \begin{pmatrix} -a & 1 \\ 0 & a \end{pmatrix} + \mathcal{O}(|\sigma|), \\ Q^{-1} &= a^{-2}\begin{pmatrix} -a & 1 \\ 0 & a \end{pmatrix} + \mathcal{O}(|\sigma|). \end{aligned}$$

For the variables $V = (V_1, V_2)^T$ defined by

$$\begin{pmatrix} u \\ v \end{pmatrix} = QV \tag{10.14}$$

one obtains scalar equations

$$V_{1x} = \lambda_1 V_1 + H_1, \qquad V_{2x} = \lambda_2 V_2 + H_2,$$

where

$$H = Q^{-1}\begin{pmatrix} 0 \\ F \end{pmatrix}, \qquad \|H\|_1 \le C_a \|F\|_1.$$

Note that the boundary condition $u(0) = 0$ transforms to

$$\lambda_1 V_1(0) + \frac{a\lambda_2}{\sigma} V_2(0) = 0. \tag{10.15}$$

By (10.13), the coefficients are bounded away from zero and infinity.

Case 1: $a < 0$, thus $\operatorname{Re}\lambda_1 > 0 > \operatorname{Re}\lambda_2$.
By Lemma 10.2 we have

$$|V_1(0)| \le \|F\|_1, \qquad \|V_1\|^2 \le \frac{1}{\operatorname{Re}\lambda_1}\, \|F\|_1^2. \tag{10.16}$$

Lemma 10.1, (10.16) and (10.15) yield

$$\|V_2\|^2 \le \frac{K_a}{|\operatorname{Re}\lambda_2|}\, \|F\|_1^2.$$

Finally, by Lemma 10.3,

$$\frac{1}{|\mathrm{Re}\,\lambda_2|} \le \frac{2}{\tau}$$

and, therefore,

$$\|V\|^2 \le K_a \left(1 + \frac{1}{\tau}\right) \|F\|_1^2.$$

Because of (10.14) and $v = u_x + au$ the desired bound follows.

Case 2: $a > 0$, thus $\mathrm{Re}\,\lambda_1 < 0 < \mathrm{Re}\,\lambda_2$.
By Lemma 10.2 we have

$$|V_2(0)| \le \|F\|_1, \qquad \|V_2\|^2 \le \frac{1}{\mathrm{Re}\,\lambda_2} \|F\|_1^2. \tag{10.17}$$

Lemma 10.1, (10.17) and (10.15) yield

$$\|V_1\|^2 \le K_a \|F\|_1^2$$

since $|\mathrm{Re}\,\lambda_1| \approx |a|$. The remaining arguments are the same as in Case 1. □

Lemma 10.6 Let $a > 0$ and let $\delta = \delta(a) > 0$ be determined as in Lemma 10.5. Consider the BVP

$$u_{xx} + au_x - \sigma u = F_x(x), \quad 0 \le x < \infty; \quad u(0) = 0,$$

under the assumptions

$$\mathrm{Re}\,\sigma \ge 0, \quad 0 < |\sigma| \le \delta, \quad F \in C^1 \cap L_1.$$

There is a constant $K = K_a$, independent of σ and F, such that

$$\|u_x\|^2 + \|u\|^2 \le K\left(\|F\|^2 + \|F\|_1^2\right).$$

Proof. Define $v(x)$ to be the solution of

$$v_x + av = F, \quad v(0) = 0.$$

Using Lemma 9.1 we find that

$$\|v_x\|^2 + \|v\|^2 \le K\|F\|^2,$$

and the estimate $\|v\|_1 \le \frac{1}{a}\|F\|_1$ follows easily from

$$v(x) = \int_0^x \mathrm{e}^{-a(x-\xi)} F(\xi)\,\mathrm{d}\xi.$$

The difference $w = u - v$ satisfies

$$w_{xx} + aw_x - \sigma w = \sigma v, \quad w(0) = 0.$$

By Lemma 10.5,

$$\|w_x\|^2 + \|w\|^2 \le K\left(1 + \frac{1}{\tau}\right) |\sigma|^2 \|v\|_1^2,$$

where

$$\tau = \frac{\operatorname{Re}\sigma}{a} + \frac{|\operatorname{Im}\sigma|^2}{a}.$$

If $\sigma = x + iy \quad (x, y \in \mathbb{R})$, then

$$|\sigma|^2 = x^2 + y^2 \le \delta x + y^2 \le C\tau.$$

Thus we find

$$\|w_x\|^2 + \|w\|^2 \le K\|v\|_1^2,$$

and, together with the estimates for v, the lemma is proved. □

Lemma 10.7 Let $a < 0$ and let $\delta = \delta(a) > 0$ be determined as in Lemma 10.5. Consider the BVP

$$u_{xx} + au_x - \sigma u = F_x(x), \quad 0 \le x < \infty; \quad u(0) = 0,$$

under the assumptions

$$\operatorname{Re}\sigma \ge 0, \quad 0 < |\sigma| \le \delta, \quad F \in C^1 \cap L_1.$$

There is a constant $K = K_a$, independent of σ and F, such that

$$\|u_x\|^2 + \|u\|^2 \le K\|F\|^2 + \frac{K}{|\operatorname{Re}\lambda_2|}\|F\|_1^2.$$

Here $\lambda_2 = \frac{\sigma}{a} + \mathcal{O}(|\sigma|^2)$ is small.

Proof. Define $v(x)$ to be the L_2-solution of

$$v_x + av = F, \quad 0 \le x < \infty;$$

thus

$$v(x) = -\int_x^\infty \mathrm{e}^{-a(x-\xi)} F(\xi)\,\mathrm{d}\xi. \tag{10.18}$$

Using Lemma 9.2 we find that

$$\|v_x\|^2 + \|v\|^2 \le K\|F\|^2,$$

and the estimates

$$|v(0)| \le \|F\|_1, \quad \|v\|_1 \le \frac{1}{|a|}\|F\|_1$$

follow easily from (10.18).

The difference $w = u - v$ satisfies

$$w_{xx} + aw_x - \sigma w = \sigma v, \quad w(0) = -v(0).$$

We write $w = w_1 + w_2$ where w_1 and w_2 solve

$$\begin{aligned} w_{1xx} + aw_{1x} - \sigma w_1 &= \sigma v, \quad w_1(0) = 0, \\ w_{2xx} + aw_{2x} - \sigma w_2 &= 0, \quad w_2(0) = -v(0). \end{aligned}$$

The estimate of w_1 proceeds in the same way as the estimate of w in the proof of the previous lemma; thus

$$\|w_{1x}\|^2 + \|w_1\|^2 \le K\|F\|_1^2.$$

Finally, we have $w_2(x) = -v(0)\mathrm{e}^{\lambda_2 x}$ and, therefore,

$$\|w_2\|^2 = \frac{|v(0)|^2}{2|\operatorname{Re}\lambda_2|} \le \frac{\|F\|_1^2}{2|\operatorname{Re}\lambda_2|}.$$

From these estimates the lemma follows. □

10.2. Weak resolvent estimate

Consider the initial-boundary value problem

$$u_t = \Delta u + \sum_{j=1}^{d} a_j D_j u + \sum_{j=1}^{d} D_j F_j(x,t), \quad x \in \mathcal{H}^d, \quad t \ge 0, \tag{10.19}$$

$$u(x,0) = 0, \quad x \in \mathcal{H}^d; \qquad u(x,t) = 0, \quad x \in \partial\mathcal{H}^d, \quad t \ge 0, \tag{10.20}$$

with real constants a_j and assume $a_1 \ne 0$.

For the one-dimensional hyperbolic problem

$$v_t = a_1 D_1 v + F(x_1,t), \quad 0 \le x_1 < \infty, \quad t \ge 0,$$

the cases $a_1 > 0$ and $a_1 < 0$ are quite different. If $a_1 > 0$, then v is an outgoing characteristic variable, no boundary condition can be given at $x_1 = 0$, and one can easily estimate v by F. In contrast, if $a_1 < 0$, then v is an in-going characteristic variable, one needs a boundary condition at $x_1 = 0$, and estimates of v by F are more restrictive. As we will see, the two cases $a_1 > 0$ and $a_1 < 0$ also lead to different stability conditions for (10.19), (10.20).

Fourier transformation in the tangential variables $x_2, \ldots, x_d$ and Laplace transformation in time leads to the following family of ordinary BVPs for $\tilde{u} = \tilde{u}(x_1, \omega, s)$:

$$D_1^2\tilde{u} + a_1 D_1\tilde{u} - \sigma\tilde{u} = -D_1\tilde{F}_1 - \mathrm{i}\sum_{j=2}^{d}\omega_j\tilde{F}_j =: \Phi(x_1,\omega,s), \quad 0 \le x_1 < \infty, \tag{10.21}$$

with boundary condition

$$\tilde{u}(0,\omega,s) = 0. \tag{10.22}$$

Here

$$\sigma = s + |\omega|^2 - \mathrm{i}\sum_{j=2}^{d}\omega_j a_j, \quad \omega \in \mathbb{R}^{d-1}, \quad \operatorname{Re} s \ge 0. \tag{10.23}$$

The next lemma gives the essential estimate of $\tilde{u}$ by $\tilde{F}$ in the case $a_1 > 0$. Its proof follows by collecting the relevant auxiliary results of Sections 9.1 and 10.1.

Lemma 10.8 Let $a_1 > 0, \operatorname{Re} s \geq 0, \omega \in \mathbb{R}^{d-1}$, and let $\delta = \delta(a_1) > 0$ be determined as in Lemma 10.5. If $|\sigma| \geq \delta$, then

$$\begin{aligned}\int_0^\infty \Big(|D_1\tilde{u}(x_1,\omega,s)|^2 + (|\omega|^2+1)|\tilde{u}(x_1,\omega,s)|^2\Big)\,\mathrm{d}x_1 \\ \leq\ K\int_0^\infty |\tilde{F}(x_1,\omega,s)|^2\,\mathrm{d}x_1 \\ =\ K\|\tilde{F}(\cdot,\omega,s)\|^2. \end{aligned} \tag{10.24}$$

If $0 < |\sigma| \leq \delta$, then

$$\begin{aligned}\int_0^\infty \Big(|D_1\tilde{u}(x_1,\omega,s)|^2 + (|\omega|^2+1)|\tilde{u}(x_1,\omega,s)|^2\Big)\,\mathrm{d}x_1 \\ \leq\ K\|\tilde{F}(\cdot,\omega,s)\|^2 + K\left(\int_0^\infty |\tilde{F}(x_1,\omega,s)|\,\mathrm{d}x_1\right)^2. \end{aligned} \tag{10.25}$$

Proof. Write the right-hand side of (10.21) as $\Phi = \Phi_1 + \Phi_2$,

$$\Phi_1 = -D_1\tilde{F}_1, \qquad \Phi_2 = -\mathrm{i}\sum_{j=2}^{d}\omega_j\tilde{F}_j,$$

and decompose $\tilde{u} = \tilde{u}_1 + \tilde{u}_2$ accordingly. We treat four cases separately.

Case 1: $|\sigma| \geq \delta$; estimate of $\tilde{u}_1$.
Note that

$$|\omega|^2 + 1 \leq |\sigma| + 1 \leq \left(1 + \frac{1}{\delta}\right)|\sigma|.$$

By Lemma 10.4,

$$\|D_1\tilde{u}_1\|^2 + (|\omega|^2+1)\|\tilde{u}_1\|^2 \leq K\|\tilde{F}_1\|^2.$$

Case 2: $|\sigma| \geq \delta$; estimate of $\tilde{u}_2$.
By Theorem 9.2,

$$\|D_1\tilde{u}_2\|^2 + \left(|\omega|^2+1\right)\|\tilde{u}_2\|^2 \leq \frac{K}{|\sigma|}\|\Phi_2\|^2 \leq K\sum_{j=2}^{d}\|\tilde{F}_j\|^2. \tag{10.26}$$

Here we have used that $|\sigma| \geq |\omega|^2$.

Case 3: $0 < |\sigma| \leq \delta$; estimate of $\tilde{u}_1$.
First note that $|\omega|^2 \leq |\sigma| \leq \delta$. Therefore, by Lemma 10.6,

$$\|D_1\tilde{u}_1\|^2 + \left(|\omega|^2+1\right)\|\tilde{u}_1\|^2 \leq K\Big(\|\tilde{F}_1\|^2 + \|\tilde{F}_1\|_1^2\Big). \tag{10.27}$$

Case 4: $0 < |\sigma| \le \delta$; estimate of $\tilde{u}_2$.
By Lemma 10.5,

$$\|D_1\tilde{u}_2\|^2 + \left(|\omega|^2 + 1\right)\|\tilde{u}_2\|^2 \le K\left(1 + \frac{1}{\tau}\right)\|\Phi_2\|^2 \tag{10.28}$$

with

$$\tau \ge \frac{\operatorname{Re}\sigma}{a_1} \ge \frac{|\omega|^2}{a_1}.$$

As before, $\|\Phi_2\|^2 \le |\omega|^2\|\tilde{F}\|^2$. Therefore, the right-hand side of (10.28) is bounded by $K\|\tilde{F}\|^2$.

Collecting these estimates, we have proved the lemma. □

It will be convenient to use the following cut-off function $\chi(\omega, s)$,

$$\chi(\omega, s) = \begin{cases} 1 & \text{if } |\sigma| \le \delta, \\ 0 & \text{if } |\sigma| > \delta, \end{cases}$$

where $\sigma = \sigma(\omega, s)$ is defined by (10.23). Also, from the definition of the Fourier–Laplace transform,

$$|\tilde{F}_j(x_1, \omega, s)| \le \int_0^\infty \int_{\mathbb{R}^{d-1}} |F_j(x_1, x_-, t)|\, \mathrm{d}x_-\, \mathrm{d}t,$$

and integration over $0 \le x_1 < \infty$ yields

$$\|\tilde{F}(\cdot, \omega, s)\|_1 \le C\|F\|_{L_1(\mathcal{H}^d \times (0,\infty))}.$$

Therefore, the two estimates proved in Lemma 10.8 can be summarized by the following inequality:

$$\begin{aligned} \int_0^\infty \left(|D_1\tilde{u}(x_1, \omega, s)|^2 + (|\omega|^2 + 1)|\tilde{u}(x_1, \omega, s)|^2\right) \mathrm{d}x_1 \\ \le K\|\tilde{F}(\cdot, \omega, s)\|^2 + K\chi(\omega, s)\|F\|^2_{L_1(\mathcal{H}^d \times (0,\infty))}. \end{aligned} \tag{10.29}$$

Here $\omega \in \mathbb{R}^{d-1}$ and $s \in \mathbb{C}$ with $\operatorname{Re} s \ge 0$ are arbitrary, except that we require $(\omega, s) \ne (0, 0)$.

(If $\omega = 0, s = 0$, then $\sigma = 0$. However, for $\sigma = 0$ the solution of the BVPs treated in Lemma 10.4, *etc.*, is not in L_2, in general. In the application of Parseval's relation, the single exceptional point $\sigma = 0$ causes no problem.)

To estimate $u(x, t)$, instead of $\tilde{u}$, we apply Parseval's relation with $\eta = 0$. First, Parseval's relation yields

$$\begin{aligned} &\int_0^\infty \|u(\cdot, t)\|^2_{H^1(\mathcal{H}^d)}\, \mathrm{d}t \\ &\le C\int_{-\infty}^\infty \int_{\mathbb{R}^{d-1}} \int_0^\infty \left(|D_1\tilde{u}(x_1, \omega, s)|^2 + (|\omega|^2 + 1)|\tilde{u}(x_1, \omega, s)|^2\right) \mathrm{d}x_1\, \mathrm{d}\omega\, \mathrm{d}\xi. \end{aligned} \tag{10.30}$$

Integrating (10.29) over $\omega \in \mathbb{R}^{d-1}$ and $s = \mathrm{i}\xi$, $-\infty < \xi < \infty$, we find that

$$\int_0^\infty \|u(\cdot,t)\|^2_{H^1(\mathcal{H}^d)}\,\mathrm{d}t \le K\int_0^\infty \|F(\cdot,t)\|^2\,\mathrm{d}t + KJ\|F\|^2_{L_1(\mathcal{H}^d\times(0,\infty))} \quad (10.31)$$

with

$$J = \int_{\mathbb{R}^{d-1}}\int_{-\infty}^{\infty} \chi(\omega,\mathrm{i}\xi)\,\mathrm{d}\omega\,\mathrm{d}\xi. \quad (10.32)$$

It is clear that J is finite since $\chi(\omega,\xi) = 0$ if $|\omega| \ge c_\delta$ or $|\xi| \ge c_\delta$.

Therefore, we have proved the following resolvent estimate.

Theorem 10.1 Consider the initial-boundary value problem (10.19) and (10.20) with $a_1 > 0$ and

$$F \in L_2(\mathcal{H}^d \times (0,\infty)) \cap L_1(\mathcal{H}^d \times (0,\infty)).$$

Then we have

$$\int_0^\infty \|u(\cdot,t)\|^2_{H^1(\mathcal{H}^d)}\,\mathrm{d}t \le C\left(\int_0^\infty \|F(\cdot,t)\|^2\,\mathrm{d}t + \left(\int_0^\infty \|F(\cdot,t)\|_1\,\mathrm{d}t\right)^2\right) \quad (10.33)$$

where C does not depend on F.

In case $a_1 < 0$ we obtain the same result if the number of space dimensions is $d \ge 3$.

Theorem 10.2 Consider the initial-boundary value problem (10.19) and (10.20) with $a_1 < 0$. Under the same assumptions on F as in Theorem 10.1 the estimate (10.33) holds if $d \ge 3$.

Proof. If $a_1 < 0$, Lemma 10.8 needs to be modified. Using the notation of the proof of Lemma 10.8, we have, by Lemma 10.7,

Case 3a: $0 < |\sigma| \le \delta$; estimate of $\tilde{u}_1$ if $a_1 < 0$:

$$\|D_1\tilde{u}_1\|^2 + (|\omega|^2+1)\|\tilde{u}_1\|^2 \le K\|\tilde{F}_1\|^2 + \frac{K}{|\mathrm{Re}\,\lambda_2|}\|\tilde{F}_1\|^2.$$

All other estimates in the proof of Lemma 10.8 remain unchanged. Therefore, in the estimate corresponding to (10.29) the factor $\chi(\omega,s)$ has to be replaced by

$$\frac{\chi(\omega,s)}{|\mathrm{Re}\,\lambda_2|}$$

and, consequently, instead of the integral J (see (10.32)) one has to consider

$$J' = \int_{\mathbb{R}^{d-1}}\int_{-\infty}^{\infty} \frac{\chi(\omega,\mathrm{i}\xi)}{|\mathrm{Re}\,\lambda_2(\omega,\mathrm{i}\xi)|}\,\mathrm{d}\omega\,\mathrm{d}\xi. \quad (10.34)$$

By Lemma 10.3,

$$|\operatorname{Re}\lambda_2| \geq \frac{1}{2}\left(\frac{|\omega|^2}{|a_1|} + \frac{(\xi-\rho)^2}{|a_1|^3}\right), \quad \rho = \sum_{j=2}^{d} \omega_j a_j.$$

Since the integral

$$\int_{|\omega|\leq 1}\int_{-\infty}^{\infty} \frac{d\xi\, d\omega}{|\omega|^2+\xi^2} \quad (\omega \in \mathbb{R}^{d-1})$$

is finite for $d-1 \geq 2$, the integral J' is also finite, and the theorem is proved. □

As in Section 6, the basic estimates of Theorem 10.1 and 10.2 can be extended. Assuming compatibility conditions are satisfied, one may assume $F_j(x,0) \equiv 0$. Then u_t satisfies

$$u_{tt} = Pu_t + \sum_j D_j F_{jt}, \quad u_t = 0 \quad \text{at} \quad t = 0,$$

and one obtains an estimate for u_t and its first space derivatives, *etc.* The perturbation terms for which one obtains nonlinear stability are described in Section 8.3.

11. Eigenvalue and spectral conditions for parabolic problems on the line

In Section 6 we have considered parabolic problems

$$u_t = Pu + \varepsilon f(x,t,u,u_x,u_{xx}) + F(x,t)$$

in a finite interval $0 \leq x \leq 1$ and have shown that the eigenvalue condition for P is necessary and sufficient for nonlinear stability. Here the eigenvalue condition for P requires that the problem

$$P\phi = \lambda\phi, \qquad R\phi = 0,$$

has no eigenvalue λ with $\operatorname{Re}\lambda \geq 0$. (With $R\phi = 0$ we denote homogeneous boundary conditions; see Section 6.) Such a result is not restricted to bounded intervals, but can be generalized to parabolic problems in bounded domains in any number of space dimensions.

On the other hand, the result cannot be extended to unbounded domains without further restrictions. For example, consider the problem on the half-line,

$$u_t = u_{xx} + au_x + G(x,t), \quad 0 \leq x < \infty,$$

with initial and boundary conditions

$$u(x,0) = 0, \quad x \geq 0; \qquad u(0,t) = 0, \quad t \geq 0,$$

and $a \in \mathbb{R}$. It is easy to see that the corresponding eigenvalue problem

$$\phi_{xx} + a\phi_x = \lambda\phi, \quad \phi(0) = 0, \quad \phi \in L_2(0,\infty),$$

has no eigenvalue λ with $\operatorname{Re}\lambda \geq 0$. (Use Lemma 9.3.) However, by Theorem 10.1, we can only obtain a *weak* resolvent estimate, and need the extra assumptions $a > 0$ and $G(x,t) = F_x(x,t)$.

We shall now discuss cases for parabolic problems on the real line $-\infty < x < \infty$ for which the eigenvalue condition for P does imply a *strong* resolvent estimate. The results can be extended to multi-dimensions.

The main principle is to reduce the problem on the infinite line to a problem on a finite interval $-L \leq x \leq L$ by expressing the 'tails' of the problem as boundary conditions at $x = \pm L$. (In numerical computations a similar process is often used when one introduces an artificial boundary for infinite domain problems; see, for example, Hagstrom and Keller (1986).)

To be more specific, consider a parabolic system

$$u_t = Pu + F(x,t), \quad x \in \mathbb{R}, \quad t \geq 0,$$

with initial condition

$$u(x,0) = 0, \quad x \in \mathbb{R}.$$

Here

$$Pu = (A(x)u_x)_x + B(x)u_x + C(x)u \tag{11.1}$$

with smooth matrix functions $A(x), B(x), C(x)$ taking values in $\mathbb{C}^{n\times n}$, and parabolicity of $u_t = Pu$ requires

$$A(x) = A^*(x) \geq \alpha I > 0, \quad x \in \mathbb{R}.$$

Also, we assume that the coefficients $A(x), B(x), C(x)$ converge exponentially fast to constant matrices $A_\pm, B_\pm, C_\pm$ as $x \to \pm\infty$. For example,

$$|A(x) - A_+| \leq K\mathrm{e}^{-\gamma x}, \quad x \geq 0,$$

with $\gamma > 0$. The properties of the resulting constant coefficient equations

$$u_t = P_+u \quad \text{and} \quad u_t = P_-u$$

are very important. (We set $P_+u = A_+u_{xx} + B_+u_x + C_+u$, and P_- is defined correspondingly.) The following lemma relates properties of the symbols $\hat{P}_\pm(\mathrm{i}\omega)$, $\omega \in \mathbb{R}$, to decay and growth properties of the solutions of the homogeneous equations

$$P_\pm u - su = 0.$$

Lemma 11.1 Let $A_0, B_0, C_0 \in \mathbb{C}^{n\times m}$ denote constant matrices with $A_0 = A_0^* \geq \alpha I > 0$, and let

$$\hat{P}_0(\kappa) = \kappa^2 A_0 + \kappa B_0 + C_0, \quad \kappa \in \mathbb{C},$$

denote the symbol of

$$P_0 u = A_0 u_{xx} + B_0 u_x + C_0 u.$$

The following two conditions are equivalent.

(1) There is a constant $\delta_1 > 0$, such that

$$\det(\hat{P}_0(\mathrm{i}\omega) - \lambda I) = 0 \quad \text{and} \quad \omega \in \mathbb{R} \tag{11.2}$$

implies $\mathrm{Re}\,\lambda \leq -\delta_1 < 0$.

(2) There are constants $\gamma > 0$ and $\delta_2 > 0$, such that

$$\det(\hat{P}_0(\kappa) - sI) = 0 \quad \text{and} \quad \mathrm{Re}\, s \geq -\delta_2 \tag{11.3}$$

implies $|\mathrm{Re}\,\kappa| \geq \gamma > 0$.

Proof. First assume condition (1) and let $\mathrm{Re}\, s \geq -\delta_1/2$. If κ solves (11.3), then (1) implies $\mathrm{Re}\,\kappa \neq 0$. Fix any $c > 0$. If $|s| \leq c$ and $\mathrm{Re}\, s \geq -\delta_1/2$ then, by continuity, there exists $\gamma = \gamma(c) > 0$ with $|\mathrm{Re}\,\kappa| \geq \gamma > 0$ for all solutions κ of (11.3). For large s, $|s| \geq c$, the roots κ of (11.3) satisfy to leading order

$$\det(\kappa^2 A - sI) = 0.$$

If $\alpha_1, \ldots, \alpha_n$ denote the eigenvalues of A, then $\alpha_j \geq \alpha$ and

$$\kappa \approx \pm\sqrt{\frac{s}{\alpha_j}}. \tag{11.4}$$

Therefore, if $\mathrm{Re}\, s \geq -1$ and $|s|$ is large, then $|\mathrm{Re}\,\kappa| \geq 1$. This shows that (1) implies (2).

Now assume condition (2) and let ω and λ satisfy (11.2). By (2) we have $\mathrm{Re}\,\lambda < -\delta_2$, and the lemma is proved. □

Under the assumptions of the lemma, equation (11.3) for κ, which is sometimes called the dispersion relation, has $2n$ roots κ_j with

$$\mathrm{Re}\,\kappa_j \leq -\gamma, \quad j = 1, \ldots, n \quad \text{and} \quad \mathrm{Re}\,\kappa_j \geq \gamma, \quad j = n+1, \ldots, 2n.$$

This follows from (11.4) and continuous dependence of the roots on s. In the language of dynamical systems, the homogeneous equation $P_0 u - su = 0$ has an exponential dichotomy on $\mathbb{R}$, which is uniform for $\mathrm{Re}\, s \geq -\delta_2$.

Now consider the variable coefficient operator P given by (11.1) and recall $A(x) \to A_\pm$ as $x \to \pm\infty$, *etc.* We say that P satisfies the *strong spectral condition* if both operators, P_+ and P_-, satisfy the conditions of the previous lemma; that is, there exists $\delta_1 > 0$ such that, for all $\omega \in \mathbb{R}$,

$$\lambda \in \sigma(\hat{P}_+(\mathrm{i}\omega)) \cup \sigma(\hat{P}_-(\mathrm{i}\omega)) \quad \text{implies} \quad \operatorname{Re}\lambda \le -\delta_1 < 0. \tag{11.5}$$

Also, P satisfies the *weak spectral condition* if, for all $\omega \in \mathbb{R}$,

$$\lambda \in \sigma(\hat{P}_+(\mathrm{i}\omega)) \cup \sigma(\hat{P}_-(\mathrm{i}\omega)) \quad \text{implies} \quad \operatorname{Re}\lambda \le 0. \tag{11.6}$$

To give a simple example, consider the scalar operator

$$Pu = u_{xx} + au_x + bu, \quad a, b \in \mathbb{R},$$

with $\hat{P}(\mathrm{i}\omega) = -\omega^2 + \mathrm{i}a\omega + b$. We see that P satisfies the strong spectral condition if $b < 0$, but only the weak spectral condition if $b = 0$. Neither condition holds for $b > 0$.

The following result can be proved.

Theorem 11.1 Assume that P has the form (11.1) and satisfies the strong spectral condition. If the eigenvalue problem

$$P\phi = \lambda\phi, \quad \phi \in L_2(\mathbb{R}), \tag{11.7}$$

has no eigenvalue λ with $\operatorname{Re}\lambda \ge 0$, then a strong resolvent estimate (as in Section 6) holds. One obtains nonlinear stability with perturbations $\varepsilon f(x, t, u, u_x, u_{xx})$ as in Section 6.

The main part of the proof consists of reducing the infinite interval to a finite one; Lemma 11.1 is one step in showing that such a reduction is possible. For the finite-interval problem, the techniques of Section 6 apply. We refer to Kreiss, Kreiss and Petersson (1994*a*) for details.

Theorem 11.1 can be used to discuss stability of travelling; see Kreiss et al. (1994*a*) and the literature cited there. (There is a technical problem resulting from a zero eigenvalue of P, which corresponds to shifting the travelling wave. Using a projection as in Henry (1981) and Kreiss et al. (1994*a*), this problem can be overcome, however.)

In many interesting cases only the weak spectral condition (11.6) is satisfied. Then, for scalar equations, one can often use exponentially weighted norms as pioneered by Sattinger (1976). Instead of using these norms, one can also make a change of the dependent variable u so that, in the new variable, the strong spectral condition (11.5) is satisfied.

For systems of equations, however, this approach does not generally give the desired results. Nevertheless, stability has recently been shown for rather general travelling shock waves governed by systems of conservation laws (Kreiss and Kreiss 1997). Essentially, the only requirements are the weak spectral condition and the condition that (11.7) has no eigenvalue λ with $\operatorname{Re}\lambda \ge 0$ except $\lambda = 0$. The inhomogeneous term and the nonlinear perturbation must satisfy the restrictions of Section 8.

Acknowledgement
This collaboration was made possible by the visit of the second author to the Mittag-Leffler Institute in Djorsholm, Sweden, which was supported by the Swedish Academy of Science. In particular, we would like to thank Professors B. Engquist and K. O. Widman for the excellent arrangements.

REFERENCES

T. Hagstrom and H. B. Keller (1986), 'Exact boundary conditions at an artificial boundary for partial differential equations in cylinders', *SIAM J. Math. Anal.* **17**, 322–341.

T. Hagstrom and J. Lorenz (1995), 'All-time existence of smooth solutions to PDEs of mixed type and the invariant subspace of uniform states', *Adv. Appl. Math.* **16**, 219–257.

D. Henry (1981), *Geometric Theory of Semilinear Parabolic Equations*, Springer, Berlin.

S. Kawashima (1987), 'Large-time behaviour of solutions to hyperbolic-parabolic systems of conservation laws and applications', *Proceedings of the Royal Society of Edinburgh* **106A**, 169–194.

S. Klainerman and G. Ponce (1983), 'Global, small amplitude solutions to nonlinear evolution equations', *Comm. Pure Appl. Math.* **36**, 133–141.

G. Kreiss and H.-O. Kreiss (1997), Stability of systems of viscous conservation laws, Technical Report 97-54, UCLA CAM Report.

G. Kreiss, H.-O. Kreiss and J. Lorenz (1998*a*), 'On stability of conservation laws', *SIAM J. Math. Anal.* To appear.

G. Kreiss, H.-O. Kreiss and J. Lorenz (1998*b*), 'On the use of symmetrizers for stability analysis'. In preparation.

G. Kreiss, H.-O. Kreiss and A. Petersson (1994*a*), 'On the convergence to steady state of solutions of nonlinear hyperbolic-parabolic systems', *SIAM J. Numer. Anal.* **31**, 1577–1604.

G. Kreiss, A. Lundbladh and D. Henningson (1994*b*), 'Bounds for threshold amplitudes in subcritical shear flow', *J. Fluid Mech.* **270**, 175–198.

H.-O. Kreiss and J. Lorenz (1989), *Initial-Boundary Value Problems and the Navier–Stokes Equations*, Academic, New York.

H.-O. Kreiss and S. H. Lui (1997), Nonlinear stability of time dependent differential equations, Technical Report 97–43, UCLA CAM Report.

H.-O. Kreiss, O. Ortiz and O. Reula (1998*c*), 'Stability of quasi-linear hyperbolic dissipative systems', *J. Diff. Eqns.* To appear.

A. M. Lyapunov (1956), *General Problem of the Stability of Motion, Collected Works*, Vol. 2, Izdat. Akad. Nauk SSSR, Moscow. (Russian).

A. Matsumura and T. Nishida (1979), 'The initial value problem for the equations of motion of compressible viscous and heat-conductive fluids', *Proc. Jap. Acad., Ser. A* **55**, 337–341.

R. Racke (1992), *Lectures on Nonlinear Evolution Equations: Initial Value Problems*, Vieweg Verlag, Braunschweig, Wiesbaden.

V. A. Romanov (1973), 'Stability and plane-parallel Couette flow', *Funktsional'nyi Analizi Ego Prilozhaniya* **7**, 62–73. Translated in *Funct. Anal. Appl.* **7**, 137–146.

D. H. Sattinger (1976), 'On the stability of waves of nonlinear parabolic systems', *Adv. Math.* **22**, 312–355.

J. Shatah (1982), 'Global existence of small solutions to nonlinear evolution equations', *J. Diff. Eqns* **46**, 409–425.

W. Strauss (1981), 'Nonlinear scattering theory at low energy', *J. Funct. Anal.* **41**, 110–133.

L. N. Trefethen (1997), 'Pseudospectra of linear operators', *SIAM Rev.* **39**, 383–406.

Acta Numerica (1998), pp. 287–336

Direct search algorithms for optimization calculations

M. J. D. Powell
Department of Applied Mathematics and Theoretical Physics,
University of Cambridge, Silver Street,
Cambridge CB3 9EW, England.
E-mail: mjdp@damtp.cam.ac.uk

Many different procedures have been proposed for optimization calculations when first derivatives are not available. Further, several researchers have contributed to the subject, including some who wish to prove convergence theorems, and some who wish to make any reduction in the least calculated value of the objective function. There is not even a key idea that can be used as a foundation of a review, except for the problem itself, which is the adjustment of variables so that a function becomes least, where each value of the function is returned by a subroutine for each trial vector of variables. Therefore the paper is a collection of essays on particular strategies and algorithms, in order to consider the advantages, limitations and theory of several techniques. The subjects addressed are line search methods, the restriction of vectors of variables to discrete grids, the use of geometric simplices, conjugate direction procedures, trust region algorithms that form linear or quadratic approximations to the objective function, and simulated annealing. We study the main features of the methods themselves, instead of providing a catalogue of references to published work, because an understanding of these features may be very helpful to future research.

CONTENTS

1. Introduction

I have contributed several Fortran subroutines for optimization calculations to IMSL (International Mathematical and Statistical Libraries), assuming that the gradient of the objective function is available. Then, in all cases, IMSL produced versions of them that make difference approximations to derivatives automatically, as many of its customers would have gone elsewhere if they had been asked to specify the first derivatives. I did not include the difference approximations myself, because I was aware that they could cause disastrous loss of accuracy in pathological cases, nor did I object strongly to the expediency of IMSL, because I wanted my software to be employed for a wide range of applications. Thus, about ten years ago, the demand from many computer users for numerical methods for optimization was met in an imperfect way that was usually adequate. The main reason for our work is to show that there are still many opportunities for improvements to optimization algorithms that do not require derivatives.

Another fundamental objection to difference approximations, in addition to loss of accuracy due to cancellation and division by small numbers, is that each gradient vector is replaced by a tight cluster of at least $n+1$ function values, where n is the number of variables. Instead, it seems more suitable, intuitively, to spread out the points at which the objective function is calculated, especially if sample values have to be taken from many parts of the space of the variables. Furthermore, when the values include some random noise, then the contribution from the noise to predicted rates of change is less if the evaluation points are widely spaced. Therefore this paper addresses algorithms for optimization that are basically different from the ones that employ gradients. On the other hand, the computer user who requires the best algorithm for his or her application should keep in mind the possibilities of difference approximations and automatic differentiation.

We let $\underline{x} \in \mathbb{R}^n$ denote a typical vector of variables, and we reserve the notation

$$F(\underline{x}), \qquad \underline{x} \in \mathcal{S} \subseteq \mathbb{R}^n, \tag{1.1}$$

for the objective function, where $\mathcal{S}$ may be a subset of $\mathbb{R}^n$ if constraints are present. Further, letting $\mathcal{X}$ be the set of points in $\mathbb{R}^n$ at which the constraints are satisfied, we assume $\mathcal{X} \subseteq \mathcal{S}$. The optimization problem under consideration is to seek a vector $\underline{x}^* \in \mathcal{X}$ that has the property

$$F(\underline{x}^*) \le F(\underline{x}), \qquad \underline{x} \in \mathcal{X}, \tag{1.2}$$

using calculated values of $F(\underline{x})$ and any constraint functions at points $\underline{x}$ that are chosen automatically. Most of the available algorithms, however, are for the case when the variables are unconstrained. Therefore we follow the usual recourse of mentioning that penalty and barrier function techniques have been developed that replace constrained calculations by unconstrained ones, good descriptions of them being given in the books by Fletcher (1987) and Gill, Murray and Wright (1981). This approach, unfortunately, usually reduces the precision of the information on the final values of the variables that is provided by explicit constraints. Therefore research on new algorithms for nonlinear constraints without the calculation of derivatives could be important to a wide range of applications. We note also that there are some serious deficiencies in the current methods for unconstrained optimization.

Three main techniques are employed for trying to achieve global convergence, namely line searches, trust regions and discrete grids. The suitability of line searches and discrete grids for this purpose is addressed in Sections 2 and 3, respectively, but trust region algorithms are not studied until Sections 6 and 7, as they require approximations to $F(\underline{x})$, $\underline{x} \in \mathbb{R}^n$. Section 4 considers simplex methods, including the procedure of Nelder and Mead (1965), because citation indexes show that it is the method of choice for many applications. Fast convergence is often achieved by algorithms that are designed to be highly efficient when the objective function is quadratic. Therefore the use of conjugate directions is the subject of Section 5. In Section 6, linear approximations are made to $F(\underline{x})$, $\underline{x} \in \mathbb{R}^n$, and to any constraint functions, which give a highly convenient and slow procedure for small numbers of variables. Some of the inefficiencies can be removed by quadratic approximations, as described in Section 7, but current procedures of this kind do not allow nonlinear constraints explicitly. Moreover, some methods make random changes to the variables, including simulated annealing, which is considered in Section 8. Finally, in Section 9, we discuss the convergence properties, the limitations, and some possible developments of the given techniques for minimization without derivatives.

2. Line search methods

Line search methods for unconstrained optimization are iterative, a starting vector of variables $\underline{x}_1 \in \mathbb{R}^n$ has to be given, and, for $k = 1, 2, 3, \ldots$, the kth iteration derives $\underline{x}_{k+1}$ from $\underline{x}_k$ in the following way. A nonzero search direction $\underline{d}_k \in \mathbb{R}^n$ is chosen. Then the function of one variable $\phi(\alpha) = F(\underline{x}_k + \alpha\, \underline{d}_k)$, $\alpha \in \mathbb{R}$, receives attention, in order to pick a new vector of variables of the form

$$\underline{x}_{k+1} = \underline{x}_k + \alpha_k\, \underline{d}_k. \tag{2.1}$$

For example, an 'exact line search' would set the step-length α_k to an α that minimizes $\phi(\alpha)$. In practice, however, one tries to choose α_k in a way that requires very few values of $F(\underline{x}_k + \alpha\, \underline{d}_k)$, $\alpha \in \mathbb{R}$, on each iteration, and it is usual to satisfy the condition

$$F(\underline{x}_{k+1}) \leq F(\underline{x}_k), \qquad k = 1, 2, 3, \ldots . \tag{2.2}$$

Of course the search directions should be able to explore the full space of the variables. Therefore line search methods should have the property that, for some integer $\ell \geq n$, any ℓ consecutive search directions span $\mathbb{R}^n$ in a strict sense. If this condition failed, then a nonzero $\underline{v} \in \mathbb{R}^n$ would be (nearly) orthogonal to the directions. Therefore a convenient form of the strict sense is that the bound

$$\max\left\{|\underline{v}^T\underline{d}_j| \,/\, \|\underline{d}_j\|_2 : j=k-\ell+1, k-\ell+2, \ldots, k\right\} \geq c\,\|\underline{v}\|_2, \quad \underline{v} \in \mathbb{R}^n, \quad (2.3)$$

is satisfied for $k \geq \ell$, where c is a positive constant. For example, a way of achieving this condition, which gives $\ell = n$ and $c = n^{-1/2}$, is to let each $\underline{d}_k$ be a coordinate direction in $\mathbb{R}^n$, and to cycle round the n coordinate directions recursively as k increases. Rosenbrock (1960) provides an extension of this technique that is sometimes useful. His first n directions are also the coordinate directions, but, when k is any positive integer multiple of n, then, before starting the $(k+1)$th iteration, he generates $\underline{d}_{k+1}, \underline{d}_{k+2}, \ldots, \underline{d}_{k+n}$ in sequence, by applying the Gram–Schmidt procedure to the differences $\underline{x}_{k+1} - \underline{x}_{k-n+j}$, $j=1,2,\ldots,n$. Further, he ensures that every step-length is nonzero, although condition (2.2) may have to fail. Some other choices of search directions are given in Section 5.

Unfortunately, condition (2.3) and exact line searches do not guarantee that limit points of the sequence $\underline{x}_k$, $k = 1, 2, 3, \ldots$, are good estimates of optimal vectors of variables, even if the objective function is continuously differentiable, and the level set $\{\underline{x} : F(\underline{x}) \leq F(\underline{x}_1)\}$ is bounded. Indeed, Powell (1973) gives an example of bad behaviour, with $n=3$ and exact line searches, where the sequence $\underline{d}_k$, $k=1,2,3,\ldots$, is generated by cycling round the coordinate directions, as mentioned in the previous paragraph. Here, for each integer i in $[1,6]$, the infinite sequence $\underline{x}_{6j+i}$, $j = 1, 2, 3, \ldots$, tends to one vertex of a cube, and the path from $\underline{x}_k$ to $\underline{x}_{k+6}$ tends to be a cycle along six edges of the cube. Further, the objective function is constant on each of these edges, which implies that two components of the gradient $\underline{\nabla} F$ are zero at each limiting vertex. The other component of $\underline{\nabla} F(\underline{x}_k)$, however, is bounded away from zero for each k. It follows that the calculated vectors of variables do not approach a stationary point of F. Therefore it is easy to modify the algorithm so that the objective function becomes less than the actual limit of the decreasing sequence $F(\underline{x}_k)$, $k \to \infty$. Specifically, we replace $\underline{d}_j$ by a difference approximation to $-\underline{\nabla} F(\underline{x}_j)$ for any integer j that is sufficiently large. Furthermore, we find in the remainder of this section that there is another remedy that does not require an estimate of $\underline{\nabla} F$.

The kind of ingredient that avoids the bad behaviour above is imposing the condition that, if $\|\underline{x}_{k+1} - \underline{x}_k\|$ is bounded away from zero, then $F(\underline{x}_k) - F(\underline{x}_{k+1})$ is bounded away from zero too. Hence, in the usual case when $F(\underline{x}_k)$,

$k=1,2,3,\ldots$, converges monotonically, we have the limit

$$\|\underline{x}_{k+1}-\underline{x}_k\| \to 0 \quad \text{as} \quad k \to \infty, \tag{2.4}$$

which prevents the cycling round the edges of the cube in the example of the previous paragraph. Further, if the directions $\underline{d}_j$, $j=1,2,3,\ldots$, satisfy inequality (2.3), and if $\underline{x}^*$ is any limit point of the infinite sequence $\underline{x}_k$, $k=1,2,3,\ldots$, then $\underline{\nabla} F(\underline{x}^*)=0$ can be obtained by a suitable line search, provided that F is continuously differentiable and bounded below. We are going to prove this assertion, not only because the restriction on $\|\underline{x}_{k+1}-\underline{x}_k\|$ may be valuable to future algorithms, but also because the method of proof provides a demonstration of the kind of analysis that can establish convergence properties. A way of achieving the restriction, due to Lucidi and Sciandrone (1997), will be given after the proof.

We aim to deduce a contradiction from the assumption $\|\underline{\nabla} F(\underline{x}^*)\|_2=\eta$, where η is a positive constant and where $\underline{x}^*$ is a limit point of the sequence $\underline{x}_k$, $k=1,2,3,\ldots$, as stated already. We seek some integers j such that $\underline{\nabla} F(\underline{x}_j)^T\underline{d}_j/\|\underline{d}_j\|_2$ is bounded away from zero, because then the step-length α_j of the equation $\underline{x}_{j+1}=\underline{x}_j+\alpha_j\,\underline{d}_j$ can be chosen so that $F(\underline{x}_j)-F(\underline{x}_{j+1})$ is also bounded away from zero, which gives the required contradiction if this happens an infinite number of times. Now, by setting $\underline{v}=\underline{\nabla} F(\underline{x}^*)$ in expression (2.3), we deduce that the inequality

$$|\underline{\nabla} F(\underline{x}^*)^T\underline{d}_j|\,/\,\|\underline{d}_j\|_2 \geq c\,\|\underline{\nabla} F(\underline{x}^*)\|_2 = c\eta \tag{2.5}$$

is achieved at least once for every ℓ consecutive positive integers j. Further, because $\underline{\nabla} F$ is continuous, this inequality implies $|\underline{\nabla} F(\underline{x}_j)^T\underline{d}_j|/\|\underline{d}_j\|_2 \geq \frac{1}{2}c\eta$, provided that $\underline{x}_j$ is sufficiently close to $\underline{x}^*$. Specifically, $\underline{x}_j$ is close enough to $\underline{x}^*$ if it satisfies $\|\underline{x}_j-\underline{x}^*\|_2\leq\varepsilon$, where ε is a positive constant that provides the property

$$\|\underline{\nabla} F(\underline{x})-\underline{\nabla} F(\underline{x}^*)\|_2 \leq \tfrac{1}{2}c\eta \quad \text{if} \quad \|\underline{x}-\underline{x}^*\|_2\leq\varepsilon, \tag{2.6}$$

because then the Cauchy–Schwarz inequality and condition (2.5) give the bound

$$\frac{|\underline{\nabla} F(\underline{x}_j)^T\underline{d}_j|}{\|\underline{d}_j\|_2} \geq \frac{|\underline{\nabla} F(\underline{x}^*)^T\underline{d}_j| - |\{\underline{\nabla} F(\underline{x}_j)-\underline{\nabla} F(\underline{x}^*)\}^T\underline{d}_j|}{\|\underline{d}_j\|_2} \geq \tfrac{1}{2}c\eta. \tag{2.7}$$

Therefore it remains to show that, on an infinite number of occasions, ℓ consecutive positive integers j satisfy $\|\underline{x}_j-\underline{x}^*\|_2\leq\varepsilon$. The limit (2.4) is helpful, because it admits an integer $j_0>0$ such that $\|\underline{x}_{j+1}-\underline{x}_j\|_2\leq\frac{1}{2}\varepsilon/(\ell-1)$ holds for all $j\geq j_0$. Hence, if $\|\underline{x}_k-\underline{x}^*\|_2\leq\frac{1}{2}\varepsilon$ occurs for some integer $k\geq j_0$, then $\|\underline{x}_j-\underline{x}^*\|_2\leq\varepsilon$ is obtained by every integer j in $[k,k+\ell-1]$. This does happen an infinite number of times, because $\underline{x}^*$ is a limit point of $\underline{x}_k$, $k=1,2,3,\ldots$, even if we require the differences between the chosen integers k to be at least ℓ. The proof is complete.

The line search procedure in Section 5 of Lucidi and Sciandrone (1997) is suitable for the above analysis, although some parameters are required that may be difficult to choose well in practice. They are numbers γ and δ that satisfy $\gamma > 0$ and $0 < \delta < 1$, and a positive sequence $\{\beta_k : k = 1, 2, 3, \ldots\}$ that tends to zero as $k \to \infty$. Then, on each iteration, there is a search for a step-length $\alpha_k = \alpha$ that has the properties

$$\left.\begin{array}{r} F(\underline{x}_k + \alpha\, \underline{d}_k) \leq F(\underline{x}_k) - \gamma\alpha^2 \|\underline{d}_k\|_2^2 \quad \text{and} \\ \min\left[F(\underline{x}_k + \hat{\alpha}\, \underline{d}_k), F(\underline{x}_k - \hat{\alpha}\underline{d}_k)\right] \geq F(\underline{x}_k) - \gamma\hat{\alpha}^2 \|\underline{d}_k\|_2^2 \end{array}\right\}, \tag{2.8}$$

where $\hat{\alpha} = \alpha/\delta$. If the first line of expression (2.8) holds for a trial $\alpha > 0$, then either α is acceptable or the second line shows that a step-length of larger modulus is allowed by the first line, namely α/δ or $-\alpha/\delta$. Thus the modulus of α is increased if necessary, and the second line is tested for the new α. This procedure is continued recursively until α is acceptable, which happens eventually because we are assuming that F is bounded below. Alternatively, if the first line of expression (2.8) fails, not only for the initial α but also for $-\alpha$, then α is replaced by $\alpha\delta$ and these tests are tried again. Thus the second inequality of expression (2.8) is achieved by the new α. Again recursion is applied, either until an acceptable step-length is found or until $\|\alpha\, \underline{d}_k\|_2 < \beta_k$ occurs, the choice $\alpha_k = 0$ being made in the latter case. Moreover, the search directions $\underline{d}_k$, $k = 1, 2, 3, \ldots$, have to satisfy the strict linear independence condition (2.3). These constructions provide the conclusion $\underline{\nabla} F(\underline{x}^*) = 0$ of the previous paragraph, as shown below.

The first line of expression (2.8) and equation (2.1) imply the bound

$$F(\underline{x}_k) - F(\underline{x}_{k+1}) \geq \gamma\, \|\underline{x}_k - \underline{x}_{k+1}\|_2^2, \qquad k = 1, 2, 3, \ldots, \tag{2.9}$$

when α_k is positive, and the bound is trivial when α_k is zero. Therefore the limit (2.4) at the beginning of the given analysis is valid, and the conclusion $\underline{\nabla} F(\underline{x}^*) = 0$ of the analysis holds, provided that inequality (2.7) causes $F(\underline{x}_j) - F(\underline{x}_{j+1})$ to be bounded away from zero. Further, the method of analysis allows us to restrict attention to values of j that satisfy two more conditions. Firstly, we assume $j \geq j_0$, where j_0 is any fixed positive integer, which may be larger than the j_0 introduced earlier. Thus we allow for the zero step-lengths in the line search procedure under consideration. Secondly, we assume $\|\underline{x}_j - \underline{x}^*\|_2 \leq \varepsilon$, although our previous use of this bound was only to establish the existence of integers j that have the property (2.7). Thus the uniform continuity of $\underline{\nabla} F$ in any neighbourhood of $\underline{x}^*$ provides the condition

$$\|\underline{\nabla} F(\underline{x}) - \underline{\nabla} F(\underline{x}_j)\|_2 \leq \tfrac{1}{4} c\eta \quad \text{if} \quad \|\underline{x} - \underline{x}_j\|_2 \leq \hat{\varepsilon}, \tag{2.10}$$

for all of the values of j that are retained, where $\hat{\varepsilon}$ is a positive number that is independent of j.

Now it follows from expressions (2.7) and (2.10) that the gradient of the line search function $\phi(\alpha)=F(\underline{x}_j+\alpha\,\underline{d}_j)$, $\alpha\in\mathbb{R}$, is bounded by the inequality

$$\begin{aligned} |\phi'(\alpha)| &= |\underline{d}_j^T\underline{\nabla}F(\underline{x}_j+\alpha\,\underline{d}_j)| \\ &\geq |\underline{d}_j^T\underline{\nabla}F(\underline{x}_j)| - \|\underline{d}_j\|_2\,\|\underline{\nabla}F(\underline{x}_j+\alpha\,\underline{d}_j)-\underline{\nabla}F(\underline{x}_j)\|_2 \\ &\geq \tfrac{1}{4}c\eta\,\|\underline{d}_j\|_2 \quad \text{if} \quad \|\alpha\,\underline{d}_j\|_2\leq\hat{\varepsilon}. \end{aligned} \tag{2.11}$$

Therefore, by choosing the sign of α to be opposite to the sign of $\phi'(0)$ and by applying $\phi(\alpha)=\int_0^\alpha\phi'(\theta)\,\mathrm{d}\theta$, we find the relation

$$F(\underline{x}_j+\alpha\,\underline{d}_j)\leq F(\underline{x}_j)-\tfrac{1}{4}c\eta\,\|\alpha\,\underline{d}_j\|_2 \quad \text{if} \quad \|\alpha\,\underline{d}_j\|_2\leq\hat{\varepsilon}. \tag{2.12}$$

Thus the first line of expression (2.8) is achieved by every α of the appropriate sign that satisfies $\|\alpha\,\underline{d}_j\|_2\leq\hat{\varepsilon}$ and $\gamma\,\|\alpha\,\underline{d}_j\|_2\leq\frac{1}{4}c\eta$. It follows that, if the parameter β_j of the line search procedure is at most $\delta\min[\hat{\varepsilon},\frac{1}{4}c\eta/\gamma]$, and if the first trial value of α on the jth iteration is at least β_j, then the procedure provides a step-length α_j that is positive. The first of these conditions is irrelevant if j_0 is sufficiently large, as assumed in the previous paragraph, and any sensible implementation observes the second condition. Therefore both of the inequalities (2.8) hold for $k=j$ with $\alpha=\alpha_j>0$. We deduce from the second one and from the property (2.12) that $\|\hat{\alpha}\,\underline{d}_j\|_2=\|\alpha_j\,\underline{d}_j\|_2/\delta$ is no less than $\min[\hat{\varepsilon},\frac{1}{4}c\eta/\gamma]$, which gives the inequality

$$\|\underline{x}_{j+1}-\underline{x}_j\|_2=\|\alpha_j\,\underline{d}_j\|_2\geq\delta\min\,[\hat{\varepsilon},\tfrac{1}{4}c\eta/\gamma]. \tag{2.13}$$

Thus condition (2.9) provides a positive lower bound on $F(\underline{x}_j)-F(\underline{x}_{j+1})$ as required. Therefore line search methods without derivatives can provide convergence properties of the kind that are acclaimed by theoreticians when $F(\underline{x})$, $\underline{x}\in\mathbb{R}^n$, need not be convex. Some further work on these questions, including new line search procedures, can be found in Grippo, Lampariello and Lucidi (1988) and in Lucidi and Sciandrone (1997).

3. Discrete grid methods

We introduce discrete grid methods by considering a simple procedure in the case when the components of $\underline{x}\in\mathbb{R}^n$ are bounded by the constraints

$$a_i\leq x_i\leq b_i, \qquad i=1,2,\ldots,n, \tag{3.1}$$

for given values of a_i and b_i that satisfy $a_i<b_i$, $i=1,2,\ldots,n$. For each i, a mesh size $h_i=(b_i-a_i)/\nu_i$ is chosen for some positive integer ν_i, and we let $\mathcal{G}$ be the finite rectangular grid

$$\mathcal{G}=\{\underline{x}:h_i^{-1}(x_i-a_i)\in\mathbb{Z}\cap[0,\nu_i],\ i=1,2,\ldots,n\}, \tag{3.2}$$

where $\mathbb{Z}$ denotes the set of integers. Let an iterative algorithm that seeks the least value of $F(\underline{x})$, $\underline{x} \in \mathcal{G}$, be given a starting point $\underline{x}_1 \in \mathcal{G}$, and let the kth iteration for each k calculate $\underline{x}_{k+1} \in \mathcal{G}$ in a way that satisfies the condition

$$F(\underline{x}_{k+1}) \leq F(\underline{x}_k), \qquad k=1,2,3,\ldots . \tag{3.3}$$

Further, let the algorithm terminate if any ℓ consecutive iterations fail to reduce the objective function, where ℓ is a prescribed positive integer. Then the finiteness of $\mathcal{G}$ guarantees that termination occurs. A reason for making this remark is to provide a contrast with the details of the analysis of convergence in the previous section.

An obvious algorithm of this type generates $\underline{x}_{k+1}$ by trying to improve only one component of $\underline{x}_k$ on the kth iteration, so it makes searches along the coordinate directions $\underline{e}_i$, $i=1,2,\ldots,n$, where the ith component of $\underline{e}_i$ is one and all the other components are zero. It is usual to cycle round these directions recursively as before, letting the search direction of the kth iteration be $\underline{e}_i$, where $i \in [1,n]$ is defined by the condition that $(k-i)/n$ is an integer. Further, let the step-length of the kth iteration be chosen so that the strict reduction $F(\underline{x}_{k+1}) < F(\underline{x}_k)$ occurs if $\underline{x}_{k+1} \neq \underline{x}_k$, and so that $\underline{x}_{k+1} \in \mathcal{G}$ has the properties

$$F(\underline{x}_{k+1}) \leq F(\underline{x}_{k+1} - h_i\,\underline{e}_i) \quad \text{and} \quad F(\underline{x}_{k+1}) \leq F(\underline{x}_{k+1} + h_i\,\underline{e}_i), \tag{3.4}$$

except that the first or second of these conditions is dropped if $\underline{x}_{k+1} - h_i\,\underline{e}_i$ or $\underline{x}_{k+1} + h_i\,\underline{e}_i$, respectively, is outside the feasible region, due to the ith component of $\underline{x}_{k+1}$ being a_i or b_i. Thus, if $\underline{x}^* \in \mathcal{G}$ is the vector of variables that is provided by the algorithm at termination, because $\ell = n$ iterations have not reduced F, then the inequality

$$F(\underline{x}^*) \leq F(\underline{x}), \qquad \underline{x} \in \mathcal{N}(\underline{x}^*), \tag{3.5}$$

is satisfied, where $\mathcal{N}(\underline{x}^*)$ is the set of points of $\mathcal{G}$ that are neighbours of $\underline{x}^*$ along coordinate directions, and we include $\underline{x}^*$ in $\mathcal{N}(\underline{x}^*)$ too.

Another grid search algorithm that achieves condition (3.5) is proposed by Hooke and Jeeves (1961). It is usually more efficient than the method of the previous paragraph, because it can generate helpful changes to the variables that are not along coordinate directions. Again $\underline{x}_1 \in \mathcal{G}$ is given, but the kth iteration derives $\underline{x}_{k+1}$ from $\underline{x}_k$ in the following way. If $k \geq 2$ and if the vector $\underline{y}_k = 2\underline{x}_k - \underline{x}_{k-1}$ satisfies the constraints (3.1), which implies $\underline{y}_k \in \mathcal{G}$, then the algorithm sets $\underline{y}_k^*$ to a point in $\mathcal{N}(\underline{y}_k)$ that provides the least value of $F(\underline{x})$, $\underline{x} \in \mathcal{N}(\underline{y}_k)$, where $\mathcal{N}(\cdot)$ has been defined already. Further, if $F(\underline{y}_k^*)$ is strictly less than $F(\underline{x}_k)$, then the choice $\underline{x}_{k+1} = \underline{y}_k^*$ is made. In all other cases, however, the algorithm sets $\underline{x}_k^*$ to a point in $\mathcal{N}(\underline{x}_k)$ that minimizes $F(\underline{x})$, $\underline{x} \in \mathcal{N}(\underline{x}_k)$, and $\underline{x}_{k+1} = \underline{x}_k^*$ is chosen if the strict inequality $F(\underline{x}_k^*) < F(\underline{x}_k)$ holds. Otherwise, termination occurs with $\underline{x}^*$ equal to $\underline{x}_k$. Thus the final vector of variables has the property (3.5) as claimed. We note that the method is based on the hypothesis that, because the step from $\underline{x}_{k-1}$ to $\underline{x}_k$ reduces the objective function, the displacement $\underline{x}_k - \underline{x}_{k-1}$ from $\underline{x}_k$ is likely to provide a further reduction, especially if the new point is chosen to be the best one in the set $\mathcal{N}(2\underline{x}_k - \underline{x}_{k-1}) = \mathcal{N}(\underline{y}_k)$.

When the objective function has continuous first derivatives, the property (3.5) allows $\underline{x}^*$ to be related to the first order conditions for a minimum of $F(\underline{x})$, $\underline{x} \in \mathcal{X}$, where $\mathcal{X}$ is the set of points in $\mathbb{R}^n$ that satisfy the constraints (3.1). These conditions are that, for each integer i in $[1, n]$, the derivative $\partial F(\underline{x})/\partial x_i$ is zero, nonnegative or nonpositive in the case $a_i < x_i < b_i$, $x_i = a_i$ or $x_i = b_i$, respectively. Now, if $\underline{x}^* + h_i \underline{e}_i$ is in the set $\mathcal{N}(\underline{x}^*)$, then the property (3.5) includes the inequality $F(\underline{x}^*) \le F(\underline{x}^* + h_i \underline{e}_i)$. Moreover, the mean value theorem gives the equation

$$F(\underline{x}^* + h_i \underline{e}_i) = F(\underline{x}^*) + h_i \, \partial F(\underline{x}^* + \zeta_i \underline{e}_i)/\partial x_i, \tag{3.6}$$

for some ζ_i in $[0, h_i]$. Thus the derivative $\partial F(\underline{x}^* + \zeta_i \underline{e}_i)/\partial x_i$ is nonnegative, which implies the condition

$$\partial F(\underline{x}^*)/\partial x_i \ge \partial F(\underline{x}^*)/\partial x_i - \partial F(\underline{x}^* + \zeta_i \underline{e}_i)/\partial x_i \ge -\omega_i(h_i), \tag{3.7}$$

where ω_i is the modulus of continuity of $\partial F(\underline{x})/\partial x_i$, $\underline{x} \in \mathcal{X}$. Similarly, if $\underline{x}^* - h_i \underline{e}_i$ is in $\mathcal{N}(\underline{x}^*)$, we deduce the bound $\partial F(\underline{x}^*)/\partial x_i \le \omega_i(h_i)$. It follows that the modulus of the difference between the ith component of $\nabla F(\underline{x}^*)$ and the ith component of a hypothetical gradient vector that satisfies the first order conditions for optimality at $\underline{x}^*$ is at most $\omega_i(h_i)$ for $i = 1, 2, \ldots, n$.

This result suggests a useful extension of the grid methods described already. It is to apply one of them until termination, and to repeat this procedure recursively, where the mesh sizes are reduced before each new step of the recursion. Specifically, after calculating the vector $\underline{x}^* = \underline{x}^*(h_1, h_2, \ldots, h_n)$ that achieves the property (3.5) for the original mesh sizes, we might halve all the mesh sizes, and then employ the algorithm to generate a new vector $\underline{x}^*$, using the old $\underline{x}^*$ as the starting point for the new calculation. If the number of recursions is infinite, and if $\hat{\underline{x}}^*$ is any limit point of the sequence $\underline{x}^*(h_1, h_2, \ldots, h_n)$ as $\max\{h_i : i = 1, 2, \ldots, n\}$ tends to zero, then it follows from the previous paragraph that $\hat{\underline{x}}^*$ satisfies the first order conditions for the least value of $F(\underline{x})$, $\underline{x} \in \mathcal{X}$.

Of course there are analogous procedures for unconstrained calculations and for the case when there are bounds on some but not all of the variables. We continue to employ a rectangular grid $\mathcal{G}$ in these cases, and to let h_i denote the mesh size in the ith coordinate direction for each integer i in $[1, n]$. Further, the starting point $\underline{x}_1$ is still required to be a grid point, and we let its components be ξ_i, $i = 1, 2, \ldots, n$. Therefore $\underline{x}$ is a point of $\mathcal{G}$ if and only if it satisfies any constraints and if all the ratios $(x_i - \xi_i)/h_i$, $i = 1, 2, \ldots, n$, are integers. These parameters should be chosen so that any bound on $\underline{x}$ holds as an equation at some of the grid points. Then condition

(3.5) can be achieved as before when the level set $\{\underline{x} : F(\underline{x}) \le F(\underline{x}_1),\ \underline{x} \in \mathcal{X}\}$ is bounded.

On the other hand, rectangular grid methods are hardly ever suitable for constraints that are more general than the simple bounds (3.1). For example, let $n=2$, let $F(\underline{x}) = x_1 + 2x_2$, $\underline{x} \in \mathbb{R}^2$, let the constraints be $-10 \le x_1 \le 10$, $x_2 \le 10$ and $x_1 + x_2 + 10 \ge 0$, let $h_1 = h_2 = 1$, and let the starting point $\underline{x}_1$ be at the origin. Then $\mathcal{G}$ includes many points on the boundary of $\mathcal{X}$, and the objective function can be reduced by decreases in x_1 and x_2. Thus it is likely that a point with the coordinates $(x_1, -x_1-10)$ will be calculated for some integer x_1 in $[-10, 0]$, because this point is on the boundary of the last constraint. Here a decrease in x_1 or x_2 would violate the constraint, while an increase in x_1 or x_2 would increase the value of the objective function. Therefore termination is likely to occur with $\underline{x}_1 \le 0$, although the optimal vector of variables is the grid point with the coordinates $(10, -20)$. The deficiency happens because further progress requires a search direction that is along or close to the boundary of the last constraint, but the direction of that boundary in $\mathbb{R}^2$ is very different from both of the coordinate directions.

If $x_1 + x_2 + 10 \ge 0$ were the only constraint on the variables in the above example, then a linear change of variables $\underline{y} = B\underline{x}$ could be made, where B is an $n \times n$ nonsingular matrix that provides $y_1 = x_1 + x_2$. Then any of the methods described already could be applied to seek the least value of $F(B^{-1}\underline{y})$ subject to $y_1 \ge -10$. The notation B for the matrix of the change of variables is taken from Torczon (1997), because she develops a general convergence theory for discrete grid methods.

That work allows both increases and decreases in mesh sizes, assuming that the variables are unconstrained and that the level set $\{\underline{x} : F(\underline{x}) \le F(\underline{x}_1)\}$ is bounded. We consider the technique when B is the unit matrix and when the mesh sizes satisfy $h_1 = h_2 = \cdots = h_n = h$, say. The value of h can be adjusted automatically, and we let its initial value be $h=1$. Then the technique includes the following four important features. Firstly, every mesh size is an integral power of a rational number $\tau = \beta/\alpha < 1$, where α and β are prescribed positive integers that are relatively prime. Secondly, if the kth iteration gives $\underline{x}_{k+1} \ne \underline{x}_k$, then every component of the difference $\underline{x}_{k+1} - \underline{x}_k$ is an integral multiple of the current h and the strict inequality $F(\underline{x}_{k+1}) < F(\underline{x}_k)$ is achieved. Thirdly, $\underline{x}_{k+1} = \underline{x}_k$ occurs only if $\underline{x}_k$ has the property

$$F(\underline{x}_k) \le F(\underline{x}), \qquad \underline{x} \in \mathcal{N}_h(\underline{x}_k), \tag{3.8}$$

where $\mathcal{N}_h$ is the $\mathcal{N}$ that has been defined already, the subscript denoting the mesh size. Fourthly, the kth iteration does not reduce h if $F(\underline{x}_{k+1})$ is less than $F(\underline{x}_k)$, but otherwise the mesh size of the $(k+1)$th iteration is the mesh size of the kth iteration multiplied by τ.

The main conclusion of Torczon (1997) is that, if the objective function has continuous first derivatives and if the number of iterations is infinite, then the given conditions imply the convergence property

$$\liminf_{k\to\infty} \|\underline{\nabla} F(\underline{x}_k)\|_2 = 0. \tag{3.9}$$

It can be proved by letting $\mathcal{K}$ be the set of integers k such that the mesh size of the $(k{+}1)$th iteration is less than all previous mesh sizes, and by deducing that the number of elements of $\mathcal{K}$ is infinite. Indeed, if we suppose that $|\mathcal{K}|$ is finite, then we can let ℓ and m be integers such that every mesh size is in the set $\{\tau^i : i{=}\ell, \ell{+}1, \ldots, m\}$, where $\ell{\le}0{\le}m$, the existence of ℓ being due to the boundedness of the level set $\{\underline{x} : F(\underline{x}){\le}F(\underline{x}_1)\}$. It follows from $\tau{=}\beta/\alpha$ that, for every iteration number k, the components of the difference $\underline{x}_{k+1}{-}\underline{x}_1$ are integral multiples of the positive constant $\beta^\ell\alpha^{-m}$. Thus, due to the boundedness of the level set, the monotonically decreasing sequence $\{F(\underline{x}_k) : k{=}1,2,3,\ldots\}$ satisfies $F(\underline{x}_{k+1}){<}F(\underline{x}_k)$ for only a finite number of values of k. Therefore there exists k_0 such that the kth iteration multiplies h by τ for all $k{\ge}k_0$, which contradicts the hypothesis $|\mathcal{K}|{<}\infty$. Hence the sequence $\{\underline{x}_k : k{\in}\mathcal{K}\}$ is infinite. Further, the given conditions and choices imply that, for every positive integer j, the jth element of this sequence has the property (3.8) with $h{=}\tau^{j-1}$. Thus the argument of the paragraph containing expressions (3.6) and (3.7) gives the bound

$$|\partial F(\underline{x}_k)/\partial x_i| \le \omega_i(\tau^{j-1}), \qquad i{=}1,2,\ldots,n. \tag{3.10}$$

Therefore the infinite sequence $\{\|\underline{\nabla} F(\underline{x}_k)\|_2 : k{\in}\mathcal{K}\}$ converges to zero, which completes the proof of the assertion (3.9).

This method of proof does not show, however, that a small value of $\|\underline{\nabla} F(\underline{x}_k)\|_2$ is achieved by a number of iterations that is suitable for practical computation. For example, if $\tau{=}2/3$, if $h{\approx}10^{-6}$ is needed to provide adequate accuracy, and if $h \approx 100$ is the greatest mesh size that occurs, then the formulae $(2/3)^{34} = 1.03{\times}10^{-6}$ and $(3/2)^{11} = 86.5$ imply that the components of $\underline{x}_{k+1} - \underline{x}_1$ should be integral multiples of $(1/3)^{34}(1/2)^{11} = 2.93 \times 10^{-20} = \eta_{\min}$, say. Then it is crucial to the last paragraph that $\eta_{\min}$ is a positive constant, although its value is less than the usual relative precision of computer arithmetic. Therefore it is hoped that there is no close relation between performance in practice and the importance of the rationality of τ to the given analysis. A strong advantage of discrete grid methods is emphasized by Torczon (1997). It is that any trial step from $\underline{x}_k$ to another grid point is allowed by the convergence theory if it provides the strict inequality $F(\underline{x}_{k+1}){<}F(\underline{x}_k)$. On the other hand, proofs of convergence of several other algorithms require the reduction in the objective function to be 'sufficiently large'.

4. Simplex methods

A simplex is the convex hull of $n+1$ points in $\mathbb{R}^n$, where the points satisfy the nondegeneracy condition that the volume of the hull is nonzero. This condition is equivalent to the linear independence of the vectors $\underline{v}_i - \underline{v}_1$, $i = 2,3,\ldots,n+1$, where $\underline{v}_i \in \mathbb{R}^n$, $i = 1,2,\ldots,n+1$, are the $n+1$ points. Simplex methods for unconstrained optimization are iterative. A simplex is available at the beginning of each iteration, and we let $\underline{v}_i$, $i=1,2,\ldots,n+1$, be its vertices. Further, the values $F(\underline{v}_i)$, $i=1,2,\ldots,n+1$, of the objective function are known. The iteration picks an integer m from $[1,n+1]$ that is usually defined by the property

$$F(\underline{v}_m) \geq F(\underline{v}_i), \qquad i=1,2,\ldots,n+1, \tag{4.1}$$

in order that $F(\underline{v}_m)$ is the worst of the function values. Further, a new vector of variables, $\hat{\underline{v}}_m$ say, is generated, a frequent choice being the point

$$\hat{\underline{v}}_m = (2/n) \sum_{\substack{i=1 \\ i\neq m}}^{n+1} \underline{v}_i - \underline{v}_m, \tag{4.2}$$

because $\hat{\underline{v}}_m$ is on the straight line from $\underline{v}_m$ through the centroid of the other vertices. Then $F(\hat{\underline{v}}_m)$ is calculated. Its value may allow the iteration to be completed by letting the simplex for the next iteration be the current one, except that $\underline{v}_m$ is replaced by $\hat{\underline{v}}_m$. Another possibility is that $\hat{\underline{v}}_m$ may be revised in a way that depends on the new function value, and then $F(\hat{\underline{v}}_m)$ is required at the revised point. Moreover, it happens occasionally that the new information causes the current simplex to shrink, in which case $\underline{v}_i$ is overwritten by $\frac{1}{2}(\underline{v}_i + \underline{v}_\ell)$ for $i = 1,2,\ldots,n+1$, where ℓ is an integer from $[1,n+1]$ such that $F(\underline{v}_\ell)$ is the least of the numbers $F(\underline{v}_i)$, $i=1,2,\ldots,n+1$. Thus the new current simplex has n new vertices, and the objective function has to be calculated at all of them before the next iteration is begun.

The first algorithm of this kind was proposed by Spendley, Hext and Himsworth (1962). We consider a version of their method that employs formula (4.2) on every iteration. If $F(\hat{\underline{v}}_m) \geq F(\underline{v}_m)$ occurs, then the calculation is terminated if the simplex has become sufficiently small, and otherwise the iteration shrinks the simplex. Alternatively, when $F(\hat{\underline{v}}_m)$ is less than $F(\underline{v}_m)$, then the only modification to the current simplex for the next iteration is the replacement of $\underline{v}_m$ by $\hat{\underline{v}}_m$. A small departure from the procedure of the previous paragraph assists the achievement of the condition $F(\hat{\underline{v}}_m) < F(\underline{v}_m)$. Specifically, if the usual choice of m would cause the vector (4.2) to be the vertex that was deleted from the previous simplex by the previous iteration, then m is chosen to be an integer in $[1,n+1]$ such that $F(\underline{v}_m)$ is the second largest of the numbers $F(\underline{v}_i)$, $i=1,2,\ldots,n+1$. Thus the same value of m is not picked by two consecutive iterations, unless a shrink is applied by the earlier iteration. Furthermore, this algorithm has the property that every simplex is regular if the initial simplex is regular, where a simplex is regular if and only if all its edges have the same length.

The last remark is relevant to an analysis of convergence in the general case. Indeed, let the method of the previous paragraph be applied for several iterations to the function $F(\underline{x})$, $\underline{x} \in \mathbb{R}^n$, where the simplex at the beginning of the calculation has the vertices $\underline{v}_i^{(0)}$, $i=1,2,\ldots,n+1$. Further, let $\underline{w}_i^{(0)}$, $i=1,2,\ldots,n+1$, be the vertices of any regular simplex in $\mathbb{R}^n$, and then define the $n \times n$ nonsingular matrix B by the equations

$$\underline{w}_i^{(0)} - \underline{w}_1^{(0)} = B\,(\underline{v}_i^{(0)} - \underline{v}_1^{(0)}), \qquad i=2,3,\ldots,n+1. \tag{4.3}$$

We now consider applying the method of the last paragraph to the function $G(\underline{y}) = F(\underline{v}_1^{(0)} + B^{-1}\{\underline{y} - \underline{w}_1^{(0)}\})$, $\underline{y} \in \mathbb{R}^n$, starting with the regular simplex just mentioned. The identities $G(\underline{w}_i^{(0)}) = F(\underline{v}_i^{(0)})$, $i=1,2,\ldots,n+1$, imply that the initial choice of m is the same as before. Thus, for the new calculation, formula (4.2) provides the vector

$$\hat{\underline{w}}_m^{(0)} = (2/n) \sum_{\substack{i=1 \\ i \neq m}}^{n+1} \underline{w}_i^{(0)} - \underline{w}_m^{(0)}. \tag{4.4}$$

Further, if $\hat{\underline{v}}_m^{(0)}$ is the vector (4.2) that occurs on the first iteration of the original calculation, then $\hat{\underline{w}}_m^{(0)}$ is related to $\hat{\underline{v}}_m^{(0)}$ by the identity

$$\begin{aligned} &\underline{v}_1^{(0)} + B^{-1}\left\{\hat{\underline{w}}_m^{(0)} - \underline{w}_1^{(0)}\right\} \\ &\quad = \underline{v}_1^{(0)} + B^{-1}\Big\{ (2/n) \sum_{\substack{i=1 \\ i \neq m}}^{n+1} (\underline{w}_i^{(0)} - \underline{w}_1^{(0)}) - (\underline{w}_m^{(0)} - \underline{w}_1^{(0)}) \Big\} \\ &\quad = \underline{v}_1^{(0)} + (2/n) \sum_{\substack{i=1 \\ i \neq m}}^{n+1} (\underline{v}_i^{(0)} - \underline{v}_1^{(0)}) - (\underline{v}_m^{(0)} - \underline{v}_1^{(0)}) \;=\; \hat{\underline{v}}_m^{(0)}. \end{aligned} \tag{4.5}$$

It follows that the function values $G(\hat{\underline{w}}_m^{(0)})$ and $F(\hat{\underline{v}}_m^{(0)})$ are the same. Hence the simplex at the beginning of the second iteration of the new calculation is related to the simplex at the beginning of the second iteration of the old calculation by the change of variables $\underline{y} = \underline{w}_1^{(0)} + B(\underline{x} - \underline{v}_1^{(0)})$, $\underline{x} \in \mathbb{R}^n$. Further, it can be shown by induction that this relation is inherited by the simplices of all iterations. Therefore, when analysing convergence, there is no loss of generality in assuming that every simplex is regular. It can be shown similarly that there is also no loss of generality in making any nondegenerate choice of the initial simplex.

One fundamental convergence question is whether the method of the second paragraph of this section can continue indefinitely without any shrinks, when the level set $\{\underline{x} : F(\underline{x}) \le F_0\}$ is bounded, where F_0 is the largest of the function values at the vertices of the initial simplex. The grid argument of the previous section provides a negative answer when $n=2$. Indeed, in this case there is no loss of generality in letting the initial vertices have the components $\underline{v}_1^{(0)}=(0,0)$, $\underline{v}_2^{(0)}=(1,0)$ and $\underline{v}_3^{(0)}=(0,1)$. Thus it follows from formula (4.2) with $n=2$ that, until the first shrink occurs, the components of every vertex of every simplex are integers. Moreover, every iteration before a shrink provides a strict reduction in the sum $\sum_{i=1}^{n+1} F(\underline{v}_i)$, where the points $\underline{v}_i$, $i=1,2,\ldots,n+1$, are still the vertices of the current simplex. Therefore the shrink is guaranteed by the remark that the number of different values of this sum subject to the conditions $F(\underline{v}_i) \le F_0$ and $\underline{v}_i \in \mathbb{Z}^n$ is finite, where $\mathbb{Z}^n$ is the set of vectors in $\mathbb{R}^n$ whose components are integers.

When $n \ge 3$, however, then the recursive use of formula (4.2) can generate an infinite number of different vertices of simplices in bounded regions of $\mathbb{R}^n$. For example, let $n=3$, let the initial simplex be a regular tetrahedron, and let two of its vertices have the components $\underline{v}_1=(0,0,0)$ and $\underline{v}_2=(1,0,0)$. We consider the replacement of $\underline{v}_m$ by $\hat{\underline{v}}_m$ recursively in the case when $\underline{v}_1$ and $\underline{v}_2$ remain as vertices of every simplex that occurs. We know that all these simplices are regular tetrahedra and that any two consecutive ones have a common face. Therefore each replacement is equivalent to rotating the current simplex about the x-axis through an angle θ that satisfies $\cos\theta=1/3$ and $-\pi/2 < \theta < \pi/2$. We let every θ be positive. Hence all the simplices are different if and only if the ratio $k\theta/(2\pi)$ is not an integer for every positive integer k. This happens because such values of k are excluded by the conditions $\cos\theta = 1/3$ and $\cos(k\theta)=1$. Indeed, if $\cos(k\theta)$ is expressed as a polynomial of degree k in $\cos\theta$, then the coefficient of $(\cos\theta)^k$ is 2^{k-1}, which is not divisible by three, and all the coefficients are integers. Therefore $\cos\theta = 1/3$ cannot be an exact solution of $\cos(k\theta) = 1$, which completes the analysis of the example. I do not know the answer to the question whether, for $n \ge 3$, the simplex method under consideration always includes a shrink in exact arithmetic, provided that the objective function has bounded level sets and suitable smoothness properties. On the other hand, a shrink is guaranteed in practice, because the number of different values of $F(\underline{x})$, $\underline{x} \in \mathbb{R}^n$, in computer arithmetic is finite.

Another simplex algorithm is proposed by Dennis and Torczon (1991). It has the property that, due to a grid argument of the kind given already, shrinks occur for any number of variables, provided that level sets of the objective function are bounded. Each iteration of this algorithm begins as before with a simplex whose vertices are $\underline{v}_i$, $i = 1,2,\ldots,n+1$, and with the function values $F(\underline{v}_i)$, $i=1,2,\ldots,n+1$. Then the basic version of an iteration finds an integer ℓ in $[1,n+1]$ such that $F(\underline{v}_\ell)$ is the least of these function values. Now the simplex that has the vertices

$$\hat{\underline{v}}_i = 2\,\underline{v}_\ell - \underline{v}_i, \qquad i=1,2,\ldots,n+1, \tag{4.6}$$

is congruent to the current one, and a switch to this simplex is likely to reduce the least function value at a vertex, because, for each i, the inequality $F(\underline{v}_\ell) \le F(\underline{v}_i)$ is satisfied, and $\underline{v}_\ell$ is the mid-point of the line segment from $\underline{v}_i$ to $\hat{\underline{v}}_i$. Therefore the algorithm calculates the objective function at the n vertices of the new simplex that are different from $\underline{v}_\ell = \hat{\underline{v}}_\ell$. It is possible, however, that the new function values fail to achieve the strict reduction

$$\min\{F(\hat{\underline{v}}_i) : i=1,2,\ldots,n+1\} < F(\underline{v}_\ell). \tag{4.7}$$

When this unsuccessful case occurs, then, as before, either the procedure is terminated because the current simplex is sufficiently small, or there is a shrink that overwrites $\underline{v}_i$ by $\frac{1}{2}(\underline{v}_i+\underline{v}_\ell)$ for $i=1,2,\ldots,n+1$. Alternatively, when condition (4.7) holds, then the basic iteration includes a test for doubling the size of the simplex by choosing the vertices $\check{\underline{v}}_i = 2\,\hat{\underline{v}}_i - \underline{v}_\ell$, $i=1,2,\ldots,n+1$, so the n function values $F(\check{\underline{v}}_i)$, $i \ne \ell$, are calculated too. Then the inequality

$$\min\{F(\check{\underline{v}}_i) : i=1,2,\ldots,n+1\} < \min\{F(\hat{\underline{v}}_i) : i=1,2,\ldots,n+1\} \tag{4.8}$$

is tried. If the conditions (4.7) and (4.8) are both satisfied, the points $\check{\underline{v}}_i$, $i=1,2,\ldots,n+1$, become the vertices of the simplex of the next iteration, but, in the remaining case when inequalities (4.7) and (4.8) hold and fail, respectively, the vertices of the new simplex are $\hat{\underline{v}}_i$, $i=1,2,\ldots,n+1$. The fact that each iteration requires $2n$ evaluations of the objective function may be an advantage in practice, because the algorithm is designed to be suitable for parallel machines.

When analysing the convergence of this method, there is no loss of generality in selecting every vertex of the initial simplex from $\mathbb{Z}^n$. Hence, if σ_k and τ_k are the number of times that the size of the simplex is shrunk or doubled, respectively, during the first k iterations, and if μ_k is the greatest of the differences $\sigma_j - \tau_j$, $j=0,1,\ldots,k-1$, where $\sigma_0=\tau_0=0$, then, at the beginning of the kth iteration, every component of every vertex of the current simplex is an integer multiple of $2^{-\mu_k}$. Thus, under the usual assumption that the set $\{\underline{x} : F(\underline{x}) \le F_0\}$ is bounded, a finite upper bound on the increasing sequence of integers $\{\mu_k : k=1,2,3,\ldots\}$ would provide a contradiction if the number of iterations were infinite. Therefore a simplex that is small enough to give termination occurs eventually. When this happens, the algorithm provides the relations

$$F(\underline{v}_\ell) \le F(\underline{v}_i), \quad F(\underline{v}_\ell) \le F(\hat{\underline{v}}_i) \quad \text{and} \quad \underline{v}_\ell = \tfrac{1}{2}(\underline{v}_i+\hat{\underline{v}}_i), \quad i=1,2,\ldots,n+1. \tag{4.9}$$

Further, the shape of the current simplex does not degenerate during the calculation because the shapes of all the simplices are the same. Thus the conditions (4.9) are analogous to expression (3.8). It follows that, if the objective function is continuously differentiable, and if the property $\|\underline{\nabla}F(\underline{v}_\ell)\|_2 \leq \varepsilon$ is required in exact arithmetic, where ε is any prescribed positive number, then it is suitable to let the final simplex be sufficiently small in the test for termination.

The algorithm in the second paragraph of this section possesses a similar property, except that, for $n \geq 3$, it is an assumption that enough shrinks are made to achieve the termination condition for any choice of the final size of the simplex. The following analysis gives an explanation of this property in the case when every simplex is regular. Let termination occur because $F(\hat{\underline{v}}_m) \geq F(\underline{v}_m)$ holds, and because the lengths of the edges of the current simplex are at most δ, which is a prescribed positive number. We may assume that the inequalities (4.1) are satisfied, because, if the alternative choice of m were made, then a switch to the usual choice would cause the new $\hat{\underline{v}}_m$ to be the vertex of the previous simplex that was replaced by $\underline{v}_m$, so the condition $F(\hat{\underline{v}}_m) \geq F(\underline{v}_m)$ is retained. We let $\underline{v}_*$ and $\underline{g}_*$ be the vectors $n^{-1}\sum_{i=1,\,i\neq m}^{n+1} \underline{v}_i$ and $\underline{\nabla}F(\underline{v}_*)$, respectively. Therefore, because the objective function has a continuous gradient in a bounded level set, the formula

$$F(\underline{x}) = F(\underline{v}_*) + (\underline{x}-\underline{v}_*)^T \underline{g}_* + o(\|\underline{x}-\underline{v}_*\|_2), \tag{4.10}$$

is valid for all points $\underline{x}$ that are of interest. Hence, by letting $\underline{x}$ run through the vertices of the current simplex that are different from $\underline{v}_m$ and averaging, we deduce the equation

$$F(\underline{v}_*) = (1/n) \sum_{\substack{i=1 \\ i\neq m}}^{n+1} F(\underline{v}_i) + o(\delta). \tag{4.11}$$

It follows from expression (4.1) that $F(\underline{v}_m) - F(\underline{v}_*)$ is at least $o(\delta)$, so expression (4.10) implies $(\underline{v}_m - \underline{v}_*)^T \underline{g}_* \geq o(\delta)$. Similarly, $F(\hat{\underline{v}}_m) \geq F(\underline{v}_m)$ implies $(\hat{\underline{v}}_m - \underline{v}_*)^T \underline{g}_* \geq o(\delta)$. Therefore, by writing equation (4.2) in the form $\hat{\underline{v}}_m - \underline{v}_* = -(\underline{v}_m - \underline{v}_*)$, we establish $(\underline{v}_m - \underline{v}_*)^T \underline{g}_* = o(\delta)$. Furthermore, by combining this result with conditions (4.1) and (4.10), we find the bounds

$$\begin{aligned}(\underline{v}_i - \underline{v}_*)^T \underline{g}_* &= F(\underline{v}_i) - F(\underline{v}_*) + o(\delta) \leq F(\underline{v}_m) - F(\underline{v}_*) + o(\delta) \\ &= (\underline{v}_m - \underline{v}_*)^T \underline{g}_* + o(\delta) = o(\delta), \quad i=1,2,\ldots,n+1.\end{aligned} \tag{4.12}$$

Let j be any integer in $[1, n+1]$ that is different from m, and let $\hat{\underline{v}}_j$ be the point where the straight line from $\underline{v}_j$ through $\underline{v}_*$ leaves the current simplex. Then $\|\hat{\underline{v}}_j - \underline{v}_*\|_2$ is of magnitude δ and, because $\hat{\underline{v}}_j$ is in the convex hull of the vertices, it has the form $\hat{\underline{v}}_j = \sum_{i=1}^{n+1} \theta_i \underline{v}_i$, where the multipliers θ_i,

$i=1,2,\ldots,n+1$, are nonnegative and sum to one. Thus the bounds (4.12) give the inequality

$$(\underline{\hat{v}}_j-\underline{v}_*)^T\underline{g}_* = \sum_{i=1}^{n+1}\theta_i\,(\underline{v}_i-\underline{v}_*)^T\underline{g}_* \le o(\delta). \tag{4.13}$$

Further, because the directions $\underline{v}_j-\underline{v}_*$ and $\underline{\hat{v}}_j-\underline{v}_*$ are exactly opposite to each other, expressions (4.12) and (4.13) imply the stronger condition $(\underline{v}_j-\underline{v}_*)^T\underline{g}_* = o(\delta)$, which holds for every integer j in $[1,n+1]$. Therefore $\underline{g}_*$ satisfies the equations

$$(\underline{v}_i-\underline{v}_1)^T\underline{g}_* = o(\delta), \qquad i=2,3,\ldots,n+1. \tag{4.14}$$

Thus all of the ratios $(\underline{v}_i-\underline{v}_1)^T\underline{g}_* \,/\, \|\underline{v}_i-\underline{v}_1\|_2$, $i=2,3,\ldots,n+1$, tend to zero if the lengths of the edges of the simplices are shrunk to zero. It follows from the uniform linear independence of the vectors $\underline{v}_i-\underline{v}_1$, $i=2,3,\ldots,n+1$, that $\|\underline{g}_*\|_2$ tends to zero too. Hence, for any positive number ε, the termination condition implies $\|\underline{g}_*\|_2\le\varepsilon$ in exact arithmetic, provided that δ is sufficiently small.

We consider also the algorithm of Nelder and Mead (1965), because it has become the most popular simplex method in practice for unconstrained optimization. The introduction to that paper states 'In the method to be described the simplex adapts itself to the local landscape, elongating down long inclined planes, changing direction on encountering a valley at an angle, and contracting in the neighbourhood of a minimum'. These properties are often, but not always, achieved in numerical experiments. An iteration of the algorithm, with typical parameter values, has all the features mentioned in the opening paragraph of this section, the other details being as follows. Let $F(\underline{v}_\ell)$ and $F(\underline{v}_{\hat{m}})$ be the least and second largest of the numbers $F(\underline{v}_i)$, $i=1,2,\ldots,n+1$. If the function value at the point (4.2) satisfies the conditions $F(\underline{v}_\ell)\le F(\underline{\hat{v}}_m)<F(\underline{v}_{\hat{m}})$, then $\underline{v}_m$ is overwritten by $\underline{\hat{v}}_m$ and the next iteration is begun. Otherwise, if $F(\underline{\hat{v}}_m)<F(\underline{v}_\ell)$ occurs, then elongation is attempted by calculating $F(\underline{\check{v}}_m)$, where $\underline{\check{v}}_m$ is the point $\underline{\hat{v}}_m+\frac{1}{2}(\underline{\hat{v}}_m-\underline{v}_m)$. Further, $\underline{v}_m$ is replaced by $\underline{\check{v}}_m$ or $\underline{\hat{v}}_m$ in the case $F(\underline{\check{v}}_m)<F(\underline{\hat{v}}_m)$ or $F(\underline{\check{v}}_m)\ge F(\underline{\hat{v}}_m)$, respectively, which provides the simplex for the next iteration. In the remaining situation $F(\underline{\hat{v}}_m)\ge F(\underline{v}_{\hat{m}})$, the inequality $F(\underline{\hat{v}}_m)<F(\underline{v}_m)$ is tested. If it holds or fails then $\underline{\hat{v}}_m$ is overwritten by $\frac{3}{4}\underline{\hat{v}}_m+\frac{1}{4}\underline{v}_m$ or $\frac{1}{4}\underline{\hat{v}}_m+\frac{3}{4}\underline{v}_m$, respectively, and $F(\underline{\hat{v}}_m)$ is calculated at the new point. Then $\underline{\hat{v}}_m$ replaces $\underline{v}_m$ as before if and only if the new $F(\underline{\hat{v}}_m)$ is strictly less than $F(\underline{v}_m)$, the alternative being to shrink the simplex in the way described already. The calculation may be terminated when the lengths of the edges of the current simplex become less than a prescribed positive number. Some of these details are taken from Wright (1996).

That paper includes an excellent discussion of the limitations, disadvantages, successes and developments of the Nelder and Mead algorithm. The fact that literature searches show that it is the most used method for unconstrained optimization in practice is remarkable, because some severe cases of failure have been found. One source of trouble is that a sequence of iterations can cause the volume of the current simplex to tend to zero, when the length of the longest edge of the simplex is bounded below by a positive constant. In particular, let $n = 2$, let the vertices of the current simplex have the coordinates $(\alpha, 0)$, $(0, -1)$ and $(0, 1)$, where α is nonzero, and let F be any smooth function with the following properties. The form of F on the x-axis, namely $F(x, 0)$, $x \in \mathbb{R}$, is strictly convex and is least at $x = 0$. Further, the inequalities

$$F(0, -1) \le F(0, 0) \quad \text{and} \quad F(0, 1) \le F(0, 0) \tag{4.15}$$

are satisfied. Then formula (4.2) provides $\hat{\underline{v}}_m = (-\alpha, 0)$, which gives the $F(\hat{\underline{v}}_m) \ge F(\underline{v}_{\hat{m}})$ situation, so the components of $\hat{\underline{v}}_m$ are altered to $(-\frac{1}{2}\alpha, 0)$ or $(\frac{1}{2}\alpha, 0)$. Now the choice between these alternatives by the algorithm, the convexity condition and $F(0, 0) < F(\alpha, 0)$ imply that the new value of $F(\hat{\underline{v}}_m)$ is strictly less than $F(\underline{v}_m)$. Thus the outcome of the iteration is the replacement of α by $-\frac{1}{2}\alpha$ or $\frac{1}{2}\alpha$. Such iterations can continue indefinitely, and then the current simplex tends to have the vertices $(0, 0)$, $(0, -1)$ and $(0, 1)$. This result is particularly unfortunate, because the modulus of the second component of the gradient $\underline{\nabla} F(\underline{x})$ can be large at all the vertices of all the simplices that occur during the calculation.

The function $F(\underline{x})$, $\underline{x} \in \mathbb{R}^2$, can be convex in the example, provided that both of the conditions (4.15) hold as equations. Then the assumptions imply that the least value of the objective function occurs at the origin, so the performance of the algorithm is adequate. On the other hand, McKinnon (1997) constructs a more interesting case with $n = 2$, where the objective function is strictly convex and has continuous second derivatives, where the gradient vector $\underline{\nabla} F(0, 0)$ is nonzero, where the number of iterations of the Nelder and Mead (1965) algorithm is infinite, and where all the vertices of the current simplex tend to the origin. Further, this unacceptable behaviour occurs without any shrinks, which is possible because the directions of the edges of the current simplex tend to be orthogonal to $\underline{\nabla} F(0, 0)$. These examples are disturbing, and they provide some very strong reasons for questioning the current use of simplex methods for unconstrained optimization calculations.

5. Conjugate direction methods

As mentioned already, conjugate direction methods apply line searches, and they are designed to be efficient for unconstrained minimization when $F(\underline{x})$, $\underline{x} \in \mathbb{R}^n$, is a strictly convex quadratic function. Further, it is hoped that some of the efficiency of the quadratic case will be inherited when the objective function is general. We assume the quadratic form

$$F(\underline{x}) = F_0 + \underline{x}^T\underline{g}_0 + \tfrac{1}{2}\,\underline{x}^T A\,\underline{x}, \qquad \underline{x}\in\mathbb{R}^n, \tag{5.1}$$

when defining conjugate directions, where A is a constant positive definite symmetric matrix. Specifically, the nonzero vectors $\underline{d}_i \in \mathbb{R}^n$ and $\underline{d}_j \in \mathbb{R}^n$ are conjugate if and only if they satisfy the equation

$$\underline{d}_i^T A\,\underline{d}_j = 0. \tag{5.2}$$

The most useful consequence of conjugacy is that, if F is the quadratic function (5.1), if the kth iteration of an algorithm is making a line search from $\underline{x}_k$ along the direction $\underline{d}_k$, and if $\underline{d}$ is any direction conjugate to $\underline{d}_k$, then the scalar product $\underline{d}^T\underline{\nabla}F(\underline{x}_k{+}\alpha\,\underline{d}_k)$ is independent of the step-length $\alpha\in\mathbb{R}$ of the line search, which can be verified easily. In particular, if $\underline{d}^T\underline{\nabla}F(\underline{x}_k)=0$ holds, then it follows from formula (2.1) that $\underline{d}^T\underline{\nabla}F(\underline{x}_{k+1})=0$ is achieved too. Moreover, an exact line search along $\underline{d}_k$ provides $\underline{d}_k^T\underline{\nabla}F(\underline{x}_{k+1})=0$. Thus, if each iteration makes an exact line search, and if each search direction is conjugate to all the previous search directions, then $\underline{x}_{k+1}$ enjoys the property

$$\underline{d}_j^T\underline{\nabla}F(\underline{x}_{k+1}) = 0, \qquad j=1,2,\ldots,k, \tag{5.3}$$

for every positive integer k. Further, if $\underline{\nabla}F(\underline{x}_{k+1})$ is nonzero, then the possible choice $\underline{d}_{k+1} = -A^{-1}\underline{\nabla}F(\underline{x}_{k+1})$ shows that there exists a $\underline{d}_{k+1}$ that is conjugate to $\underline{d}_j$, $j=1,2,\ldots,k$. On the other hand, it follows from the positive definiteness of A that directions that are conjugate to each other are linearly independent. Therefore $\underline{\nabla}F(\underline{x}_{k+1})=0$ occurs in equation (5.3) if k attains the value n, and then $\underline{x}_{k+1}$ is the optimal vector of variables.

The remarks made so far are well known, and they are the basis of some very successful algorithms for unconstrained optimization when the gradient $\underline{\nabla}F(\underline{x})$, $\underline{x}\in\mathbb{R}^n$, can be calculated. We are assuming, however, that no derivatives are available. Therefore we turn our attention to the construction of search directions from function values in ways that provide some useful conjugacy properties if F happens to be quadratic.

The following technique is highly useful. If F is a strictly convex quadratic function, and if $\underline{x}_p$ and $\underline{x}_q$ are different points of $\mathbb{R}^n$ that satisfy $\underline{d}^T\underline{\nabla}F(\underline{x}_p)=0$ and $\underline{d}^T\underline{\nabla}F(\underline{x}_q)=0$, for some nonzero vector $\underline{d}$, then the choice $\underline{d}_q=\underline{x}_q-\underline{x}_p$ provides a search direction that is conjugate to $\underline{d}$. It is true because equation (5.1) implies the identity

$$\underline{d}^T\underline{\nabla}F(\underline{x}_q)-\underline{d}^T\underline{\nabla}F(\underline{x}_p) = \underline{d}^T(A\,\underline{x}_q{+}\underline{g}_0)-\underline{d}^T(A\,\underline{x}_p{+}\underline{g}_0) = \underline{d}^TA\,(\underline{x}_q{-}\underline{x}_p), \tag{5.4}$$

and it allows sequences of mutually conjugate directions to be formed. Indeed, let the directions $\underline{d}^{(j)}$, $j=1,2,\ldots,\ell$, be mutually conjugate, where ℓ is an integer from $[1,n-1]$, and suppose that we require a direction $\underline{d}^{(\ell+1)}$ that is conjugate to $\underline{d}^{(j)}$, $j=1,2,\ldots,\ell$. Then we pick integers p and q that satisfy $p \ge \ell+1$ and $q \ge p+\ell+1$, and we let the sequence of points $\underline{x}_k$, $k=1,2,3,\ldots$, be generated by an algorithm that makes exact line searches on every iteration. Further, we include the search directions

$$\underline{d}_{p-\ell-1+j} = \underline{d}_{q-\ell-1+j} = \underline{d}^{(j)}, \qquad j=1,2,\ldots,\ell. \tag{5.5}$$

Then, corresponding to equation (5.3), the line searches and conjugacy provide the identities

$$\underline{d}_{p-\ell-1+j}^T \underline{\nabla} F(\underline{x}_p)=0 \quad \text{and} \quad \underline{d}_{q-\ell-1+j}^T \underline{\nabla} F(\underline{x}_q)=0, \qquad j=1,2,\ldots,\ell, \tag{5.6}$$

in the quadratic case. Therefore the remark at the beginning of this paragraph is applicable if $\underline{d}$ is any of the vectors $\underline{d}^{(j)}$, $j=1,2,\ldots,\ell$. It follows that the direction $\underline{d}^{(\ell+1)} = \underline{x}_q - \underline{x}_p$ satisfies all the required conjugacy conditions, provided that $\underline{x}_q$ is different from $\underline{x}_p$. A generalization of this method helps to achieve $\underline{x}_q \neq \underline{x}_p$. It is that the step from $\underline{x}_p$ to $\underline{x}_{p+1}$ can be arbitrary, because $q \ge p+\ell+1$ implies that $\underline{d}_p$ is not included in expression (5.5).

This construction allows the least value of a strictly convex quadratic function to be calculated using $\frac{1}{2}n(n+1)$ line searches and $n-1$ displacements, as suggested by Smith (1962). A version of his method lets $\underline{s}^{(j)}$, $j=1,2,\ldots,n$, be any linearly independent vectors in $\mathbb{R}^n$, and it forms a sequence of mutually conjugate directions $\underline{d}^{(j)}$, $j=1,2,\ldots,n$, such that, for each integer ℓ in $[1,n]$, the direction $\underline{d}^{(\ell)}$ is in the linear space spanned by $\underline{s}^{(j)}$, $j=1,2,\ldots,\ell$. Therefore $\underline{d}^{(1)}=\underline{s}^{(1)}$ and $\ell=1$ are set initially, and $\underline{x}_1$ is any given point in $\mathbb{R}^n$. Then $\underline{x}_2$ is calculated by an exact line search from $\underline{x}_1$ along the direction $\underline{d}^{(1)}$. These starting conditions are a special case of the fact that, when $\underline{d}^{(j)}$, $j=1,2,\ldots,\ell$, are known for any ℓ in $[1,n]$, then a point $\underline{x}_p$ is available that satisfies the first part of expression (5.6) for the search directions $\underline{d}_{p-\ell-1+j}$, $j=1,2,\ldots,\ell$, the value of p being $\frac{1}{2}\ell(\ell+3)$, which includes $p=2$ when $\ell=1$. For each $\ell<n$, the method generates a point $\underline{x}_q$, different from $\underline{x}_p$, that achieves the second part of expression (5.6), in order that the choice $\underline{d}^{(\ell+1)}=\underline{x}_q-\underline{x}_p$ has the required conjugacy properties. Specifically, $\underline{x}_{p+1}$ is any point of the form $\underline{x}_p+\alpha_p\,\underline{s}^{(\ell+1)}$, where α_p is nonzero, and then, for $k=p+1,p+2,\ldots,p+\ell$, the new vector of variables $\underline{x}_{k+1}$ is obtained by an exact line search from $\underline{x}_k$ in the direction $\underline{d}_k=\underline{d}^{(k-p)}$, which provides $q=p+\ell+1$. There is also an exact line search from $\underline{x}_q$ in the direction $\underline{d}^{(\ell+1)}=\underline{x}_q-\underline{x}_p$, giving the point $\underline{x}_{q+1}=\underline{x}_{p+\ell+2}$. The conjugacy conditions allow this point to be the new $\underline{x}_p$ when ℓ is increased by one, and the identity

$$\tfrac{1}{2}\,\ell\,(\ell+3)+\ell+2=\tfrac{1}{2}\,(\ell+1)\,(\ell+4) \tag{5.7}$$

shows that $p=\frac{1}{2}\ell(\ell+3)$ is preserved. The method is applied recursively until $\ell=n$ is attained. Indeed, $\underline{x}_p$ is optimal when $\ell=n$, because it is the result of n consecutive line searches along mutually conjugate directions.

This method can be applied when F is a general function, using a practical line search method that would be exact if F were quadratic. Then we call the technique of the last paragraph a 'cycle', because it is usual to employ several cycles, where the final point $\underline{x}_p$ of each cycle becomes the starting point $\underline{x}_1$ of the next cycle. An unsatisfactory feature of this algorithm, however, is that, for each integer ℓ in $[1,n]$, a cycle makes just $n+1-\ell$ line searches along $\underline{d}^{(\ell)}$, so the search direction $\underline{d}^{(n)}$ occurs only once in the $\frac{1}{2}n(n+1)$ line searches. Therefore Powell (1964) suggests the following procedure, where each cycle includes only n or $n+1$ line searches, using directions that span the full space of the variables.

A cycle requires a starting point $\underline{x}_1$ and n linearly independent directions $\underline{d}_j$, $j=1,2,\ldots,n$, say, that are not expected to have any conjugacy properties initially in the quadratic case. For $k=1,2,\ldots,n$, the point $\underline{x}_{k+1}$ is generated by a line search from $\underline{x}_k$ along $\underline{d}_k$. Then the function value $F(\underline{x}_{n+1}+\underline{d}_{n+1})$, where $\underline{d}_{n+1}=\underline{x}_{n+1}-\underline{x}_1$, is calculated for a test that is specified in the next paragraph. If the test fails, the next cycle is given the current search directions, and its starting point is the current $\underline{x}_{n+1}$. Otherwise, if the test succeeds, then $\underline{x}_{n+2}$ is generated by a line search from $\underline{x}_{n+1}$ along $\underline{d}_{n+1}$, and $\underline{x}_{n+2}$ becomes the starting point of the next cycle. Further, the search directions of the next cycle are obtained by deleting one of the first n vectors from the sequence $\underline{d}_1,\underline{d}_2,\ldots,\underline{d}_{n+1}$, without changing the order of the sequence. A reasonable way of picking the vector that is deleted is also given in the next paragraph. For the moment, however, we make the ideal assumptions that F is a strictly convex quadratic function, that all the line searches are exact, that the test that has been mentioned never fails, that the deleted vector is always $\underline{d}_1$, that the search directions $\underline{d}_j$, $j=2,3,\ldots,n+1$, of every cycle are linearly independent, and that the number of cycles is n. Then, by considering the construction of $\underline{d}^{(\ell+1)}$ in the paragraph that includes expressions (5.4)–(5.6), we deduce that, at the end of the jth cycle, the last j of the directions $\underline{d}_1,\underline{d}_2,\ldots,\underline{d}_{n+1}$ are mutually conjugate, where j is any integer in $[1,n]$. This statement is trivial for $j=1$ and, using induction, we suppose that it is true for $j\in[1,n-1]$. In this case, corresponding to equation (5.6), the points $\underline{x}_1$ and $\underline{x}_{n+1}$ of the $(j+1)$th cycle have the properties

$$\underline{d}_k^T\underline{\nabla}F(\underline{x}_1)=0 \quad\text{and}\quad \underline{d}_k^T\underline{\nabla}F(\underline{x}_{n+1})=0, \qquad k=n,n-1,\ldots,n-j+1, \quad (5.8)$$

where $\underline{d}_k$ is now the kth search direction of the $(j+1)$th cycle. Thus the vector $\underline{d}_{n+1}=\underline{x}_{n+1}-\underline{x}_1$, which does not vanish because of the linear independence assumption, is conjugate to $\underline{d}_k$, $k=n,n-1,\ldots,n-j+1$, which establishes the inductive hypothesis. It follows that the last n search directions of the nth cycle are mutually conjugate. Therefore the optimal vector of variables is calculated, the total number of line searches of all the cycles being $n(n+1)$.

Let m be the integer in $[1, n+1]$ such that the cycle of the previous paragraph removes $\underline{d}_m$ from the sequence $\underline{d}_1, \underline{d}_2, \ldots, \underline{d}_{n+1}$, in order to provide the search directions for the next cycle. Therefore $m = n+1$ occurs if and only if the test to be described fails. The purpose of the test is to make the new directions as linearly independent as possible in the quadratic case. We define the 'Euclidean measure' of linear independence of the vectors $\underline{s}_j \in \mathbb{R}^n$, $j = 1, 2, \ldots, n$, to be the modulus of the determinant of the matrix with the columns $\underline{s}_j / \|\underline{s}_j\|_2$, $j = 1, 2, \ldots, n$. Moreover, when F is the quadratic function (5.1) and when its second derivative matrix A is positive definite, we define the 'natural measure' to be the modulus of the determinant of the matrix with the columns $\underline{s}_j / (\underline{s}_j^T A\, \underline{s}_j)^{1/2}$, $j = 1, 2, \ldots, n$. The method of Powell (1964) chooses m in a way that maximizes the 'natural measure' of the directions that are retained. In order to specify this choice, we deduce from the identity

$$\underline{d}_{n+1} = \underline{x}_{n+1} - \underline{x}_1 = \sum_{k=1}^{n} \alpha_k\, \underline{d}_k \tag{5.9}$$

that, if $m \leq n$ occurs in the quadratic case, then the natural measure of the retained directions is the natural measure of $\underline{d}_j$, $j = 1, 2, \ldots, n$, multiplied by the number

$$\beta_m = \alpha_m\, (\underline{d}_m^T A\, \underline{d}_m)^{1/2} / \, (\underline{d}_{n+1}^T A\, \underline{d}_{n+1})^{1/2}, \tag{5.10}$$

but we require an expression for this quantity in terms of the available function values. Now, for each integer k in $[1, n]$, the point $\underline{x}_{k+1}$ is calculated by an exact line search along $\underline{d}_k$, which provides the formula

$$F(\underline{x}_k) - F(\underline{x}_{k+1}) = \tfrac{1}{2}\, (\underline{x}_k - \underline{x}_{k+1})^T A\, (\underline{x}_k - \underline{x}_{k+1}) = \tfrac{1}{2}\, \alpha_k^2\, \underline{d}_k^T A\, \underline{d}_k \tag{5.11}$$

in the quadratic case. We also make use of the fact that the function (5.1) has the property

$$F(\underline{x}_{n+1} - \underline{d}_{n+1}) - 2\, F(\underline{x}_{n+1}) + F(\underline{x}_{n+1} + \underline{d}_{n+1}) = \underline{d}_{n+1}^T A\, \underline{d}_{n+1}. \tag{5.12}$$

Therefore m is picked by the following procedure for all functions $F(\underline{x})$, $\underline{x} \in \mathbb{R}^n$. We let $\hat{m}$ be a value of k in $[1, n]$ that maximizes the difference $F(\underline{x}_k) - F(\underline{x}_{k+1})$, $k = 1, 2, \ldots, n$. The choice $m = \hat{m}$ is made if the test

$$F(\underline{x}_{\hat{m}}) - F(\underline{x}_{\hat{m}+1}) > \tfrac{1}{2}\, F(\underline{x}_1) - F(\underline{x}_{n+1}) + \tfrac{1}{2}\, F(\underline{x}_{n+1} + \underline{d}_{n+1}) \tag{5.13}$$

is satisfied, which provides $\beta_m > 1$ in the quadratic case, but $m = n+1$ is selected otherwise. It can be deduced from the condition $F(\underline{x}_{\hat{m}}) - F(\underline{x}_{\hat{m}+1}) \leq F(\underline{x}_1) - F(\underline{x}_{n+1})$ that the test (5.13) is equivalent to the one that is used in Powell (1964).

A property of the 'natural measure' makes it appropriate for the present application. The property is that, if F is the strictly convex quadratic function (5.1), then the natural measure of the directions $\underline{d}_k \in \mathbb{R}^n$, $k = 1, 2, \ldots, n$, is greatest if and only if the directions are mutually conjugate. In order to prove this assertion, we assume without loss of generality that the lengths of the directions satisfy $\underline{d}_k^T A\, \underline{d}_k = 1$, $k=1,2,\ldots,n$. Therefore the natural measure is $|\det D|$, where D is the $n\times n$ matrix with the columns $\underline{d}_k$, $k=1,2,\ldots,n$. Further, if the directions are mutually conjugate, then D satisfies $D^T A\, D = I$, which implies $|\det D| = (\det A)^{-1/2}$. Thus the natural measure is the same for all sets of mutually conjugate directions. It follows that the assertion is true, provided that lack of conjugacy allows the natural measure to be increased. Therefore we assume $\underline{d}_1^T A\, \underline{d}_2 \neq 0$. If $\underline{d}_1$ and $\underline{d}_2$ are overwritten by $\underline{s}_1 = \cos\theta\, \underline{d}_1 - \sin\theta\, \underline{d}_2$ and $\underline{s}_2 = \sin\theta\, \underline{d}_1 + \cos\theta\, \underline{d}_2$, respectively, for any real θ, then $\det D$ does not change, but the natural measure of the directions is multiplied by $\{(\underline{s}_1^T A\, \underline{s}_1)(\underline{s}_2^T A\, \underline{s}_2)\}^{-1/2}$. This factor is greater than one when $\theta = \pi/4$, say, due to the elementary relation

$$\begin{aligned}(\underline{s}_1^T A\, \underline{s}_1)(\underline{s}_2^T A\, \underline{s}_2) &= \tfrac{1}{4}\{(\underline{d}_1 - \underline{d}_2)^T A\, (\underline{d}_1 - \underline{d}_2)\}\{(\underline{d}_1 + \underline{d}_2)^T A\, (\underline{d}_1 + \underline{d}_2)\} \\ &= (1 - \underline{d}_1^T A\, \underline{d}_2)(1 + \underline{d}_1^T A\, \underline{d}_2) = 1 - (\underline{d}_1^T A\, \underline{d}_2)^2 < 1, \quad (5.14)\end{aligned}$$

which completes the proof.

On the other hand, the 'Euclidean measure' of linear independence of the directions $\underline{d}_k$, $k=1,2,\ldots,n$, is greatest if and only if the directions are mutually orthogonal. It has the advantage over the 'natural measure' that it cannot become misleading in general calculations due to pathological features of the objective function, but it is less suitable as an aid for achieving conjugacy. It is possible, however, to maximize both measures simultaneously when F is the function (5.1), by letting the search directions be mutually orthogonal eigenvectors of the symmetric matrix A. The 'Praxis' algorithm of Brent (1973) takes advantage of this remark in the following way. The natural measure of linear independence is employed throughout, and also, after about every n^2 line searches, the matrix A is defined by the equation $D^T A\, D = I$, where the columns of D are still the vectors $\underline{d}_k$, $k=1,2,\ldots,n$. Further, these vectors are normalized so that the second derivative of the line search function $F(\underline{x}_k + \alpha\, \underline{d}_k)$, $\alpha \in \mathbb{R}$, is approximately one at $\alpha = \alpha_k$ for each k. Then the current search directions are replaced by mutually orthogonal eigenvectors of A. Another interesting use of eigenvectors occurs in the algorithm of Brodlie (1975). Here the technique for achieving the conjugacy condition $\underline{d}_p^T A\, \underline{d}_q = 0$, where p and q are any integers that satisfy $1 \le p < q \le n$, is analogous to the annihilation of the (p,q) off-diagonal matrix element in Jacobi's method for calculating the eigenvalues of an $n\times n$ symmetric matrix. Thus the search directions can remain mutually orthogonal, but, when F is quadratic and $n \geq 3$, it is usually not possible to calculate the least value of F in a finite number of iterations.

There is some convergence theory for the algorithms of this section when $F(\underline{x})$, $\underline{x} \in \mathbb{R}^n$, is a convex function that need not be quadratic. In particular, the analysis of Toint and Callier (1977) allows some freedom in the steplengths of the line searches.

6. Linear approximation methods

The changes to the variables in the simplex methods of Section 4 depend on the positions $\underline{v}_i$, $i=1,2,\ldots,n+1$, of the vertices of the current simplex, and on an integer m in $[1,n+1]$ that is usually defined by the conditions $F(\underline{v}_m)\geq F(\underline{v}_i)$, $i=1,2,\ldots,n+1$. These methods make no other use of $F(\underline{v}_i)$, $i=1,2,\ldots,n+1$, however, when choosing the next vector of variables for the calculation of the objective function, although the function values at the vertices can provide highly useful information when F is smooth. In particular, there is a unique linear polynomial from $\mathbb{R}^n$ to $\mathbb{R}$, Φ say, that satisfies the interpolation conditions

$$\Phi(\underline{v}_i) = F(\underline{v}_i), \qquad i=1,2,\ldots,n+1, \tag{6.1}$$

and often $\underline{\nabla}\Phi$ is very helpful for reducing the least calculated value of F. Therefore we will consider changes to the variables that are derived from Φ. The given procedures also allow constraints on the variables of the form

$$c_p(\underline{x}) \geq 0, \qquad p=1,2,\ldots,m, \tag{6.2}$$

where m denotes the number of constraints from now until the end of the section. The constraint functions have to be specified by a subroutine that calculates $c_p(\underline{x})$, $p=1,2,\ldots,m$, at points $\underline{x}\in\mathbb{R}^n$ that are generated automatically. These points include the vertices of the current simplex, in order that, for each p, we can let γ_p be the linear polynomial from $\mathbb{R}^n$ to $\mathbb{R}$ whose coefficients are defined by the equations $\gamma_p(\underline{v}_i)=c_p(\underline{v}_i)$, $i=1,2,\ldots,n+1$. Then the inequalities (6.2) are approximated by the linear conditions

$$\gamma_p(\underline{x}) \geq 0, \qquad p=1,2,\ldots,m. \tag{6.3}$$

Most of the techniques of this section are taken from Powell (1994).

We restrict attention to iterative algorithms, where, for $k=1,2,3,\ldots$, the kth iteration employs the current simplex and the linear approximations of the previous paragraph. The initial simplex is constructed from data that include a recommended distance between vertices, namely $\Delta_1>0$. For each k, we let $\underline{x}_k$ be the best of the vertices $\underline{v}_i$, $i=1,2,\ldots,n+1$, which implies the conditions

$$\underline{x}_k\in\{\underline{v}_i : i=1,2,\ldots,n+1\} \quad \text{and} \quad F(\underline{x}_k)\leq F(\underline{v}_i), \quad i=1,2,\ldots,n+1, \tag{6.4}$$

in the unconstrained case. A suitable extension of the last inequality for $m\geq 1$ is given in the final paragraph of this section. Each iteration until termination generates a new vector of variables, $\underline{v}_{n+2}$ say, where the difference $\underline{v}_{n+2}-\underline{x}_k$ is either a 'minimization step' or a 'simplex step'. The values of F and any constraint functions are calculated at $\underline{v}_{n+2}$. Then the $n+1$ vertices of the simplex of the next iteration are chosen by deleting one point from the set $\{\underline{v}_i : i=1,2,\ldots,n+2\}$. Further, $\underline{x}_{k+1}$ is defined in the way mentioned earlier, any ties being broken by retaining $\underline{x}_{k+1}=\underline{x}_k$, unless a change provides a strict improvement according to the criterion for the best vertex. An iteration also sets the parameters Δ_{k+1} and ρ_{k+1} before increasing k, where $\rho_1=\Delta_1$. All of these operations receive further consideration below.

The minimization of $\Phi(\underline{x})$, $\underline{x}\in\mathbb{R}^n$, subject to the constraints (6.3), is a linear programming problem that usually fails to have a finite solution in the case $m<n$. Further, it is likely that the linear approximations are too inaccurate to be useful when $\underline{x}$ is far from the current simplex. Therefore we consider algorithms that employ trust region bounds. Specifically, the vector $\underline{v}_{n+2}$ of the kth iteration has to satisfy the inequality

$$\|\underline{v}_{n+2}-\underline{x}_k\| \le \rho_k, \tag{6.5}$$

where ρ_k is a positive number that is available at the beginning of the iteration, but it may be reduced occasionally. On most iterations, $\underline{v}_{n+2}$ is the vector $\underline{x}$ that minimizes $\Phi(\underline{x})$ subject to $\|\underline{x}-\underline{x}_k\|\le\rho_k$ and the conditions (6.3), and then $\underline{v}_{n+2}-\underline{x}_k$ is the 'minimization step' of the previous paragraph, provided that $\|\underline{v}_{n+2}-\underline{x}_k\|$ is as small as possible if the solution to this subproblem is not unique. It can happen, however, that the constraints of the subproblem are inconsistent, and then the 'minimization step' is defined by minimizing the greatest violation of a linear constraint, namely $\max\{-\gamma_p(\underline{v}_{n+2}):p=1,2,\ldots,m\}$, subject to inequality (6.5), where again any nonuniqueness is taken up by reducing $\|\underline{v}_{n+2}-\underline{x}_k\|$. Powell (1994) addresses these calculations when the vector norm is Euclidean, and recommends a procedure that generates the path $\underline{v}_{n+2}(\alpha)$, $0\le\alpha\le\rho_k$, in $\mathbb{R}^n$, where $\underline{v}_{n+2}(\alpha)$ is the $\underline{v}_{n+2}$ that would be required if ρ_k were equal to α. This path begins at the point $\underline{v}_{n+2}(0)=\underline{x}_k$, and is continuous and piecewise linear. Further, the different pieces of the path correspond to different indices of critical constraints, the qth constraint being critical if and only if the conditions $\gamma_q(\underline{x})\le 0$ and $\gamma_q(\underline{x})\le\gamma_p(\underline{x})$, $p=1,2,\ldots,m$, hold. Sometimes the length $\|\underline{v}_{n+2}-\underline{x}_k\|$ of the 'minimization step' is 'too small', and then it is usual to replace $\underline{v}_{n+2}-\underline{x}_k$ by a 'simplex step'. There are also iterations that calculate only a 'simplex step'. The reasons for these alternatives are as follows.

We consider the case when there are no given constraints on the variables, when $\underline{v}_{n+2}-\underline{x}_k$ is a 'minimization step', when $\|\underline{v}_{n+2}-\underline{x}_k\|$ is large enough for $F(\underline{v}_{n+2})$ to be calculated, and when the new function value has the property

$$F(\underline{v}_{n+2}) \ge F(\underline{x}_k). \tag{6.6}$$

Then, because the definition of the minimization step implies $\Phi(\underline{v}_{n+2})<F(\underline{x}_k)$, the approximation $\Phi(\underline{v}_{n+2})\approx F(\underline{v}_{n+2})$ is inadequate. There are two main causes of the inadequacy, and it is important to distinguish between them. Firstly, $\underline{v}_{n+2}$ may be so far from $\underline{x}_k$ that very good linear approximations to F in a neighbourhood of $\underline{x}_k$ may be unsuitable at $\underline{v}_{n+2}$, due to second and higher order terms or lack of smoothness of the objective func-

tion. Secondly, although the bound (6.5) may ensure that any one of these very good approximations provides a minimization step that is successful at reducing the least calculated value of F, the interpolation conditions (6.1) may define a linear polynomial Φ that is unhelpful. This can happen if one or more of the distances $\|\underline{v}_i-\underline{x}_k\|$, $i=1,2,\ldots,n+1$, is much greater than ρ_k or if the current simplex is nearly degenerate. The appropriate remedy in the first case is to shorten the length of the minimization step on the next iteration by choosing $\rho_{k+1}<\rho_k$, which is a standard technique in trust region algorithms. In the second case, however, the remedy is to choose a better simplex. When $\underline{v}_{n+2}$ is calculated for the latter purpose, we call $\underline{v}_{n+2}-\underline{x}_k$ a 'simplex step', the actual choice of $\underline{v}_{n+2}$ being the subject of the following paragraph. The choice is independent of Φ, except for a plus or minus sign, and $\underline{v}_{n+2}$ becomes one of the vertices of the simplex of the next iteration. The need for such steps is clear if a given constraint on the variables is linear, and if $\underline{v}_{n+2}$ satisfies the constraint as an equation for all minimization steps. Indeed, if there were no simplex steps in this case, and if all of the vertices of the initial simplex have been removed from the current simplex by earlier iterations, then all of the current vertices $\underline{v}_i$, $i=1,2,\ldots,n+1$, are on the boundary of the linear constraint. Thus the equations (6.1) fail to define the coefficients of Φ, because the matrix of the equations is singular. Therefore a reason for the 'simplex steps' is to oppose any tendencies for the current simplex to become degenerate.

It has been mentioned that $\Delta_1>0$ is a prescribed parameter that controls the size of the initial simplex. Most of the later iterations employ $\Delta_k=\Delta_{k-1}$, and each Δ_k is an acceptable length for the edges of the current simplex, the length being relevant to the suitability of the linear polynomial Φ defined by the equations (6.1). Specifically, it is assumed that the nonlinearities of the objective function may damage the usefulness of the approximation $\Phi\approx F$, if any of the distances $\|\underline{v}_i-\underline{x}_k\|$, $i=1,2,\ldots,n+1$, is much greater than Δ_k. On the other hand, when 'minimization steps' are successful at improving the best vector of variables so far, then there is no need for any 'simplex steps'. Thus 20 consecutive iterations, say, may make changes to the variables that are minimization steps, and all of the changes may be in roughly the same direction in $\mathbb{R}^n$, which causes $\max\{\|\underline{v}_i-\underline{x}_k\| : i=1,2,\ldots,n+1\}$ to become large. Eventually, however, we expect the sequence of successful iterations to be interrupted by a minimization step that makes $\underline{v}_{n+2}$ no better than $\underline{x}_k$, which means that inequality (6.6) occurs in the unconstrained case. Then the next iteration employs a 'simplex step'. When the kth iteration tries to take a simplex step, an integer ℓ in $[1,n]$ is calculated that has the property

$$\|\underline{v}_\ell-\underline{x}_k\| = \max\{\|\underline{v}_i-\underline{x}_k\| : i=1,2,\ldots,n+1\}. \tag{6.7}$$

Further, the condition $\|\underline{v}_\ell-\underline{x}_k\|\le\beta\,\Delta_k$ is tested, where $\beta>1$ is a prescribed constant that has the value $\beta=2.1$ in the work of Powell (1994). If the test fails, then $\underline{v}_{n+2}$ is chosen in a way that makes it suitable to delete $\underline{v}_\ell$ from the set $\{\underline{v}_i : i=1,2,\ldots,n+2\}$, when generating the vertices of the simplex of the next iteration. Specifically, letting $\underline{\omega}_\ell \in \mathbb{R}^n$ be a vector of unit length that is orthogonal to the face of the current simplex that is without $\underline{v}_\ell$, we let $\underline{v}_{n+2}$ be the point

$$\underline{v}_{n+2} = \underline{x}_k \pm \Delta_k \underline{\omega}_\ell, \tag{6.8}$$

where the $\pm$ sign is negative if and only if $\underline{x}_k - \Delta_k \underline{\omega}_\ell$ is better than $\underline{x}_k + \Delta_k \underline{\omega}_\ell$, according to a criterion in the last paragraph of this section. Otherwise, if $\|\underline{v}_\ell - \underline{x}_k\| \le \beta \Delta_k$ is achieved, the algorithm seeks a different integer ℓ in $[1, n+1]$. Indeed, letting σ_i be the distance from $\underline{v}_i$ to the plane in $\mathbb{R}^n$ that contains the vertices of the current simplex that are different from $\underline{v}_i$, the new ℓ minimizes σ_ℓ subject to $\underline{v}_\ell \neq \underline{x}_k$. Therefore a very small value of σ_ℓ indicates that the simplex is nearly degenerate. The inequality $\sigma_\ell \ge \alpha \Delta_k$ is tried, where $\alpha < 1$ is another positive constant, for instance $\alpha = 1/4$. If the inequality fails, then $\underline{v}_{n+2}$ is defined by formula (6.8) for the new ℓ, where the $\pm$ sign and $\underline{\omega}_\ell$ are as before. Further, the new simplex is generated by replacing $\underline{v}_\ell$ by $\underline{v}_{n+2}$ in the list of vertices, which increases the volume of the current simplex by the factor Δ_k / σ_ℓ. If $\sigma_\ell \ge \alpha \Delta_k$ holds, however, then the positions of the vertices $\underline{v}_i$, $i = 1, 2, \ldots, n+1$, are assumed to be adequate for the equations (6.1) that define Φ, and we say that the simplex is 'acceptable'. Then the iteration tries to generate $\underline{v}_{n+2}$ by a 'minimization step' instead of by a 'simplex step'.

We are now ready to consider the choices between the minimization and simplex step alternatives, the values of Δ_k and ρ_k, $k = 1, 2, 3, \ldots$, and a condition for terminating the calculation. Simple rules are recommended for adjusting Δ_k and for termination. Specifically, Δ_1 is given, and, until termination, the kth iteration sets $\Delta_{k+1} = \Delta_k$, where k is still the iteration number. The value of Δ_k at the start of the kth iteration is provisional, however, in order that a few iterations can reduce Δ_k, although no increases are allowed. A positive parameter, Δ_* say, has to be prescribed that satisfies $\Delta_* \le \Delta_1$, because it is a lower bound on every Δ_k. The changes in Δ_k have to be such that $\Delta_k = \Delta_*$ occurs after a finite number of reductions. The situation that causes a reduction is described in the paragraph after next. The calculation terminates when this situation occurs and Δ_k has already reached the value Δ_*. These rules afford the following useful properties. Every iteration until termination picks a vector of variables $\underline{v}_{n+2}$ that satisfies the inequality

$$\|\underline{v}_{n+2} - \underline{x}_k\| \ge \Delta_k, \tag{6.9}$$

and Δ_k is not reduced until this condition seems to prevent further improvements to the variables. The user can pick a value of Δ_1 that causes substantial adjustments to the variables to be tried at the beginning of the calculation, which can alleviate the damage from any random noise in the function values. Then the bound (6.9) can be refined gradually by the decreases in Δ_k. Further, when the given functions are smooth, good accuracy can usually be achieved at termination by letting Δ_* be sufficiently small.

Powell (1994) sets $\rho_k = \Delta_k$ throughout the calculation, but changes to the variables that are much greater than Δ_k are sometimes necessary for efficiency. Indeed, there are unconstrained calculations with quadratic objective functions such that, when Δ_k is reduced, the distance from $\underline{x}_k$ to the optimal vector of variables is of magnitude $M\Delta_k$, where M is the condition number of the second derivative matrix $\nabla^2 F$. Therefore it may be helpful to allow ρ_k to be much larger than Δ_k. Moreover, the initial choice $\rho_1 = \Delta_1$ has been mentioned already, and it is reasonable to set ρ_k to the new value of Δ_k when Δ_k is decreased, because ρ_k should become less than the old value of Δ_k for the moment, but condition (6.9) excludes $\rho_k < \Delta_k$. These remarks suggest the following guidelines for the choice of ρ_k, $k=1,2,3,\ldots$. We pick $\rho_k = \Delta_k$ for each new Δ_k, which causes ρ_k to be less than its value at the beginning of any iteration that decreases Δ_k, but there are no other changes to ρ_k during an iteration. The value $\rho_{k+1} = \rho_k$ is often set at the end of the kth iteration, and it always occurs when $\underline{v}_{n+2} - \underline{x}_k$ is a 'simplex step'. On the other hand, $\rho_{k+1} > \rho_k$ is allowed when $\underline{v}_{n+2} - \underline{x}_k$ is a 'minimization step' that provides $\underline{x}_{k+1} \neq \underline{x}_k$. If a minimization step fails to improve the best vector of variables so far, however, then the next minimization step is required to be substantially shorter than the present one, except that the bound (6.9) is preserved. Therefore the value

$$\rho_{k+1} = \max\,[\,\Delta_k,\; \tfrac{1}{2}\,\|\underline{v}_{n+2} - \underline{x}_k\|\,], \tag{6.10}$$

for example, may be suitable. Some choices of ρ_k after successful minimization steps can be found in Chapter 5 of Fletcher (1987), for instance, and in other published descriptions of trust region methods.

Each iteration until termination has to choose a 'minimization step' or a 'simplex step'. The method that fixes the choice is specified below, using the nomenclature that a minimization step is 'long enough' if it satisfies inequality (6.9), and is 'questionable' unless its length is exactly Δ_k and the current simplex is 'acceptable'. When the iteration does not reduce Δ_k, then the choice between the alternatives is determined by the following four rules, which are given in order of priority.

(1) A minimization step is preferred if it is long enough and if either $k=1$ or the previous iteration improved the best vector of variables so far.

(2) A minimization step is preferred if it is long enough and if the previous iteration applied a simplex step.

(3) A simplex step is preferred if neither (1) nor (2) apply and if the current simplex is not acceptable.

(4) A minimization step is preferred if it is long enough, if the current simplex is acceptable and if the previous iteration employed a minimization step that is questionable.

Thus the remaining possibilities are the following two situations.

(5) The current simplex is acceptable and the minimization step is not long enough.
(6) The current simplex is acceptable, the minimization step is long enough, the previous iteration applied a minimization step that is not questionable, but that iteration did not improve the best vector of variables.

In these cases the time has come to reduce Δ_k and ρ_k. Therefore termination occurs if Δ_k has attained the value Δ_*. Otherwise, after reducing Δ_k and ρ_k, the required choice is determined by three more rules.

(7) The minimization step is preferred if it is long enough for the new Δ_k.
(8) The simplex step is chosen if (7) fails and if the current simplex is not acceptable for the new Δ_k.
(9) In all other cases, Δ_k is still too large, so we introduce a recursion by branching back to the part of the algorithm that either causes termination or reduces Δ_k.

Thus each iteration before termination picks just one vector $\underline{v}_{n+2}$ at which the values of the given functions from $\mathbb{R}^n$ to $\mathbb{R}$ are calculated.

Another question that requires an answer is the choice of the n+1 vertices of the simplex for the next iteration from $\{\underline{v}_i : i = 1, 2, \ldots, n+2\}$. We let $\underline{v}_\ell$ be the point that is not retained, which agrees with equation (6.8) when $\underline{v}_{n+2} - \underline{x}_k$ is a 'simplex step'. We propose a new choice of ℓ when $\underline{v}_{n+2} - \underline{x}_k$ is a 'minimization step', however, because the technique in Powell (1994) assumes $\rho_k = \Delta_k$ for every k. Let $\underline{x}_{k+1} \in \{\underline{v}_i : i = 1, 2, \ldots, n+2\}$ be determined before ℓ is selected, which is possible because we require the best vector of variables so far. Further, let the real multipliers θ_i, $i = 1, 2, \ldots, n$+2, satisfy the equation

$$\sum_{i=1}^{n+2} \theta_i \,(\underline{v}_i - \underline{x}_{k+1}) = 0, \tag{6.11}$$

where θ_{i_*} is zero for the integer i_* that is defined by $\underline{x}_{k+1} = \underline{v}_{i_*}$, but some of the other multipliers are nonzero. It follows from the nondegeneracy of the current simplex that the values of the multipliers are determined uniquely except for a scaling factor. Now, if i and j are different integers in $[1, n+2]$ such that θ_i and θ_j are nonzero, and if $\mathcal{S}_i$ and $\mathcal{S}_j$ are the new simplices for $\ell = i$ and $\ell = j$, respectively, then equation (6.11) implies the property

$$|\,\mathrm{Vol}\,\mathcal{S}_i \,/\, \mathrm{Vol}\,\mathcal{S}_j\,| = |\,\theta_i \,/\, \theta_j\,|\,. \tag{6.12}$$

Therefore it may be suitable to pick ℓ by satisfying the condition $|\theta_\ell| = \max\{|\theta_i| : i=1,2,\ldots,n+2\}$. This method, however, would favour the retention of any points $\underline{v}_i$ that are far from $\underline{x}_{k+1}$, and we do not want the new simplex to have a large volume because the lengths of some of its sides are much greater than Δ_k. Instead we take the view for the moment that, for every i in $[1,n+2]$ such that $\|\underline{v}_i - \underline{x}_{k+1}\|$ exceeds Δ_k, the point $\underline{v}_i$ is replaced by the point on the line segment from $\underline{x}_{k+1}$ to $\underline{v}_i$ that is distance Δ_k from $\underline{x}_{k+1}$, but $\underline{v}_i$ is unchanged for the other values of i. Then we choose ℓ by applying the procedure just described to these new points. Specifically, for each integer i in $[1,n+2]$, we find that θ_i in equation (6.11) has to be scaled by $\max[1, \|\underline{v}_i - \underline{x}_{k+1}\|/\Delta_k]$, because of the temporary change to $\underline{v}_i$. Therefore we let ℓ be an integer in $[1,n+2]$ that has the property

$$|\theta_\ell| \max[\Delta_k, \|\underline{v}_\ell - \underline{x}_{k+1}\|] \geq |\theta_i| \max[\Delta_k, \|\underline{v}_i - \underline{x}_{k+1}\|], \quad i=1,2,\ldots,n+2. \tag{6.13}$$

Thus, if the current simplex has a vertex that is far from the best vector of variables, there is a tendency to exclude it from the simplex of the next iteration.

The merit function, Ψ say, of the calculation provides a balance between the value of the objective function and any constraint violations, in order to determine the best vertex of the current simplex. Specifically, Ψ is the same as F when there are no constraints, and, for $m \geq 1$, the form

$$\Psi(\underline{x}) = F(\underline{x}) + \mu[\max\{-c_p(\underline{x}) : p=1,2,\ldots,m\}]_+, \qquad \underline{x} \in \mathbb{R}^n, \tag{6.14}$$

is taken from Powell (1994). Here μ is a parameter that is zero initially and that may be increased automatically as described below. Further, the subscript '+' indicates that the expression in square brackets is replaced by zero if and only if its value is negative. Thus $\Psi(\underline{x}) = F(\underline{x})$ occurs whenever $\underline{x}$ is feasible, and it is helpful to scale the constraint functions so that the values $c_p(\underline{x})$, $p=1,2,\ldots,m$, have similar magnitudes for typical vectors $\underline{x}$. Expression (6.4) is extended to $m \geq 0$ by requiring the best vertex $\underline{x}_k$ to satisfy the conditions

$$\underline{x}_k \in \{\underline{v}_i : i=1,2,\ldots,n+1\} \quad \text{and} \quad \Psi(\underline{x}_k) \leq \Psi(\underline{v}_i), \quad i=1,2,\ldots,n+1. \tag{6.15}$$

After choosing $\underline{x}_k$, both the minimization and the simplex steps are independent of Ψ and μ, but Ψ is usually important to what happens next. Indeed, if the new vector of variables of the kth iteration, namely $\underline{v}_{n+2}$, is generated by a minimization step, then usually another minimization step is chosen by the $(k+1)$th iteration if and only if the strict inequality $\Psi(\underline{v}_{n+2}) < \Psi(\underline{x}_k)$ holds. Further, this inequality should be achieved if all the linear approximations of the first paragraph of this section are exact. Therefore we require the value of μ to provide the property

$$\Upsilon(\underline{v}_{n+2}) < \Upsilon(\underline{x}_k), \qquad \text{if } \underline{v}_{n+2} - \underline{x}_k \text{ is a minimization step}, \tag{6.16}$$

where Υ is the piecewise linear approximation

$$\Upsilon(\underline{x}) = \Phi(\underline{x}) + \mu\,[\max\{-\gamma_p(\underline{x}) : p=1,2,\ldots,m\}]_+, \qquad \underline{x}\in\mathbb{R}^n, \qquad (6.17)$$

to the merit function (6.14). Now a minimization step either reduces the contribution from the constraints to expression (6.17), or the contribution is zero and $\Phi(\underline{v}_{n+2})<\Phi(\underline{x}_k)$ occurs, where we are excluding steps that are zero, because they are abandoned automatically, due to the failure of inequality (6.9). It follows that condition (6.16) can be achieved whenever it is required by choosing a sufficiently large value of μ. Therefore Powell (1994) proposes the following technique for increasing μ. Whenever a minimization step is calculated that has the property (6.9), we let $\bar{\mu}$ be the least nonnegative value of μ that provides $\Upsilon(\underline{v}_{n+2})\leq\Upsilon(\underline{x}_k)$. Further, μ is unchanged in the case $\mu\geq\frac{3}{2}\bar{\mu}$, but otherwise it is increased to $2\bar{\mu}$. A possible consequence of an increase in μ is that $\underline{x}_k$ is no longer the optimal vertex, and then the calculated minimization step would be incorrect. Therefore $\underline{x}_k$ is changed if necessary to another vertex that satisfies the conditions (6.15). Then the minimization step is recalculated, so μ may have to be increased again, which may cause a further change to the optimal vertex. Fortunately, this procedure does not cycle, because each change to $\underline{x}_k$ causes a strict reduction in $\max\{-\gamma_p(\underline{x}_k) : p=1,2,\ldots,m\}$. Another use of Υ is that the $\pm$ sign of expression (6.8) is negative if and only if $\Upsilon(\underline{x}_k-\Delta_k\underline{\omega}_\ell)$ is less than $\Upsilon(\underline{x}_k+\Delta_k\underline{\omega}_\ell)$. Some remarks on the convergence properties of the algorithm of this section are made in Section 9.

7. Quadratic approximation methods

Many of the techniques of this section are similar to those of Section 6, but now we let the approximation $\Phi(\underline{x})$, $\underline{x}\in\mathbb{R}^n$, to the objective function $F(\underline{x})$, $\underline{x}\in\mathbb{R}^n$, be a quadratic polynomial instead of a linear polynomial. Therefore Φ has $\frac{1}{2}(n+1)\,(n+2)=\hat{n}$, say, independent coefficients, that may be defined by the interpolation conditions

$$\Phi(\underline{v}_i) = F(\underline{v}_i), \qquad i=1,2,\ldots,\hat{n}, \qquad (7.1)$$

where the vectors $\underline{v}_i$, $i=1,2,\ldots,\hat{n}$, are points in $\mathbb{R}^n$. These points should have the property that, if expression (7.1) is written as a system of linear equations, the unknowns being the coefficients, then the matrix of the system is nonsingular. The Lagrange functions of the interpolation problem will be useful later. Therefore we reserve the notation χ_i, $i=1,2,\ldots,\hat{n}$, for the quadratic polynomials from $\mathbb{R}^n$ to $\mathbb{R}$ that satisfy the equations

$$\chi_i(\underline{v}_j) = \delta_{ij}, \qquad 1\leq i,j\leq\hat{n}, \qquad (7.2)$$

where δ_{ij} is the Kronecker delta. It follows that Φ is the function

$$\Phi(\underline{x}) = \sum_{i=1}^{n} F(\underline{v}_i)\, \chi_i(\underline{x}), \qquad \underline{x} \in \mathbb{R}^n. \tag{7.3}$$

The main advantage of quadratic over linear polynomials is that quadratics include some second derivative information, which allows the development of algorithms that have useful superlinear convergence properties. We are going to consider some of the ideas proposed for constructing and applying quadratic approximations to F when there are no constraints on the variables.

The algorithm of Winfield (1973) not only employs the interpolation equations (7.1) to define Φ, but also it includes some of the earliest work on trust regions. The kth iteration of that method is given all the values of the objective function calculated so far, $\hat{n}$ of them being obtained before the first iteration. Let these values be $F(\underline{v}_i)$, $i=1,2,\ldots,\check{n}$, where $\check{n} \geq \hat{n}$, let $\underline{x}_k$ be a best vector of variables, which means that it satisfies the conditions

$$\underline{x}_k \in \{\underline{v}_i : i=1,2,\ldots,\check{n}\} \quad \text{and} \quad F(\underline{x}_k) \leq F(\underline{v}_i), \quad i=1,2,\ldots,\check{n}, \tag{7.4}$$

and let the current data be ordered so that the sequence of distances $\|\underline{v}_i - \underline{x}_k\|$, $i=1,2,\ldots,\check{n}$, increases monotonically. Then the kth iteration generates the quadratic polynomial Φ by trying to interpolate the function values of only the first $\hat{n}$ terms of the sequence, in accordance with the notation (7.1). Further, the iteration calculates the vector $\underline{x} \in \mathbb{R}^n$ that minimizes $\Phi(\underline{x})$ subject to the bound $\|\underline{x} - \underline{x}_k\| \leq \rho_k$, where the trust region radius is chosen automatically and satisfies $\rho_k \leq 0.99\, \|\underline{v}_{\hat{n}} - \underline{x}_k\|$, in order that the value of F at the new point will be included in the interpolation conditions of the $(k+1)$th iteration. One reason for mentioning the algorithm is that it acts in an enterprising way when the system (7.1) is degenerate. Specifically, the degeneracy is ignored, it is assumed that the calculation of Φ is sufficiently robust to provide a quadratic function that allows the trust region subproblem to be solved, and the resultant $\underline{x}$ receives no special treatment. Thus some unpredictable changes to the variables occur that may remove the degeneracy after a few iterations. Indeed, Winfield (1973) states that 'This natural cure of ill-conditioning is more efficient than restarting the algorithm by evaluating $F(\underline{x})$ at the points of a grid'. The other methods that we study, however, ensure that each Φ is well defined.

The Lagrange functions that have been mentioned provide a convenient way of avoiding singularity in the equations (7.1). The technique suggests itself if one tries to modify the algorithm of Powell (1994) for unconstrained optimization, described in the previous section, so that the linear polynomial Φ of expression (6.1) is replaced by the quadratic polynomial that is defined by the equations (7.1). We retain from Section 6 the parameters Δ_k and

ρ_k, $k = 1, 2, 3, \ldots$, and the rules that give their values. Moreover, in the quadratic case, the points $\underline{v}_i$, $i = 1, 2, \ldots, \hat{n}$, for the first iteration can be the vertices and the mid-points of the edges of a nondegenerate simplex in $\mathbb{R}^n$, where the lengths of the edges are still of magnitude Δ_1. Otherwise, for $k \geq 2$, these points are chosen by the previous iteration, and $\underline{x}_k$ satisfies the conditions

$$\underline{x}_k \in \{\underline{v}_i : i=1,2,\ldots,\hat{n}\} \quad \text{and} \quad F(\underline{x}_k) \leq F(\underline{v}_i), \quad i=1,2,\ldots,\hat{n}. \qquad (7.5)$$

Further, $\underline{v}_{\hat{n}+1} - \underline{x}_k$ is still a 'minimization step' if $\underline{v}_{\hat{n}+1}$ is the vector $\underline{x} \in \mathbb{R}^n$ that minimizes $\Phi(\underline{x})$ subject to $\|\underline{x} - \underline{x}_k\| \leq \rho_k$, which is the trust region subproblem of the previous paragraph. On the other hand, a 'simplex step' is usually required if the previous iteration generated a minimization step that failed to reduce the least calculated value of F. In this case, guided by equation (6.7), we let ℓ be an integer in $[1, \hat{n}]$ that maximizes $\|\underline{v}_\ell - \underline{x}_k\|$. If this distance is unacceptably large, then we have to pick a point $\underline{v}_{\hat{n}+1}$ that will replace $\underline{v}_\ell$ in the system (7.1) on the next iteration. Therefore we require a formula that is analogous to expression (6.8), and that is suitable when Φ is a quadratic polynomial.

Now the main property of the point (6.8) is that it maximizes the volume of the simplex of the next iteration subject to $\|\underline{v}_{n+2} - \underline{x}_k\| \leq \Delta_k$. Further, the volume of the simplex is a constant multiple of the modulus of the determinant of the matrix of the system (6.1), when the usual basis of the space of linear polynomials is employed. Therefore an analogous choice of the 'simplex step' when Φ is quadratic would maximize the modulus of the determinant of the $\hat{n} \times \hat{n}$ system (7.1), after $\underline{v}_\ell$ is replaced by $\underline{v}_{\hat{n}+1}$, where $\underline{v}_{\hat{n}+1}$ has to satisfy $\|\underline{v}_{\hat{n}+1} - \underline{x}_k\| \leq \Delta_k$. We write $\underline{x} = \underline{v}_{\hat{n}+1}$ for the moment, we regard the new determinant as a function of $\underline{x} \in \mathbb{R}^n$, and we find that it is a quadratic polynomial in $\underline{x}$. Further, the determinant must vanish if $\underline{x}$ is any point of the set $\{\underline{v}_i : i=1,2,\ldots,\hat{n}\}$ that is different from $\underline{v}_\ell$. Thus an elementary normalization provides the identity

$$\text{New determinant / Old determinant} = \chi_\ell(\underline{x}), \qquad \underline{x} \in \mathbb{R}^n. \qquad (7.6)$$

Therefore we define $\underline{v}_{\hat{n}+1} - \underline{x}_k$ to be a 'simplex step' for the chosen integer $\ell \in [1, \hat{n}]$ if and only if $\underline{v}_{\hat{n}+1}$ is a vector of variables $\underline{x}$ that maximizes $|\chi_\ell(\underline{x})|$ subject to $\|\underline{x} - \underline{x}_k\| \leq \Delta_k$. This definition has the advantage of being independent of the choice of basis of the space of quadratic polynomials. Further, the simplex step can be calculated by solving two trust region subproblems of the type that has been encountered already. Indeed, if two vectors of variables are generated by minimizing the quadratic functions χ_ℓ and $-\chi_\ell$ subject to the trust region bound, then the required $\underline{v}_{\hat{n}+1}$ is the vector that gives the larger value of $|\chi_\ell|$.

When the kth iteration tries to take a 'simplex step', the algorithm may find that all of the points $\underline{v}_i$, $i = 1, 2, \ldots, \hat{n}$, are sufficiently close to $\underline{x}_k$, which corresponds to the condition $\|\underline{v}_\ell - \underline{x}_k\| \le \beta\,\Delta_k$ in Section 6. Then a test for near-degeneracy of the system (7.1) is required, that is analogous to the use of σ_ℓ and α in the paragraph that includes equations (6.7) and (6.8). There the replacement of $\underline{v}_\ell$ by the point (6.8) increases the modulus of the determinant of the system (6.1) by the factor Δ_k/σ_ℓ. Therefore we continue to let $\alpha < 1$ be a positive constant, for instance $\alpha = 1/4$, and we seek an integer ℓ in $[1, \hat{n}]$ such that the replacement of $\underline{v}_\ell$ by $\underline{v}_{\hat{n}+1}$ increases the modulus of the determinant of the system (7.1) by a factor of more than $1/\alpha$, where $\underline{v}_{\hat{n}+1}$ is defined at the end of the previous paragraph, because this choice maximizes the modulus of the new determinant. Specifically, the test for near-degeneracy in the quadratic case is as follows. The integer ℓ runs through the set $\{1, 2, \ldots, \hat{n}\}$, but similar tests on recent iterations may make it advantageous not to begin with $\ell = 1$. For each ℓ, the maximum value of $|\chi_\ell(\underline{x})|$, $\|\underline{x} - \underline{x}_k\| \le \Delta_k$, is calculated. If $|\chi_\ell(\underline{x})| > 1/\alpha$ occurs, the task of searching for a suitable ℓ is complete, because the replacement of $\underline{v}_\ell$ by the vector $\underline{v}_{\hat{n}+1}$ that has been mentioned provides a substantial improvement to the positions of the interpolation points. Then $\underline{v}_{\hat{n}+1} - \underline{x}_k$ is a 'simplex step', and the function value $F(\underline{v}_{\hat{n}+1})$ is required for the system (7.1) of the next iteration. Otherwise, if no integer ℓ in $[1, \hat{n}]$ provides $|\chi_\ell(\underline{v}_{\hat{n}+1})| > 1/\alpha$, then, as before, we say that the current interpolation points are 'acceptable', and the iteration may generate $\underline{v}_{\hat{n}+1}$ by a 'minimization step'. We also retain the rule that the minimization step is abandoned if it fails to satisfy $\|\underline{v}_{\hat{n}+1} - \underline{x}_k\| \ge \Delta_k$, which is important to the criteria for reducing Δ_k and for termination, as described in Section 6.

We let each choice between a 'minimization' and a 'simplex' step in the quadratic case be the same as in Section 6, the rules listed on pages 314–315 being applied as before. A modification is needed, however, to the technique that selects the interpolation points for the $(k+1)$th iteration, after $F(\underline{v}_{\hat{n}+1})$ has been calculated and $\underline{v}_{\hat{n}+1} - \underline{x}_k$ is a minimization step. These points are all but one of the vectors $\underline{v}_i$, $i = 1, 2, \ldots, \hat{n}+1$, and again we let $\underline{v}_\ell$ denote the point that is rejected. Here it is important to note that, in contrast to the previous two paragraphs, $\underline{v}_{\hat{n}+1}$ is now independent of ℓ, because it is generated by the minimization step before ℓ is chosen. In order to retain a best vector of variables so far, we let i_* be an integer in $[1, \hat{n}+1]$ such that $F(\underline{v}_{i_*})$ is the least of the function values $F(\underline{v}_i)$, $i = 1, 2, \ldots, \hat{n}+1$. Then the value $\ell = i_*$ is prohibited, because $\underline{x}_{k+1}$ is going to be the point $\underline{v}_{i_*}$. It would be straightforward to pick the ℓ that maximizes the modulus of the determinant of the system (7.1) on the next iteration if we wished to do so. Indeed, if $\ell \in [1, \hat{n}]$, then it follows from the identity (7.6) that the determinant of the new system is the determinant of the present one multiplied by $\chi_\ell(\underline{v}_{\hat{n}+1})$. Therefore, after defining $\theta_{\hat{n}+1} = 1$ and $\theta_i = \chi_i(\underline{v}_{\hat{n}+1})$, $i = 1, 2, \ldots, \hat{n}$, and then replacing θ_{i_*} by zero, we would let ℓ satisfy the equation $|\theta_\ell| = \max\{|\theta_i| \,:\, i = 1, 2, \ldots, \hat{n}+1\}$. This notation provides an analogy with the statement made immediately after expression (6.12). Again, however, we prefer to take the distances $\|\underline{v}_i - \underline{x}_{k+1}\|$, $i = 1, 2, \ldots, \hat{n}$, into account, so our choice of ℓ is derived from a condition that is analogous to inequality (6.13).

Specifically, if i is any integer in $[1, \hat{n}]$ such that $\|\underline{v}_i - \underline{x}_{k+1}\| > \Delta_k$ occurs, we make a notional shift of $\underline{v}_i$ to $\hat{\underline{v}}_i$, say, which is a point on the line segment from $\underline{x}_{k+1}$ to $\underline{v}_i$ that is within distance Δ_k of $\underline{x}_{k+1}$. Further, we let $\hat{\chi}_i$ be the quadratic polynomial that satisfies the Lagrange conditions $\hat{\chi}_i(\hat{\underline{v}}_i) = 1$ and $\hat{\chi}_i(\underline{v}_j) = 0$, for every integer j in $[1, \hat{n}]$ that is different from i. Hence $\hat{\chi}_i$ is the function

$$\hat{\chi}_i(\underline{x}) = \chi_i(\underline{x}) \,/\, \chi_i(\hat{\underline{v}}_i), \qquad \underline{x} \in \mathbb{R}^n. \tag{7.7}$$

Now, because of the inequality $\|\underline{v}_i - \underline{x}_{k+1}\| > \Delta_k$, we assume that the temporary replacement of $\underline{v}_i$ by $\hat{\underline{v}}_i$ would make the determinant of the system (7.1) more relevant to our consideration of possible near-degeneracy. Therefore we change the value of θ_i, given in the previous paragraph, to the number $\hat{\chi}_i(\underline{v}_{\hat{n}+1}) = \chi_i(\underline{v}_{\hat{n}+1}) / \chi_i(\hat{\underline{v}}_i)$, but there is still some freedom in the position of $\hat{\underline{v}}_i$. We have to avoid positions that are too close to other interpolation points, and it is easy to make $|\chi_i(\hat{\underline{v}}_i)|$ as large as possible, because χ_i is a quadratic function of one variable on the line segment from $\underline{x}_{k+1}$ to $\underline{v}_i$. On the other hand, it would be unsuitable to allow $|\chi_i(\hat{\underline{v}}_i)|$ to exceed one, because then $\|\underline{v}_i - \underline{x}_{k+1}\| > \Delta_k$ would assist the retention of $\underline{v}_i$ in the set of interpolation points. These remarks lead to the formula

$$\theta_i = \chi_i(\underline{v}_{\hat{n}+1}) \,/\, \min\left[1, \max\left\{|\chi_i(\underline{x}_{k+1} + \alpha\,[\underline{v}_i - \underline{x}_{k+1}])| : 0 \le \alpha \le \bar{\alpha}\right\}\right], \tag{7.8}$$

where $\bar{\alpha} = \Delta_k / \|\underline{v}_i - \underline{x}_{k+1}\|$. Further, this choice is just $\theta_i = \chi_i(\underline{v}_{\hat{n}+1})$, as before, when i is any integer in $[1, \hat{n}]$ that satisfies $i \ne i_*$ and $\|\underline{v}_i - \underline{x}_{k+1}\| \le \Delta_k$. Moreover, $\theta_{\hat{n}+1} = 1$ is the most reasonable scaling factor to apply to the determinant when there is no change to the interpolation points. Therefore we recommend these values of θ_i, and, after replacing θ_{i_*} by zero, we let ℓ be an integer in $[1, \hat{n}+1]$ that maximizes $|\theta_\ell|$.

We have found that, due to the identity (7.6), Lagrange functions are highly useful for selecting points $\underline{v}_i$, $i = 1, 2, \ldots, \hat{n}$, such that the quadratic polynomial Φ is well defined by the equations (7.1). I presented a paper on this technique at the '5th Stockholm Optimization Days' in 1994, but I did not write a report on it then, because I had hoped that the technique would be developed further by a research student. Later I encouraged another research student, namely Evan Jones, to study the subject, and he proposed the two trust region radii, namely Δ_k and ρ_k, that are introduced in Section 6, although there is only one trust region radius in the algorithm of Powell (1994). He investigated numerically whether or not the number of iterations is reduced by including a second trust region radius, and usually some improvements occur. In any case, the idea is attractive, because it allows some large changes to the variables to co-exist with the security of never increasing Δ_k as k advances. Therefore I described the method, and showed some preliminary numerical results, at the '5th SIAM Conference on Optimization' in 1996. I had expected Jones to write a paper on his work, but unfortunately that has not happened either. Therefore one of the reasons for the present contribution to *Acta Numerica* is to catch up on the recording of this work.

Another description of the use of Lagrange functions is given in Section 4 of Conn, Scheinberg and Toint (1997*b*), which includes a generous acknowledgement to the conference talks mentioned in the previous paragraph. That paper also addresses the idea of employing 'Newton fundamental polynomials' instead of Lagrange functions, where these polynomials in the quadratic case are a constant Lagrange polynomial, n linear Lagrange polynomials, and $\frac{1}{2}n(n+1)$ quadratic Lagrange polynomials that are derived from one, $n+1$ and all of the interpolation points $\underline{v}_i$, $i=1,2,\ldots,\hat{n}$, respectively. They provide a different basis of the $\hat{n}=\frac{1}{2}(n+1)\,(n+2)$-dimensional space of quadratic polynomials, which is helpful when fewer than $\hat{n}$ values of F are available to determine Φ. An outline of a trust region algorithm for unconstrained minimization without derivatives is given too. A major departure from the work of this section is that, if the kth iteration takes a 'minimization step' that reduces F by an amount that compares favourably with the corresponding reduction in Φ, then Δ_{k+1} is allowed to be larger than Δ_k. Nevertheless, this trust region radius is reduced only when the positions of the interpolation points satisfy acceptability conditions that are similar to the ones specified in the complete paragraph that follows equation (7.6). Therefore, in comparison with the technique of Jones that employs both Δ_k and ρ_k, several extra function values may occur if the larger trust region radius is successful for only a small number of iterations. An earlier paper by Conn, Scheinberg and Toint (1997*a*) also considers Newton fundamental polynomials and presents an outline of a similar trust region algorithm. Further, the convergence property of the algorithm is studied under certain assumptions, including the uniform boundedness of the second derivative matrices $\nabla^2\Phi$. It is proved that, if the number of iterations is infinite, then the property $\liminf_{k\to\infty}\|\underline{\nabla}F(\underline{x}_k)\|=0$ is achieved.

The last topic of this section is the algorithm of Elster and Neumaier (1995), which is designed for optimization calculations subject to the simple bounds (3.1) on the variables. The algorithm is remarkable, because it combines quadratic approximations to F and trust regions with some of the properties of discrete grids that are considered in Section 3. Thus termination is achieved, even if the values of the objective function are distorted by noise. There is a close analogy with the two trust region idea of Evan Jones, because it is appropriate to let ρ_k be the trust region radius and Δ_k be the grid size. The algorithm retains all the calculated values of F. Then each quadratic approximation Φ is formed by least squares fitting to some of them, using a technique that is interesting, because it begins by generating a Hessian approximation G, and then it restricts attention to only about $2n+2$ function values, in order to fit the parameters $a \in \mathbb{R}$, $\underline{g} \in \mathbb{R}^n$ and $\kappa \in \mathbb{R}$ of the approximation

$$\Phi(\underline{x}) = a + \underline{g}^T(\underline{x}-\underline{x}_k) + \tfrac{1}{2}\kappa\,(\underline{x}-\underline{x}_k)^T G\,(\underline{x}-\underline{x}_k), \qquad \underline{x} \in \mathbb{R}^n, \tag{7.9}$$

where $\underline{x}_k$ is still the vector of variables that provides the least value of F so far. The algorithm requires Φ, ρ_k and $\underline{x}_k$ for the calculation of a 'minimization step'. Then the new vector of variables at the end of this step is shifted to the nearest grid point, $\underline{x}_+$ say. The use of grids ensures that, after only a finite number of iterations, the function value $F(\underline{x}_+)$ will have been found by an earlier iteration. When this happens, or when three consecutive minimization steps fail to achieve $F(\underline{x}_+) < F(\underline{x}_k)$, a procedure is invoked that is similar to a 'simplex step'. The procedure derives and may apply a linear polynomial approximation to F, using values of the objective function at grid points that have to be neighbours of $\underline{x}_k$. Thus the decision is taken whether or not to reduce Δ_k before resuming the minimization steps. Alternatively, as in Section 6, termination occurs if a reduction in Δ_k is required but Δ_k is already at a prescribed lower bound. Several numerical experiments in Elster and Neumaier (1995) show that this algorithm compares favourably with the method of Nelder and Mead (1965), and with a finite difference implementation of a quasi-Newton algorithm.

8. Simulated annealing

The algorithms that we have studied so far are designed to converge to a local minimum of the objective function $F(\underline{x})$, $\underline{x} \in \mathbb{R}^n$, subject to any constraints on the variables. Many practical optimization calculations, however, have several local minima that are not optimal. Therefore some methods that select vectors of variables using random number generators have become very popular for a wide range of applications. Usually they possess the property that, if there are no constraints, if the objective function is continuous and has bounded level sets, if its least value is $F(\underline{x}^*)$, and if ε is any positive number, then, as the number of random vectors tends to infinity, there is probability one of choosing an $\underline{x}$ that satisfies $F(\underline{x}) \le F(\underline{x}^*)+\varepsilon$. Two approaches of this kind, namely 'simulated annealing' and 'genetic algorithms' are highly active fields of research, and the procedures that have been developed are employed often in practice. Indeed, the books by van Laarhoven and Aarts (1987) and by Goldberg (1989), respectively, both contain more than 200 references.

Genetic algorithms require a one-to-one correspondence between strings of binary digits and vectors of variables in finite precision arithmetic. Thus there is a value of $F(\underline{x})$, $\underline{x} \in \mathbb{R}^n$, for each string. An iteration constructs a new set of m strings from a given set of m strings, where m is a parameter, the value $m=100$ being typical. New strings are generated in pairs, where each pair is derived from two 'parents' that are picked randomly from the given set. The random choice is biased towards better values of the objective function, so a string can be a parent more than once during the iteration, which provides some 'natural selection'. Further, some randomness also occurs in the procedure that breeds two children from the binary digits of their parents. Hence it is difficult to relate the actual changes of variables to the original optimization calculation. On the other hand, several properties of the simulated annealing method have interesting explanations in terms of the objective function and the sequence of iterations. Therefore that method will be considered in the remainder of this section.

The simulated annealing procedure is analogous to the cooling of a liquid to a solid state, starting at a high temperature. When the liquid is hot, very many changes of state can occur, which correspond to many changes of the variables of an optimization calculation, that are allowed to make the objective function worse. Then the energy of the system is lowered, and, if the rate of cooling is sufficiently slow, the liquid should become frozen in the state of least energy, the analogy in the optimization calculation being the required global minimum. An algorithm requires a weighted list of possible changes from one state to another. Moreover, the probability that a change is made depends on the objective functions of the states and on a parameter that corresponds to temperature. The following paragraph gives a method that applies these ideas in the discrete case when $\underline{x}$ is restricted to a finite set $\mathcal{X} \subset \mathbb{R}^n$ that has m elements, say. In practice, m may be the number of different routes that can occur in the travelling salesman problem, but it is instructive to consider much smaller values of m.

Let $\mathcal{X}$ be the set $\{\underline{x}^{(i)} : i = 1, 2, \ldots, m\}$, and let W be a very sparse $m \times m$ matrix of nonnegative real numbers, such that W_{ij} is positive if and only if a transition from state $\underline{x}^{(i)}$ to $\underline{x}^{(j)}$ is allowed for high enough temperatures. The transition is more likely if W_{ij} is larger, and we assume the normalization

$$\sum_{j=1}^{m} W_{ij} = 1, \qquad i=1,2,\ldots,m. \tag{8.1}$$

The method requires these weights, a starting point $\underline{x}_1 \in \mathcal{X}$, and a positive parameter T, which is the initial value of the temperature. For $k=1,2,3,\ldots$, the vector of variables $\underline{x}_k$ is revised to $\underline{x}_{k+1}$ in the following way. Let q be the integer in $[1,m]$ that satisfies $\underline{x}^{(q)} = \underline{x}_k$. An integer p is picked at random from $[1,m]$, where the probability of choosing p is W_{qp}. The new vector $\underline{x}_{k+1}$ is always $\underline{x}^{(p)}$ in the case $F(\underline{x}^{(p)}) \le F(\underline{x}^{(q)})$. Otherwise, the new vector is either $\underline{x}^{(p)}$ or $\underline{x}^{(q)}$, the selection being random with the probabilities

$$\exp\left\{-[F(\underline{x}^{(p)})-F(\underline{x}^{(q)})]/T\right\} \quad \text{or} \quad 1-\exp\left\{-[F(\underline{x}^{(p)})-F(\underline{x}^{(q)})]/T\right\}, \tag{8.2}$$

respectively. Occasionally an iteration may reduce the value of T, by observing some rules that are considered later. These rules also provide a condition for terminating the sequence of iterations.

We consider the probability that $\underline{x}_{k+1}$ is equal to $\underline{x}^{(i)}$, when the positive number T is not altered in the algorithm that has just been described. For convenience of notation, we assume without loss of generality that the sequence $F(\underline{x}^{(i)})$, $i=1,2,\ldots,m$, increases monotonically. Therefore, if $\underline{x}_k=\underline{x}^{(j)}$, then $\underline{x}_{k+1}=\underline{x}^{(i)}$ occurs with probability P_{ij}, where P is the $m\times m$ matrix with the elements

$$P_{ij} = \begin{cases} W_{ji}, & i<j, \\ W_{jj}+\sum_{\ell=j+1}^{m} W_{j\ell}\left(1-\exp\{-[F(\underline{x}^{(\ell)})-F(\underline{x}^{(j)})]/T\}\right), & i=j, \\ W_{ji}\exp\{-[F(\underline{x}^{(i)})-F(\underline{x}^{(j)})]/T\}, & i>j, \end{cases} \tag{8.3}$$

except that the sum is suppressed in the case $i=j=m$. It follows that, if $\underline{v}_k$ is the vector in $\mathbb{R}^m$ whose jth component is the probability that $\underline{x}_k=\underline{x}^{(j)}$ holds, where j is any integer in $[1,m]$, then the ith component of the product $\underline{v}_{k+1}=P\underline{v}_k$ is the probability that $\underline{x}_{k+1}$ is equal to $\underline{x}^{(i)}$. Further, if s is the integer in $[1,m]$ such that $\underline{x}_1$ is $\underline{x}^{(s)}$, and if $\underline{e}_s$ is the sth coordinate vector in $\mathbb{R}^m$, then we deduce the formula

$$\underline{v}_k = P^{k-1}\underline{e}_s, \qquad k=1,2,3,\ldots. \tag{8.4}$$

Now the property

$$\|P\|_1 = \max\{\textstyle\sum_{i=1}^{m}|P_{ij}| = \sum_{i=1}^{m}P_{ij}=1 : j=1,2,\ldots,m\} = 1 \tag{8.5}$$

implies that the spectral radius of P is at most one, and in fact it equals one, because of the eigenvalue equation $\underline{e}^TP=\underline{e}^T$, where the components of $\underline{e}\in\mathbb{R}^m$ are all one. Moreover, we make a nondegeneracy assumption that is suitable for the present application, namely that there is only one eigenvalue of P with modulus equal to the spectral radius. Hence, if $\underline{v}_*\in\mathbb{R}^m$ is a nonzero solution of $P\underline{v}_*=\underline{v}_*$, then the sequence of vectors (8.4) converges to a multiple of $\underline{v}_*$ as $k\to\infty$. Further, because every vector in this sequence is nonnegative and has components that sum to one, the limit of the sequence has these properties too. It follows that we can normalize $\underline{v}_*$ so that its components also sum to one, and then formula (8.4) provides $\lim_{k\to\infty}\underline{v}_k=\underline{v}_*$, for every choice of the integer s. Therefore the limiting behaviour of the algorithm of the previous paragraph is that the states $\underline{x}^{(i)}$, $i=1,2,\ldots,m$, occur with probabilities that are equal to the corresponding components of the nonnegative vector $\underline{v}_*$.

Let $(\underline{v})_i$ denote the ith component of $\underline{v}\in\mathbb{R}^n$. When T is small, it is usual for $(\underline{v}_*)_i$ to be of magnitude $\exp\{-[F(\underline{x}^{(i)})-F(\underline{x}^{(1)})]/T\}$ for $i=1,2,\ldots,n$. It follows from the previous paragraph that the simulated annealing algorithm is likely to pick the state $\underline{x}^{(1)}$, which is the optimal vector of variables. A way of making the assertion about magnitudes plausible depends on the $m\times m$ diagonal matrix D with the diagonal elements $\exp\{-F(\underline{x}^{(i)})/T\}$, $i=1,2,\ldots,m$. Specifically, we define $\underline{\hat{v}}_*\in\mathbb{R}^m$ to be the positive multiple of $D^{-1}\underline{v}_*$ that satisfies $\|\underline{\hat{v}}_*\|_2=1$, we deduce from the eigenvalue equation $P\underline{v}_*=\underline{v}_*$ that $\underline{\hat{v}}_*$ is an eigenvector of the matrix $Q=D^{-1}PD$ with eigenvalue one, and we find that expression (8.3) provides the elements

$$Q_{ij}=\begin{cases} W_{ji}\exp\left\{-[F(\underline{x}^{(j)})-F(\underline{x}^{(i)})]/T\right\}, & i<j,\\ W_{jj}+\sum_{\ell=j+1}^{m}W_{j\ell}\left(1-\exp\left\{-[F(\underline{x}^{(\ell)})-F(\underline{x}^{(j)})]/T\right\}\right), & i=j,\\ W_{ji}, & i>j. \end{cases} \tag{8.6}$$

Thus, if $T\to 0$, the matrix Q remains bounded, and each element Q_{ij} with $i<j$ tends to zero in the usual case when the inequality $F(\underline{x}^{(i)})\le F(\underline{x}^{(j)})$ is strict. Further, the first row of Q tends to be a multiple of the coordinate vector $\underline{e}_1^T$ if $F(\underline{x}^{(1)})$ is less than $F(\underline{x}^{(2)})$. Now, because Q is similar to P, the nondegeneracy assumption of the previous paragraph implies that all but one of the eigenvalues of Q have modulus less than one, so, apart from scaling by a constant, $\underline{\hat{v}}_*$ is the only eigenvector of Q with eigenvalue one. These remarks suggest that $(\underline{\hat{v}}_*)_1$ may remain bounded away from zero as $T\to 0$. Further, the nearly upper triangular structure of Q and the equation $Q\,\underline{\hat{v}}_*=\underline{\hat{v}}_*$ make it possible that none of the components of $\underline{\hat{v}}_*$ tends to zero as $T\to 0$. Then the magnitudes of the components of $\underline{v}_*$ can be derived from the fact that $\underline{v}_*$ is the multiple of $D\,\underline{\hat{v}}_*$ whose components sum to one, which yields the assertion under consideration.

When F is a function of one variable, and when $\mathcal{X}$ contains m different real numbers $\underline{x}^{(i)}$, $i=1,2,\ldots,m$, then a typical choice of the matrix W lets W_{ij} be nonzero if and only if $i\neq j$ and there are no elements of $\mathcal{X}$ between the numbers $\underline{x}^{(i)}$ and $\underline{x}^{(j)}$. Further, it lets the nonzero elements in each row of W be the same. Hence, if α and β are the integers in $[1,m]$ such that $\underline{x}^{(\alpha)}$ and $\underline{x}^{(\beta)}$ are the least and greatest elements of $\mathcal{X}$, respectively, then the αth and βth rows of W each contain only one nonzero element, which has the value 1, while the other rows of W each contain two nonzero elements, and they have the value 1/2, in accordance with equation (8.1). The reason for mentioning this example is that it provides excellent corroboration for the suggestions in the previous paragraph. Indeed, it is straightforward to show that the vector $\underline{v}\in\mathbb{R}^m$ that has the positive components

$$v_i=\left(1-\tfrac{1}{2}\delta_{i\alpha}-\tfrac{1}{2}\delta_{i\beta}\right)\exp\left\{-F(\underline{x}^{(i)})/T\right\},\qquad i=1,2,\ldots,m, \tag{8.7}$$

satisfies the eigenvalue equation $P\underline{v}=\underline{v}$. Therefore $\underline{v}_*$ is the multiple of $\underline{v}$ whose components sum to one, which agrees completely with the suggestions.

It can also happen that the limit (8.4) is unhelpful to the calculation of a global minimum. For example, we consider the case when $n=1$, $m=4$ and $\underline{x}^{(i)}=i$, $i=1,2,3,4$, but we depart from the previous ordering of points of $\mathcal{X}$ by picking the function values

$$F(\underline{x}^{(1)})=0, \quad F(\underline{x}^{(2)})=3, \quad F(\underline{x}^{(3)})=1 \quad \text{and} \quad F(\underline{x}^{(4)})=2. \tag{8.8}$$

Further, we let each row of W have two elements of zero and two elements of 1/2, where $W_{ij}=1/2$ occurs if and only if $\underline{x}^{(j)}$ is one of the two nearest neighbours of $\underline{x}^{(i)}$. Thus P is the matrix

$$P=\begin{pmatrix} 1-\frac{1}{2}\theta-\frac{1}{2}\theta^3 & \frac{1}{2} & 0 & 0 \\ \frac{1}{2}\theta^3 & 0 & \frac{1}{2}\theta^2 & \frac{1}{2}\theta \\ \frac{1}{2}\theta & \frac{1}{2} & 1-\frac{1}{2}\theta-\frac{1}{2}\theta^2 & \frac{1}{2} \\ 0 & 0 & \frac{1}{2}\theta & \frac{1}{2}-\frac{1}{2}\theta \end{pmatrix}, \tag{8.9}$$

where θ denotes $\exp(-1/T)$. Hence straightforward calculation gives the exact solution

$$\underline{v}=\begin{pmatrix} (2+\theta)\,\theta \\ (2+\theta)\,(1+\theta^2)\,\theta^2 \\ (2+\theta^2)\,(1+\theta) \\ (2+\theta^2)\,\theta \end{pmatrix} \tag{8.10}$$

of the eigenvalue equation $P\underline{v}=\underline{v}$. Therefore, when T is small, the given algorithm tends to provide the vector of variables $\underline{x}^{(3)}$, although it is clear that $F(\underline{x}^{(1)})$ is the least of the function values (8.8). This pathological behaviour occurs because a transition from the $\underline{x}^{(1)}$ state to the $\underline{x}^{(3)}$ state is more likely than a return from the $\underline{x}^{(3)}$ state to the $\underline{x}^{(1)}$ state, although $F(\underline{x}^{(1)})$ is less than $F(\underline{x}^{(3)})$. Indeed, the return from $\underline{x}^{(3)}$ to $\underline{x}^{(1)}$ has to be via $\underline{x}^{(2)}$, and the large value of $F(\underline{x}^{(2)})$ is obstructive. Therefore it is helpful if the sparsity structure of the matrix W is symmetric.

The rate of cooling is very important to efficiency, especially when the starting vector of variables $\underline{x}_1 \in \mathcal{X}$ is near or at a local minimum of the objective function that is not optimal. We address some of the questions that arise when $\mathcal{X}$ is the finite set $\{\underline{x}^{(i)} : i=1,2,\ldots,m\}$ as before. Further, we suppose for convenience that the function values $F(\underline{x}^{(i)})$, $i=1,2,\ldots,m$, are all different. Therefore it is suitable to define $\underline{x}^{(j)}$ to be a local minimum if, for every integer i in $[1,m]$ that satisfies $i \neq j$ and $W_{ij}>0$, the strict inequality $F(\underline{x}^{(i)}) > F(\underline{x}^{(j)})$ holds. Because the simulated annealing method is likely to pick a vector of variables that is a local minimum, and because we can regard any $\underline{x}_{k+1}$ as a starting point for the subsequent iterations, we address the case when the initial vector $\underline{x}_1 = \underline{x}^{(s)}$ is a local minimum. Then the probability that $\underline{x}_2$ is different from $\underline{x}_1$ has the value $\sum_{i=1,\, i\neq s}^{m} W_{is} \exp\{-[F(\underline{x}^{(i)})-F(\underline{x}^{(s)})]/T\}$, which tends to zero if $T\to 0$. Further, when $\underline{x}_2$ is different from $\underline{x}_1$, then the most likely choice of $\underline{x}_3$ may be $\underline{x}^{(s)}$ again. These remarks indicate why a fairly large value of T should be chosen initially.

A suitable way of making this choice automatically is as follows. For each iteration number k, let the integers p and q be taken from expression (8.2), so q is defined by $\underline{x}^{(q)}=\underline{x}_k$, while $\underline{x}^{(p)}$ is the new vector of variables if $\underline{x}_{k+1}$ is different from $\underline{x}_k$. We make use of the remark that $F(\underline{x}^{(p)})-F(\underline{x}^{(q)})=\Delta_k$, say, is independent of T, and we aim for a probability of about 50% that an iteration will change the vector of variables when a change would increase the objective function. Therefore, if k is the index of the first iteration that gives $\Delta_k>0$, we define a provisional value of T by the condition $\exp(-\Delta_k/T)=1/2$. Then, on subsequent iterations, the provisional T is revised to an approximate solution of the equation

$$\text{Average value of } \{\exp(-\Delta_j/T) : j\in[1,k],\ \Delta_j>0\} \;=\; 1/2, \qquad (8.11)$$

until the number of values of j in the braces reaches a prescribed amount. For instance, we may fix the initial T by condition (8.11) when 50 of the numbers Δ_j, $j=1,2,\ldots,k$, are positive, and this T may be used for hundreds of iterations.

The analogy with the cooling of a liquid motivates some automatic reductions in T, but it has been mentioned that the temperature should not be decreased too rapidly. Specifically, we require the property that, if $\underline{x}_1 = \underline{x}^{(s)}$ is a 'strict local minimum', then the probability of choosing $\underline{x}_{k+1} = \underline{x}_k$ on every iteration is zero for $k\to\infty$. Here 'strict local minimum' means that, if a single iteration can change the variables from $\underline{x}^{(s)}$ to $\underline{x}^{(\ell)}$, then $F(\underline{x}^{(\ell)}) > F(\underline{x}^{(s)})$ holds. The probability of $\underline{x}_{k+1}=\underline{x}_k$ when $\underline{x}_k=\underline{x}^{(s)}$ is the P_{ss} element of expression (8.3), assuming the usual ordering $F(\underline{x}^{(i)})\leq F(\underline{x}^{(i+1)})$, $i=1,2,\ldots,m-1$. We also assume the strict inequality $F(\underline{x}^{(s)})<F(\underline{x}^{(s+1)})$. Therefore, letting T_k be the current temperature, this probability has the lower bound

$$\begin{aligned} P_{ss} &\geq W_{ss} + \textstyle\sum_{\ell=s+1}^{m} W_{s\ell} \left(1-\exp\left\{-[F(\underline{x}^{(s+1)})-F(\underline{x}^{(s)})]/T_k\right\}\right) \\ &\geq 1-\exp\left\{-[F(\underline{x}^{(s+1)})-F(\underline{x}^{(s)})]/T_k\right\} \\ &= 1-\exp(-\Delta_*/T_k), \qquad (8.12) \end{aligned}$$

say, the second inequality being valid because the local minimum property of $\underline{x}^{(s)}$ provides $\sum_{i=s}^{m} W_{si}=1$. Thus there is zero probability of no change to the variables only if the sequence of temperatures T_k, $k=1,2,3,\ldots$, satisfies the equation

$$\prod_{k=1}^{\infty}\{1-\exp(-\Delta_*/T_k)\}=0, \tag{8.13}$$

which is equivalent to the condition

$$\sum_{k=1}^{\infty}\exp(-\Delta_*/T_k)=\infty. \tag{8.14}$$

It follows that the cooling rate is too fast if we let each T_k be of magnitude $1/k$. Indeed, in this case the analytic formula for the sum of a geometric progression shows that the left-hand side of expression (8.14) is finite.

On the other hand, if T_k is of magnitude $1/\log k$ for large k, then condition (8.14) can hold. Specifically, the choice $T_k = c/\log k$, where c is a positive constant, has the property (8.14) if and only if the integral

$$\int_1^{\infty}\exp(-\Delta_* c^{-1}\log\theta)\,\mathrm{d}\theta = \int_0^{\infty}\exp\{(1-\Delta_* c^{-1})\,t\}\,\mathrm{d}t \tag{8.15}$$

is infinite, where the second integral is due to change of variables $t=\log\theta$. Therefore we require c to be at least Δ_*, and a greater lower bound on c can allow for the inequalities of expression (8.12). Thus $T_k \sim 1/\log k$ is the cooling rate that is usually recommended for a simulated annealing algorithm, as mentioned in van Laarhoven and Aarts (1987).

In practice, however, it is easy to avoid $\underline{x}_k=\underline{x}^{(s)}$ for every k by not decreasing T until $\underline{x}_{k+1}$ is different from $\underline{x}_k$. Indeed, a typical 'cooling schedule' would begin by generating T by the method described earlier that satisfies equation (8.11) for a suitable choice of k. Then, whenever 50 iterations, say, have provided increases in the objective function for the current T, either the temperature is multiplied by 0.9 or the calculation is terminated, the last action being taken when T is sufficiently small. Moreover, in the usual case when the range of the variables $\underline{x}\in\mathbb{R}^n$ is a continuum instead of a discrete set $\mathcal{X}$, a technique is required for choosing the vector $\underline{x}^{(p)}$ of the method in the paragraph containing equation (8.1). It is often suitable to pick $\underline{x}^{(p)}$ at random from the set $\{\underline{x} : \|\underline{x}-\underline{x}_k\|_2\le\rho\}$, where ρ is a prescribed positive constant. Then the rules for deciding between the alternatives $\underline{x}_{k+1}=\underline{x}^{(q)}$ and $\underline{x}_{k+1}=\underline{x}^{(p)}$ are the same as before. There are several computer programs that apply the simulated annealing method. For example, a program for the solution of the travelling salesman problem is given in Press, Flannery, Teukolsky and Vetterling (1986).

9. Discussion

Usually it is very difficult to discover general convergence theorems for algorithms for nonlinear optimization that perform well in practice, especially when first derivatives are not available. Therefore I have always supported the view that new algorithms should be developed from ideas for techniques that may provide improvements to actual calculations. Further, as well as trying the techniques in numerical experiments, in order to find any advantages over other methods, it is important to study what can go wrong. Such research is hardly ever conclusive, and the outcome may be an algorithm that compares favourably with other procedures on a range of test problems, but that fails occasionally. Many papers on such work are published in journals and even more are rejected. On the other hand, the literature also includes several descriptions of algorithms that are designed so that convergence properties can be proved, which rules out methods that allow the construction of examples that show failure. In particular, both the procedure that makes exact line searches along coordinate directions recursively and the method of Nelder and Mead (1965) are excluded, because of the counter-examples that are mentioned in Sections 2 and 4, respectively. Indeed, there seems to be hardly any correlation between the algorithms that are in regular use for practical applications and the algorithms that enjoy guaranteed convergence in theory, although advances on the theoretical side should influence the development of software for optimization calculations. We will consider the methods of Sections 2–8 in the light of these comments.

The convergence theory of Section 2 presents a result that is not well known. It is that the conditions (2.3) and (2.8) on the search directions and step-lengths, respectively, can ensure that $\liminf\{\|\underline{\nabla} F(\underline{x}_k)\| : k=1,2,3,\ldots\}$ is zero in exact arithmetic, provided that the objective function has continuous first derivatives and the sequence $\underline{x}_k$, $k=1,2,3,\ldots$, has a limit point. This property is achieved by a method that requires the positive parameters β_k, $k=1,2,3,\ldots$, that are introduced just before expression (2.8), but the need for these data is objectionable. Indeed, if the parameters are not superfluous, then they fix some important decisions about step-lengths before all or most of the values of the objective function are calculated. Instead it would be reasonable to generate the parameters automatically as the iterations proceed, which would also assist the use of the algorithm. We explain this remark by assuming that $F(\underline{x}_p)-F(\underline{x}^*)$ is less than $\gamma\beta_p^2$ at the beginning of the pth iteration, where $F(\underline{x}^*)$ is the least value of the objective function. Then the first line of expression (2.8) implies $\|\alpha\,\underline{d}_p\|_2<\beta_p$, but the procedure in Section 2 sets the step-length α_p to zero in this case. Further, if q is the number of the next iteration that changes the variables, then $\underline{x}_q=\underline{x}_p$ and $\gamma\beta_q^2\leq F(\underline{x}_p)-F(\underline{x}^*)$ must hold, while the iterations with numbers $k\in[p,q-1]$ each calculate at least one value of $F(\underline{x}_k+\alpha\,\underline{d}_k)$, although they cannot alter the current vector of variables until $\gamma\beta_k^2$ is reduced to at most $F(\underline{x}_p)-F(\underline{x}^*)$. Therefore it is suitable to pick $\beta_{k+1}<\beta_k$ during these iterations, and to set $\beta_{k+1}=\beta_k$ in the case $\underline{x}_{k+1}\neq\underline{x}_k$, but $\beta_{k+1}=\beta_k$ may be safer than $\beta_{k+1}<\beta_k$ if the reason for $\underline{x}_{k+1}=\underline{x}_k$ is that $\underline{d}_k$ is practically orthogonal to the unknown gradient $\underline{\nabla} F(\underline{x}_k)$. Thus the initial choice of β_k, $k=1,2,3,\ldots$, may be avoided, by letting β_{k+1} depend on β_k and available values of F, which is like the adjustment of trust region radii in trust region algorithms. Further, if the differences $\beta_k - \beta_{k+1}$ are sufficiently parsimonious, then it may be possible to replace both parts of expression (2.8) by the single condition

$$F(\underline{x}_k + \alpha\, \underline{d}_k) \le F(\underline{x}_k) - \gamma \beta_k^2 \quad \text{if} \quad \alpha \ne 0, \tag{9.1}$$

with the proviso that the step-length is set to zero only if inequality (9.1) fails in the case $\alpha = \beta_k / \|\underline{d}_k\|_2$. These remarks suggest that some useful convergence properties can be achieved in a line search method, by using search directions that satisfy expression (2.3), and by introducing one or two trust region ideas into the control of step-lengths.

The discrete grid methods, considered in Section 3, are both an inspiration and a target for adverse criticism. The analysis of convergence given there justifies the title 'Direct search methods: once scorned, now respectable' of Wright (1996). Moreover, if the set $\{\underline{x} : F(\underline{x}) \le F(\underline{x}_1)\}$ is bounded and if the mesh size does not change, then eventually a vector of variables is generated that has occurred already during the calculation, which is useful, because this situation can activate operations like reductions in the mesh size, as stated in Section 3 and in the last paragraph of Section 7. On the other hand, examples like the minimization of the function

$$F(\underline{x}) = F(x_1, x_2) = (x_1 - x_2 + 0.2)^2 + 10^{-4}(x_1 + x_2 + 77)^2, \qquad \underline{x} \in \mathbb{R}^2, \tag{9.2}$$

cause some consternation. Specifically, if the points of the grid are all vectors whose components are integers, and if $F(0,0) = 0.6329$ has been calculated, then the best grid point so far has to satisfy $x_1 = x_2$. Therefore it cannot be one of the four neighbours of $(0,0)$. Further, the least value of the continuous function (9.2) occurs at $(x_1, x_2) = (-38.6, -38.4)$, and the nearest grid point is $(-39, -38)$, which gives a function value that exceeds $F(0,0)$. Thus the use of discrete grids can impair the efficiency of the algorithm of Elster and Neumaier (1995), for example, even when F is quadratic. These disadvantages of restricting the vectors of variables to discrete grids are well known, so 'respectability' has been achieved by the recent work on convergence theory.

I have never liked the simplex methods of Section 4, because it seems very wasteful to make changes to the variables in ways that depend only on the signs of differences between calculated function values, the magnitudes of the differences being ignored. Therefore I was surprised to be told more than ten years ago that the Nelder and Mead (1965) algorithm is the method for unconstrained optimization that is used most often. I was even more surprised to learn later about the cases of failure that are reported at the end of Section 4. These remarks provide a clear demonstration that there is a strong need for easy-to-use algorithms that are more reliable. New procedures are proposed occasionally. For example, Kelley (1997) has developed an 'oriented restart' technique that replaces the current simplex by a more suitable one automatically, when the current simplex is found to be nearly degenerate. Another possible way of avoiding collapse is to confine all the vertices of all the simplices to a discrete grid. Then it may be possible to generalize equation (4.2) to the formula

$$\hat{\underline{v}}_m = 2 \sum_{\substack{i=1 \\ i \neq m}}^{n+1} \theta_i \underline{v}_i - \underline{v}_m, \tag{9.3}$$

where the multipliers θ_i are nonnegative and sum to one. Thus the next vector of variables can depend on the actual values of the objective function at the vertices of the current simplex, which may provide better efficiency in typical applications.

The conjugate direction method of Powell (1964), described in Section 5, is another algorithm that has been in regular use for more than 30 years. When I developed it, I was concerned about the choice of m in the paragraph containing expressions (5.9)–(5.13), because the given procedure may not retain those search directions that have already been given the required conjugacy properties in the quadratic case. Then I took the view that it was possible to secure convergence by using search directions that span the full space of the variables. However, we now know, from the example in the third paragraph of Section 2, that even condition (2.3) on an infinite sequence of search directions does not guarantee $\liminf_{k\to\infty} \|\underline{\nabla} F(\underline{x}_k)\| = 0$, when the line searches are exact, although F can be very smooth with bounded level sets. Therefore, because sufficiently accurate line searches are vital to the given techniques that achieve conjugacy, the convergence properties of conjugate direction methods deserve further attention. It is possible that our remarks on inequality (9.1), that suggest a zero step-length if the inequality fails, may be helpful.

The first version of the method in Section 6 was developed for Westland Helicopters, in order to solve a range of optimization problems that each have about 4 variables and 10 constraints. It was anticipated correctly that, due to the tiny number of variables, the inefficiencies that arise from linear approximations to the objective and constraint functions would be tolerable. An easy-to-use Fortran program was written that implements the algorithm of Powell (1994), and it is available from the author. The program is very slow, as expected, when there are no constraints, because of the importance of the curvature of the objective function. On the other hand, linear approximations to constraints are usually excellent for reducing the freedom in the variables in a suitable way. We address one convergence question, namely whether the method of Section 6 terminates if only a finite number of iterations cause $\underline{x}_{k+1}$ to be different from $\underline{x}_k$. This assumption is reasonable because of the limited precision of the computer arithmetic, and because $\underline{x}_{k+1} \neq \underline{x}_k$ requires the strict inequality $\Psi(\underline{x}_{k+1}) < \Psi(\underline{x}_k)$, where

Ψ is the merit function (6.14). We give careful attention to the rules listed on pages 314–315, remembering that the number of reductions in Δ_k is also finite. Thus we find that termination does not occur if and only if the situations (5) and (6) are not reached during an infinite number of consecutive iterations, and we suppose that termination fails in this way. Then Δ_k is independent of k, and the above assumption allows $\underline{x}_k$ to be independent of k too, without loss of generality. In other words, no iteration improves the best vector of variables so far, which excludes rule (1). The number of minimization steps is infinite, however, because otherwise every step would be a simplex step eventually, which would give a contradiction, due to the increases in the volume of the simplex after the inequality

$$\max\{\|\underline{v}_i-\underline{x}_k\| : i=1,2,\ldots,n+1\} \le \beta\Delta_k \tag{9.4}$$

is achieved. Therefore the trust region radius ρ_k is reduced until $\rho_k=\Delta_k$ holds for all sufficiently large k. Then $\beta>1$ implies that a minimization step does not increase the number of integers i in $[1,n+1]$ that satisfy $\|\underline{v}_i-\underline{x}_k\|>\beta\Delta_k$, nor does it diminish the volume of the simplex, although a nonoptimal vertex of the simplex may be changed. It follows that the volume of the simplex would become unbounded if the number of simplex steps were infinite, so rules (2) and (3) are also irrelevant for large k. Therefore every simplex becomes acceptable. Further, we deduce from $\rho_k=\Delta_k$ that there are no more 'questionable' minimization steps. Thus rule (4) is excluded too, which completes the proof of termination. One purpose of this analysis is to justify the given rules. In particular, lack of termination would be prevalent if the last proviso of rule (4) were deleted, because it is usual for the minimization step to satisfy $\|\underline{v}_{n+2}-\underline{x}_k\|=\rho_k$ when Φ is a linear function.

The analysis of termination in the previous paragraph applies also to the algorithm that is the main subject of Section 7. Further, we can generalize that algorithm by letting Φ be an approximation to F from any prescribed finite-dimensional linear space of functions from $\mathbb{R}^n$ to $\mathbb{R}$, say $\mathcal{A}$. Then $\hat{n}$ is the dimension of $\mathcal{A}$, and each iteration begins with points $\underline{v}_i$, $i=1,2,\ldots,\hat{n}$, such that the equations (7.1) define Φ uniquely. Therefore $\mathcal{A}$ includes cardinal functions χ_i, $i=1,2,\ldots,\hat{n}$, derived from the conditions (7.2) as before. Further, all the uses of the cardinal functions in Section 7 are preserved, including the definition of the 'simplex step'. In particular, if $\mathcal{A}$ is the $(n+1)$-dimensional space of linear polynomials, then the generalization reduces to the method of Section 6 when there are no constraints. A choice of $\mathcal{A}$ that is between the linear polynomial case and the $\hat{n}=\frac{1}{2}(n+1)\,(n+2)$ quadratic polynomial case has been suggested by Philippe Toint (private communication), and I expect it to be highly useful. It is relevant when F has the form

$$F(\underline{x}) = \sum_{j=1}^{t} F_j(\underline{x}), \qquad \underline{x}\in\mathbb{R}^n, \tag{9.5}$$

where each of the functions F_j, $j=1,2,\ldots,t$, is independent of most of the components of $\underline{x}$. Then we let $\mathcal{S}$ be the set of pairs of integers $\{p,q\}$ from $[1,n]$, such that $\{p,q\}$ is in $\mathcal{S}$ if and only if no F_j depends on both x_p and x_q. Alternatively, if F is twice differentiable, then we pick the set

$$\mathcal{S} = \{\{p,q\} : \partial^2 F(\underline{x}) \,/\, \partial x_p \, \partial x_q \equiv 0\}, \tag{9.6}$$

which may be larger than before. Of course, we let $\mathcal{A}$ be the linear space of quadratic polynomials with the zero second derivatives $\partial^2 \Phi(\underline{x}) \,/\, \partial x_p \, \partial x_q = 0$, $\{p,q\} \in \mathcal{S}$, for every Φ in $\mathcal{A}$. Thus the elements of $\mathcal{A}$ are given some of the sparsity properties of F, and $\hat{n}$ may be much less than $\frac{1}{2}(n+1)\,(n+2)$, which would provide large reductions in the amount of routine computation. Another question that is important to applications is whether the method of this paragraph can be extended to allow constraints on the variables. It is straightforward to approximate the constraints $c_p(\underline{x}) \geq 0$, $p=1,2,\ldots,m$, by the inequalities $\gamma_p(\underline{x}) \geq 0$, $p=1,2,\ldots,m$, where each γ_p is the element of $\mathcal{A}$ that is defined by the equations $\gamma_p(\underline{v}_i) = c_p(\underline{v}_i)$, $i=1,2,\ldots,\hat{n}$, at the beginning of the iteration. The calculation of a minimization step would be difficult, however, if we had to minimize a quadratic function subject to quadratic constraints. Therefore it may be best to give most attention to linear constraints for a while. The algorithm of Elster and Neumaier (1995), for instance, allows simple bounds on the variables.

The simulated annealing method of Section 8 is attractive to many computer users, because it is easy to apply, the amount of routine work of each iteration is only linear in n, and some theory suggests that it is possible to find the global solution of several optimization problems. Further, in many applications, any reduction in the value of the objective function may be financially rewarding, which does not favour algorithms that stop at a local minimum. Indeed, the failures of the method of Nelder and Mead (1965) may seem to be no worse than termination at a local minimum. Of course these remarks depend on the nature of the application, partly because high accuracy would be inappropriate for many models of real situations. On the other hand, there is also a wide range of precise optimization problems in science and engineering, so there are strong reasons for further developments of the techniques of Sections 2–7, and it would be good to have some efficient software for such problems that is attractive to computer users. My opinion of simulated annealing is that, for small and moderate values of n, it is highly inefficient to take decisions randomly. Therefore one may be able to construct procedures that provide similar results much more quickly, in cases when most of the computing time is spent on calculations of values of the objective function. Indeed, the importance of global optimization in practice makes an excellent case for much more work on general algorithms that need not be trapped by local minima.

REFERENCES

R. P. Brent (1973), *Algorithms for Minimization without Derivatives*, Prentice-Hall, Englewood Cliffs, NJ.

K. W. Brodlie (1975), 'A new direction set method for unconstrained minimization without evaluating derivatives', *J. Inst. Math. Appl.* **15**, 385–396.

A. R. Conn, K. Scheinberg and Ph. L. Toint (1997*a*), 'On the convergence of derivative-free methods for unconstrained optimization', in *Approximation Theory and Optimization* (M. D. Buhmann and A. Iserles, eds), Cambridge University Press, Cambridge, pp. 83–108.

A. R. Conn, K. Scheinberg and Ph. L. Toint (1997*b*), 'Recent progress in unconstrained nonlinear optimization without derivatives', *Math. Prog.* **79**, 397–414.

J. E. Dennis and V. Torczon (1991), 'Direct search methods on parallel machines', *SIAM J. Optim.* **1**, 448–474.

C. Elster and A. Neumaier (1995), 'A grid algorithm for bound constrained optimization of noisy functions', *IMA J. Numer. Anal.* **15**, 585–608.

R. Fletcher (1987), *Practical Methods of Optimization*, Wiley, Chichester.

P. E. Gill, W. Murray and M. H. Wright (1981), *Practical Optimization*, Academic Press, London.

D. E. Goldberg (1989), *Genetic Algorithms in Search, Optimization and Machine Learning*, Addison-Wesley, Reading, MA.

L. Grippo, F. Lampariello and S. Lucidi (1988), 'Global convergence and stabilization of unconstrained minimization methods without derivatives', *J. Optim. Theory Appl.* **56**, 385–406.

R. Hooke and T. A. Jeeves (1961), 'Direct search solution of numerical and statistical problems', *J. Assoc. Comput. Mach.* **8**, 212–229.

C. T. Kelley (1997), 'Detection and remediation of stagnation in the Nelder–Mead algorithm using a sufficient decrease condition', North Carolina State University Report CRSC-TR97-2.

P. J. M. van Laarhoven and E. H. L. Aarts (1987), *Simulated Annealing: Theory and Applications*, Reidel, Dordrecht.

S. Lucidi and M. Sciandrone (1997), 'On the global convergence of derivative free methods for unconstrained optimization', preprint, Università di Roma 'La Sapienza', Italy.

K. I. M. McKinnon (1997), 'Convergence of the Nelder–Mead simplex method to a nonstationary point', preprint (to be published in *SIAM J. Optim.*).

J. A. Nelder and R. Mead (1965), 'A simplex method for function minimization', *Comput. J.* **7**, 308–313.

M. J. D. Powell (1964), 'An efficient method for finding the minimum of a function of several variables without calculating derivatives', *Comput. J.* **7**, 155–162.

M. J. D. Powell (1973), 'On search directions for minimization algorithms', *Math. Prog.* **4**, 193–201.

M. J. D. Powell (1994), 'A direct search optimization method that models the objective and constraint functions by linear interpolation', in *Advances in Optimization and Numerical Analysis* (S. Gomez and J-P. Hennart, eds), Kluwer Academic, Dordrecht, pp. 51–67.

W. H. Press, B. P. Flannery, S. A. Teukolsky and W. T. Vetterling (1986), *Numerical Recipes: The Art of Scientific Computation*, Cambridge University Press, Cambridge.

H. H. Rosenbrock (1960), 'An automatic method for finding the greatest or least value of a function', *Comput. J.* **3**, 175–184.

C. S. Smith, (1962), 'The automatic computation of maximum likelihood estimates', N.C.B. Sci. Dept. Report SC 846/MR/40.

W. Spendley, G. R. Hext and F. R. Himsworth (1962), 'Sequential application of simplex designs in optimisation and evolutionary operation', *Technometrics* **4**, 441–461.

Ph. L. Toint and F. M. Callier (1977), 'On the accelerating property of an algorithm for function minimization without calculating derivatives', *J. Optim. Theory Appl.* **23**, 531–547.

V. Torczon (1997), 'On the convergence of pattern search algorithms', *SIAM J. Optim.* **7**, 1–25.

D. Winfield (1973), 'Function minimization by interpolation in a data table', *J. Inst. Math. Appl.* **12**, 339–347.

M. H. Wright (1996), 'Direct search methods: once scorned, now respectable', in *Numerical Analysis 1995* (D. F. Griffiths and G. A. Watson, eds), Addison Wesley Longman, Harlow, pp. 191–208.

Acta Numerica (1998), pp. 337–377

Choice of norms for data fitting and function approximation

G. A. Watson
Department of Mathematics,
University of Dundee,
Dundee DD1 4HN, Scotland
E-mail: gawatson@mcs.dundee.ac.uk

For the approximation of functions and data, it is often appropriate to minimize a norm. Many norms have been considered, and a review is presented of methods for solving a range of problems using a wide variety of norms.

CONTENTS

1. Introduction

Two central problems in approximation theory are the approximation of data and the approximation of functions. Given a set of data or a function for which an approximation is required, two fundamental issues are the kind of approximation which should be chosen, and the measure of goodness of fit which should be used. These will of course depend very much on the precise nature of the underlying problem, and our main interest here is a consideration of methods for obtaining solutions to a wide range of such problems which are of practical importance.

As far as the approximation of functions is concerned, the goal is to replace the original function by one that is simpler or more manageable. The most useful measure is the Chebyshev norm: as Meinardus says in the preface to his book (Meinardus 1967), '... it has by far the greatest practical importance'. There has of course been much interest in the use of the L_2

norm, and one reason for this is that the analysis is relatively straightforward, with linear approximation leading directly to the solution of a linear system of equations. There are obvious advantages if the basis functions are orthogonal or orthonormal, so that this is in some ways a natural measure to use when approximating by orthogonal polynomials, Fourier series or orthogonal wavelets. The emphasis here is on methods for computing approximations, and so we will not pursue this further, but confine attention almost entirely to the Chebyshev norm: some relevant material is covered in Section 5. The L_1 norm has also attracted some interest, and we will deal briefly with it in Section 6.

For the approximation of data the situation is quite different, and there are many criteria that are of practical value. There are two main features of data approximation. Firstly, there is a wide range of characteristics which the data may possess. For example, the data may arise from the sampling of a known function, or they may be observed or measured data for which the underlying form is known: a simple example is data, generated experimentally, which might be expected to lie on a straight line. At the other extreme, the data may be irregularly positioned, with no discernible pattern. Secondly, observed data generally contain errors, and the nature of these is an important consideration in deciding a measure for goodness of fit.

The simplest and most direct way of choosing the unknown parameters in data-fitting problems is by interpolation. This may be appropriate, for example, if the model is linear, there is a large number of parameters, and it is known that a nonsingular system will occur. Other more sophisticated measures may indeed be unsuitable or impracticable because of the sheer size of the problem. This is the case in the approximation of scattered data using radial basis functions. Also, if there is significantly different behaviour of the data in different regions, so that considerable flexibility is required, then spline functions may be appropriate, and again interpolation may be the natural thing to use.

This article is, however, not concerned with interpolation, and thus, in the data-fitting context, it will be assumed that the data can be modelled by a function containing a number of free parameters, and minimizing a norm is appropriate. Perhaps the most commonly occurring criterion in such cases is the least squares norm. Its use has a long and distinguished history, it is relatively well understood, and there are good algorithms available. Yet there are often situations where it is not ideal. For example, a statistical justification for least squares requires certain assumptions about the error pattern in the data, and if these are not satisfied there may be bias in the estimate.

Therefore there are many other norms which are of interest in data fitting, and which have been studied from both a theoretical and a practical point of view. We will use this statement as an excuse for giving in Section 2 a very general theoretical treatment of the conventional linear problem for arbitrary norms, and we will consider in particular the question of characterization of solutions of such problems. The analysis is of course contained in a treatment of approximation problems in completely general normed linear spaces, and this is well known to approximation theorists. However, whereas that requires some very sophisticated functional analysis, very powerful results can be obtained for the present problem in a comparatively straightforward manner, and this should be accessible to the general readership of this book. The main tool is the subdifferential of the norm, which extends the idea of the derivative to the non-differentiable case. Therefore we will give some attention to this, and then go on to use some of the results in special cases.

In Section 3, an important modification of the usual linear problem is addressed in some (though not complete) generality. In Section 4, nonlinear problems are briefly considered, again with some emphasis on a general treatment.

2. Approximation to data by linear models

Suppose that a relationship exists between variables so that one of the variables can be expressed as a linear combination of n functions of the others. Then, if a set of values of the variables is generated which is assumed to satisfy this relationship, the result is a system of linear equations, say

$$A\mathbf{x} = \mathbf{b}, \tag{2.1}$$

where $\mathbf{x} \in \mathbb{R}^n$ represents the unknown coefficients, $\mathbf{b} \in \mathbb{R}^m$ is a vector of values of the dependent variable, and $A \in \mathbb{R}^{m\times n}$ is formed from the values of the other variables. If the available data are subject to errors, and $m > n$, then (2.1) is an over-determined linear system which normally has no (exact) solution. If it is assumed that the errors are only present in the data values forming the vector $\mathbf{b}$, then we can introduce a vector $\mathbf{r}$ of perturbations of $\mathbf{b}$, representing these errors, so that

$$\mathbf{r} = A\mathbf{x} - \mathbf{b}, \tag{2.2}$$

and choose $\mathbf{x}$ to make the components of $\mathbf{r}$ small in some sense.

The problem of the solution of an overdetermined system of linear equations in this form has attracted enormous interest. Typically, this is done by solving the following problem:

$$\text{find } \mathbf{x} \in \mathbb{R}^n \text{ to minimize } \|\mathbf{r}\|, \tag{2.3}$$

where $\|\cdot\|$ is a given norm on $\mathbb{R}^m$. A solution always exists, and if the norm is differentiable, then we can easily characterize a minimum by zero-derivative conditions. Such conditions are readily extended to the general case through the use of the subdifferential. We will consider this next.

2.1. Characterization of solutions

Let $\|\cdot\|$ be a norm on $\mathbb{R}^m$. Then, for any $\mathbf{v} \in \mathbb{R}^m$, the *dual* norm is the norm on $\mathbb{R}^m$ defined by

$$\|\mathbf{v}\|^* = \max_{\|\mathbf{r}\| \le 1} \mathbf{r}^T\mathbf{v}. \tag{2.4}$$

The relationship between a norm on $\mathbb{R}^m$ and its dual is symmetric, so that, for any $\mathbf{r} \in \mathbb{R}^m$,

$$\|\mathbf{r}\| = \max_{\|\mathbf{v}\|^* \le 1} \mathbf{r}^T\mathbf{v}. \tag{2.5}$$

Important special cases are the l_p norms,

$$\begin{aligned} \|\mathbf{r}\|_p &= \left(\sum_{i=1}^m |r_i|^p\right)^{1/p}, \quad 1 \le p < \infty, \\ \|\mathbf{r}\|_\infty &= \max_{1\le i\le m} |r_i|. \end{aligned}$$

Then the dual norm is the l_q norm, where $1/p + 1/q = 1$.

Definition 1 The *subdifferential*, or set of subgradients of $\|\mathbf{r}\|$, is given by

$$\partial\|\mathbf{r}\| = \{\mathbf{v} \in \mathbb{R}^m : \|\mathbf{s}\| \ge \|\mathbf{r}\| + (\mathbf{s}-\mathbf{r})^T\mathbf{v}, \quad \text{for all } \mathbf{s} \in \mathbb{R}^m\}. \tag{2.6}$$

This set is closed, bounded and convex. It is also easily seen that it is just the set of vectors $\mathbf{v}$ such that equality holds in (2.5): in other words we have the following very useful result.

Lemma 1 Let $\mathbf{r} \in \mathbb{R}^m$. Then

$$\partial\|\mathbf{r}\| = \{\mathbf{v} \in \mathbb{R}^m : \|\mathbf{r}\| = \mathbf{r}^T\mathbf{v}, \ \|\mathbf{v}\|^* \le 1\}. \tag{2.7}$$

Proof. Let $\mathbf{v}$ be in the set defined by (2.6). Then, setting $\mathbf{s} = 0$, and $\mathbf{s} = 2\mathbf{r}$, it follows that $\|\mathbf{r}\| = \mathbf{r}^T\mathbf{v}$, and further $\|\mathbf{v}\|^* \le 1$, from the definition. Thus $\mathbf{v}$ lies in the set defined by (2.7). The reverse implication is immediate. □

If $\|\mathbf{r}\|$ is differentiable at $\mathbf{r}$, then the subdifferential is a singleton with

$$\partial\|\mathbf{r}\| = \left\{\mathbf{v} \in \mathbb{R}^m : v_i = \frac{\partial\|\mathbf{r}\|}{\partial r_i}, \quad i = 1, \ldots, m\right\}.$$

This follows from the inequality in (2.6), using convexity. If $\mathbf{r} = 0$, then obviously

$$\partial\|\mathbf{r}\| = \{\mathbf{v} \in \mathbb{R}^m : \|\mathbf{v}\|^* \le 1\}.$$

If $\mathbf{r} \ne 0$, then it is a consequence of (2.5) that $\|\mathbf{v}\|^* = 1$. For the l_p norms, we have the sets

$$\begin{aligned} \partial\|\mathbf{r}\|_1 &= \{\mathbf{v} \in \mathbb{R}^n : v_i = \operatorname{sign}(r_i), r_i \ne 0; \ |v_i| \le 1, r_i = 0\}, & (2.8) \\ \partial\|\mathbf{r}\|_p &= \{\mathbf{v} \in \mathbb{R}^n : v_i = \operatorname{sign}(r_i)|r_i|^{p-1}\|\mathbf{r}\|_p^{1-p}\}, \quad 1 < p < \infty, & (2.9) \end{aligned}$$

and, defining

$$J = \{i : |r_i| = \|\mathbf{r}\|_\infty\},$$

$$\partial\|\mathbf{r}\|_\infty = \left\{\mathbf{v} \in \mathbb{R}^m : \text{sign}(v_i) = \text{sign}(r_i), \quad i \in J; \quad v_i = 0, \quad i \notin J; \right.$$
$$\left. \sum_{i \in J} |v_i| = 1 \right\}. \quad (2.10)$$

The concept of the subdifferential enables us to characterize readily the solutions of linear best approximation problems set in $\mathbb{R}^m$. Two preliminary lemmas are required. Lemma 2 can in fact be given in a much stronger form, but this version is sufficient for our purposes; Lemma 3 gives an expression for the directional derivative of the norm in terms of the subdifferential. With the assumption that the reader knows that the minimum of a continuous function over a closed and bounded set is attained, the rest of this section should be entirely self-contained.

Lemma 2 Let $\mathbf{r}, \mathbf{s} \in \mathbb{R}^m$, let $\gamma \in \mathbb{R}$, and let $\mathbf{v}(\gamma) \in \partial\|\mathbf{r} + \gamma\mathbf{s}\|$. Then the limit points of $\mathbf{v}(\gamma)$ as $\gamma \to 0$ lie in $\partial\|\mathbf{r}\|$.

Proof. Clearly $\{\mathbf{v}(\gamma)\}$ has limit points; let one of these be $\mathbf{v}$. Then $\|\mathbf{v}\|^* \leq 1$. Further,

$$\begin{aligned} \mathbf{v}(\gamma)^T\mathbf{r} &= \mathbf{v}(\gamma)^T(\mathbf{r} + \gamma\mathbf{s}) - \gamma\mathbf{v}(\gamma)^T\mathbf{s} \\ &= \|\mathbf{r} + \gamma\mathbf{s}\| - \gamma\mathbf{v}(\gamma)^T\mathbf{s}. \end{aligned}$$

Letting $\gamma \to 0$, the result follows. □

Lemma 3 Let $\mathbf{r}, \mathbf{s} \in \mathbb{R}^m$. Then

$$\lim_{\gamma \to 0+} \frac{\|\mathbf{r} + \gamma\mathbf{s}\| - \|\mathbf{r}\|}{\gamma}. \quad (2.11)$$

Proof. For all $\mathbf{v} \in \partial\|\mathbf{r}\|$, using Definition 1,

$$\|\mathbf{r} + \gamma\mathbf{s}\| \geq \|\mathbf{r}\| + \gamma\mathbf{v}^T\mathbf{s}.$$

Also, for all $\mathbf{v}(\gamma) \in \partial\|\mathbf{r} + \gamma\mathbf{s}\|$, again using Definition 1,

$$\|\mathbf{r}\| \geq \|\mathbf{r} + \gamma\mathbf{s}\| - \gamma\mathbf{v}(\gamma)^T\mathbf{s}.$$

Combining these inequalities shows that, for all $\mathbf{v} \in \partial\|\mathbf{r}\|$, $\mathbf{v}(\gamma) \in \partial\|\mathbf{r} + \gamma\mathbf{s}\|$, we have for $\gamma > 0$

$$\mathbf{v}^T\mathbf{s} \leq \frac{\|\mathbf{r} + \gamma\mathbf{s}\| - \|\mathbf{r}\|}{\gamma} \leq \mathbf{v}(\gamma)^T\mathbf{s}.$$

Letting $\gamma \to 0+$ and using Lemma 2, the result follows. □

These results enable a general, and simple, characterization result to be established.

Theorem 1 The problem (2.3) is solved by $\mathbf{x}$, with $\mathbf{r} = A\mathbf{x} - \mathbf{b}$, if and only if there exists $\mathbf{v} \in \partial\|\mathbf{r}\|$ with

$$A^T\mathbf{v} = 0. \tag{2.12}$$

Proof. Let $\mathbf{x}$ be a solution but assume that (2.12) is not satisfied. Consider the problem:

$$\text{find } \mathbf{v} \in \partial\|\mathbf{r}\| \text{ to minimize } \|A^T\mathbf{v}\|_2.$$

Then a solution exists, at $\mathbf{w}$, say. By convexity, $\lambda\mathbf{v} + (1-\lambda)\mathbf{w} \in \partial\|\mathbf{r}\|$, for $0 \le \lambda \le 1$, where $\mathbf{v} \in \partial\|\mathbf{r}\|$ is arbitrary. Thus

$$\begin{aligned} 0 &\le \|A^T\Big(\lambda\mathbf{v} + (1-\lambda)\mathbf{w}\Big)\|_2^2 - \|A^T\mathbf{w}\|_2^2 \\ &= \lambda^2\|A^T(\mathbf{v}-\mathbf{w})\|_2^2 + 2\lambda(\mathbf{v}-\mathbf{w})^T AA^T\mathbf{w}. \end{aligned}$$

The last term on the right-hand side will actually be negative for small positive λ if the coefficient of λ is negative, which would lead to a contradiction, and so

$$\mathbf{v}^T AA^T\mathbf{w} \ge \mathbf{w}^T AA^T\mathbf{w} > 0.$$

Thus, setting $\mathbf{s} = -AA^T\mathbf{w}$ in Lemma 3, and using the fact that $\mathbf{v}$ is arbitrary, contradicts the assumption that $\mathbf{x}$ gives a minimum of the norm.

Now let the conditions hold for some $\mathbf{x} \in \mathbb{R}^n$, with $\mathbf{w} \in \partial\|\mathbf{r}\|$ satisfying (2.12). Then, if $\mathbf{y} \in \mathbb{R}^n$ is arbitrary,

$$\begin{aligned} \|A\mathbf{y} - \mathbf{b}\| &\ge \mathbf{w}^T(A\mathbf{y} - \mathbf{b}) \\ &= \mathbf{w}^T(A\mathbf{x} - \mathbf{b}) \\ &= \|A\mathbf{x} - \mathbf{b}\|. \end{aligned}$$

The proof is completed. □

This result, in conjunction with appropriate sets $\partial\|\mathbf{r}\|$, can be used to obtain specific characterization results.

2.2. l_1 approximation

There is particular interest in approximation using the l_1 norm because it has the property of de-emphasizing the effect of wild points or gross errors in $\mathbf{b}$. We will expand on that after giving a characterization result. For any $\mathbf{x} \in \mathbb{R}^n$, let I denote the set of indices where $r_i = 0$, and let I^c denote its complement. From Theorem 1 and (2.8), we have the following theorem.

Theorem 2 The vector $\mathbf{x}$ is a solution to the l_1 problem if and only if there exists $\boldsymbol{\lambda} \in \mathbb{R}^m$ with

$$\lambda_i = \text{sign}(r_i), \quad i \in I^c, \quad \text{and} \quad |\lambda_i| \le 1, \quad i \in I,$$

such that

$$A^T \boldsymbol{\lambda} = 0. \tag{2.13}$$

The l_1 problem is said to be *primal nondegenerate* if, at any point $\mathbf{x}$, the rows of A corresponding to $i \in I$ are linearly independent; it is *dual nondegenerate* if, for any $\boldsymbol{\lambda}$ satisfying (2.13), at most $m - n$ components of $\boldsymbol{\lambda}$ are equal to 1 in modulus.

Suppose that, at a solution, one of the values of b_i, $i \in I^c$, is perturbed in such a way that sign(r_i) is unchanged. Then clearly the characterization conditions are unaffected, so that the solution is unchanged. It is this property of robustness, or insensitivity to possibly large errors in the data, that makes the l_1 norm important.

Because $\|\mathbf{r}\|_1$ is a piecewise linear function, the most commonly used algorithms have been based on movement between intersections of points having sets of zero components of $\mathbf{r}$. In addition to methods based explicitly on linear programming (for example, Barrodale and Roberts 1973), variants based on reduced gradients (Osborne 1987, Shi and Lukas 1996), or projected gradients (Bartels, Conn and Sinclair 1978) are popular: for some relationships, see Osborne (1985). These methods are finite and, with efficient implementation, including line searches, have for a long time appeared to represent the right way to tackle the problem. Recently, however, there has been interest in iterative methods, which effectively smooth the problem. Partly, this has been due to the success of interior point methods for linear programming problems, and attempts have been made to use such ideas in the l_1 problem. We will examine first a method of this type, based on the idea of affine scaling.

Let A have rank n and let the QR decomposition of A be given by

$$A = YR,$$

where R is $n \times n$ upper triangular, and $[Y : Z]$ is an $m \times m$ orthogonal matrix. Then (2.13) is equivalent to

$$\boldsymbol{\lambda} = Z\mathbf{w}, \tag{2.14}$$

where $\mathbf{w} \in \mathbb{R}^{m-n}$ and (2.2) is equivalent to

$$Z^T(\mathbf{r} + \mathbf{b}) = 0. \tag{2.15}$$

Define D_r to be the diagonal matrix with (i,i) element r_i, and let $\mathbf{g} \in \mathbb{R}^m$ be the vector defined by

$$g_i = \begin{cases} \text{sign}(r_i), & i \in I^c, \\ 0, & i \in I. \end{cases}$$

It follows that $\mathbf{x}$ solves the l_1 problem if and only if there exists $\mathbf{r} \in \mathbb{R}^m$, $\mathbf{w} \in \mathbb{R}^{m-n}$, such that

$$D_r(\mathbf{g} - Z\mathbf{w}) = 0 \tag{2.16}$$

$$Z^T(\mathbf{r} + \mathbf{b}) = 0, \tag{2.17}$$

and, in addition,

$$-1 \leq (Z\mathbf{w})_i \leq 1, \quad i = 1, \ldots, m. \tag{2.18}$$

An affine scaling method attempts to solve (2.16), (2.17), (2.18) by an iterative method that computes a direction of progress from the current vector $\mathbf{r}$ by solving the following subproblem:

$$\begin{aligned} &\text{minimize}_{\mathbf{d} \in \mathbb{R}^m} \ \mathbf{g}^T\mathbf{d} \\ &\text{subject to} \ \ Z^T\mathbf{d} = 0, \\ &\qquad \|D^{-1}\mathbf{d}\|_2 \leq \tau, \end{aligned} \tag{2.19}$$

where D is a given positive definite diagonal matrix, and τ is a given positive number that restricts the size of $\mathbf{d}$: (2.19) can be thought of as scaling the solution. It is a straightforward exercise to show that $\mathbf{d}^* = \alpha\mathbf{d}$ is the solution, where α is a suitably chosen scalar and $\mathbf{d}$ is given by

$$\mathbf{d} = -A\left(A^T D^{-2} A\right)^{-1} A^T \mathbf{g}. \tag{2.20}$$

Assume that, at the current $\mathbf{r}$, (2.17) is satisfied (and so remains satisfied for subsequent $\mathbf{r}$), and also $I = \phi$. Then $\mathbf{g}$ is just the gradient of $\|\mathbf{r}\|_1$ at the current point, and it follows that the solution $\mathbf{d}$ is a descent direction for the l_1 norm (since $\mathbf{d} = 0$ implies that $A^T\mathbf{g} = 0$). A line search to minimize the piecewise linear function $\|\mathbf{r} + \alpha\mathbf{d}\|$ with respect to α can readily be incorporated, and, if we stop short of the optimal step length, then we can start again from a point with $I = \phi$. Note that no explicit value of τ is actually required.

Now consider the alternative of applying Newton's method to (2.16) and (2.17). The Newton step in $\mathbf{r}$ and $\mathbf{w}$ is given by solving the system

$$\begin{bmatrix} D_\beta & -D_r Z \\ Z^T & 0 \end{bmatrix} \begin{bmatrix} \delta\mathbf{r} \\ \delta\mathbf{w} \end{bmatrix} = \begin{bmatrix} -D_r(\mathbf{g} - Z\mathbf{w}) \\ 0 \end{bmatrix}, \tag{2.21}$$

where $D_\beta = \text{diag}\{\beta_i, \ldots, \beta_m\}$, with $\beta_i = g_i - (Z\mathbf{w})_i$, $i = 1, \ldots, m$. It follows from this system that

$$\boldsymbol{\delta}\mathbf{r} = -A\left(A^T D_r^{-1} D_\beta A\right)^{-1} A^T \mathbf{g}. \tag{2.22}$$

Note that (2.20) and (2.22) have the same general form. Let D be chosen by

$$D = \text{diag}\{|r_i|^{1/2}, \quad i = 1, \ldots, n\},$$

and write

$$\boldsymbol{\delta}_\theta \mathbf{r} = -A(A^T W_\theta A)^{-1} A^T \mathbf{g}, \tag{2.23}$$

where

$$W_\theta = \text{diag}\{|r_i^{-1}(g_i - (1-\theta)(Z\mathbf{w})_i)|, \quad i = 1, \ldots, m\}.$$

Then the choice $\theta = 1$ in (2.23) gives (2.20), and the choice $\theta = 0$ gives (2.22), provided that $D_r^{-1} D_\beta$ is positive definite. That this is true for a set of points $\mathbf{r}$ arbitrarily close to a solution but having no component of $\mathbf{r}$ zero is a key observation in the hybrid method of Coleman and Li (1992*a*). By providing a criterion for choosing θ so that $\theta \to 0$ as $(\mathbf{r}, \boldsymbol{\lambda})$ tend to optimal values, and working with (2.23), they develop a method which is globally convergent to a solution, with a quadratic convergence rate, if the l_1 problem is primal and dual nondegenerate. The full Newton step is not taken asymptotically because of the damping required to maintain differentiability; however, sufficiently accurate approximations to the full Newton step are achieved to permit quadratic convergence.

Note that (2.23) may be solved by first calculating the l_2 solution of the system

$$W_\theta^{1/2} A\, \mathbf{d}_1 = W_\theta^{-1/2} \mathbf{g},$$

followed by setting

$$\boldsymbol{\delta}_\theta \mathbf{r} = -A\mathbf{d}_1. \tag{2.24}$$

The new value of $\mathbf{r}$ is then obtained by a line search in the direction $\boldsymbol{\delta}_\theta \mathbf{r}$. To obtain a new value of $\boldsymbol{\lambda}$, it helps to observe that use of the hybrid method corresponds to solving the system analogous to (2.21), namely

$$\begin{bmatrix} W_\theta & -Z \\ Z^T & 0 \end{bmatrix} \begin{bmatrix} \boldsymbol{\delta}_\theta \mathbf{r} \\ \boldsymbol{\delta}\mathbf{w} \end{bmatrix} = \begin{bmatrix} -(\mathbf{g} - Z\mathbf{w}) \\ 0 \end{bmatrix}. \tag{2.25}$$

It follows from this that the current value of $\boldsymbol{\lambda}$ can be updated to the value

$$Z(\mathbf{w} + \boldsymbol{\delta}\mathbf{w}) = \mathbf{g} + W_\theta \boldsymbol{\delta}_\theta \mathbf{r},$$

with $\boldsymbol{\delta}_\theta \mathbf{r}$ given by (2.24). Of course, (2.13) remains satisfied. Although Z appears as an aid to the theoretical development of the method, note that its actual computation is unnecessary, and we need only work with $\mathbf{r}$ and $\boldsymbol{\lambda}$.

An initial approximation may be obtained by choosing the $\mathbf{r}$ satisfying (2.2) of minimum l_2 norm, and taking $\boldsymbol{\lambda}$ to be a multiple of that $\mathbf{r}$.

It may appear that difficulties are inevitable in practice as components of $\mathbf{r}$ tend to zero. However, Coleman and Li (1992*a*) show that theoretically this is not a problem, and neither is it in practice if the method is implemented carefully. They give numerical results which show improvement over an earlier attempt to provide a method based on an interior point approach. Although comparisons with other methods do not seem to be available, the affine scaling method seems promising for large problems, since it appears to be insensitive to problem size.

Another method that attempts to smooth the l_1 problem has been developed recently by Madsen and Nielsen (1993). It is based on the use of the Huber M-estimator, defined by

$$\psi_\gamma \equiv \psi_\gamma(\mathbf{r}) = \sum_{i=1}^{m} \rho(r_i), \tag{2.26}$$

where

$$\rho(t) = \begin{cases} t^2/2, & |t| \le \gamma, \\ \gamma(|t| - \gamma/2), & |t| > \gamma, \end{cases} \tag{2.27}$$

$\mathbf{r}$ is given by (2.2) and γ is a scale factor or tuning constant. The function (2.26) is convex and once continuously differentiable, but has discontinuous second derivatives at points where $|r_i| = \gamma$. The mathematical structure of the Huber M-estimator is considered by Clark (1985). Clearly, if γ is chosen large enough, then ψ_γ is just the least squares function; in addition, if γ tends to zero, then limit points of the set of solutions minimize the l_1 norm (see Theorem 3). It is the latter property that concerns us here. It has been suggested by Madsen and Nielsen (1993) and also Li and Swetits (1998) that the preferred method for solving the l_1 problem is via a sequence of Huber problems for a sequence of scale values $\gamma \to 0$. This algorithmic development has led to increased interest in the relationship between the Huber M-estimator and the l_1 problem; see, for example, Madsen, Nielsen and Pinar (1994), Li and Swetits (1998).

Let a partition be defined by an index set σ and its complement σ^c as follows:

$$\sigma = \{i : |r_i| \le \gamma\}, \qquad \sigma \cup \sigma^c = \{1, 2, \ldots, m\}. \tag{2.28}$$

Then the system of equations determined by the necessary conditions for $\mathbf{x}$ to be a solution is

$$A_\sigma^T A_\sigma \mathbf{x} = A_\sigma^T \mathbf{b}_\sigma - \gamma \sum_{i \in \sigma^c} g_i \mathbf{a}_i, \tag{2.29}$$

where $\mathbf{a}_i^T$ denotes the ith row of A, $g_i = \text{sign}(r_i)$ as before, A_σ is obtained from A by deleting rows corresponding to indices $i \in \sigma^c$, and $\mathbf{b}_\sigma$ is defined similarly.

For given γ, the Huber problem can be solved by Newton's method, or a variant, using a line search. There are other possibilities, and a comparison of eight algorithms for this problem (*inter alia*) is given by Ekblom and Nielsen (1996). Algorithms based on continuation are also given by Clark and Osborne (1986), Boncelet and Dickinson (1984). For given $\mathbf{x} \in \mathbb{R}^n$, define W as a diagonal matrix with elements 1 if $|r_i| \leq \gamma$ and 0 otherwise. Then, assuming that no value of $|r_i|$ is equal to γ, and letting $\mathbf{s} \in \mathbb{R}^m$ be defined by

$$s_i = \begin{cases} 0, & i \in \sigma, \\ \text{sign}(r_i), & i \in \sigma^c, \end{cases}$$

it is easily seen by differentiating the Huber function that ψ_γ is minimized if and only if

$$A^T \left[\frac{1}{\gamma} W\mathbf{r} + \mathbf{s} \right] = 0. \tag{2.30}$$

The formal Newton step $\mathbf{d}$ for solving this system of equations ignores the discontinuity in derivative. It satisfies

$$\frac{1}{\gamma} A^T W A\mathbf{d} = -A^T \left[\frac{1}{\gamma} W\mathbf{r} + \mathbf{s} \right], \tag{2.31}$$

or

$$A^T W A\mathbf{d} = -A^T (W\mathbf{r} + \gamma\mathbf{s}), \tag{2.32}$$

If A has full rank n, then the rank of W can always be taken to be at least n at the solution of the M-estimation problem: Osborne (1985). However, this does not ensure that W has rank at least n in a step of the Newton iteration. Problems with singularity of the linear system can be avoided by inserting additional 1s into the diagonal positions of W, or indeed the unit matrix can be used. The solution to (2.32) is most efficiently obtained through LU factorization of the matrix on the left-hand side; one step of iterative refinement is recommended in Madsen and Nielsen (1993). A line search may be needed to ensure descent.

As the iteration proceeds, the partition σ will change, until the partition valid at the solution is obtained. Then the iteration terminates in one further Newton step (because a quadratic is being minimized). Changes in the partition translate into changes in W (and $\mathbf{s}$), and corresponding changes to the LU factors of $A^T WA$. It is here that the efficiency of the algorithm is achieved, because changes of a single index in the partition mean a rank 1 change so that updating of the factors is all that is required, a typical iteration costing $O(n^2)$ operations. Changes of more than one index need refactorization, at a cost of $O(n^3)$ operations: however, this is in practice needed only occasionally (see, for example, Table 2 of Madsen and Nielsen 1993). Once a solution has been obtained for a particular γ, the value of γ is reduced and the process repeated, using warm starts. Again, only a rank 1 change to A^TWA may be needed.

A key observation here is that it is not necessary to let γ go to zero, but the method can be terminated at a nonzero value. The relevant result is as follows.

Theorem 3 Let $\mathbf{x}_\delta$ minimize ψ_δ, and suppose that $\mathbf{s}$ and W remain constant for $0 < \gamma < \delta$. Then $\mathbf{x}_\delta + \delta\mathbf{v}$ solves the l_1 problem, where

$$\delta A^T W A\mathbf{v} = -A^T W\mathbf{r}(\mathbf{x}_\delta). \tag{2.33}$$

Proof. For $0 < \gamma < \delta$, define

$$\mathbf{x}_\gamma = \mathbf{x}_\delta + (\delta - \gamma)\mathbf{v}.$$

Then

$$\begin{aligned} &A^T W\mathbf{r}(\mathbf{x}_\gamma) + \gamma A^T\mathbf{s} \\ &\quad= A^T W\Big(\mathbf{r}(\mathbf{x}_\delta) + (\delta-\gamma)A\mathbf{v}\Big) + \gamma A^T\mathbf{s} \\ &\quad= A^T W\mathbf{r}(\mathbf{x}_\delta) + (\delta-\gamma)A^T W A\mathbf{v} + \gamma A^T\mathbf{s} \\ &\quad= A^T W\mathbf{r}(\mathbf{x}_\delta) - \left(\tfrac{\delta-\gamma}{\delta}\right) A^T W\mathbf{r}(\mathbf{x}_\delta) + \gamma A^T\mathbf{s} \quad \text{using (2.33)} \\ &\quad= (\gamma/\delta)A^T W\mathbf{r}(\mathbf{x}_\delta) + \gamma A^T\mathbf{s} \\ &\quad= (\gamma/\delta)\Big(A^T W\mathbf{r}(\mathbf{x}_\delta) + \delta A^T\mathbf{s}\Big) \\ &\quad= 0, \end{aligned}$$

using (2.30) and the definition of $\mathbf{x}_\delta$. It follows from (2.30) that $\mathbf{x}_\gamma$ minimizes the Huber function ψ_γ.

Now, for $0 < \gamma < \delta$,

$$A^T W\mathbf{r}(\mathbf{x}_\gamma) + \gamma A^T\mathbf{s} = 0$$

is equivalent to

$$\sum_{i\in\sigma} r_i(\mathbf{x}_\gamma)\mathbf{a}_i + \gamma\sum_{i\in\sigma^c} g_i(\mathbf{x}_\gamma)\mathbf{a}_i = 0,$$

or

$$\sum_{i\in\sigma} \frac{r_i(\mathbf{x}_\gamma)}{\gamma}\mathbf{a}_i + \sum_{i\in\sigma^c} g_i(\mathbf{x}_\gamma)\mathbf{a}_i = 0. \tag{2.34}$$

Using (2.28), continuity implies that

$$r_i(\mathbf{x}_0) = 0, \quad i \in \sigma,$$

where $\mathbf{x}_0 = \mathbf{x}_\delta + \delta\mathbf{v}$. Thus

$$\sigma \subset I = \{i : r_i(\mathbf{x}_0) = 0\}.$$

In addition,

$$g_i(\mathbf{x}_\gamma) = g_i(\mathbf{x}_0), \quad i \in \sigma^c.$$

Now,

$$\left|\frac{r_i(\mathbf{x}_\gamma)}{\gamma}\right| \leq 1, \quad i \in \sigma.$$

Thus there exist numbers v_i, $|v_i| \leq 1$, $i \in I$ such that

$$\sum_{i \in I} v_i \mathbf{a}_i + \sum_{i \in I^c} g_i(\mathbf{x}_0)\mathbf{a}_i = 0.$$

Thus $\mathbf{x}_0$ solves the l_1 problem and the proof is complete. □

Because the algorithm cannot return to the same sign vector, the condition $0 < \gamma < \delta$ will eventually be satisfied. Note that the matrix on the left-hand side of equation (2.33) is such that no new factorization is needed to compute $\mathbf{v}$.

Numerical results given by Madsen and Nielsen (1993) show that careful implementation of the method can make it superior to the algorithm of Barrodale and Roberts (1973). The larger scale calculations, however, are carried out only for randomly generated problems with $m/n = 2$, which does not seem appropriate for problems of practical interest. More efficient implementations of simplex-based methods, with careful attention to line search performance and scaling, are available for solving the l_1 problem: see Bloomfield and Steiger (1983), Osborne and Watson (1996). Thus, although the Huber-based approach seems to be a promising one, further investigation would be valuable.

2.3. l_p approximation, $1 < p < \infty$

For given $\mathbf{x} \in \mathbb{R}^n$, let $\mathbf{r} = \mathbf{r}(\mathbf{x})$, let D_r be as in the previous section, and let $D_{|r|}$ denote the matrix

$$D_{|r|} = \text{diag}\,\{|r_1|, \dots, |r_m|\}.$$

Then, applying Theorem 1, or by direct differentiation, it follows that $\mathbf{x}$ minimizes $\|\mathbf{r}\|_p$ if and only if

$$A^T D_{|r|}^{p-1} \mathbf{g} = 0, \tag{2.35}$$

where, as before, $g_i = \text{sign}(r_i)$, $i = 1, \dots, m$. When $p = 2$, this gives the usual normal equations; otherwise it is a nonlinear system of equations for a solution $\mathbf{x}$. If the value $p = 2$ is not satisfactory because of the error pattern, there may be merit in moving p towards 1. The range $1 < p < 2$ is of particular interest computationally because there is reduced smoothness: problems with $p \geq 2$ are twice differentiable, problems with $1 < p < 2$ are once differentiable, and the case $p = 1$ is non-differentiable.

One way to proceed to find a solution is to recognize that (2.35) can be written as

$$A^T D_{|r|}^{p-2} \mathbf{r} = 0, \tag{2.36}$$

and this can be viewed as a weighted system of normal equations with weighting matrix $W = D_{|r|}^{p-2}$. This matrix is of course only defined for $1 < p < 2$ if no component of $\mathbf{r}$ is zero, and this will be assumed at present. Fixing this matrix at the current value of $\mathbf{x}$ and solving this weighted least squares problem for the new approximation gives the technique known as *iteratively reweighted least squares* (IRLS). This is attractive since least squares problems are easy to solve. If $\mathbf{x}$ is the current approximation, and $\mathbf{x} + \Delta\mathbf{x}$ is the new approximation, we have

$$A^T D_{|r|}^{p-2} A(\mathbf{x} + \Delta\mathbf{x}) = A^T D_{|r|}^{p-2} \mathbf{b}. \tag{2.37}$$

This simple iteration process will converge if started from close enough to a solution and also if p is close enough to 2. In fact, a stronger result is available for the special case when $1 < p < 2$. The following lemma, which is readily proved, is helpful.

Lemma 4 Let scalars a, b be given with $b \neq 0$. Then, if $1 < p < 2$,

$$|a|^p - |b|^p \leq \tfrac{p}{2}|b|^{p-2}\left(|a|^2 - |b|^2\right), \tag{2.38}$$

with equality only if $|a| = |b|$.

Theorem 4 If $1 < p < 2$, the method of IRLS is convergent from any starting point to a point satisfying (2.36).

Proof. Let $\mathbf{r}^+$ be the vector $\mathbf{r}$ evaluated at $\mathbf{x}+\Delta\mathbf{x}$ defined by (2.37). Setting $a = \mathbf{r}_i^+$, $b = \mathbf{r}_i$ in (2.38) and summing from $i = 1$ to m gives

$$\begin{aligned} \|\mathbf{r}^+\|_p^p - \|\mathbf{r}\|_p^p &\leq \tfrac{p}{2}\left\{\sum_{i=1}^m |r_i^+|^2 |r_i|^{p-2} - \sum_{i=1}^m |r_i|^p\right\} \\ &\leq 0, \end{aligned}$$

using the definition of $\mathbf{r}^+$. Thus there is strict reduction unless (2.36) is satisfied, and the result is proved. □

Unfortunately, the properties of guaranteed convergence provided by this theorem, coupled with the simplicity of the iteration process, are offset by the fact that the process is accompanied by a potentially unsatisfactory convergence rate: this is linear, with convergence constant $|p-2|$ (see Osborne 1985). Thus, as p approaches 1, for example, convergence can be intolerably slow.

Consider now the alternative of using Newton's method. Let

$$\mathbf{f}(\mathbf{x}) = A^T D_{|r|}^{p-2}\mathbf{r}.$$

Then it is readily seen that

$$\nabla\mathbf{f}(\mathbf{x}) = (p-1)A^T D_{|r|}^{p-2}A,$$

so the Newton step $\mathbf{d}$ satisfies the linear system of equations

$$\begin{aligned}(p-1)A^T D_{|r|}^{p-2}A\mathbf{d} &= -A^T D_{|r|}^{p-2}\mathbf{r}\\ &= -A^T D_{|r|}^{p-2}(A\mathbf{x}-\mathbf{b})\end{aligned}$$

or

$$A^T D_{|r|}^{p-2}A(\mathbf{x}+(p-1)\mathbf{d}) = A^T D_{|r|}^{p-2}\mathbf{b}.$$

Comparing this with (2.37) shows that the IRLS step is just $(p-1)$ times the Newton step. It is easily seen that $\mathbf{d}^T\mathbf{f}(\mathbf{x}) < 0$, so that $\mathbf{d}$ is a descent direction for $\|\mathbf{r}\|_p^p$, and so it makes sense to incorporate a line search. If this is done in both methods, then essentially the same method is obtained.

As already indicated, one of the difficulties of the range $1 < p < 2$ is the fact that second derivatives do not always exist if any component of $\mathbf{r}$ becomes zero. Different strategies have been proposed to get round this difficulty. However, not just zero components but *nearly zero* components are potentially troublesome. There is some evidence, however, that these phenomena are not *by themselves* a major problem, but only if they are accompanied by p being close to 1. The main difficulty appears to be due to the fact that, as p approaches 1, we are coming closer to a discontinuous problem, effectively to a constrained problem. It seems necessary to recognize this in a satisfactory algorithm, and consider some of the elements of the l_1 problem in devising an approach that will deal with small values of p in a satisfactory way. This is the philosophy in the recent approach of Li (1993), which we will now describe.

Let Z be the matrix defined previously whose rows form a basis for the null space of A^T, so that

$$Z^T A = 0.$$

Define

$$\mathbf{g}_p = pD_{|r|}^{p-1}\mathbf{g},$$

the derivative of $\|\mathbf{r}\|_p^p$ with respect to $\mathbf{r}$. Then (2.35) is equivalent to

$$\mathbf{g}_p - Z\mathbf{w} = 0. \qquad (2.39)$$

Of course, when $p = 1$, $\mathbf{g}_p$ is just $\mathbf{g}$. If D_r is nonsingular, then $\mathbf{x}$ solves the l_p problem if and only if there exists $\mathbf{r} \in \mathbb{R}^m$, $\mathbf{w} \in \mathbb{R}^{m-n}$ such that

$$D_r(\mathbf{g}_p - Z\mathbf{w}) = 0, \tag{2.40}$$

$$Z^T(\mathbf{r} + \mathbf{b}) = 0. \tag{2.41}$$

If D_r is singular, then an additional requirement is that if $r_i = 0$, then $(Z\mathbf{w})_i = 0$. Notice that (2.40) and (2.41) are just the system of equations (2.16) and (2.17) in the l_1 case, so that (2.40) incorporates the complementary slackness conditions which are an essential part of the l_1 characterization.

Applying Newton's method to (2.40) and (2.41) gives, as before, the step in $\mathbf{r}$ as

$$\boldsymbol{\delta}\mathbf{r} = -A\left(A^T D_r^{-1} D_\beta A\right)^{-1} A^T \mathbf{g}_p,$$

where in this case

$$\boldsymbol{\beta} = p\mathbf{g}_p - Z\mathbf{w}.$$

Clearly, when $p = 1$, this is just the matrix D_β defined before. In order to globalize the method, one can use a technique similar to that used for the l_1 method. In this case a suitable matrix W_θ is given by

$$W_\theta = \operatorname{diag}\{|r_i^{-1}(p(\mathbf{g}_p)_i - (1-\theta)(Z\mathbf{w})_i)|,\ i = 1, \ldots, m\},$$

and the step in the direction $\mathbf{r}$ is then

$$\boldsymbol{\delta}_\theta \mathbf{r} = -A\left(A^T W_\theta A\right)^{-1} A^T \mathbf{g}_p.$$

Clearly $\theta = 0$ gives

$$W_\theta = D_r^{-1} D_\beta,$$

and the Newton step. If $\theta = 1$, then

$$\boldsymbol{\delta}_1 \mathbf{r} = p^{-1} A \Delta \mathbf{x},$$

or $1/p$ times the IRLS step. It is therefore possible to develop an algorithm for the l_p problem which is essentially equivalent to the method for $p = 1$ described above. A line search is of course required, and the details are given by Li (1993). For $1 < p < 2$, the method is globally convergent to $\mathbf{r}^*$ satisfying (2.35) if the rows of A corresponding to zero components of $\mathbf{r}^*$ are linearly independent; the convergence is superlinear if $\mathbf{r}^*$ has no zero component.

Numerical results given by Li (1993) show that the new method is clearly superior to IRLS (with the same line search) for values of p close to 1, with the gap between the two methods widening as p approaches 1. There is little difference for values of $p \geq 1.5$ or so. As with the l_1 case, the number of iterations appears to be independent of the problem size.

Finally we mention briefly the cases when $p > 2$. These are perhaps of less interest in practice, and also in theory, since second derivatives always exist. Newton's method (with line search) is perfectly satisfactory in most cases. Clearly, for very large p, scaling problems may well be a factor. We will not dwell on this, but will move on to consider the limiting case, the Chebyshev approximation problem.

2.4. Chebyshev approximation

For any $\mathbf{x} \in \mathbb{R}^n$, and

$$\mathbf{r} = A\mathbf{x} - \mathbf{b},$$

let J denote the set of indices

$$J = \{i : |r_i| = \|r\|_\infty\},$$

and let J^c denote its complement. Then, using Theorem 1 and (2.10) we have the following theorem.

Theorem 5 The vector $\mathbf{x}$ is a solution to (2.3) in the Chebyshev case if and only if there exists $\boldsymbol{\lambda} \in \mathbb{R}^m$ with

$$\lambda_i = 0, \quad i \in J^c, \quad \text{and} \quad \lambda_i \geq 0, \quad i \in J,$$

such that

$$A^T D_g \boldsymbol{\lambda} = 0, \tag{2.42}$$

where

$$D_g = \text{diag}\,\{g_1, \ldots, g_m\},$$

with $g_i = \text{sign}(r_i),\ i = 1, \ldots, m$, as before.

A solution $\mathbf{x}$ is unique if A satisfies the Haar condition, that is, if every $n \times n$ submatrix is nonsingular. This condition is sufficient only if $m = n+1$ (see, for example, Watson 1980). The function $\|\mathbf{r}\|_\infty$ is piecewise linear, and the most popular methods have been based on the simplex method of linear programming or variants: see, for example, Barrodale and Phillips (1975), Bartels, Conn and Li (1989). These are finite, moving through sets of points with J containing $n + 1$ indices until optimality is achieved: the fact that there is always a solution at a point with $n + 1$ indices in J if A has rank n is just a restatement of the fact that the solution to a linear programming problem occurs at a basic feasible solution.

Recently, as in the l_1 case, there has been interest in the possibility of smoothing the problem. We describe here an approach that is the analogue of the methods previously described, due to Coleman and Li (1992*b*). Recall that their l_1 approach was able to cross lines of non-differentiability, avoided derivative discontinuities except in the limit, and combined descent steps based on derivatives with Newton steps in an automatic way. In the current problem, $\|\mathbf{r}\|_\infty$ is differentiable provided that the norm is attained at just one component of $\mathbf{r}$. Let this be the jth component. The region of non-differentiability is therefore defined by the hyperplanes

$$|r_i| = |r_j|, \quad i \neq j.$$

Because j will change from iteration to iteration, it is not obvious how the l_1 method extends to this case; in particular, there is no global transformation that will result in new variables corresponding to the distances to these hyperplanes.

The approach used in Coleman and Li (1992*b*) is to define *local* transformations. At each step a transformation is defined which transforms $\mathbf{r}$ to a new variable $\mathbf{s}$ whose components measure the distance to the hyperplanes of non-differentiability. At the current point $\mathbf{x}$, with $\mathbf{r}$ defined as usual, let J be a singleton with

$$|r_j| = \|r\|_\infty.$$

Then we require $\mathbf{s} \in \mathbb{R}^m$ to be defined by

$$s_i = \begin{cases} g_j r_j - g_i r_i, & i \neq j, \\ g_j r_j, & i = j. \end{cases}$$

Alternatively, defining $T \in \mathbb{R}^{m \times m}$ by

$$T = (\mathbf{g} + g_j \mathbf{e}_j)\mathbf{e}_j^T - D_g,$$

then it is easily seen that

$$T^{-1} = g_j(\mathbf{e} + \mathbf{e}_j)\mathbf{e}_j^T - D_g,$$

and in addition

$$\mathbf{r} = T\mathbf{s}.$$

Of course, T depends on $\mathbf{r}$, and so the transformation is a local one.

Let Z be defined as before so that $A^T Z = 0$. Then the Chebyshev problem can be expressed in the form

$$\begin{aligned} &\text{minimize}_{\mathbf{s} \in \mathbb{R}^m} \quad \|T\mathbf{s}\|_\infty \\ &\text{subject to } Z^T(T\mathbf{s} + \mathbf{b}) = 0. \end{aligned} \tag{2.43}$$

Let $\mathbf{r}$ be such that $\|\mathbf{r}\|_\infty$ is differentiable, with $J = \{j\}$. Then the gradient vector is given by $D_g \mathbf{e}_j$. A descent direction $\mathbf{d}_s$ for the current $\mathbf{s}$, analogous to that defined for the l_1 case, can be obtained by solving the following subproblem:

$$\begin{aligned} &\text{minimize}_{\mathbf{d} \in \mathbb{R}^n} \quad \mathbf{e}_j^T D_g T \mathbf{d} \\ &\text{subject to } Z^T T \mathbf{d} = 0, \quad \text{and} \quad \|D^{-1}\mathbf{d}\|_2 \leq \tau. \end{aligned} \tag{2.44}$$

Again, D is a positive definite diagonal scaling matrix, and τ is a positive number that restricts the size of $\mathbf{d}_s$. As before, up to a scalar multiple, we can easily see that

$$\mathbf{d}_s = -T^{-1}A\left(A^TT^{-T}D^{-2}T^{-1}A\right)^{-1}A^TD_g\mathbf{e}_j,$$

so that the corresponding descent direction for $\mathbf{r}$ analogous to (2.20) is

$$\mathbf{d} = -A\left(A^TT^{-T}D^{-2}T^{-1}A\right)^{-1}A^TD_g\mathbf{e}_j. \tag{2.45}$$

This vector can be obtained via the solution of a least squares problem, and a line search enables $\mathbf{r}$ to be updated. Details of this and a suitable line search are given in Coleman and Li (1992*b*). This affine scaling algorithm may be slowly convergent, and we consider next the application of Newton's method to the problem. This requires suitable reinterpretation of the characterization conditions as a nonlinear system, analogous to (2.16) and (2.17). The following theorem is key to this.

Theorem 6 Necessary and sufficient conditions for $\mathbf{r}$ to solve the Chebyshev problem are that there exists $\mathbf{w} \in \mathbb{R}^m$ such that

$$\begin{aligned} D_sT^T(D_g\mathbf{e}_j - Z\mathbf{w}) &= 0, & (2.46)\\ Z\mathbf{r} + Z\mathbf{b} &= 0, & (2.47)\\ \lambda_i &\geq 0, \quad i \in I - \{j\}, &\\ 1 - \sum_{i \in J - \{j\}} \lambda_i &\geq 0, & \end{aligned}$$

where

$$\lambda_i = g_i(Zw)_i, \quad i \in I - \{j\}.$$

Proof. Let $\mathbf{r}$ solve the Chebyshev problem. Then, obviously,

$$Z^T\mathbf{r} + Z^T\mathbf{b} = 0.$$

Further, from the characterization result, there exist numbers λ_i, $i \in I$, $\mathbf{w} \in \mathbb{R}^m$, such that

$$\sum_{i \in I} g_i\lambda_i\mathbf{e}_i = Z\mathbf{w},$$

where

$$\lambda_i \geq 0, \quad i \in J,$$

$$\sum_{i \in J} \lambda_i = 1.$$

It follows that

$$\sum_{i\in I} g_i\lambda_i T^T\mathbf{e}_i = T^T Z\mathbf{w},$$

and so

$$T^T D_g\mathbf{e}_j = \sum_{i\in I-\{j\}} \lambda_i e_i + T^T Z\mathbf{w}. \tag{2.48}$$

Thus

$$D_s T^T(D_g\mathbf{e}_j - Z\mathbf{w}) = 0,$$

and necessity is established.

Now let the stated conditions be satisfied. Clearly this implies that (2.48) is satisfied in the ith component, for $i \in J^c \cup \{j\}$. For $i \in J\backslash\{j\}$,

$$\left(T^T D_g\mathbf{e}_j\right)_i - \left(T^T Z\mathbf{w}\right)_i = g_i(Z\mathbf{w})_i,$$

and so, by the definition of λ_i, (2.48) also holds in these components. Thus (2.48) is satisfied, and reversing the argument of the proof of necessity leads to the required result. □

Consider now the application of Newton's method to the nonlinear system consisting of (2.46) and (2.46). Define D_β as before, where

$$\boldsymbol{\beta} = T^T(\mathbf{g} - Z\mathbf{w}).$$

Then the Newton step in $\mathbf{r}$ and $\mathbf{w}$ is given by the system of equations

$$\begin{bmatrix} D_\beta T^{-1} & -D_s T^T Z \\ Z^T & 0 \end{bmatrix} \begin{bmatrix} \delta\mathbf{r} \\ \delta\mathbf{w} \end{bmatrix} = \begin{bmatrix} -D_s T^T(D_g\mathbf{e}_j - Z\mathbf{w}) \\ 0 \end{bmatrix}. \tag{2.49}$$

It follows from this system that

$$\delta\mathbf{r} = -A\left(A^T T^{-T} D_s^{-1} D_\beta T^{-1} A\right)^{-1} A^T D_g\mathbf{e}_j. \tag{2.50}$$

Coleman and Li (1992*b*) show that in a neighbourhood of a solution to (2.3), where $\|\mathbf{r}\|_\infty$ is differentiable, the matrix being inverted on the right-hand side of (2.50) is positive definite, so that the Newton direction becomes a descent direction.

Note that (2.45) and (2.50) have the same general form, and suitable definition of D enables a smooth transition to be made between the steps (2.45) and (2.50). Let D be chosen by

$$D = D_s^{1/2}.$$

Further, define

$$W_\theta = D_s^{-1}\left((1-\theta)D_{|\beta|} + \theta\mathbf{e}\right),$$

where $D_{|\beta|}$ denotes the diagonal matrix whose diagonal elements are the modulus of those of D_β. Define the search direction as

$$\mathbf{d} = -A\left(A^T T^{-T} W_\theta T^{-1} A\right)^{-1} A^T D_g \mathbf{e}_j. \tag{2.51}$$

Then, when $\theta = 1$, (2.51) just gives (2.45), and when $\theta = 0$, (2.51) gives (2.50), provided that D_β is positive definite: it is shown by Coleman and Li (1992*b*) that this holds in a neighbourhood of the solution excluding non-differentiable points.

Details of the way in which the parameter θ can be chosen, as well as other computational issues, are given by Coleman and Li (1992*b*). It is also shown that, under nondegeneracy conditions analogous to those for the l_1 problem, global convergence to a solution at a quadratic rate may be established. Numerical results show that the number of iterations is relatively insensitive to problem size.

The application of interior point methods for linear programming problems to l_1 and l_∞ problems has been considered by several other authors, for example Meketon (1987), Ruzinsky and Olsen (1989), Zhang (1993) and Duarte and Vanderbei (1994). An approach based on row (column) relaxation or proximal point methods has been used by Dax (1989, 1993), Dax and Berkowitz (1990): this may have potential for large sparse problems. It is interesting that all these smoothing methods have as a subproblem the solution of a weighted least squares problem.

The extent to which these new methods for both the l_1 and l_∞ problems will usurp methods of simplex type remains to be seen. The main methods of the latter type for which there is readily available software are not the most up to date, and in particular the issue of the provision of efficient line searches is a live one. It would seem that such issues are unlikely to be resolved until efficient codes for all these different types of method have been produced.

3. Total least norm problems

An assumption made in Section 2 was that errors were only present in $\mathbf{b}$, so that the model equations (2.2) were appropriate. However, in many situations, the independent variable values are also in error. The problem of dealing with errors in all variables is well known in the statistics literature, and solution methods go back to the end of the last century. The general situation is complicated by the need to make additional assumptions about the error structure in order to ensure identifiability (Moran 1959).

One way of taking errors also in the independent variables into account is to introduce a matrix E of perturbations of A. This leads to the relationship

$$\mathbf{r} = (A + E)\mathbf{x} - \mathbf{b}. \tag{3.1}$$

The analogue of the general problem considered in Section 2 is then the problem

$$\text{minimize } \|E : \mathbf{r}\| \text{ subject to (3.1)}, \tag{3.2}$$

where the norm is now a norm on $m \times (n+1)$ matrices. This problem, in which the norm is the Frobenius norm, was first analysed by Golub and Van Loan (1980), who referred to it as the *total least squares problem.* Since that time, problems of this kind have generated enormous interest, and there have been extensions in various directions, for example to structured problems. There are many applications, to systems identification, frequency estimation, superresolution, control theory, *etc.* In keeping with the underlying theme of this article, we will consider (3.2) for a general class of norms. This is not merely of theoretical interest, because (as in the previous section) norms other than the least squares norm are relevant in practice, and there has been algorithmic development.

It is not possible to deal with this problem in complete generality (as is the case when E is zero). There are some fundamental differences. As we show below, existence of solutions is not even guaranteed. It is not a convex problem, and general characterization results are not available. However, we can make progress in an analysis of the problem if we confine attention to a wide range of matrix norms known as separable matrix norms, a concept introduced in Osborne and Watson (1985). Before introducing these, we need the concept of the dual matrix norm: this is defined (analogous to (2.4)) by

$$\|M\|^* = \max_{\|N\| \leq 1} \operatorname{trace}(N^T M). \tag{3.3}$$

Definition 2 A matrix norm on $m \times (n+1)$ matrices is said to be *separable* if, given vectors $\mathbf{u} \in \mathbb{R}^m$ and $\mathbf{v} \in \mathbb{R}^{n+1}$, there are vector norms $\| \cdot \|_A$ on $\mathbb{R}^m$ and $\| \cdot \|_B$ on $\mathbb{R}^{n+1}$ such that

$$\|\mathbf{u}\mathbf{v}^T\| = \|\mathbf{u}\|_A \|\mathbf{v}\|_B^*, \qquad \|\mathbf{u}\mathbf{v}^T\|^* = \|\mathbf{u}\|_A^* \|\mathbf{v}\|_B.$$

A useful property of separable norms is the following. Let $Z \in \mathbb{R}^{m \times (n+1)}$, $\|\mathbf{v}\|_B \leq 1$. Then, for separable norms,

$$\begin{aligned}
\|Z\mathbf{v}\|_A &= \max_{\|\mathbf{u}\|_A^* \leq 1} \mathbf{u}^T Z \mathbf{v} \\
&\leq \max_{\|\mathbf{u}\|_A^* \leq 1, \|\mathbf{v}\|_B \leq 1} \mathbf{u}^T Z \mathbf{v} \\
&\leq \max_{\|\mathbf{u}\mathbf{v}^T\|^* \leq 1} \operatorname{trace}(\mathbf{v}\mathbf{u}^T Z) \\
&\leq \max_{\|G\|^* \leq 1} \operatorname{trace}(G^T Z) \\
&= \|Z\|,
\end{aligned}$$

so that

$$\|Z\| \geq \max_{\|\mathbf{v}\|_B \leq 1} \|Z\mathbf{v}\|_A. \tag{3.4}$$

Examples of separable matrix norms are as follows.

Example 1 Operator norms defined by

$$\|M\| = \max_{\|\mathbf{x}\|_d \leq 1} \|M\mathbf{x}\|_c,$$

are separable with $\|\cdot\|_A = \|\cdot\|_c$, and $\|\cdot\|_B = \|\cdot\|_d$; in other words, equality holds in (3.4).

Example 2 The norms defined by

$$\|M\| = \left(\sum_{i,j} |m_{i,j}|^p\right)^{1/p}, \quad 1 \leq p < \infty,$$

are separable with $\|\cdot\|_A = \|\cdot\|_p$, and $\|\cdot\|_B = \|\cdot\|_q$, where $1/p + 1/q = 1$.

Example 3 Orthogonally invariant norms such that

$$\|M\| = \|UMV\|,$$

where U and V are orthogonal, are separable. Both norms are (unweighted) l_2 norms provided that $\|\mathbf{e}_1\mathbf{e}_1^T\| = 1$. This follows because if σ_1 is the largest singular value of M, then for any vectors $\mathbf{u}$ and $\mathbf{v}$ we will have

$$\|\mathbf{u}\mathbf{v}^T\| = \|\sigma_1\mathbf{e}_1\mathbf{e}_1^T\| = \sigma_1,$$

by assumption, where σ_1 is the largest singular value of $\mathbf{u}\mathbf{v}^T$. But

$$\sigma_1 = \|\mathbf{u}\|_2\|\mathbf{v}\|_2,$$

and the result is established. For example, a class of such orthogonally invariant norms is the class of Schatten p norms, defined by

$$\|M\|_{C_p} = \left(\sum_{i=1}^{n} \sigma_i^p\right)^{1/p}, \quad 1 \leq p < \infty,$$

where $\sigma_1, \ldots, \sigma_n$ are the singular values of M.

To solve the total least norm problem it is necessary to minimize $\|E : \mathbf{r}\|$ subject to (3.1). This is greatly facilitated by replacing the problem by an equivalent but generally much more tractable problem:

$$\text{minimize } \|Z\mathbf{v}\|_A \text{ subject to } \|\mathbf{v}\|_B = 1, \tag{3.5}$$

where $Z = [A : -\mathbf{b}]$.

Theorem 7 Let $\mathbf{v}$ solve (3.5) with $v_{n+1} \neq 0$. Then (3.2) is solved by

$$[E : -\mathbf{r}] = -Z\mathbf{v}\mathbf{w}^T,$$

where $\mathbf{w} \in \partial\|\mathbf{v}\|_B$, and $\mathbf{x}$ is such that

$$\mathbf{v}^T = \alpha(\mathbf{x}^T, 1), \quad \alpha \in \mathbb{R}. \tag{3.6}$$

Proof. For E, $\mathbf{r}$ and $\mathbf{x}$ as defined,

$$\begin{aligned} \|[E : -\mathbf{r}]\| &= \|Z\mathbf{v}\|_A \|\mathbf{w}\|_B^* \\ &= \|Z\mathbf{v}\|_A. \end{aligned}$$

Now let E, $\mathbf{r}, \mathbf{x}$ be *any* feasible set for (3.1), with $\mathbf{v}$ defined by (3.6). Then

$$\begin{aligned} \|Z\mathbf{v}\|_A &= \|[E : -\mathbf{r}]\mathbf{v}\|_A \\ &\leq \|[E : -\mathbf{r}]\|, \end{aligned}$$

using (3.4). The result follows. □

Notice that the matrix $[E : -\mathbf{r}]$ which gives a solution is a rank 1 matrix. However, if there is no $\mathbf{v}$ solving (3.5) with last component nonzero, then there is no solution to the total least norm problem.

Example 4 Let

$$A = \begin{bmatrix} 1 & 0 \\ 0 & 1 \\ 1 & 1 \end{bmatrix}, \quad \mathbf{b} = \begin{bmatrix} 1 \\ 1 \\ 1 \end{bmatrix}.$$

Clearly $\mathbf{v} = (1, 1, 0)^T$ (suitably normalized) gives a zero value for the minimum in (3.5). Further, for any norm, $\|E : \mathbf{r}\|$ can be made arbitrarily small; however, it can never be zero because $\mathbf{b}$ is not in the range of A.

Consider the case when the matrix norm is the Frobenius norm. Then both vector norms occurring in the separable norm definition are least squares norms and the problem (3.5) becomes that of determining the smallest singular value of the matrix Z. The total least squares problem was first analysed by Golub and Van Loan (1980). It may happen that certain columns of A are known to be exact, so that the corresponding columns of E should be zero. This is easily dealt with by fixing the corresponding components of $\mathbf{v}$ in (3.5) to be zero: see Watson (1983), Osborne and Watson (1985). The solution of (3.5) for other norms is less straightforward, not least because the problem is not a convex one, and so local solutions are possible.

Necessary conditions for a solution of (3.5) can be given in the following form. (In the case when $\|\cdot\|_B$ is polyhedral, then these conditions are also sufficient for a local solution; see Watson 1983.)

Theorem 8 Let $\mathbf{v}$ solve (3.5). Then, for every $\mathbf{w} \in \partial\|\mathbf{v}\|_B$, there exists $\mathbf{g} \in \partial\|Z\mathbf{v}\|_A$ such that

$$Z^T\mathbf{g} = \|Z\mathbf{v}\|_A\mathbf{w}.$$

An algorithm for the minimization of the norm defined by

$$\|M\| = \sum_{i,j} |m_{i,j}|,$$

the total l_1 problem ($\|\cdot\|_A$ is the l_1 norm, and $\|\cdot\|_B$ is the l_∞ norm), is given by Osborne and Watson (1985). Variants of these problems, such as the total l_p problem and the orthogonal l_1 problem, have also been considered in Watson (1984), Späth and Watson (1987). The idea of introducing structure into the matrix E has also been investigated. The case of zero columns has already been mentioned, but there are important applications when perturbations of A should preserve other sparsity patterns, or Toeplitz, Vandermonde or Hankel forms. While (3.5) is useful for certain types of structure (see, for example, Watson 1988, 1991), for others it seems necessary to work with the original problem (3.2).

For example, Rosen, Park and Glick (1996, 1997) develop a method concerned with retaining given structure, and permitting the use of general norms. We will illustrate this in the important special case when A has Toeplitz structure that must be preserved, as occurs in system identification. Then

$$A_{ij} = \alpha_{n+i-j}, \quad i = 1, \dots, m, \quad j = 1, \dots, n,$$

so that we can define A entirely by its first row

$$\rho_1(A) = [\alpha_n, \alpha_{n-1}, \dots, \alpha_1],$$

and its first column

$$\kappa_1(A) = [\alpha_n, \alpha_{n+1}, \dots, \alpha_{n+m-1}]^T.$$

Thus, in (3.1) we can define E to have the same form, with unique unknown elements, say β_i, $i = 1, \dots, n+m-1$. Assume that the matrix norm is an l_p norm defined for any matrix M by

$$\|M\| = \left(\sum |m_{ij}|^p\right)^{1/p}, \quad 1 \le p \le \infty.$$

Then

$$\|E : \mathbf{r}\| = \left\| \begin{array}{c} \mathbf{r}(\boldsymbol{\beta}, \mathbf{x}) \\ W\boldsymbol{\beta} \end{array} \right\| \tag{3.7}$$

where the vector norm is the l_p norm, where

$$\mathbf{r}(\boldsymbol{\beta}, \mathbf{x}) = (A + E)\mathbf{x} - \mathbf{b},$$

and where W is an $(m+n-1) \times (m+n-1)$ diagonal weighting matrix which accounts for repetitions of elements in E. The problem of minimizing (3.7) is of course nonlinear in $\boldsymbol{\beta}$ and $\mathbf{x}$, and methods for the l_1, l_2 and l_∞ norms based on linearization are given in Rosen et al. (1996). Extensions to problems where A depends nonlinearly on parameters which have to be estimated are given in Rosen et al. (1997). The techniques required are those for nonlinear problems, so we will not deal with this further but proceed to a more general study of this class.

4. Approximation to data by nonlinear models

When the model contains free parameters which occur nonlinearly, and it is assumed that errors are only present in the dependent variable, we obtain a problem which can be posed in the form

$$\text{minimize}_{\mathbf{x}\in\mathbb{R}^n} \quad \|\mathbf{f}(\mathbf{x})\|, \tag{4.1}$$

where $\mathbf{f} : \mathbb{R}^n \to \mathbb{R}^m$, $m > n$, the dependence of $\mathbf{f}$ on $\mathbf{x}$ is nonlinear, and the norm is a norm on $\mathbb{R}^m$. Provided that $\mathbf{f}$ is differentiable, we can write $A(\mathbf{x})$ for the $m \times n$ matrix of partial derivatives of $\mathbf{f}$ with respect to the components of $\mathbf{x}$. Allowing an arbitrary norm, we have the following result.

Theorem 9 Let $\mathbf{x}$ solve (4.1), with $\mathbf{f}$ such that

$$\mathbf{f}(\mathbf{z}) = \mathbf{f}(\mathbf{x}) + A(\mathbf{x})(\mathbf{z}-\mathbf{x}) + o(\|\mathbf{z}-\mathbf{x}\|_p)$$

for all $\mathbf{z}$ in a neighbourhood of $\mathbf{x}$, where $\|\cdot\|_p$ is a norm on $\mathbb{R}^n$. Then there exists $\mathbf{v} \in \partial\|\mathbf{f}(\mathbf{x})\|$ such that

$$A(\mathbf{x})^T\mathbf{v} = 0.$$

This result may be established along the lines of the proof of necessity in Theorem 1; the condition is not sufficient except in the special case when the norm is a convex function of $\mathbf{x}$. See Watson (1980), for example, for the details.

A general class of methods can be given for finding a point satisfying the conditions of this theorem (a stationary point). The basis is the solution of a sequence of linearized subproblems defined at the current approximation $\mathbf{x}$ to a stationary point, which enables an improved approximation to be obtained. A typical subproblem has the form

$$\begin{aligned} &\text{minimize}_{\mathbf{d}\in\mathbb{R}^n} \quad \|\mathbf{f}(\mathbf{x}) + A(\mathbf{x})\mathbf{d}\| \\ &\text{subject to } \|\mathbf{d}\|_A \le \tau, \end{aligned}$$

where τ is a suitably chosen positive scalar, and $\|\cdot\|_A$ is a suitably chosen norm. An analysis of the use of this subproblem is quite straightforward. Because $\mathbf{d} = 0$ is a candidate for a solution, we must have

$$\|\mathbf{f}(\mathbf{x}) + A(\mathbf{x})\mathbf{d}\| \le \|\mathbf{f}(\mathbf{x})\|.$$

But, for γ such that $0 < \gamma \leq 1$,

$$\begin{aligned} \mathbf{f}(\mathbf{x}+\gamma\mathbf{d}) &= \mathbf{f}(\mathbf{x}) + \gamma A(\mathbf{x})\mathbf{d} + o(\gamma) \\ &= \gamma(\mathbf{f}(\mathbf{x}) + A(\mathbf{x})\mathbf{d}) + (1-\gamma)\mathbf{f}(\mathbf{x}) + o(\gamma), \end{aligned}$$

and so

$$\begin{aligned} \|\mathbf{f}(\mathbf{x}+\gamma\mathbf{d})\| &\leq \gamma\|\mathbf{f}(\mathbf{x}) + A(\mathbf{x})\mathbf{d}\| + (1-\gamma)\|\mathbf{f}(\mathbf{x})\| + o(\gamma) \\ &= \|\mathbf{f}(\mathbf{x})\| + \gamma\Big(\|\mathbf{f}(\mathbf{x}) + A(\mathbf{x})\mathbf{d}\| - \|\mathbf{f}(\mathbf{x})\|\Big) + o(\gamma). \end{aligned}$$

Thus $\mathbf{d}$ is a descent direction for $\|\mathbf{f}\|$ at $\mathbf{x}$ unless

$$\|\mathbf{f}(\mathbf{x}) + A(\mathbf{x})\mathbf{d}\| = \|\mathbf{f}(\mathbf{x})\|.$$

Thus $\mathbf{d} = 0$ is a solution to the subproblem. Hence, using the characterization result given in Theorem 1 (noting that the bound constraint is inactive) and applying Theorem 9 shows that $\mathbf{x}$ is a stationary point.

In practice, a line search can be avoided: either $\mathbf{x}$ is replaced by $\mathbf{x}+\mathbf{d}$ to give a better approximation, or the subproblem can be re-solved with a reduced value of τ. If care is taken with the rules for choosing τ, this process can give convergence to a stationary point.

For the important cases of the l_2, l_1 and l_∞ norms, methods of this type are well known. A second-order rate of convergence is possible for polyhedral norm problems (which includes l_1 and l_∞). For example, consider the l_∞ norm, and suppose that $\mathbf{x}^*$ is a stationary point, with the current approximation $\mathbf{x}$ in a neighbourhood of $\mathbf{x}^*$. Suppose also that $\|\mathbf{f}(\mathbf{x}^*)\|$ is attained at exactly $n+1$ components of $\mathbf{f}(\mathbf{x}^*)$, say $j_1, \ldots, j_{n+1}$. Then, if the bound constraints on the solution of the linearized subproblems stay inactive, solutions can be interpreted as steps of Newton's method applied to the solution of

$$f_{j_i}(\mathbf{x}) - \sigma_i h = 0, \quad i = 1, \ldots, n+1, \tag{4.2}$$

where $\sigma_i = \pm 1$. A similar analogy is possible with the l_1 norm, with the corresponding requirement being that, at a stationary point $\mathbf{x}^*$, there are exactly n zeros of $\mathbf{f}(\mathbf{x}^*)$. In general, however, such conditions do not hold: there are too few nonlinear equations like (4.2) to determine the unknowns, and convergence can be slow. This has led to the incorporation of second derivative information into the subproblems. This can be done in different ways, for example by adding $\frac{1}{2}\mathbf{d}^T H\mathbf{d}$ to the objective function of the linearized subproblem, where H is the Hessian matrix (or some symmetric approximation to the Hessian) of the Lagrangian function at the current point for the problem posed as an optimization problem. The subproblems can usually be solved as quadratic programming problems. Good methods for the l_1 and l_∞ problems have been available for some time, and some references can be found in the review paper by Watson (1987).

Such methods are good for small problems. They are quite sophisticated, and relatively heavy computationally. In addition, sparsity in A, say, cannot easily be exploited. For the l_1 and l_∞ norms, with a suitable choice of $\|\cdot\|_A$, the linearized subproblems can usually be posed as linear programming problems, and efficient techniques are available for large sparse problems. Indeed, for very large problems, there may be little choice but to use the linearized subproblem as the basis for an algorithm, with, if possible, simple modifications introduced to help speed up convergence. Such ideas for l_∞ problems are proposed by Jonasson (1993), Jonasson and Madsen (1994). It remains to be seen how effective such methods will become, but, in any event, the basic subproblem will remain a very important and robust tool.

As already indicated, the problem (4.1) most often arises from the data-fitting problem analogous to (2.3) in the linear case, where we can write the ith component of $\mathbf{f}$ as

$$f_i(\mathbf{x}) = F(\mathbf{x}, \xi_i) - b_i, \quad i = 1, \dots, m,$$

and where F depends nonlinearly on the parameters forming the vector $\mathbf{x}$. Here the data consists of sets of points (ξ_i, b_i), $i = 1, \dots, m$, with only $\mathbf{b}$ containing errors. There is of course also the possibility of errors in the values of the variables ξ_i, and in this case the model equations can be written

$$f_i(\mathbf{x}) = F(\mathbf{x}, \xi_i + \delta_i) - b_i, \quad i = 1, \dots, m. \tag{4.3}$$

Then it is appropriate to minimize some norm of the vector $\mathbf{v}$ in $\mathbb{R}^{2m}$ whose components are f_i, $i = 1, \dots, m$, δ_i, $i = 1, \dots, m$, where we assume for simplicity that ξ_i (and hence δ_i) are in $\mathbb{R}$ (although these can in practice have many components). In this form, the problem is referred to as an errors-in-variables problem. For example, we could minimize the (square of the) l_2 norm of all the errors,

$$\mathbf{v}^T\mathbf{v} = \sum_{i=1}^{n} (f_i^2 + \delta_i^2), \tag{4.4}$$

subject to (4.3). This problem is referred to as *orthogonal distance regression*, the reason for this name being that we are minimizing the sum of squares of the distances from the data points (ξ_i, b_i) to the model curve. An efficient method for minimizing (4.4) is given by Boggs, Byrd and Schnabel (1987), with software available in Boggs, Byrd, Donaldson and Schnabel (1989). The point here is that there is considerable structure which can be exploited.

The minimization of the l_1 norm of $\mathbf{v}$ has also been considered, again exploiting the structure; see, for example, Watson and Yiu (1991), Watson (1997).

5. Chebyshev approximation of functions

Let $f(x)$, $\phi_i(x)$, $i = 1, \ldots, n$, be continuous functions defined on a compact set X. Then the problem of approximating f in the Chebyshev sense by a linear combination of the functions ϕ_i can be stated as: find coefficients c_i, $i = 1, \ldots, n$, to minimize

$$\left\| f - \sum_{i=1}^{n} c_i \phi_i \right\| = \max_{x \in X} \left| f(x) - \sum_{i=1}^{n} c_i \phi_i(x) \right|.$$

The analogue of (2.42) is that there exist points x_j, $j = 1, \ldots, t$ in X where the norm is attained with sign g_i, $i = 1, \ldots, t$, and corresponding numbers μ_i, $i = 1, \ldots, t$ such that $\mu_i g_i > 0$, $i = 1, \ldots, t$ and

$$A^T \boldsymbol{\mu} = 0, \tag{5.1}$$

where A is the $t \times n$ matrix with (i, j) element $\phi_j(x_i)$. Suppose that $X = [a, b] \subset \mathbb{R}$. Then, if the system of functions ϕ_i, $i = 1, \ldots, n$ forms a Chebyshev set on $[a, b]$, so that no nontrivial linear combination of the functions has more than $(n - 1)$ zeros in $[a, b]$, then it is readily shown from (5.1) that $t \geq n + 1$, and the values of μ_i alternate in sign. This gives us the well-known classical alternation characterization property: there are $n + 1$ points in $[a, b]$ where the norm is attained with the error alternating in sign as we go from left to right, or briefly

$$\mathcal{A}(f - \phi)_{[a,b]} \geq n + 1, \tag{5.2}$$

where ϕ denotes the approximation. The exchange algorithms of Remes (both single-point exchange and multi-point exchange) are effective ways of computing the (unique) best approximation.

Some general alternation theorems are also available for problems with constraints. For example, Brosowski and da Silva (1992) consider the problem of approximation on $[a, b]$ by a linear combination of functions forming a Chebyshev set on $[a, b]$, subject to certain side-conditions. Their results contain as a special case the classical theorem, and other known results for one-sided and restricted range approximation.

Once the Chebyshev set condition is dropped, then life becomes much more complicated. There may be nonuniqueness of solutions, but a more serious problem in practice is that, at a solution, there may be fewer than $n + 1$ points where the norm is attained. If this possibility is ignored, then exchange methods can be applied, although convergence can be slow and ill-conditioning can occur (because of coalescing points in the set where the norm is attained). Multivariate problems are particularly susceptible to this difficulty. Methods of Newton type are available, with the Newton method used as a local method when information about the number of points where the norm is attained at a solution is known, along with associated information. Methods of this type also apply to nonlinear problems (for example, Watson 1976).

A method for solving a wide class of continuous Chebyshev approximation problems, linear as well as nonlinear, is given by Jing and Fam (1987). The algorithm is shown to be convergent (possibly to a local minimum), and convergence is quadratic in nondegenerate cases. The method has some similarities to a method due to Jonasson and Watson (1982). Both approaches solve a sequence of linearized subproblems on the current set of points where the error function attains its local maxima, followed by a line search to obtain an improved approximation. Under appropriate conditions, both methods are equivalent locally to the Remes (multi-point) exchange method.

Most emphasis, however, has been on the use of particular approximating functions whose special properties can be exploited, and we consider some examples of these.

5.1. Chebyshev approximation by spline functions

Aside from interpolation, the use of spline functions for approximation has mainly been concerned with the use of the Chebyshev norm. Consider now the problem of approximating from the space of spline functions defined as follows.

Definition 3 Let integers m and k be given, and let $a = x_0 < x_1 < \cdots < x_{k+1} = b$. Then

$$S_m = \{s \in C^{m-1}[a,b] : s(x) \in \Pi_m \text{ on } [x_i, x_{i+1}], \quad i = 0, \dots, k\},$$

is the space of polynomial splines of degree m with k fixed knots, where Π_m denotes the space of polynomials of degree m. S_m is a linear space with dimension $m + k + 1$.

The theory of approximation by Chebyshev sets does not apply to approximation from S_m. However, S_m is an example of a *weak Chebyshev space*: there exists at least one best approximation from S_m to any continuous function that has the classical alternation property (although there may be others that do not). The theory of Chebyshev approximation by splines with fixed knots is fully developed, and a characterization of best approximation goes back to Rice (1967) and Schumaker (1968). What is required is the existence of an interval $[x_p, x_{p+q}] \subset [a,b]$, with $q \geq 1$ such that there are at least $q + m + 1$ alternating extrema on $[x_p, x_{p+q}]$ or, in the notation introduced in (5.2),

$$\mathcal{A}(f - s)_{[x_p, x_{p+q}]} \geq q + m + 1,$$

where $s \in S_m$.

In addition to characterization of solutions, there has been interest in conditions for uniqueness and *strong uniqueness* of best approximations.

Definition 4 A function $s_f \in S_m$ is called a strongly unique best approximation to $f \in C[a,b]$ if there is a constant $K_f > 0$ such that, for all $s \in S_m$,

$$\|f - s\| \geq \|f - s_f\| + K_f \|s - s_f\|.$$

In general, best approximations are not unique. However, the uniqueness (and strong uniqueness) of best spline approximations is characterized by the fact that *all* knot intervals contain sufficiently many alternating extrema (see Nürnberger 1989).

An iterative algorithm for computing best Chebyshev approximations from spline spaces is due to Nürnberger and Sommer (1983). As in the classical Remes method, a substep at each iteration is the computation of a spline $s \in S_m$ such that

$$(-1)^i \Big(f(\xi_i) - s(\xi_i) \Big) = h, \quad i = 1, \ldots, m + k + 2,$$

for some some real number h, and given points $\xi_1, \ldots, \xi_{m+k+2}$ in $[a,b]$. The number of equations reflects the fact that S_m has dimension $m+k+1$. Then one of the points ξ_i is replaced by a point where $\|f - s\|$ is attained in $[a,b]$ to get a new set of points $\{\xi_i\}$. The usual Remes exchange rule can result in a singular system of equations, so a modified exchange rule is needed. Such a rule is given by Nürnberger and Sommer (1983), which ensures that the new system has a unique solution. Because of possible nonuniqueness of best approximations, the proof of convergence is fairly complicated. However, a convergence result can be established. A multiple exchange procedure can also be implemented, and quadratic convergence is possible. The above results can be extended to more general spline spaces, where the polynomials are replaced by linear combinations of functions forming Chebyshev sets: see, for example, Nürnberger, Schumaker, Sommer and Strauss (1985).

To permit the full power of splines, one should allow the knots to vary, rather than be fixed in advance. The corresponding approximation problem is then a difficult nonlinear problem. To guarantee existence of best approximations, multiple knots have to be allowed. There may be local solutions; a characterization of best approximations is not known. For the case of k free knots, necessary and (different) sufficient conditions of the alternation kind given above may be proved. Let q' denote the sum of the knot multiplicities at the points $x_{p+1}, \ldots, x_{p+q-1}$. Then it is *necessary* for $s \in S_m$ to be a best Chebyshev approximation with k free knots to f in $[a,b]$ that there exists an interval $[x_p, x_{p+q}] \subset [a,b]$ with $q \geq 1$ such that

$$\mathcal{A}(f - s)_{[x_p, x_{p+q}]} \geq m + q + q' + 1$$

(Nürnberger, Schumaker, Sommer and Strauss 1989); it is *sufficient* for $s \in S_m$ to be a best Chebyshev approximation with k free knots to f in

$[a, b]$ that there exists an interval $[x_p, x_{p+q}] \subset [a, b]$ with $q \geq 1$ such that

$$\mathcal{A}(f - s)_{[x_p, x_{p+q}]} \geq m + k + q' + 2$$

(Braess 1971). The necessary condition is strengthened to a possibly longer alternant by Mulansky (1992). Other results on this topic are given by Kawasaki (1994), Nürnberger (1994*a*).

Although a characterization of best spline approximations with free knots is not known, a characterization of strongly unique best spline approximations with free simple knots is available: what is required is that *all* knot intervals contain sufficiently many alternating extrema (Nürnberger 1987; see also Nürnberger 1994*b*).

Some algorithms for computing best Chebyshev approximations by free knot splines are available. For example, Nürnberger, Sommer and Strauss (1986) (see also Meinardus, Nürnberger, Sommer and Strauss 1989) give an algorithm that converges through sequences of knot sets from an arbitrary set of knots. For each set of k knots, best Chebyshev degree m polynomial approximations to f are obtained on each subinterval using the classical Remes algorithm. The knots are then adjusted by a 'levelling' process, so that the maximum errors of the polynomial best approximations are equalized. Finally, the algorithm for fixed knots described above is applied on the levelled knot set.

Generalizations to multivariate splines have mainly been concerned with interpolation problems. But consider bivariate splines on $[a_1, b_1] \times [a_2, b_2]$. This region can be divided into rectangles by knot lines $x = x_i$, $y = y_i$, $i = 1, \ldots, s$, and a tensor product spline space can be defined. As in the univariate problem, partitions can be defined and improved systematically in such a way that best Chebyshev approximations are obtained in the limit. Some recent work on this problem is given by Meinardus, Nürnberger and Walz (1996), Nürnberger (1997). However, there are many unsolved problems: see Nürnberger (1996).

5.2. *Chebyshev approximation by rational functions*

Another important class of approximation problems is the best Chebyshev approximation of continuous functions by rational functions. The basic problem is as follows: define R_{nm} by

$$R_{nm} = \left\{ \frac{P(x)}{Q(x)} : P(x) = \sum_{j=0}^{n} a_j p_j(x), \quad Q(x) = \sum_{k=0}^{m} b_k q_k(x), \right.$$
$$\left. Q(x) > 0 \text{ on } [a, b] \right\}.$$

Then, given $f(x) \in C[a, b]$, we need to determine $R \in R_{nm}$ to minimize $\|f - R\|$, using the Chebyshev norm on $C[a, b]$. For the special case when

$P(x)$ and $Q(x)$ are polynomials of degree n and m, respectively, existence of a best approximation is guaranteed, is unique (up to a normalization), and is characterized by an alternating set of $n + m + 2 - d$ points,

$$\mathcal{A}(f - R)_{[a,b]} \geq n + m + 2 - d,$$

where d is the defect of the approximation, that is, the minimum difference between the *actual* degree of $P(x)$ and $Q(x)$ and n and m, respectively. If $d > 0$, the best approximation is said to be degenerate. For more general quotients, existence is no longer guaranteed, although characterization results are available (though not necessarily of alternation type), and uniqueness results may be extended.

For rational approximation by quotients of polynomials on an interval, the analogue of the Remes exchange method may be applied. It assumes nondegeneracy of the best approximation, and second-order convergence can be obtained. The system of linear equations that needs to be solved in the linear problem is replaced by a nonlinear system in the rational problem, equivalent to an eigenvalue problem, and various methods were proposed for this in the 1960s. Breuer (1987) has suggested a different direct approach to this subproblem, which uses continued fraction interpolation, and which, it is claimed, can lead to a considerable increase in efficiency, and also accuracy and robustness.

If attention is restricted to a discrete subset of $[a, b]$, with positivity of $Q(x)$ only required on the discrete set, then existence of best approximations is no longer guaranteed, even in the polynomial case, and characterization and uniqueness results are no longer valid. The Remes algorithm nevertheless may be applied, although a serious competitor is the differential correction algorithm, first proposed by Cheney and Loeb (1961), and further analysed by Barrodale, Powell and Roberts (1972), Cheney and Powell (1987). The method, which consists of a sequence of linear programming problems, has guaranteed convergence from any starting point, with quadratic convergence in the absence of degeneracy. It may in theory be applied to problems on intervals (Dua and Loeb 1973), but the solution of the subproblems is not straightforward.

Let the discrete subset on which a solution is required be x_i, $i = 1, \ldots, t$, and let R_{nm}^D be the set

$$R_{nm}^D = \left\{ \frac{P(x)}{Q(x)} : P(x) = \sum_{j=0}^{n} a_j p_j(x), \quad Q(x) = \sum_{k=0}^{m} b_k q_k(x), \right.$$
$$\left. Q(x_i) > 0, \quad i = 1, \ldots, t \right\}.$$

Let $P/Q \in R^D_{nm}$, and let Δ satisfy

$$|f(x_i) - P(x_i)/Q(x_i)| \le \Delta, \quad i = 1, \dots, t.$$

Then

$$|f(x_i)Q(x_i) - P(x_i)| \le \Delta Q(x_i), \quad i = 1, \dots, t. \tag{5.3}$$

Expanding the right-hand side, regarded as the function $g(\Delta, Q(x_i)) = \Delta Q(x_i)$, in a Taylor series about $(\Delta_k, Q_k(x_i))$ gives for each i

$$\Delta Q(x_i) = \Delta_k Q_k(x_i) + (\Delta - \Delta_k)Q_k(x_i) + (Q(x_i) - Q_k(x_i))\Delta_k + \dots,$$

so to first-order terms the ith term of (5.3) can be written

$$\begin{aligned} &|f(x_i)Q(x_i) - P(x_i)| \\ &\quad \le \Delta_k Q_k(x_i) + (\Delta - \Delta_k)Q_k(x_i) + (Q(x_i) - Q_k(x_i))\Delta_k \\ &\quad = (\Delta - \Delta_k)Q_k(x_i) + Q(x_i)\Delta_k. \end{aligned}$$

Thus, to first-order terms,

$$|f(x_i)Q(x_i) - P(x_i)| - \Delta_k Q(x_i) \le (\Delta - \Delta_k)Q_k(x_i), \quad i = 1, \dots, t,$$

or

$$\max_{1 \le i \le t} \left\{ \frac{|f(x_i Q(x_i) - P(x_i)| - \Delta_k Q(x_i)}{Q_k(x_i)} \right\} + \Delta_k \le \Delta. \tag{5.4}$$

The differential correction algorithm is as follows.

(1) Choose an initial approximation $R_1 = P_1/Q_1 \in R^D_{nm}$; set $k = 1$.

(2) Determine $P(x)$ and $Q(x)$ to minimize the left-hand side of (5.4), where

$$\Delta_k = \max_{1 \le i \le t} |f(x_i) - P_k(x_i)/Q_k(x_i)|,$$

and Q_k is not identically zero.

(3) Set $P_{k+1} = P$, $Q_{k+1} = Q$, $k = k + 1$ and continue unless there is convergence.

This algorithm generates a monotonic sequence of numbers that decrease to the minimum error. Starting with $Q_1(x_i) > 0$, $i = 1, \dots, t$, subsequent denominators retain this property. When the solution is unique, Cheney and Powell (1987) show that convergence is at least superlinear. Barrodale et al. (1972) show that quadratic convergence is obtained in the polynomial case in the absence of degeneracy.

A potentially unsatisfactory feature of approximation from R_{nm} (or R^D_{nm}) is that the denominator, although positive, can become arbitrarily close to zero at certain points. It is not sufficient simply to impose a lower bound on Q, because of the possibility of multiplying both numerator and denominator by an arbitrary constant. A modification of the differential correction algorithm that applies to problems with a lower bound on the denominator and upper bounds on the absolute values of the coefficients b_i is given by Kaufman and Taylor (1981). It is more natural, however, to impose upper and lower bounds on the denominators themselves ('constrained denominators'). Some aspects of uniqueness are considered by Li and Watson (1997). A modified differential correction algorithm for this problem has been given by Gugat (1996*a*). This involves a constraint of the form

$$\mu(x) \leq Q(x) \leq \nu(x), \tag{5.5}$$

over the appropriate set, where μ and ν are continuous functions. (In fact the algorithm applies to much more general problems than the one considered here, possibly defined on intervals, even having nonlinear expressions P and Q.) The subproblem corresponding to (5.5) above differs in that the additional conditions

$$\mu(x_i) \leq Q(x_i) \leq \nu(x_i), \quad i = 1, \ldots, t \tag{5.6}$$

are imposed. However, it differs also in that, whereas the original method starts with an arbitrary approximation, with denominator Q_1 positive on $x_i, i = 1, \ldots, t$, and with error Δ_1, the method of Gugat (1996*a*) can start with an arbitrary number Δ_1 that is allowed to be smaller than the current error, and an arbitrary (feasible) denominator Q_1. This flexibility turns out to be an important advantage: for example, numerical results show that the choice $\Delta_1 = 0$ is a good one. Subsequent Δ_k are defined as in the original algorithm, but with the constraints (5.6) included in step (2). It is shown by Gugat (1996*a*) that convergence results for the original version carry over. The development outlined above shows that the differential correction algorithms have links with Newton's method. Other methods using variants of Newton's method are those of Hettich and Zenke (1990) and Gugat (1996*b*). However, in contrast to the methods considered here, these do not generate a monotonic sequence.

6. L_1 approximation of functions

While the theory of best Chebyshev approximation to functions has (perhaps quite naturally) received considerable attention, the same cannot be said for best L_1 approximation. Given the same setting as at the start of Section 5, the problem is

$$\text{minimize} \int_X |f(x) - \sum_{i=1}^{n} c_i\phi_i(x)| \, dx. \tag{6.1}$$

For given $\mathbf{c}$, let $Z(\mathbf{c})$ denote the zeros of $f(x) - \sum_{i=1}^{n} c_i\phi_i(x)$ in X, and for points where this is nonzero, let $g(x, \mathbf{c})$ denote the sign. (We may define g to be zero at other points.) Define

$$V(\mathbf{c}) = \{v(x) : \|v\|_\infty \leq 1, \;\; v(x) = g(x, \mathbf{c}), x \notin Z\}.$$

Then it may be shown (for example, Watson 1980) that $\mathbf{c}$ is a solution to (6.1) if and only if there exists $v \in V(\mathbf{c})$ such that

$$\int_X v(x)\phi_j(x)\,\mathrm{d}x = 0, \quad j = 1,\ldots,n. \tag{6.2}$$

Further, if the system of functions $\{\phi_i(x),\ldots,\phi_n(x)\}$ forms a Chebyshev set on $[a,b]$, the best approximation is unique.

If the measure of the set Z is zero (for example if Z just consists of a finite set of points) then clearly (6.2) can be written as

$$\int_X g(x,\mathbf{c})\phi_j(x) = 0, \quad j = 1,\ldots,n.$$

This corresponds to the case when the norm is differentiable, and the above equations are just zero-derivative conditions with respect to the components of $\mathbf{c}$. The likelihood of these being appropriate in practice means that usually the problem is a smooth one. It also means that great store is placed on the points where there are sign changes, or equivalently where the approximation interpolates f. If these points were known, and were exactly n in number, then we could compute the best approximation by interpolation, *provided that there were no other changes of sign in the error of the resulting approximation.*

Definition 5 The points $x_1 < \cdots < x_t \in (a,b) = (x_0, x_{t+1})$, where $1 \le t \le n$, are called *canonical points* if

$$\sum_{i=0}^{t}(-1)^i \int_{x_i}^{x_{i+1}} \phi_j(x)\,\mathrm{d}x = 0, \quad j = 1,\ldots,n.$$

In the Chebyshev set case, existence and uniqueness of such a set of points were established by Micchelli (1977). For approximation by polynomials of degree $n-1$ in $[a,b]$, $t = n$ and the location of those canonical points is known – they lie at the zeros of the Chebyshev polynomial of the second kind of degree n (shifted if necessary). Thus their location is independent of f. Interpolation at these points can quite frequently result in the best polynomial approximation. (For further analysis of the L_1 problem, see Pinkus 1989.)

Example 5 Consider the approximation of $f(x) = 5 + 6\mathrm{e}^{2x} + 2\sin(4x)$ by polynomials of degree $n-1$ on $[-1,\ 1]$. Table 1 gives the outcome of determining a polynomial by interpolation at the zeros of the second kind Chebyshev polynomial $U_n(x)$. Shown are the number of zeros of the error in $[-1,\ 1]$, the value of the l_1 norm for the approximation given by the interpolant, and the minimum value of the norm. Clearly, when the number of zeros equals n, the best l_1 approximation is obtained and the norm is the minimum norm, otherwise it is not.

Table 1. *Interpolating polynomials*

n	no of zeros	norm	minimum norm
2	2	7.658748	7.658748
3	5	1.022593	0.816405
4	5	0.986141	0.812920
5	5	0.704263	0.704263
6	7	0.081135	0.063107
7	7	0.061799	0.061799
8	9	0.005072	0.003947
9	9	0.003937	0.003937
10	11	0.000192	0.000150
11	11	0.000150	0.000150

An algorithm for computing best L_1 approximations from general linear subspaces is given by Watson (1981). It is essentially of exchange type, based on the calculation of the zeros of the error at each iteration, and the construction of descent directions. It is also of Newton type, since it constructs the Hessian matrix of the error when it exists, and can have a second-order convergence rate. In a sense, it can be thought of as analogous to the second algorithm of Remes for Chebyshev problems, where a sequence of sets of zeros plays the role of a sequence of sets of extreme points; the connection with Newton's method under appropriate circumstances is also something the methods have in common. An algorithm for nonlinear problems is given by Watson (1982).

For best L_1 approximation by splines with fixed knots, it is known that every continuous function has a unique best approximation. Further, under certain assumptions, the best approximation can be determined by interpolation at canonical points. These results go back to Micchelli (1977). Little, if any, practical work has been done on this or more general problems.

Acknowledgement
I am grateful to Yuying Li, Kaj Madsen, Gunther Nürnberger, Mike Osborne and Mike Powell, who were kind enough to read parts of this paper and make constructive comments.

REFERENCES

I. Barrodale and C. Phillips (1975), An improved algorithm for discrete Chebyshev linear approximation, in *Proc. 4th Manitoba Conf. on Numer. Math.* (B. L. Hartnell and H. C. Williams, eds), University of Manitoba (Winnipeg, Canada), pp. 177–190.

I. Barrodale and F. D. K. Roberts (1973), 'An improved algorithm for discrete l_1 linear approximation', *SIAM J. Numer. Anal.* **10**, 839–848.

I. Barrodale, M. J. D. Powell and F. D. K. Roberts (1972), 'The differential correction algorithm for rational l_∞ approximation', *SIAM J. Numer. Anal.* **9**, 493–504.

R. Bartels, A. R. Conn and Y. Li (1989), 'Primal methods are better than dual methods for solving overdetermined linear systems in the l_∞ sense?', *SIAM J. Numer. Anal.* **26**, 693–726.

R. Bartels, A. R. Conn and J. W. Sinclair (1978), 'Minimisation techniques for piecewise differentiable functions: the l_1 solution to an overdetermined linear system', *SIAM J. Numer. Anal.* **15**, 224–241.

P. Bloomfield and W. L. Steiger (1983), *Least Absolute Deviations*, Birkhäuser, Boston.

P. T. Boggs, R. H. Byrd and R. B. Schnabel (1987), 'A stable and efficient algorithm for nonlinear orthogonal distance regression', *SIAM J. Sci. Statist. Comput.* **8**, 1052–1078.

P. T. Boggs, R. H. Byrd, J. R. Donaldson and R. B. Schnabel (1989), 'ODRPACK, Software for weighted orthogonal distance regression', *ACM Trans. Math. Software* **15**, 348–364.

C. G. Boncelet, Jr. and B. W. Dickinson (1984), 'A variant of Huber robust regression', *SIAM J. Sci. Statist. Comput.* **5**, 720–734.

D. Braess (1971), 'Chebyshev approximation by spline functions with free knots', *Numer. Math.* **17**, 357–366.

P. T. Breuer (1987), A new method for real rational uniform approximation, in *Algorithms for Approximation* (J. C. Mason and M. G. Cox, eds), Clarendon Press, Oxford, pp. 265–283.

B. Brosowski and A. R. da Silva (1992), A general alternation theorem, in *Approximation Theory* (G. A. Anastassiou, ed.), Marcel Dekker, Inc., New York, pp. 137–150.

E. W. Cheney and H. L. Loeb (1961), 'Two new algorithms for rational approximation', *Numer. Math.* **3**, 72–75.

E. W. Cheney and M. J. D. Powell (1987), 'The differential correction algorithm for generalized rational functions', *Constr. Approx.* **3**, 249–256.

D. I. Clark (1985), 'The mathematical structure of Huber's M-estimator', *SIAM J. Sci. Statist. Comput.* **6**, 209–219.

D. I. Clark and M. R. Osborne (1986), 'Finite algorithms for Huber's M-estimator', *SIAM J. Sci. Statist. Comput.* **7**, 72–85.

T. Coleman and Y. Li (1992*a*), 'A globally and quadratically convergent affine scaling algorithm for l_1 problems', *Math. Prog.* **56**, 189–222.

T. Coleman and Y. Li (1992*b*), 'A globally and quadratically convergent method for linear l_∞ problems', *SIAM J. Numer. Anal.* **29**, 1166–1186.

A. Dax (1989), 'The minimax solution of linear equations subject to linear constraints', *IMA J Numer. Anal.* **9**, 95–109.

A. Dax (1993), 'A row relaxation method for large minimax problems', *BIT* **33**, 262–276.

A. Dax and B. Berkowitz (1990), 'Column relaxation methods for least norm problems', *SIAM J. Sci. Statist. Comput.* **11**, 975–989.

S. N. Dua and H. L. Loeb (1973), 'Further remarks on the differential correction algorithm', *SIAM J. Numer. Anal.* **10**, 123–126.

A. M. Duarte and R. J. Vanderbei (1994), Interior point algorithms for lsad and lmad estimation, Technical Report SOR-94-07, Programs in Operations Research and Statistics, Princeton University.

H. Ekblom and H. B. Nielsen (1996), A comparison of eight algorithms for computing M-estimates, Technical Report 1996-15, Technical University of Denmark.

G. H. Golub and C. F. Van Loan (1980), 'An analysis of the total least squares problem', *SIAM J. Numer. Anal.* **17**, 883–893.

M. Gugat (1996*a*), 'An algorithm for Chebyshev approximation by rationals with constrained denominators', *Constr. Approx.* **12**, 197–221.

M. Gugat (1996*b*), 'The Newton differential correction algorithm for uniform rational approximation with constrained denominators', *Numer. Algorithms* **13**, 107–122.

R. Hettich and P. Zenke (1990), 'An algorithm for general restricted rational Chebyshev approximation', *SIAM J. Numer. Anal.* **27**, 1024–1033.

Z. Jing and A. T. Fam (1987), 'An algorithm for computing continuous Chebyshev approximations', *Math. Comp.* **48**, 691–710.

K. Jonasson (1993), 'A projected conjugate gradient method for sparse minimax problems', *Numer. Algorithms* **5**, 309–323.

K. Jonasson and K. Madsen (1994), 'Corrected sequential linear programming for sparse minimax optimization', *BIT* **34**, 372–387.

K. Jonasson and G. A. Watson (1982), A Lagrangian method for multivariate continuous Chebyshev approximation problems, in *Multivariate Approximation Theory 2* (W. Schempp and K. Zeller, eds), Birkhäuser, Basel, pp. 211–221.

E. H. Kaufman and G. D. Taylor (1981), 'Uniform approximation by rational functions having restricted denominators', *J. Approx. Theory* **32**, 9–26.

H. Kawasaki (1994), 'A second-order property of spline functions with one free knot', *J. Approx. Theory* **78**, 293–297.

C. Li and G. A. Watson (1997), 'Strong uniqueness in restricted rational approximation', *J. Approx. Theory* **89**, 96–113.

W. Li and J. J. Swetits (1998), 'Linear l_1 estimator and Huber M-estimator', *SIAM J. Optim.* To appear.

Y. Li (1993), 'A globally convergent method for L_p problems', *SIAM J. Optim.* **3**, 609–629.

K. Madsen and H. B. Nielsen (1993), 'A finite smoothing algorithm for linear l_1 estimation', *SIAM J. Optim.* **3**, 223–235.

K. Madsen, H. B. Nielsen and M. C. Pinar (1994), 'New characterizations of l_1 solutions of overdetermined linear systems', *Operations Research Lett.* **16**, 159–166.

G. Meinardus (1967), *Approximation of Functions: Theory and Numerical Methods*, Springer, Berlin.

G. Meinardus, G. Nürnberger and G. Walz (1996), 'Bivariate segment approximation and splines', *Adv. Comput. Math.* **6**, 25–45.

G. Meinardus, G. Nürnberger, M. Sommer and H. Strauss (1989), 'Algorithms for piecewise polynomials and splines with free knots', *Math. Comp.* **53**, 235–247.

M. S. Meketon (1987), Least absolute value regression, Technical report, AT&T Bell Laboratories, Murray Hill, New Jersey.

C. A. Micchelli (1977), 'Best L^1 approximation by weak Chebyshev systems and the uniqueness of interpolating perfect splines', *J. Approx. Theory* **19**, 1–14.

P. A. P. Moran (1959), 'Random processes in economic theory and analysis', *Sankhya* **21**, 99–126.

B. Mulansky (1992), 'Chebyshev approximation by spline functions with free knots', *IMA J. Numer. Anal.* **12**, 95–105.

G. Nürnberger (1987), 'Strongly unique spline approximation with free knots', *Constr. Approx.* **3**, 31–42.

G. Nürnberger (1989), *Approximation by Spline Functions*, Springer, Berlin.

G. Nürnberger (1994*a*), Approximation by univariate and bivariate splines, in *Second International Colloquium on Numerical Analysis* (D. Bainov and V. Covachev, eds), VSP, Utrecht, pp. 143–153.

G. Nürnberger (1994*b*), 'Strong unicity in nonlinear approximation and free knot splines', *Constr. Approx.* **10**, 285–299.

G. Nürnberger (1996), 'Bivariate segment approximation and free knot splines: research problems 96-4', *Constr. Approx.* **12**, 555–558.

G. Nürnberger (1997), Optimal partitions in bivariate segment approximation, in *Surface Fitting and Multiresolution Methods* (A. Le Méhauté, C. Rabut and L. L. Schumaker, eds), Vanderbilt University Press, Nashville, pp. 271–278.

G. Nürnberger and M. Sommer (1983), 'A Remez type algorithm for spline functions', *Numer. Math.* **41**, 117–146.

G. Nürnberger, L. L. Schumaker, M. Sommer and H. Strauss (1985), 'Approximation by generalized splines', *J. Math. Anal. Appl.* **108**, 466–494.

G. Nürnberger, L. L. Schumaker, M. Sommer and H. Strauss (1989), 'Uniform approximation by generalized splines with free knots', *J. Approx. Theory* **59**, 150–169.

G. Nürnberger, M. Sommer and H. Strauss (1986), 'An algorithm for segment approximation', *Numer. Math.* **48**, 463–477.

M. R. Osborne (1985), *Finite Algorithms in Optimisation and Data Analysis*, Wiley, Chichester.

M. R. Osborne (1987), The reduced gradient algorithm, in *Statistical Data Analysis Based on the L_1 Norm and Related Methods* (Y. Dodge, ed.), North Holland, Amsterdam, pp. 95–107.

M. R. Osborne and G. A. Watson (1985), 'An analysis of the total approximation problem in separable norms, and an algorithm for the total l_1 problem', *SIAM J. Sci. Statist. Comput.* **6**, 410–424.

M. R. Osborne and G. A. Watson (1996), Aspects of M-estimation and l_1 fitting problems, in *Numerical Analysis: A R Mitchell 75th Birthday Volume* (D. F. Griffiths and G. A. Watson, eds), World Scientific, Singapore, pp. 247–261.

A. Pinkus (1989), *On L^1 Approximation*, Cambridge University Press, Cambridge.

J. R. Rice (1967), 'Characterization of Chebyshev approximation by splines', *SIAM J. Math. Anal.* **4**, 557–567.

J. B. Rosen, H. Park and J. Glick (1996), 'Total least norm formulation and solution for structured problems', *SIAM J. Matrix Anal. Appl.* **17**, 110–126.

J. B. Rosen, H. Park and J. Glick (1997), Total least norm for linear and nonlinear structured problems, in *Recent Advances in Total Least Squares Techniques and Errors-in-Variables Modeling* (S. Van Huffel, ed.), SIAM, Philadelphia, pp. 203–214.

S. A. Ruzinsky and E. T. Olsen (1989), 'l_1 and l_∞ minimization via a variant of Karmarkar's algorithm', *IEEE Trans. Acoustics, Speech and Signal Processing* **37**, 245–253.

L. L. Schumaker (1968), 'Uniform approximation by Tchebycheffian spline functions', *J. Math. Mech.* **18**, 369–378.

M. Shi and M. A. Lukas (1996), 'On the reduced gradient algorithm for L_1 norm minimization with linear constraints'. Research Report, Department of Mathematics and Statistics, Murdoch University, Australia.

H. Späth and G. A. Watson (1987), 'On orthogonal linear l_1 regression', *Numer. Math.* **51**, 531–543.

G. A. Watson (1976), 'A method for calculating best nonlinear Chebyshev approximations', *J. Inst. Math. Appl.* **18**, 351–360.

G. A. Watson (1980), *Approximation Theory and Numerical Methods*, Wiley, Chichester.

G. A. Watson (1981), 'An algorithm for linear L_1 approximation of continuous functions', *IMA J. Numer. Anal.* **1**, 157–167.

G. A. Watson (1982), A globally convergent method for (constrained) nonlinear continuous L_1 approximation problems, in *Numerical Methods of Approximation Theory 1981* (L. Collatz, G. Meinardus and H. Werner, eds), Birkhäuser, Berlin, pp. 233–243. ISNM 59.

G. A. Watson (1983), The total approximation problem, in *Approximation Theory IV* (C. K. Chui, L. L. Schumaker and J. D. Ward, eds), Academic Press, New York, pp. 723–728.

G. A. Watson (1984), The numerical solution of total l_p approximation problems, in *Numerical Analysis Dundee 1983* (D. F. Griffiths, ed.), Springer, Berlin, pp. 221–238.

G. A. Watson (1987), Methods for best approximation and regression problems, in *The State of the Art in Numerical Analysis* (A. Iserles and M. J. D. Powell, eds), Clarendon Press, Oxford, pp. 139–164.

G. A. Watson (1988), 'The smallest perturbation of a submatrix which lowers the rank of the matrix', *IMA J. Numer. Anal.* **8**, 295–303.

G. A. Watson (1991), 'On a general class of matrix nearness problems', *Constr. Approx.* **7**, 299–314.

G. A. Watson (1997), The use of the L_1 norm in nonlinear errors-in-variables problems, in *Recent Advances in Total Least Squares Techniques and Errors-in-Variables Modeling* (S. Van Huffel, ed.), SIAM, Philadelphia, pp. 183–192.

G. A. Watson and K. F. C. Yiu (1991), 'On the solution of the errors in variables problem using the l_1 norm', *BIT* **31**, 697–710.

Y. Zhang (1993), 'A primal-dual interior point approach for computing the l_1 and l_∞ solutions of overdetermined linear systems', *J. Optim. Theory Appl.* **77**, 323–341.

For EU product safety concerns, contact us at Calle de José Abascal, 56–1°,
28003 Madrid, Spain or eugpsr@cambridge.org.

www.ingramcontent.com/pod-product-compliance
Ingram Content Group UK Ltd.
Pitfield, Milton Keynes, MK11 3LW, UK
UKHW050417210126
466816UK00032B/429
* 9 7 8 0 5 2 1 1 5 7 6 5 0 *